Lecture Notes in Electrical Engineering

Volume 141

Ying Zhang (Ed.)

Future Communication, Computing, Control and Management

Volume 1

Springer

Editor
Ying Zhang
Wuhan University
Wuhan
China

ISSN 1876-1100 e-ISSN 1876-1119
ISBN 978-3-642-27310-0 e-ISBN 978-3-642-27311-7
DOI 10.1007/978-3-642-27311-7
Springer Heidelberg New York Dordrecht London

Library of Congress Control Number: 2011944178

Printed on acid-free paper

Springer is part of Springer Science+Business Media (www.springer.com)

ICF4C 2011 Preface

We are pleased to announce that 2011 International Conference on Future Communication, Computing, Control and Management (ICF4C 2011) will be held on December 16–17, 2011, Phuket, Thailand. ICF4C is a leading International Conference on Communication, Computing, Control and Management. It is also an outstanding chance to meet and network with reliability colleagues from around the world.

Phuket is one of the southern provinces of Thailand. Neighbouring provinces are Phang Nga and Krabi, but as Phuket is an island there are no land boundaries. The island is served by Phuket International Airport, located in the north of the island. The ph of Phuket is an aspirated p, so the name is pronounced Poock-get. The name Phuket is apparently derived from the word Bukit in Malay which means mountain or hill, as this is what the island appears like from a distance.

Phuket is the biggest island in Thailand, located in the Andaman Sea off southern Thailand. The island is mostly mountainous with a mountain range in the west of the island from the north to the south. The mountains of Phuket form the southern end of the Phuket mountain range, which ranges for 440 km from the Kra Isthmus. The highest elevation of the island is Mai Thao Sip Song, at 529 m above sea level.

The objective of ICF4C 2011 is to provide a forum for researchers, engineers, and government officials involved in the general areas of Communication, Computing, Control and Management to disseminate their latest research results and exchange views on the future research directions of these fields.

2011 International Conference on Future Communication, Computing, Control and Management (ICF4C 2011) is sponsored by Intelligent Information Technology Application Research Association, Hong Kong and Hong Kong Education Society. The mission is to bring together innovative academics and industrial experts in the field of Communication, Computing, Control and Management. The ICF4C 2011 will also include presentations of contributed papers and state-of-the-art lectures by invited keynote speakers.

We would like to take this opportunity to thank all of the individuals who have contributed to this conference, including authors, keynote speakers, the reviewers, committees and sponsors. Special thanks go to Springer Publisher.

We hope that ICF4C 2011 will be successful and enjoyable to all participants. We wish each of you successful deliberations, stimulating discussions, new friendships and all enjoyment that Phuket, Thailand can offer. While this is a truly remarkable conference, there is more yet to come. We look forward to seeing all of you next year at the ICF4C 2012.

Ming Chen, Intelligent Information Technology Application Research Association, Hong Kong

ICF4C 2011 Organizing Committee

General Chair

Mark Zhou	Hong Kong Education Society, Hong Kong

General-Co Chairs

Mark Zhou	Hong Kong Education Society, Hong Kong
Jun Zhang	Huazhong University of Science and Technology, China

Program Co-Chairs

Ming Chen	Intelligent Information Technology Application Research Association, Hong Kong
Jin Wen	ACM Chongqing Chapter, China

Organizing Co-Chairs

Liangming Zhang	Hong Kong University, Hong Kong
Jun Hu	Chinese University of Hong Kong, Hong Kong

Publication Chair

Ying Zhang	Wuhan University, China

International Committee

Jianwei Zhang	Suzhou University, China
Nian Zhang	Suzhou University, China
Minli Dai	Suzhou Vocational University, China
Qi Luo	Wuhan Insitute of Technology, China
Wei Li	Asia Pacific Human-Computer Interaction Research Center, Hong Kong
Alireza Yazdizadeh	International Science and Engineering Center, Hong Kong
Yong Hu	Nanchang University, China

Contents

Research of Dynamic Clustering Time Synchronization Algorithm in Wireless Sensor Networks 1
Ping Wang, ShiWu Xu

Design of On-Demand Time Synchronization Algorithm in Wireless Sensor Networks 9
Ping Wang, ShiWu Xu

Power Load Forecasting Based on the Indicators of Industrialization and Urbanization 17
Gu Changzhou, Long Jie, Chen Xiaoping, Guo Ying

Quantum Bit and Square Well in Quantum Computation 25
Lu Jun

Physical Realization of Harmonic Oscillator Quantum Computer 29
Lu Jun

Intrusion Detection Model Based on Set Pair Analysis Theory 35
Yuan Kan, Wei Jiang, Xiaojuan Ji

Research on Maximum Wind Energy Capturing of Doubly Fed Induction Wind Generator Based on Fuzzy Control Technology 43
Haihui Song, Tian De

A Developed Switching Vector Control Method Using DSP TMS320F240 . . . 49
Wen Youmei

A Modeling Approach Using UML 2 for Security Protocols in Distributed System 57
Xiaoxiong Zhou

The Edge-Detection Algorithm Based on Enlarged Block-Range 65
Wang Weiqing

The Image Correction Algorithm Based on Combined Transformation 73
Wang Weiqing

Designing Network Immunization Strategies Based on Graph Partitioning Methods 83
Changxi Niu, Lemin Li, Du Xu

Researches into Semiorthogonality Quality of Multivariant Wavelet Packages with Short Support 89
Yongcai Hu

The Research of a Pair of Canonical Frames Yielded by Finitely Supported Refinable Functions 99
Hailin Gao

The Study on Private Investment Fund Risk Assessment Using Analytic Hierarchy Process 109
Li Xin

An Improved Background Modeling Method for Target Detection 117
Cheng Mingzhi, Gao Junxiang

The Intelligent Management System of the Buses on Busy Roads 125
Gu Wenyang, Chen Huanlin, Zhang Jiajian, Wu Xiang

Car Speed Measurement Based on Feature Point Tracking in Traffic Video 131
Shi-Song Zhu, Min Gu, Toshio Koga

Adaptive Edge-Detection Implementation for Video Processing Based on DM642 141
Yixuan Wu, Zhiwei Tang

Dynamic Modeling, Simulation and Experiment of the Delta Robot 149
Wang Pan Feng, Zhang Li Min, Zhao Xue Man

From CAD to Search, Rebuild the Industrial Chain of Architecture by IT 157
Jing Zhang, Jia Lang, Mingjiu Yu, Lei Qiu

Research on Development and Strategy for Hebei Wushu Schools 165
Shiru Hao

Stock Prediction with Directed Cross-Correlation Network 173
Hua Chen, Qiming Sun

Solution for Transforming Web Application to Multi-tenant SaaS Application 181
Yangpeng Zhu, Jing Zhang, Junhuai Li

Risk Acceptance Criteria of Oil and Gas Pipelines Projects in Construction Period . . . 189
Wu Wenping, Feng Xia

The SWOT Analysis on the Development of Renewable Energy Constructions in Xi'an . . . 197
Wu Wenping, Liang Xuelan

The Passenger Car Chassis Upgraded Based on Independent Suspension . . . 207
Yang Wu, Li Ming, Liang Zhi Hong

Application of Artificial Neural Network to Predict Real Estate Investment in Qingdao . . . 213
Ping Zhang, Wenjing Ma, Tiejun Zhang

Test Requirements Simplification Based on Nonlinear Data Structure . . . 221
Jianyuan Su, Xiaoming Wang

An Agent-Based Approach for Cost Evaluation Using ABC Method: Application to INOTIS Enterprise . . . 227
Noria Taghezout, Sonia Ait Ameur, Amel Bekkal, Bouabdellah Ascar

Power Engineering Investment Forecasting Based on Covering Rough Set . . . 235
Lili Zhu, Qingyou Yan, Cunbin Li

Improving Wavelet Transform for the Impact-Echo Method of Non Destructive Testing . . . 241
T. Bouden, M. Nibouche, F. Djerfi, S. Dib

A Hybrid Intelligent Algorithm for Mixed-Variable Optimization Problems . . . 249
Zhang Yunqiang, Tan Ying, Sun Chaoli, Zeng Jianchao

Annotating Taiwan Cancer Registry to caDSR for International Interoperability . . . 257
Shin-Bo Chen, Chien-Yeh Hsu, Chi-Hung Huang

A Comprehensive Behavioural Process Model for Design Synthesis . . . 265
Yi-Min Deng, Yongsheng Ma, Meiyan Wang, Liang Shi

Research and Application of Contactless Electrical Energy Transmission System . . . 271
Huang Ke, Zhang Liping, Kong Xianwang, Yang Yongkui, Wang Junwei

Rapid Manufacturing: An Industrial Revolution or a Passing Fad? . . . 279
Sarat Singamneni

X-FEM Studies on an Inclined Crack in a 2-D Finite Piezoelectric Media . . . 285
R.R. Bhargava, Kuldeep Sharma

Development of Flow Stress Model Based on Internal Variables 291
Jarosław Nowak, Dmytro S. Svyetlichny, Łukasz Łach

Invasion Detection Algorithm Research and Application 297
Zhao Yanwei

Application Layer Protocol Identification Algorithm Design 307
Liao Xiaoju, Yin Mingyong, Lu Hai

A Novel Intrusion Detection Method Base on Attribute Reduction Algorithm 313
Zhao Yanwei

Design of Coreless Winding Machine Electrical Control System 321
Wu Yaohui, Shao Pengfei, Wei Rencan

Kinematic Simulation of Rescue Robot 329
Weichun Zhang, Baohao Pei, Jie Chen, Xianbin Du, Bingbing Ma, Xiao Wang

The Optimal Design and Modal Analysis for a Truck Crane Frame 337
Weichun Zhang, Baohao Pei, Bingbing Ma, Jie Chen, Xianbin Du

Finite Element Analysis of Some Anchor Windlass Based on ABAQUS 343
Zhuang Hong, Li Qiong, Zhang Jian, Tang Wen-xian

FCM-Based Model for Network Security Assessment 349
Shihui Cheng

Energy Regeneration System for Hybrid Electric Vehicle Based on CAN Bus 357
Li Ran, Wu Junfeng, Li Gechen

The Probabilistic Approach to Modification of Seismic Linear Response Spectra 365
Cada Zdeneka, Kala Jirib, Salajka Vlastislavc, Kanicky Viktord

Study on Vibration and Noise of Hybrid Electric Vehicle 373
Xiong Jian-qiang, Huang Ju-hua, Liao Qun

Brand Logo Design Symbol Research 383
Jiaqing Guan, Qian He

A CBR-Based CAD System for Subframe Design of Aerial Work Trucks . . . 389
Li Qi, Liu Xinhua, Liu Youhui, Yin Jilin

Assembly Sequence Planning Based on Ant Colony Algorithm 397
Liu Youhui, Liu Xinhua, Li Qi

The Feasible Algorithm Based on a Class of Multiple Affine Bi-Frames with Arbitrary Real Translation Number ... 405
Yongcai Hu

Dense Sub Graph Based Video Semantic Mining ... 415
Shilin Zhang, Heping Li, Shuwu Zhang

CAM Software Products for Creation of Programs for CNC Machining ... 421
Peter Michalik, Jozef Zajac, Ján Duplák, Alexander Pivovarnik

Research on the Information Security Management in E-commerce System ... 427
Qin Qin, Ge Li

Multi-objective Optimization of Coal-Fired Boiler Efficiency and NOx Emission under Different Ecological Environment ... 433
Wang Weiqing

Ranking the Importance of Classes via Software Structural Analysis ... 441
Li Ding-wei, Li Bing, He Peng, Pan Wei-feng

Measurement Study of Self-concept and Academic Help-Seeking on Junior Middle School Students ... 451
Changxiu Shi, Jiajun Yao, Shuoqiu Gan, Fan Yu

The Description of Characters of Matrix-Block Multielement Wavelet Wraps according to a Dilation Matrix ... 457
Ruihua Hu, Xianghai Li

Mine Mechanical Equipments Fault Detection Based on Artificial Immune System ... 467
Xin Yue, Dongge Wen, Haifeng Ma, Jingfeng Song

Study on Rotor-Side Converter of Doubly Fed Induction Generator ... 473
Haihui Song, Tian De

The Study on the Multi-object Fuzzy Decision-Making of the M&A of Resource–Typed Enterprises Based on Rough Sets ... 479
Tian Liang, Ying Chu

Algorithm of License Plate Location Based on Morphological Multi-structural Elements ... 487
Xiong Chun-rong, Chen Tie-jun, Lv Jie

Medical Image Segmentation of PACS System Image Post-processing ... 495
Lv Jie, Xiong Chun-rong, Xie Miao

A New Multi-ant Colony Optimization Algorithm ... 501
He Yueshun, Du Ping

Design and Realization of Zhejiang Water Environment Information Disclosure System 509
Yanli Ma, Bin Xie

Matrix Inversion Method for Load Identification in Transfer Paths Analysis 517
Zhang Fang, Zhou Hong, Wang Erbing

Operational Modal Analysis for Automobile 525
Wang Jing, Zhou Hong, Xu Gang, Wang Erbing, Lu Xiang

Study on Control Algorithm of the Electro-hydraulic Servo System for Load Simulation Test 533
Hu Yudong, Zhou Hong, Xu Gang

Research on the Effect of Paths Coupling of Transfer Path Analysis Method 543
Mu Le, Zhou Hong, Wang Erbing, Xu Haiqing, Su Lang

Network Security Analysis Based on Graph Theory Model with Neutral Network 551
Mo Ming-zhong

Fault Diagnosis of Gearboxes Based on the Local Mean Decomposition Method 559
Jinshan Lin

Study on Stability of Continuous-Time Hopfield Network 565
Li Tu, Juanjuan Song

An Economic Model for the Evaluation of the Economic Value of Cloud Computing Federation 571
Zehua Zhang, Xuejie Zhang

Color Calibration and Color Management in Digital Camera 579
Songhua He, Zhen Liu, Lijie Wang

A New Method for CT/MR Image Fusion Based on Bandelets Transform . . . 587
Huimin Lu, Yujie Li, Lifeng Zhang, Shiyuan Yang, Seiichi Serikawa

Color Image Segmentation Using Fast Density-Based Clustering Method . . . 593
Yujie Li, Huimin Lu, Lifeng Zhang, Shiyuan Yang, Seiichi Serikawa

Modeling High-Speed Network Traffic with Truncated α-Stable Processes 599
Lin Bai, Wenyang Guan, Chen Chen, Jianhua He, Rui Wang

Author Index 605

Research of Dynamic Clustering Time Synchronization Algorithm in Wireless Sensor Networks

Ping Wang and ShiWu Xu

Key Laboratory of OptoElectronic Science and Technology for Medicine of Ministry of Education, Fujian Normal University, Fuzhou, China, 350007
pwang@fjnu.edu.cn, xushiwu@126.com

Abstract. In this paper, We propose a DCTS (Dynamic Clustering Time Synchronization) algorithm. The algorithm is divided into two phases. In level discovery phase, a clustering structure is established in network. In synchronization phase, we adopt TPSN algorithms to achieve time synchronization between the base station and cluster-head, and use DMTS algorithm to complete time synchronization between the cluster-head and cluster nodes. Simulation results show that the accuracy of DCTS is high, and the energy consume of DCTS is less than TPSN.

Keywords: LEACH, Time synchronization, DMTS, TPSN, Fusion Design.

1 Introduction

Wireless sensor networks have drawn much attention from the academia and industry as they offer an unprecedented range of potential applications. Most of these applications require sensor nodes to maintain local clocks in order to determine the timing of the events. To maintain the node the time synchronism is important in the wireless sensing network, which is also the premise of the guarantee precise of node localization, the reliable data fusion, the target tracking and examination [1].

In 2002, Jeremy Elson and Kay Romer firstly proposed the importance of Time Synchronization for Wireless Sensor Network in the International Conference, so far, there are already many quite mature synchronization algorithm. Such as TPSN (Timing-Sync Protocol for Sensor Networks) algorithm[2], which based on the sender-recipient. The synchronization algorithm based on receiver - receiver, the typical method such as RBS（Reference Broadcast Synchronization）algorithm[3]. Some people propose time synchronizations algorithms based on dynamic clustering, Such as, CHTS（Cluster-based Hierarchical Time Synchronization）algorithm[4], GCS(Gradient Clock Synchronization)[5]. PBS (Pairwise Broadcast Synchronization) [6]. BTS (Broadcast Time Synchronization)[7]. These types of synchronization algorithms have advantages and disadvantages, the precision is high, the relative energy consumption is also high, therefore in the specific wireless sensing network, the compromise between accuracy and power consumption is important, which can not only guarantee the accuracy of the actual demand, but also to minimize the power consumption of the entire network.

Y. Zhang (Ed.): Future Communication, Computing, Control and Management, LNEE 141, pp. 1–7.
springerlink.com

2 The Design of Synchronization Algorithm

2.1 The Selection Phase of the Cluster-Head Nodes and Cluster Nodes Using LEACH Routing Algorithm

In order to improve the survival time of the entire network, Wendi Rabiner Heinzelman and other scholars proposed LEACH(Low-Energy Adaptive Clustering Hierarchy)[8]. Which each sensor node has the opportunity to act as a cluster head node, and the selection of cluster head node based on the suggested percentage of cluster heads for the network and the number of times the node has been a cluster-head so far. Each node in the network randomly selected a number between 0 and 1, if the number is less than a threshold T (n), then the node act as the cluster head node. T (n) calculated as formula (1)

$$T(n)=\begin{cases}\dfrac{p}{1-p\left(r \bmod \dfrac{1}{p}\right)} & \text{if } n\in G\\ 0 & \text{otherwise}\end{cases} \tag{1}$$

Equation (1), p represents the percentage of cluster head nodes, r is the current round, G is the set of nodes that have not been cluster-heads in the last 1 / p rounds. LEACH algorithm by setting T (n) values to ensure that each node will be a cluster-head at some point within 1/p rounds to balance the node energy consumption. if the cluster head node is determined, the cluster head node for the current round broadcasts an message to the rest of the nodes. it use a CSMA MAC protocol to avoid conflict in the broadcast process. Then the non cluster head nodes in the whole network by receiving the signal strength to decide the cluster which they belong to, and select the strongest signal strength of the source node as their cluster head node, and inform the relevant cluster head nodes that themselves to be cluster members.

The network model of DCTS is shown in Fig.1, in order to reduce the interference of the other experimental conditions, we assume n sensor nodes has the following properties:

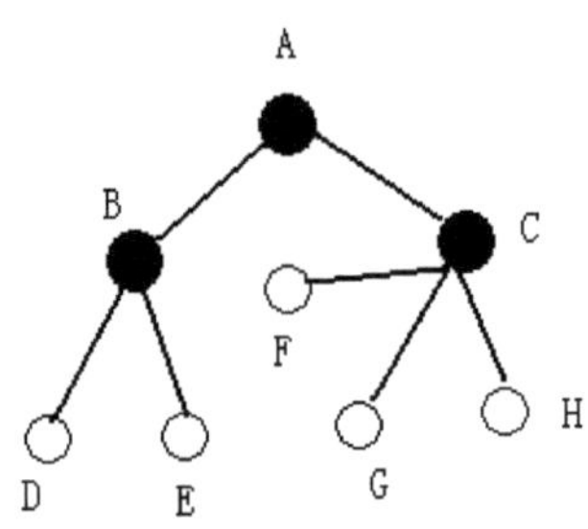

Fig. 1. Network model

Each node has a routing function and can be the cluster head node by LEACH algorithm.

All nodes are static, and know the location information in advance.

Each node uses the same kinds of materials, especially using the same crystal.

To describe conveniently, in Fig.1, A node is a coordinator, we assume that the first selection of the cluster head node is node B and node C. In Fig.1, the two sensor nodes of D, E are the cluster nodes in the cluster head node B, the three sensor nodes of F, G, H are the cluster nodes in the cluster head node C.

2.2 Time Synchronization Phase

First, the synchronization with TPSN algorithm between coordinator and cluster head node. Synchronization principle is shown in Fig.2, Coordinator use broadcast way to send time synchronization packet, if cluster-head receive this packet, at time T1, the cluster head node send a synchronization packet to coordinator, the packet contains the level of cluster-head and T1, after coordinate received the packet, record the time is T2, and immediately respond a synchronization packet to cluster-head, the packet contains the level of coordinator ,T1 ,T2 and response time T3.After cluster-head received the reply from coordinator, record the time is T4. So we can calculate the offset time m between the cluster head node and coordinator, we assumed the same propagation delay from sending and receiving messages, and we assumed this propagation delay time is d. because from T1 to T4, the time is short. We assumed the local clock time offset m is the same.

From $T2 = T1 + m + d$, $T4 = T3 - m + d$ .So we can get

$$m = \frac{(T2 - T1) - (T4 - T3)}{2} \tag{2}$$

$$d = \frac{(T2 - T1) + (T4 - T3)}{2} \tag{3}$$

As a result, the local clock on cluster head node could be synchronized with coordinator local clock by adding offset m.

After the cluster head node had revised their local clock time, second, we can use DMTS algorithm to finish the time synchronization between the cluster head node and cluster nodes. Synchronization principle is shown in Fig.3. When the cluster head nodes detect the channel idle, we can add the timestamp t0 to the broadcast packet, in order to eliminate the sending node processing delay and MAC layer access delay. Assuming the packet length is NA bits (Include the preamble and synchronization words), the cost time of send each bit is t, If receiver has been received the synchronization words, we record the local clock is t1, and before we revise the local clock, we note the time is t2, in order to let cluster nodes could be synchronized with the cluster head node ,cluster nodes local clock can be revised to t0+tNA+(t2-t1).

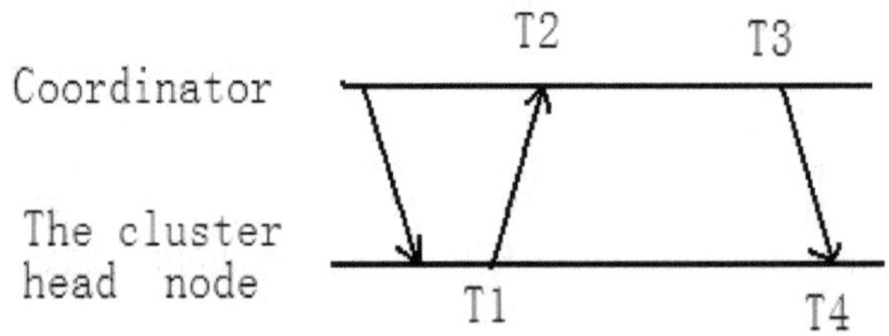

Fig. 2. The principle of time synchronization between coordinator and cluster head node

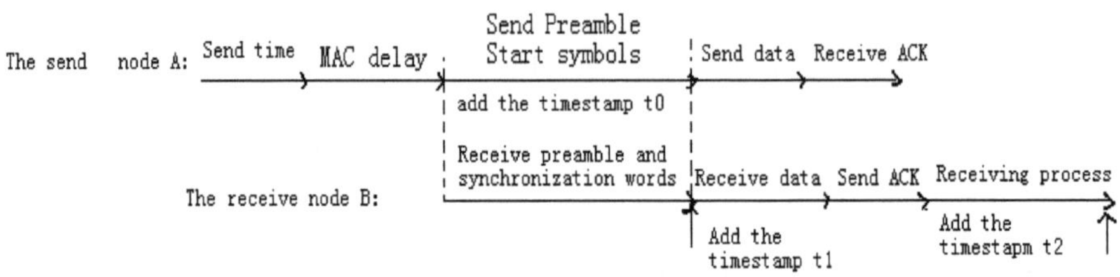

Fig. 3. The principle of time synchronization between Cluster-head and cluster node

2.3 The Analytical Error from the Design of Time Synchronization Algorithm

In order to describe convenient, we assume the coordinator is node A and the cluster head node is node B in Fig.2,because the time synchronization between coordinator and the cluster head node use TPSN algorithm, so have the same source of error. The TPSN algorithm use time mark in the MAC layer, it can eliminate time error from the send time and access time. We can get:

$$Error1=\frac{S_{UC}}{2}+\frac{P_{UC}}{2}+\frac{R_{UC}}{2}+\frac{RD_{t1\to t4}^{A\to B}}{2} \tag{4}$$

In the equation (4), $S_{UC}=S_A-S_B$, $P_{UC}=P_{A\to B}-P_{B\to A}$, $R_{UC}=R_B-R_A$, SA is the packet transmission time of node A, SB is the packet transmission time of node B, $P_{A\to B}$ is the time of node A transmit the packet to node B, $P_{B\to A}$ is the time of node B transmit the packet to node A, RA is the time of node A processing packet, RB is the time of node B processing packet, $RD_{t1\to t4}^{A\to B}$ represent the additional local clock offset between node A and node B from t1 to t4.

Second, we analysis the synchronization error between the cluster head node and the cluster nodes, To describe the convenient, we assume the cluster head node is node A and the cluster node is node B in the Fig.3,The cluster head node detected the channel idle at time T0,after the packet arrived to the cluster nodes, the cluster nodes add a timestamp T1 to the packet, and before cluster nodes revise the local time, record the time is T2, cluster nodes completed the adjustment at T3.So we can get:

$$T3=T0+T_{A\to B}+P_{A\to B}+R_B+D_{t0}^{A\to B} \tag{5}$$

$$T_{A\to B}+R_B=nt+T_{error}+\mathrm{R}_{error}+(T2-T1) \tag{6}$$

And we can get the local clock offset m of DMTS:

$$m = P_{A\to B} + T_{error} + \mathrm{R}_{error} + D_{t3}^{A\to B} + RD_{t0\to t3}^{A\to B} \tag{7}$$

The error of DMTS: $Error2 = m - D_{t3}^{A\to B} = P_{A\to B} + T_{error} + \mathrm{R}_{error} + RD_{t0\to t3}^{A\to B}$ (8)

In the equation (7), $D_{t0}^{A\to B} = D_{t3}^{A\to B} + RD_{t0\to t3}^{A\to B}$, n is the length of preamble, Terror is the error from during transmission bits. Rerror is the error from the processing time mark. $D_{t0}^{A\to B}$ is the local clock offset between node A and node B at time t0, $D_{t3}^{A\to B}$ is the local clock offset between node A and node B at time t3, $RD_{t0\to t3}^{A\to B}$ represent the additional local clock offset between node A and node B from t0 to t3.

From the equation (4), we can know that The accuracy of TPSN algorithm is high, because the TPSN algorithm use time mark in the MAC layer, it can eliminate time error from the send time and access time, The disadvantage is that TPSN is point to point synchronization algorithm, each time synchronization only a pair of nodes, and it needs send two messages and receiving two messages, the energy consumption of TPSN is high, from the equation (8), we can know the error of DMTS is high, because DMTS is the one-way time synchronization algorithm, it can not eliminate the affect from the Terror and Rerror., but doing a synchronization, it only needs send a message and receiving a message, the energy consumption of DMTS is low.

2.4 The Analysis of Algorithm Performance

Using MATLAB to analysis the performance of DCTS algorithm, and compared with TPSN algorithm. In the Fig.4, Re representative the receive packets used to synchronize, Tr representative the transmit packets used to synchronize, total representative the total packets used to synchronize. Because TPSN Re is the same as TPSN Tr, with a straight line. From the Fig.5, we can see although the Synchronization accuracy of DCTS is worse than TPSN, the four hops of synchronization accuracy is about 50.2us use DCTS. it can be used in some occasions which not need high precision synchronization. From the Fig.4, we can see the number of sent messages use DCTS is about 1/6 of TPSN, the number of receive messages use DCTS is about half of TPSN, and the total number of exchange messages use DCTS is about 1/3 of TPSN, as we know, in the wireless sensor network, the most of power consumer is in the exchange messages, especially in the sent messages, From the perspective of energy consumption, DCTS is better than TPSN. It can prolong the life time of nodes in the network. Although the synchronization accuracy of TPSN is high, the power consume is also too high, so it is not suitable for wireless sensor network.

In the industrial harmful gas detection, most of the node as terminal node to sample harmful gas data values, and only a few node as routing node, which used to store and forwarding the data from the terminal node. Through multiple hops transmission to the PC, data processing, get the result of we need. If a lot of the terminal node in wireless sensor network. Using the DCTS algorithm is proposed in this paper. The premise of keeping synchronous precision demand, it also can reduce the power consumption in the all network.

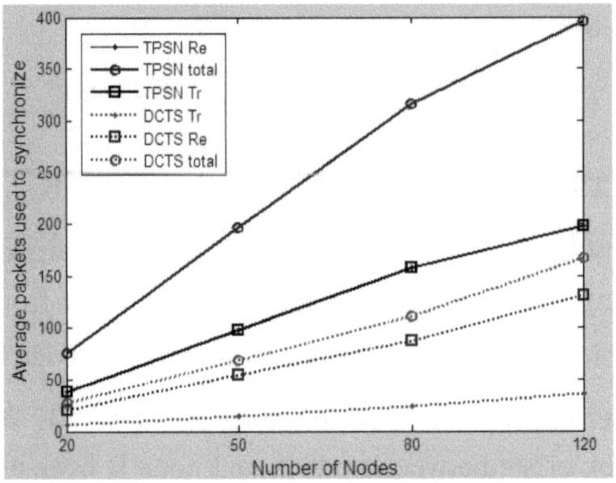

Fig. 4. Average packets used to synchronize a network from 20 nodes to 120 nodes

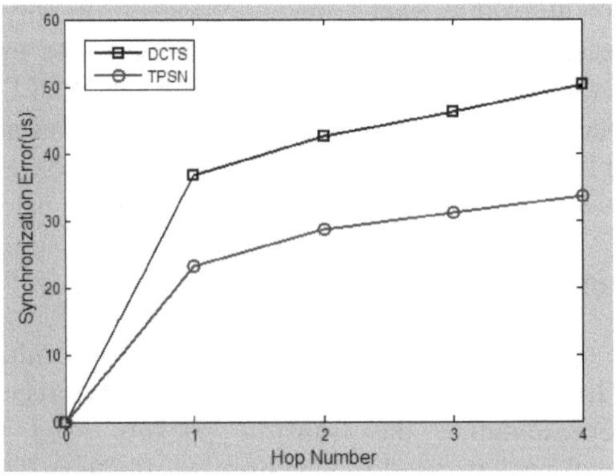

Fig. 5. The relation between synchronization error and hop number

3 Summary

The innovation of this paper is making use the advantage of LEACH routing algorithm, the whole network is divided into several clusters, thus reducing the synchronization node synchronization error cumulative impact of a single hop, meanwhile the dynamic selection of the cluster head node balances the power consumption of nodes in the network ,extends the life of the whole network and combines with the different roles of the cluster head node and cluster node in the network. Simulation results show that the accuracy of DCTS is high, and the energy consume of DCTS is less than TPSN.

Acknowledgment. The work in this paper was partially supported by Science and Technology Department of Fujian Province-The Embedded Designed of the Neural Network and E-mail Filtering System (JB0976) and The Research of The Wireless Sensor Networks in Smart Home System (JB09071);

References

1. Cheng, K., Lui, K., Wu, Y., Tam, V.: IEEE Transaction on Wireless Communications 8, 1764 (2009)
2. Ganeriwal, S., Kumar, R., Srivastava, M.B.: In: Proceedings of the 1st International Conference Embedded Networked Sensor Systems (SenSys's 2003), p. 138. ACM press (2003)
3. Elson, J., Girod, L., Estrin, D.: In: Proceedings of the 5th Symposium on Operating Systems Design and Implementation (OSDI 2002), USA, p. 147 (2002)
4. Kim, H., Kim, D., Yoo, S.-E.: 20th International Conference on Advanced Information Networking and Applications, p. 2 (2006)
5. Sommer, P., Wattenhofer, R.: In: Proc. 8th ACM/IEEE International Conference on Information Processing in Sensor Networks (IPSN), p. 37 (2009)
6. Cheng, K.-Y., Lui, K.-S., Wu, Y.-C., Tam, V.: IEEE Transactions on Wireless Communications 8, 1764 (2009)
7. Xu, C.-N., Zhao, L., Xu, Y.-J., Li, X.-W.: Chinese Journal of Computers 30, 514 (2007)
8. Heinzelman, W., Chandrakasan, A., Balakrishnan, H.: In: Proceedings of the 33rd Hawaii International Conference on System Science, Hawaii, USA, p. 3005 (2000)

Design of On-Demand Time Synchronization Algorithm in Wireless Sensor Networks

Ping Wang and ShiWu Xu

Key Laboratory of OptoElectronic Science and Technology
for Medicine of Ministry of Education, Fujian Normal University, Fuzhou, China, 350007
pwang@fjnu.edu.cn, xushiwu@126.com

Abstract. This paper briefly described the development of wireless sensor network time synchronization algorithm in recent years. More attention was paid to introduce TPSN and RBS algorithms, considering the characteristics of routing node and terminal node in the wireless network. Design a time algorithm based on RBS and TPSN, we proposed a ODTS(on-demand Time Synchronization) algorithm, comparing with TPSN algorithm, the new algorithm not only keeping the synchronization accuracy, but also reduce the power consumption of the entire network.

Keywords: Time synchronization, RBS, TPSN, Fusion Design, Error analysis.

1 Introduction

In recent years, Wireless sensor networks has been rapid development. It can be used many areas of life, for example, military, health, family, business and so on. Time synchronization is one of the basic services of WSN. It can be used in many occasions such as time-stamping sensor event, localization, energy-saving sleep mode and data fusion. NTP protocol is the Internet time synchronization protocol standard, the wireless sensor network has the limited power resources, adverse wireless channel conditions. So NTP protocol is not suitable for wireless sensor networks. GPS systems can also provide high-precision time synchronization, GPS antenna must be installed in open space and it needs high energy consumption, it is too expensive to attach on cheap node. it is also not suitable for wireless sensor networks[1,2].

2 Related Research

Currently, there are a lot of synchronization techniques have been proposed for wireless sensor networks. Such as, TPSN (Timing-Sync Protocol for Sensor Networks) algorithm[3], it based on the sender-receiver time synchronization algorithm, RBS (Reference Broadcast Synchronization) algorithm[4], it follows receiver-receiver model. DCTS (Distributed Consensus Time Synchronization) algorithm[5,6]. PBS (Pairwise Broadcast Synchronization)[7]. BTS (Broadcast Time Synchronization)[8].

Y. Zhang (Ed.): Future Communication, Computing, Control and Management, LNEE 141, pp. 9–15.
springerlink.com

2.1 The Analysis of TPSN Algorithm

TPSN (Timing-Sync Protocol for Sensor Networks) algorithm, the network structure of TPSN is hierarchical. It can be divided into two stages, the first stage is level discovery phase, the second stage is a synchronous phase. Fig.1 shows message exchange between pair of nodes, T1 and T4 are used to record local clock of Synchronization node A .T2 and T3 are used to record local clock of Reference node B. At time T1, the synchronization node A send a synchronization packet to reference node B, the packet contains the level of A and T1,after B has been received the packet, record the time is T2, and immediately respond a synchronization reply packet to the A, the packet contains the level of reference node B ,T1 ,T2 and response time T3. Node A receives the packet at T4. Assuming the same propagation delay from Sending and receiving messages, and this propagation delay time is d. At time t1, Assuming the local clock time offset between synchronization node A and reference node B is m. because from T1 to T4, the time is short, the local clock time offset is the same. From $T2 = T1 + m + d$, $T4 = T3 - m + d$. We can get

$$m = \frac{(T2 - T1) - (T4 - T3)}{2} \tag{1}$$

$$d = \frac{(T2 - T1) + (T4 - T3)}{2} \tag{2}$$

As a result, the local clock on node A could be synchronized with node B's local clock by adding offset m.

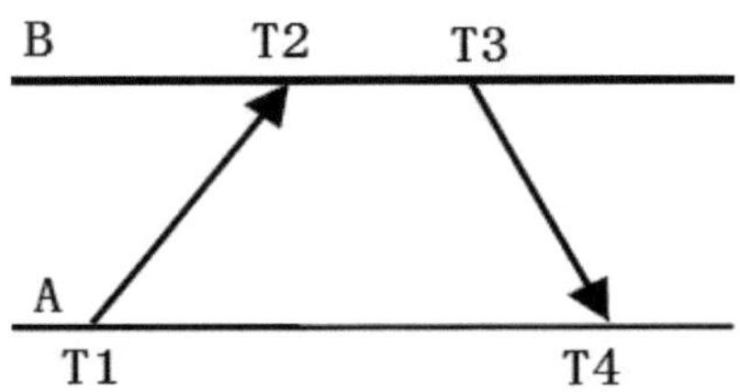

Fig. 1. The theory of TPSN algorithm

2.2 The Analysis of RBS Algorithm

RBS (Reference Broadcast Synchronization) algorithm, The simplest form of RBS is the broadcast of a single pulse to two receivers, allowing them to estimate their relative phase offsets. Which are shown in Fig.2. Reference node transmitter broadcasts a reference packet to two receivers (A and B). Each receiver records the time that the reference was received, according to its local clock. The receivers exchange their observations. Based on this single broadcast alone, the receivers have sufficient information to form a local (or relative) timescale. In order to increase the precision of synchronization statistically, by sending more than one reference. So we can get:

$$\forall i \in n, j \in n: Offset[i,j] = \frac{1}{m}\sum_{k=1}^{m}(T_{j,k} - T_{i,k}) \tag{3}$$

n represent the number of receivers, m represent the number of reference broadcast, Tr, b represent r's clock when it received broadcast b. In the multiple hops, RBS use least-squares linear regression to calculate skew.

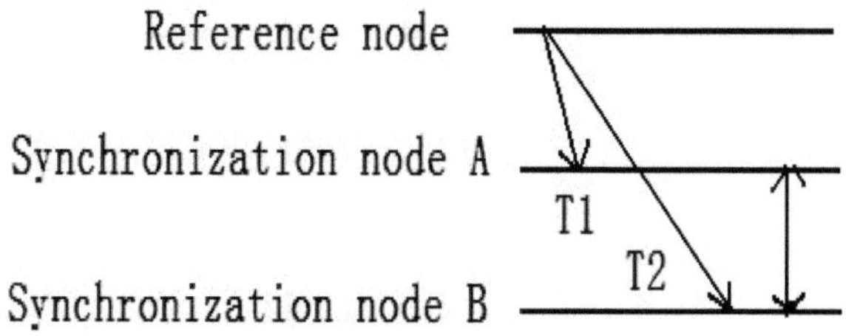

Fig. 2. The Algorithm schematic of RBS

3 Design of Time Synchronization Algorithm

In this paper, we proposed a ODTS algorithm, which based on the TPSN and RBS algorithm, using TPSN algorithm to synchronization between routing node and routing node, adopting RBS algorithm to synchronization among routing node and terminal node, not only keeping the synchronization accuracy, but also reduce the power consumption of the entire network.

3.1 Level Discovery Phase

To describe convenient, we take 4 nodes for example, which was shown is Fig.3, Solid line represents the relationship between father and son. Dotted line represents node can receive synchronization packets in the communication range. The root node A is assigned a level 0 and it initiates this phase by broadcasting a level_discovery packet. The level_discovery packet contains the identity and the level of the sender. B node receive this packet and assign themselves a level, one greater than the level they have received, and known his father node is A, After B node establishing level, B node broadcast a new level_discovery packet, containing his own level. Node C and D receive this packet, establishing their level, note their father node is B, This process is continued and eventually every node in the network is assigned a level. So we can get a routing table.

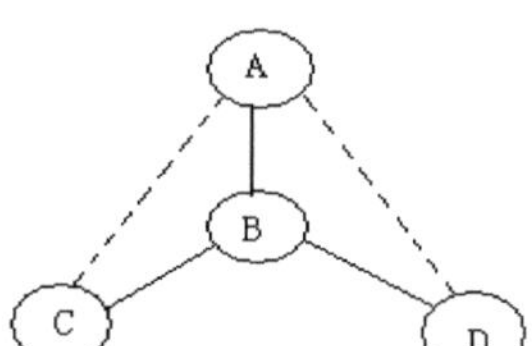

Fig. 3. The diagram of nodes network

3.2 Synchronization Phase

In Fig.4, First, we finished synchronization between root node A and routing node B, At time T1, node B send a synchronization packet to node A, the packet contains the level of B and T1, after node A received the packet, record the time is T21, and immediately respond a synchronization packet to the node B, the packet contains the level of node A, T1, T21 and response time T3.After synchronization node B received the reply from reference node A, record the time is T41. Assuming that the clock drift and the propagation delay do not change in this small span of time, node B can calculate the clock drift m and propagation delay d as:

$$m = \frac{(T21 - T1) - (T41 - T3)}{2} \tag{4}$$

$$d = \frac{(T21 - T1) + (T41 - T3)}{2} \tag{5}$$

Knowing the drift, node B can correct its clock accordingly, so that it synchronizes to node A.

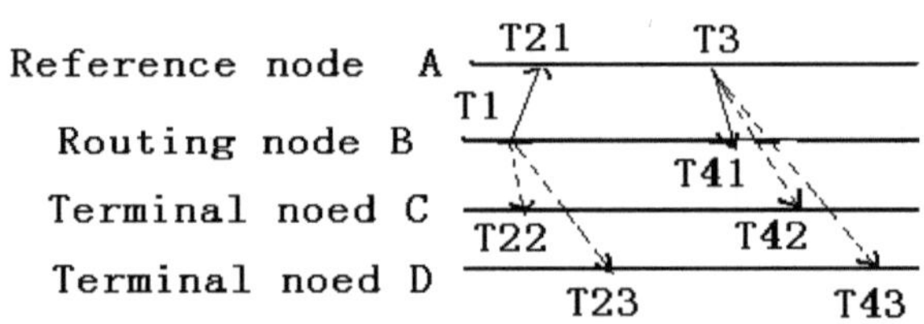

Fig. 4. The schematic of ODTS algorithm

Second, if we have finished synchronization between root node and routing node, now, we can start synchronization between routing node and terminal node. When node B send a synchronization packet to node A, because Node C and D in the communication range of B, Node C receives this packet at T22, Node D receives this packet at T23, when node A respond a synchronization packet to the node B, because Node C and D in the communication range of A, Node C receives this packet at T42, Node D receives this packet at T43, we assume the offset between A and C is m1, and the offset between A and D is m2, we can get m1=T22-T21,m2=T23-T21, knowing the drift, node C and D can correct its clock accordingly, so that it synchronizes to node A.

3.3 Analytical of Error Sources for ODTS Time Synchronization Algorithm

Error analysis of the terminal node C, according the principle of ODTS from Fig.4, we can get

$$T21 = T1 + S_B + A_B + N_B t + P_{B\to A} + R_A + D_{t1}^{B\to A} \tag{6}$$

$$T22 = T1 + S_B + A_B + N_B t + P_{B\to C} + R_C + D_{t1}^{B\to C} \tag{7}$$

so we can get the offset m1 of terminal node c is

$$m1 = T_{22} - T_{21} = (P_{B\to C} - P_{B\to A}) + (R_C - R_A) + (D_{t1}^{B\to C} - D_{t1}^{B\to A}) \tag{8}$$

$$D_{t1}^{B\to C} - D_{t1}^{B\to A} = D_{t1}^{A\to C} = D_{t42}^{A\to C} + RD_{t1\to t42}^{A\to C} \tag{9}$$

So we can get the error of node C is

$$Error1 = m1 - D_{t42}^{A\to C} = (P_{B\to C} - P_{B\to A}) + (R_C - R_A) + RD_{t1\to t42}^{A\to C} \tag{10}$$

As the same, we can get the offset of other terminal node.

Error analysis of routing node B

Because the time synchronization between coordinator and the cluster head node use TPSN algorithm, so have the same source of error. The TPSN algorithm use time mark in the MAC layer, it can eliminate time error from the send time and access time. We can get the error of node B [3]:

$$Error2 = \frac{S_{UC}}{2} + \frac{P_{UC}}{2} + \frac{R_{UC}}{2} + \frac{RD_{t1\to t4}^{A\to B}}{2} \tag{11}$$

T1 is measured by B's local clock at time t1,T22 represent the local time of node C ,T21 represent the local time of node A. SB is the packet transmission time of node B, AB is the access time of sending packet. NBt is the time of sending NB bits. $D_{t1}^{B\to A}$ is the local clock offset between node B and node A at time t1. $S_{UC} = S_A - S_B$, $P_{UC} = P_{A\to B} - P_{B\to A}$, $R_{UC} = R_B - R_A$, SA is the packet transmission time of node A, SB is the packet transmission time of node B, $P_{A\to B}$ is the time of node A transmit the packet to node B, $P_{B\to A}$ is the time of node B transmit the packet to node A, RA is the time of node A processing packet, RB is the time of node B processing packet, $RD_{t1\to t4}^{A\to B}$ represent the additional local clock offset between node A and node B from time t1 to t4.

3.4 The Analysis of Algorithm Performance

We use 4 different networks to test ODTS performance, we assuming terminal node can receive packet within two hops routing node. We compare power consumer and precision of ODTS algorithm with TPSN algorithm. Case 1 represents 1 routing node and 1 terminal node. Case 2 represents 2 routing node and 3 terminal node. Case 3 represents 3 routing node and 5 terminal node. Case 4 represents 4 routing node and 6 terminal node.

From table 1 and table 2, we know that ODTS keeping the same number of receive packets used to synchronize with TPSN, but the sent packets is less than TPSN, as we known, in the wireless sensor network, the most of power consume is sent packets, and from four different case, more terminal nodes, more to reduce the power consumption of the entire network. From table 3 and table 4, although the accuracy of ODTS is less than TPSN, the difference is not large. From the perspective of energy consumption, ODTS is better than TPSN. It can prolong the life time of nodes in the network. Although the synchronization accuracy of TPSN is high, the power consume is also too high, so it is not suitable for wireless sensor network.

Table 1. The Average packets used to synchronize using TPSN algorithm

	Case 1	Case 2	Case 3	Case 4
Number of sent packets	2 packets	8 packets	14 packets	18 ackets
Number of receive packets	2 packets	8 packets	14 packets	18 ackets
Total number of packets	4 packets	16 packets	28 packets	36 ackets

Table 2. The Average packets used to synchronize using ODTS algorithm

	Case 1	Case 2	Case 3	Case 4
Number of sent packets	2 packets	2 packets	4 packets	6 packets
Number of receive packets	2 packets	8 packets	14 packets	18 ackets
Total number of packets	4 packets	10 packets	18 packets	24 ackets

Table 3. Statistics of synchronization error over multi-hop of TPSN algorithm

	1 hop distance	2 hop distance	3 hop distance	4 hop distance
Best case error	3.2us	4.6us	5.9us	5.1us
Worst case rror	30.6us	46.8us	51.2us	50.3us
Average error	18.2us	22.3us	25.8us	25.3us

Table 4. Statistics of synchronization error over multi-hop of ODTS algorithm

	1 hop distance	2 hop distance	3 hop distance	4 hop distance
Best case error	3.2us	4.9us	6.0us	5.8us
Worst ase error	30.6us	48.8us	52.6us	53.2us
Average error	18.2us	24.5us	28.3us	27.6us

4 Summary

AS we known, the precision of TPSN is higher .but Average packets used to synchronize needs energy consumption of four packets, the innovation of this paper is combined TPSN and RBS, proposed a ODTS algorithm, we full use of the advantages of both algorithm. Compared to the TPSN, he precision declined, but the precision is very close to TPSN. In the industrial harmful gas detection, most of the node as terminal node to sample harmful gas data values, and only a few node as routing node, which used to store and forwarding the data from the terminal node. Through multiple hops transmission to the PC, data processing, get the result of we need. If a lot of the terminal node in wireless sensor network. Using the ODTS algorithm is proposed in this paper. The premise of keeping synchronous precision demand, it also can reduce the power consumption in the all network.

Acknowledgment. The work in this paper was partially supported by Science and Technology Department of Fujian Province-The Embedded Designed of the Neural Network and E-mail Filtering System (JB0976) and The Research of The Wireless Sensor Networks in Smart Home System (JB09071);

References

1. Lenzen, C., Sommer, P., Wattenhofer, R.: In: Proc. 7th ACM Conference on Embedded Networked Sensor Systems (SenSys), USA (2009)
2. Sommer, P., Wattenhofer, R.: In: Proc. 8th ACM/IEEE International Conference on Information Processing in Sensor Networks (IPSN), San Francisco, USA, p. 37 (2009)
3. Ganeriwal, S., Kumar, R., Srivastava, M.B.: In: Proceedings of the 1st International Conference Embedded Networked Sensor Systems (SenSys's 2003), p. 138. ACM press, USA (2003)
4. Elson, J., Girod, L., Estrin, D.: In: Proceedings of the 5th Symposium on Operating Systems Design and Implementation (OSDI 2002), USA, p. 147 (2002)
5. Gang, X., Kishore, S.: In: Wireless Communications and Networking Conference, p. 1. IEEE, Budapest (2009)
6. Sommer, P., Wattenhofer, R.: In: Proc. 8th ACM/IEEE International Conference on Information Processing in Sensor Networks (IPSN), San Francisco, USA, p. 37 (2009)
7. Cheng, K.-Y., Lui, K.-S., Wu, Y.-C., Tam, V.: IEEE Transactions on Wireless Communications 8, 1764 (2009)
8. Xu, C.-N., Zhao, L., Xu, Y.-J., Li, X.-W.: Chinese Journal of Computers 30, 514 (2007)

Power Load Forecasting Based on the Indicators of Industrialization and Urbanization

Gu Changzhou[1,a], Long Jie[1,b], Chen Xiaoping[2,c], and Guo Ying[2,d]

[1] Jiaozuo Power Supply Company , Henan Province, China
[2] North China Electric Power University,Beijing, China
[a] cxp4828722@126.com, [b] 88519961@qq.com,
[c] wenshi2@163.com, [d] 331543746@qq.com

Abstract. In this article, through selecting and calculating of one city's index of industrialization and urbanization from 1990 to 2005, we established multiple linear regression model by industrialization and urbanization index as a parameter based on classical linear regression model. Then test the model by historical data of power consumption from 2006 to 2008. The test error shows that the model of the city power load forecasting result is quite satisfactory, so we use it to predict electrical load in 2011-2015. The result can provide reference for the planning of the Grid.

Keywords: Load forecast, Industrial index, Urbanization index.

1 Introduction

Power Load forecasting is under the condition of considering some important factors(such as operating performance, compatibility decisions, natural conditions and social impact), it is a mathematical method that researching or using a system to handle the load of past and future to determine the future value of a particular moment of the load to under the conditions of a certain accuracy. Electric power system using load forecasting is one of the important issues for domestic and foreign scholars, and generated a lot of theory and method for power load forecasting, such as value of consumption method, electricity elasticity coefficient method, integrated power consumption, regression analysis and gray system prediction method etc.

We already know that electricity demand and economic and social development are closely related. Therefore, as a measure of social and economic development of industrialization and urbanization indicators will inevitably affect the power consumption, but the current study makes little connection between the two indicators above and electricity. In order to validate the strong correlation between them, this paper makes a city for example, after calculating the index of industrialization and urbanization, using multiple linear regression model to fit, it has high prediction accuracy by inspection. Finally, the city's electricity demand from 2011 to 2015 was forecasted based on the model, look forward to the future of network planning for the city to provide evidence.

Y. Zhang (Ed.): Future Communication, Computing, Control and Management, LNEE 141, pp. 17–23.
springerlink.com

2 Industrial Indicators

2.1 The Definition of Industrialization and the Impact on Power Demand Ease of Use

In the existing "industrial" definition can be roughly divided into the following two categories: narrow industrialization is the industry (manufacturing and machinery production) or the process of secondary industry output value (or income) in the share of GDP continued to rise, and the rising process of industrial output and employment proportion. General industrialization not only includes industrial itself mechanization and modernization, but also agricultural mechanization and modernization, including more extensive content such as urbanization, and even refers to the entire process of economic modernization. Electricity demand and industrial development are closely linked, industrial electricity consumption has a large proportion in the whole society, the rising of industrial electricity consumption will stimulate the growth of total electricity consumption directly to grow rapidly. In 2008, the city industrial output value reached 1.250683 trillion yuan, in the scale of industrial added value, the light industrial added value accounted for only 16.3%, heavy industry accounted for 83.7%. Rapid industrial development will inevitably bring about the rapid growth of electricity consumption and electricity consumption structure changes. Electricity as an important industrial energy, will play an important supporting role for industry in a long period. In the city's total electricity consumption, industrial electricity accounts for a large proportion, the rising of industrial electricity will stimulate the growth of total electricity consumption directly to grow rapidly. In 2008, the second industrial consumption in this city accounted for 76.7% in whole society. In secondary industry, textile industry, pharmaceutical industry, fabricated metal products, ferrous metal smelting industry and non-ferrous metals smelting and transportation, electronic equipment manufacturing industry etc. which are high energy consuming industries are the main factors to drive the rapid development of electric consumption. With the upgrade of consumers, automotive, housing has increasingly become a hot issue, industrial development produced more and more important effect to electricity consumption. Put the level of industrialization as an important factor taken into account in power load forecasting, it has an important practical significance for predicting correctly for the city's electricity needs and planning the development of power industry.

2.2 The Meaning of Industrial Index and Its Calculation

The level of industrialization is an important indicator to determine the stage of regional economic development. From the current situation of domestic academic research, the analysis of stage of industrialization in China is based on standard theory of industrialization. Based on the above concepts and standard theory of industrialization, and taking the actual situation of our country and data availability into account, we think we can select the following four categories of indicators to calculate industrial index: (1) income level indicators: per capita GDP; (2) industry structure indicators: the proportion of non-agricultural output value; (3) industry structure indicators: re-light ratio; (4) the structure of employment indicators: the proportion of non-agricultural labor force. we used principal component analysis to determine the index weight in determining, industrial index obtained is estimated as:

Per capita industrialization index = 0.3 * GDP +0.27 * +0.28 * the proportion of non-agricultural output value of heavy industry, light industry labor ratio 0.15 * the proportion of non-agricultural industries

According to this method, the industrial index of the city from 1990 to 2008 are calculated, the normalized results are as follows in Table1:

Table 1. Industrial index of the city from 1990 to 2008

Year	Industrial Index	Year	Industrial Index
1990	0.913	2000	4.543
1991	0.989	2001	5.011
1992	1.173	2002	5.599
1993	1.519	2003	6.687
1994	2.029	2004	8.004
1995	2.558	2005	10.739
1996	3.072	2006	11.074
1997	3.441	2007	14.521
1998	3.729	2008	15.949
1999	4.033		

2.3 The Prediction of Industrial Index of the City

In this paper, see 19-year the average history growth rate as the city's industrial index from 2011 to 2015, predicted results are as follows Table 2:

Table 2. The prediction of industrial index of the city from to 2011 to 2015

Year	2011	2012	2 013	2014	2015
Predictive Value	21.985	25.245	28.988	33.286	38.221

3 Indicators of Urbanization

3.1 The Basic Meaning of Urbanization

The so-called urbanization is a natural, social and historical process which urban population and geographical size is growing continuously, that refers to the rural population shift to cities and towns, secondary and tertiary industries to keep the town together, so that an increasing number of cities and towns. Urbanization mainly in two aspects: on the one hand, it reflected in the person's geographic location and occupational changes and the resulting evolution of modes of production and way of life; on the other hand, it showed by the increasing number of urban population and cities, towns and urban expansion and intensification of economic and social improvement in the level of modernization.

To promote the process of urbanization in rural areas, so that more farmers moved into a fully equipped, beautiful small towns and new communities. The promotion of urbanization means the changes in people's living standards and consumer attitudes. The level of household consumption in driving growth, will inevitably bring about the development of industrial and service sectors, and ultimately have a major impact on electricity consumption. Therefore, load forecasting which is necessary to the level of urbanization as an important factor taken into account.

The proportion of non-agricultural population is a widely used measure of the academic stage of an important indicator of the level of urbanization. This paper uses this indicator to measure the city's urbanization level.

Table 3. The level of urbanization from 1990 to 2008 in the city

Year	The Level of Urbanization	Year	The Level of Urbanization
1990	0.560	2000	0.584
1991	0.562	2001	0.586
1992	0.563	2002	0.589
1993	0.564	2003	0.594
1994	0.566	2004	0.596
1995	0.568	2005	0.599
1996	0.571	2006	0.602
1997	0.572	2007	0.605
1998	0.576	2008	0.607
1999	0.580		

3.2 The Forecast of Level of Urbanization

Macroeconomic policies from the city and the current development trend analysis, the steady growth of non-agricultural population trend will continue for a period, and with the continuous development of the national economy, is still a growth stage, the level of urbanization will continue to increase. To this end, on the basis of research data, a city with the gray prediction method to predict the level of urbanization, the data for the 2000~2008 year with the level of urbanization.

After calculation the prediction model is:
$x^{(1)}(k+1) = 117.1039e^{0.005k} - 116.52$

The correlation coefficient of prediction model was 0.734 by correlation test. This model predicts the level of urbanization for the city between 2011~2015, are shown in Table 4:

Table 4. Predictive value of urbanization level from 2011 to 2015

Year	2011	2012	2013	2014	2015
The Level of Urbanization	0.617	0.620	0.623	0.626	0.630

4 Multiple Linear Regression Process

Many factors affect the power consumption, especially long-term forecast which forecast period is longer. In this process, the factors that impact on electricity consumption are constantly changing and developing itself, so the analysis of factors affect the demand for electricity will help improve the prediction accuracy. With the continuous development of economy, these indicators that GDP, industrialization and urbanization are the main factors that impact on electricity consumption. Through analysis, each indicator has a strong correlation with electricity, using these three indicators multiple regression prediction model can overcome the traditional shortcomings of taking a single factor into account, it has a certain credibility and able to draw more realistic predictions. The following is using relevant datas and SPSS software for regression analysis.

GDP, Industrialization and Urbanization. Using SPSS software to obtain the following regression results in Table 5:

Table 5. Regression results

Model	R	R squared	Adjusted R square	Standard error of estimation	DW test value
1	.997	.995	.993	6.57080	.916

a. Independent variable combinations : (Constant), urbanization, industrialization, GDP
b. dependent variable: consumption

Table 6. Deviation and the F statistic

Model		Sum of Squares	df	F	Sig.
1	Regression	97225.635	3	750.626	.000
	Residual	518.104	12		
	The total deviation	97743.740	15		

a. Independent variable combinations: (Constant), urbanization, industrialization、GDP
b. dependent variable: Electricity consumption

Table 7. Regression coefficient estimate value

		Regression oefficient estimate value			t	Sig.
Model		B	Std.Error	Beta		
	(Constant)	40.114	453.927		.640	.534
	GDP	.099	.033	1.237	2.992	.011
	Urbanization	-5.515	9.342	-.187	-.590	.566
	Industrialization	-339.992	813.390	-.055	418	.683

a. dependent variable: Electricity consumption

Though the Model fitting results are good, we can see t test is not through from the estimation results of regression coefficient. so consider doing regression analysis again after eliminating independent variable GDP.

5 The Industrialization and the Urbanization

Table 8. Regression results

Model	R	R squared	Adjusted R square	Standard error of estimation	DW test value
1	.995	.991	.989	8.34127	1.386

a. Independent variable combinations: (Constant), urbanization, industrialization
b. dependent variable: consumption

Table 9. Deviation and the F statistic

Model		Sum of Squares	df	F	Sig.
1	Regression	96839.240	2	695.916	.000
	Residual	904.499	13		
	The total deviation	97743.740	15		

a. Independent variable combinations: (Constant), urbanization, industrialization
b. dependent variable: consumption

Table 10. Regression coefficient estimate value

		Regression coefficient estimate value			t	Sig.
Model		B	Std.Error	Beta		
1	(Constant)	-826.628	327.554		-2.524	.025
	Urbanization	21.643	2.797	.734	7.737	.000
	Industrialization	1662.845	586.395	.269	2.836	.014

a dependent variable: consumption

We can see from the regression results that R = 0.995, F = 695.916, t test values is satisfied, but the only drawback is that there is serial correlation (DW = 1.386), the whole model is also quite satisfactory. Multiple regression model is to be established as following:

Y=21.643*X1+1662.845*X2-826.628

Using multiple regression model which we have received to predict the electricity consumption from 2006 to 2008 and compared with actual annual consumption, the error analysis as follows:

Table 11. Error Analysis

Year	Actual consumption	Electricity consumption forecast	Error
2006	445.73	424.08	4. 85%
2007	510.68	503.67	1.37%
2008	535.26	537.90	0.49%

This table shows that using the model to predict the city's electricity consumption is an effective prediction method and has achieved high precision. Using this model to predict the city's electricity demand from 2011 to 2015, the result is shown in Table 12:

Table 12. The future demand forecast[unit: TWh]

Year	2011	2012	2013	2014	2015
Predict value	675.17	750.71	836.71	934.72	1048.18

6 Summary

Taking GDP, industrial indicators, urbanization indicators into account, we use multiple regression model to predict the city's electricity demand from 2011 to 2015 in order to provide the basis for future network planning.

References

1. Niu, D.: Load forecasting technology and its application, vol. 3-4, pp. 165–210. China Electric Power Press, Beijing (1998)
2. Duan, A.: The market economy under the conditions of load forecasting. Demand Side Management (2), 24 (2002)
3. Wang, X.: Urbanization on the Development of Overseas. Hebei Normal University (4) (2006)
4. Yu, X.: Characteristics of urbanization in Chongqing. Cooperative Economic and Technology (01) (2009)
5. Dong, H.: At this stage of urbanization in China Construction. The Legal System and Society (05) (2009)

Quantum Bit and Square Well in Quantum Computation

Lu Jun

Department of Foundational Science,
Beijing Union University, Beijing, 100101, China
lujun@buu.edu.cn

Abstract. The spectacular promise of quantum computers is to enable new algorithms which render feasible problems requiring exorbitant resources for their solution on a classical computer. Quantum computation is based on transformation of quantum states. Quantum bits are two-level quantum systems, and as the simplest elementary building blocks for a quantum computer, they provide a convenient labeling for pairs of states and their physical realizations. This paper shows how a single quantum bit can be represented by the two lowest levels in a square well potential, and how simple perturbations of the potential can effect computational operations on the quantum bit. The controlling system is just another quantum system, and it couples to the one we are trying to do quantum computation. These problems lead to decoherence.

Keywords: quantum bit, square well, quantum computation, decoherence.

1 Introduction

Quantum computation and quantum information is the study of the information processing tasks that can be accomplished using quantum mechanical systems. Like many simple but profound ideas it was a long time before anybody thought of doing information processing using quantum mechanical systems. Quantum mechanics is the most accurate and complete description of the world known. It is also the basis for an understanding of quantum computation and quantum information.

Algorithms are the key concept of computer science. An algorithm is a precise recipe for performing some task. The fundamental model for algorithms will be the Turing machine [1]. This is an idealized computer, rather like a modern personal computer, but with a simpler set of basic instructions, and an idealized unbounded memory. They can be used to execute any algorithm whatsoever, even one running on an apparently much more powerful computer.

Practically speaking, many interesting problems are impossible to solve on a classical computer, not because they are in principle insoluble, but because of the astronomical resources required to solve realistic cases of the problem [2]. The spectacular promise of quantum computers is to enable new algorithms which render feasible problems requiring exorbitant resources for their solution on a classical computer.

Y. Zhang (Ed.): Future Communication, Computing, Control and Management, LNEE 141, pp. 25–28.
springerlink.com

Quantum computation and quantum information is a field of fundamental interest because we believe quantum information processing machines can actually be realized in nature. Otherwise, the field would be just a mathematical curiosity. Nevertheless, experimental realization of quantum circuits, algorithms, and communication systems has proven extremely challenging. For building a quantum computer, the elementary units of the experimental requirements are quantum bits. To realize a quantum computer, we must not only give quantum bits some robust physical representation, but also select a system in which they can be made to evolve as desired [3, 4]. Furthermore, we must be able to prepare quantum bits in some specified set of initial states, and to measure the final output state of the system.

2 Representation of Quantum Information

Quantum computation is based on transformation of quantum states. Quantum bits are two-level quantum systems, and as the simplest elementary building blocks for a quantum computer, they provide a convenient labeling for pairs of states and their physical realizations. Thus, for example, the four states of a spin-3/2 particle, | m= +3/2 >, | m= +1/2 >, | m= -1/2 >, | m= -3/2 > could be used to represent two quantum bits.

For the purpose of computation, the crucial realization is that the set of accessible states should be finite. The position x of a particle along a one-dimensional line is not generally a good set of states for computation, even though the particle may be in a quantum state | x >, or even some superposition $\sum_x c_x |x\rangle$. This is because x has a continuous range of possibilities, and the Hilbert space has infinite size, so that in the absence of noise the information capacity is infinite. The presence of noise reduces the number of distinguishable states to a finite number.

In fact, it is generally desirable to have some aspect of symmetry dictate the finiteness of the state space, in order to minimize decoherence. For example, a spin-1/2 particle lives in a Hilbert space spanned by the | ↑ > and | ↓ > states; the spin state cannot be anything outside this two-dimensional space, and thus is a nearly ideal quantum bit when well isolated.

If the choice of representation is poor, then decoherence will result. For example, a particle in a finite square well which is just deep enough to contain two bound states would make a mediocre quantum bit, because transitions from the bound states to the continuum of unbound states would be possible. These would lead to decoherence since they could destroy quantum bit superposition states. For single quantum bits, the figure of merit is the minimum lifetime of arbitrary superposition states; a good measure, used for spin states and atomic systems, is T2 , the transverse relaxation time of states such as $(|0\rangle + |1\rangle)/2$. Note that T1 , the longitudinal relaxation time of the higher energy | 1 > state, is just a classical state lifetime, which is usually longer than T2 .

3 Square Well and Quantum Bit

A prototypical quantum system is known as the "square well", which is a particle in a one-dimensional box [5], behaving according to Schrödinger equation

$$i\hbar\frac{d\psi}{dt}=H\psi \quad . \tag{1}$$

The Hamiltonian for this system is

$$H=\frac{p^2}{2m}+V(x) \tag{2}$$

where V(x) = 0 for 0 < x < L , and V(x) = ∞ otherwise, The energy eigenstates, expressed as wavefunctions in the position basis, are

$$\psi_n=\sqrt{\frac{2}{L}}\sin\left(\frac{n\pi}{L}x\right), \tag{3}$$

where n is an integer, and

$$\psi_n(t)=\exp(-iE_nt)\psi_n \quad , \tag{4}$$

$$E_n=\frac{n^2\pi^2m}{2L^2} \quad . \tag{5}$$

These states have a discrete spectrum. In particular, suppose that we arrange matters such that only the two lowest energy levels need be considered in an experiment. We define an arbitrary wavefunction of interest as

$$\psi=a\psi_1+b\psi_2 \tag{6}$$

Since

$$\psi(t)=\exp[-i(E_1+E_2)/2t][a\exp(-i\omega t)\psi_1+b\exp(i\omega t)\psi_2] \quad . \tag{7}$$

where ω = (El - E2)/2, we can just forget about everything except a and b, and write the state as the two-component vector

$$\psi=\begin{bmatrix}a\\b\end{bmatrix} \quad . \tag{8}$$

This two-level system represents a quantum bit. Under time evolution, this quantum bit evolves under the effective Hamiltonian $H=\hbar\omega Z$, which can be disregarded by moving into the rotating flame. To perform operations to this quantum bit, we perturb H. Consider the effect of adding the additional term

$$\delta V(x)=-V_0(t)\frac{9\pi^2}{16L}\left(\frac{x}{L}-\frac{1}{2}\right), \tag{9}$$

to V(x). In the basis of the two-level system, this can be rewritten by taking the matrix elements

$$V_{nm}=\langle\psi_n|\delta V(x)|\psi_m\rangle, \tag{10}$$

giving V11= V22= 0, and V12= V21= V0 , such that, to lowest order in V0 , the perturbation to H is H1=V0(t)X . This generates rotations about the x axis. Similar techniques can be used to perform other single quantum bit operations, by manipulating the potential function.

This shows how a single quantum bit can be represented by the two lowest levels in a square well potential, and how simple perturbations of the potential can effect computational operations on the quantum bit. However, perturbations also introduce higher order effects, and in real physical systems boxes are not infinitely deep, other levels begin to enter the picture, and the two-level approximation begins to fail. Also, in reality, the controlling system is just another quantum system, and it couples to the one we are trying to do quantum computation. These problems lead to decoherence.

4 Summary

An important characteristic of the measurement process for quantum computation is the wavefunction collapse which describes what happens when a projective measurement is performed. The output from a good quantum algorithm is a superposition state which gives a useful answer with high probability when measured.

Many difficulties with measurement can be imagined, for example, inefficient photon counters and amplifier thermal noise can reduce the information obtained about measured quantum bit states in the scheme just described. Furthermore, projective measurements are often difficult to implement. They require that the coupling between the quantum and classical systems be large, and switchable. Measurements should not occur when not desired, otherwise they can be a decoherence process.

Surprisingly, however, strong measurements are not necessary, weak measurements which are performed continuously and never switched off, are usable for quantum computation. This is made possible by completing the computation in time short compared with the measurement coupling, and by using large ensembles of quantum computers. These ensembles together give an aggregate signal which is macroscopically observable and indicative of the quantum state. Use of an ensemble introduces additional problems.

Acknowledgment. This work was supported by the Funding Project for Academic Human Resources Development in Institutions of Higher Learning under the Jurisdiction of Beijing Municipality (Grant No. PHR200907120), the Science and Technology Research Program of Beijing Municipal Education Commission (Grant No. KM201111417007), and the Innovative Talent Foundation of Beijing Union University.

References

1. Turing, A.M.: Proc. Lond. Math. Soc. 42, 230 (1936)
2. Feynman, R.P.: Int. J. Theor. Phys. 12, 467 (1982)
3. Divincenzo, D.P.: Science 270, 255 (1995)
4. Divincenzo, D.P.: Proc. R. Soc. London A 454, 261 (1998)
5. Lu, J., Wang, X.M.: Advanced Materials Research 160, 625 (2011)

Physical Realization of Harmonic Oscillator Quantum Computer

Lu Jun

Department of Foundational Science,
Beijing Union University, Beijing, 100101, China
lujun@buu.edu.cn

Abstract. Algorithms are the key concept of computer science. The spectacular promise of quantum computers is to enable new algorithms which render feasible problems requiring exorbitant resources for their solution on a classical computer. The relationship between the energy eigenstates of a simple harmonic oscillator and quantum computation comes by taking a finite subset of these states to represent quantum bits. These quantum bits will have lifetimes determined by physical parameters such as the cavity quality factor, which can be made very large by increasing the reflectivity of the cavity walls. Moreover, unitary transforms can be applied by simply allowing the system to evolve in time. We begin by studying the system Hamiltonian, and then discuss how one might implement simple quantum logic gates.

Keywords: Quantum computer, harmonic oscillator, quantum bit, unitary transform.

1 Introduction

Algorithms are the key concept of computer science. An algorithm is a precise recipe for performing some task. The fundamental model for algorithms will be the Turing machine [1]. This is an idealized computer, rather like a modern personal computer, but with a simpler set of basic instructions, and an idealized unbounded memory. They can be used to execute any algorithm whatsoever, even one running on an apparently much more powerful computer.

Practically speaking, many interesting problems are impossible to solve on a classical computer, not because they are in principle insoluble, but because of the astronomical resources required to solve realistic cases of the problem [2]. The spectacular promise of quantum computers is to enable new algorithms which render feasible problems requiring exorbitant resources for their solution on a classical computer.

Quantum computation and quantum information is a field of fundamental interest because we believe quantum information processing machines can actually be realized in nature. Otherwise, the field would be just a mathematical curiosity. Nevertheless, experimental realization of quantum circuits, algorithms, and communication systems has proven extremely challenging. For building a quantum computer, the elementary units of the experimental requirements are quantum bits. To realize a quantum

Y. Zhang (Ed.): Future Communication, Computing, Control and Management, LNEE 141, pp. 29–34.
springerlink.com

computer, we must not only give quantum bits some robust physical representation, but also select a system in which they can be made to evolve as desired [3, 4]. Furthermore, we must be able to prepare quantum bits in some specified set of initial states, and to measure the final output state of the system.

The challenge of experimental realization is that these basic requirements can often only be partially met. A coin has two states, and makes a good bit, but a poor quantum bit because it cannot remain in a superposition state for very long. A single nuclear spin can be a very good quantum bit, because superposition of being aligned with or against an external magnetic field can last a long time. But it can be difficult to build a quantum computer from nuclear spins because their coupling to the world is so small that it is hard to measure the orientation of single nuclei. The observation that the constraints are opposing is general: a quantum computer has to be well isolated in order to retain its quantum properties, but at the same time its quantum bits have to be accessible so that they can be manipulated to perform a computation and to read out the results. A realistic implementation must strike a delicate balance between these constraints, so that the relevant question is not how to build a quantum computer, but rather, how good a quantum computer can be built [5-12].

An example of a simple harmonic oscillator is a particle in a parabolic potential well:

$$V(x)=\frac{1}{2}m\omega^2x^2 \quad . \tag{1}$$

In the classical world, this could be a mass on a spring, which oscillates back and forth as energy is transferred between the potential energy of the spring and the kinetic energy of the mass. It could also be a resonant electrical circuit, where the energy sloshes back and forth between the inductor and the capacitor. In these systems, the total energy of the system is a continuous parameter.

In the quantum domain, which is reached when the coupling to the external world becomes very small, the total energy of the system can only take on a discrete set of values. An example is given by a single mode of electromagnetic radiation trapped in a high Q cavity; the total amount of energy (up to a fixed offset) can only be integer multiples of $\hbar\omega$, an energy scale which is determined by the fundamental constant $\hbar$ and the frequency of the trapped radiation ω .

The set of discrete energy eigenstates of a simple harmonic oscillator can be labeled as $|n\rangle$, where $n=0,1,\cdots,\infty$. The relationship to quantum computation comes by taking a finite subset of these states to represent quantum bits. These quantum bits will have lifetimes determined by physical parameters such as the cavity quality factor Q, which can be made very large by increasing the reflectivity of the cavity walls. Moreover, unitary transforms can be applied by simply allowing the system to evolve in time. We begin by studying the system Hamiltonian, and then discuss how one might implement simple quantum logic gates such as the controlled—NOT.

2 Quantum Harmonic Oscillator

The harmonic oscillator is an extremely important and useful concept in the quantum description of the physical world, and a good way to begin to understand its properties is to determine the energy eigenstates of its Hamiltonian [13-17]. The Hamiltonian for a particle in a one-dimensional parabolic potential is

$$H = \frac{p^2}{2m} + \frac{1}{2} m\omega^2 x^2 \quad , \tag{2}$$

where p is the particle momentum operator, m is the mass, x is the position operator, and ω is related to the potential depth. Recall that x and p are operators in this expression, which can be rewritten as

$$H = \hbar\omega\left(a^+ a + \frac{1}{2}\right) \quad , \tag{3}$$

where a+ and a are creation and annihilation operators, defined as

$$a = \frac{1}{\sqrt{2m\hbar\omega}}(m\omega x + ip) \quad , \tag{4}$$

$$a^+ = \frac{1}{\sqrt{2m\hbar\omega}}(m\omega x - ip) \quad . \tag{5}$$

One way to determine the energy eigenstates of the Hamiltonian for the harmonic oscillator is simply to solve the Schrödinger equation

$$\frac{\hbar^2}{2m}\frac{d^2\psi_n(x)}{dx^2} + \frac{1}{2} m\omega^2 x^2 \psi_n(x) = E\psi_n(x) \quad , \tag{6}$$

for $\psi_n(x)$ and the eigenenergies E, subject to $\psi_n(x) \to 0$ at $x = \pm\infty$, and $\int \psi_n(x) dx = 1$, the first five solutions are sketched in Fig. 1. These wave functions describe the probability amplitudes that a particle in the harmonic oscillator will be found at different positions within the potential.

Although these pictures may give some intuition about what a physical system is doing in co-ordinate space, we will generally be more interested in the abstract algebraic properties of the states. Specifically, suppose $|\psi\rangle$ satisfies Eq. (6) with energy E. Then defining operators a and a+ as in Eq. (4) and (5), we find that since $[H, a^+] = \hbar\omega a^+$,

$$Ha^+|\psi\rangle = ([H, a^+] + a^+ H)|\psi\rangle = (\hbar\omega + E)a^+|\psi\rangle \quad , \tag{7}$$

that is, $a^+|\psi\rangle$ is an eigenstate of H, with energy $E + \hbar\omega$. Similarly, $a|\psi\rangle$ is an eigenstate with energy $E - \hbar\omega$. Because of this, a+ and a are called raising and lowering operators, It follows that $a^{+n}|\psi\rangle$ are eigenstates for any integer n, with energies $E + n\hbar\omega$. There are thus an infinite number of energy eigenstates, whose energies are equally spaced apart, by $\hbar\omega$. Moreover, since H is positive definite, there must be some $|\psi_0\rangle$ for which $a|\psi_0\rangle = 0$, this is the ground state – the eigenstate of H with lowest energy. These results efficiently capture the essence of the quantum harmonic oscillator, and allow us to use a compact notation $|n\rangle$ for the eigenstates, where n is an integer, and

$$H|n\rangle = \hbar(n+1/2)|n\rangle. \quad (8)$$

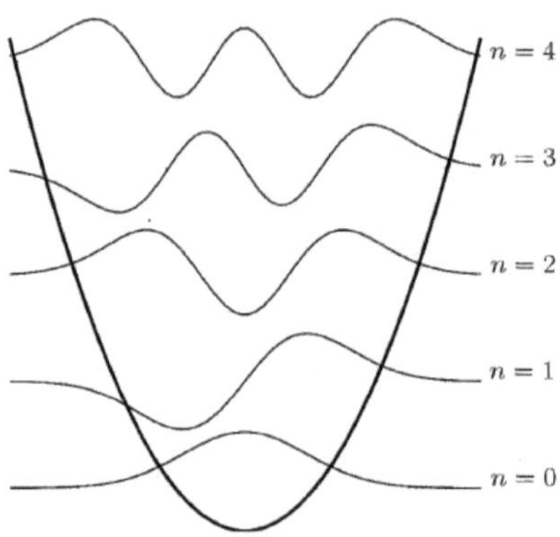

Fig. 1. The first five solutions of the Schrödinger equation (6) for the harmonic oscillator. These wave functions describe the probability amplitudes that a particle in the harmonic oscillator will be found at different positions within the potential.

3 Quantum Computation

The eigenstates $|n\rangle$ of H, where $n = 0,1,\cdots,\infty$ have the properties

$$a^+a|n\rangle = n|n\rangle, \quad (9)$$

$$a^+|n\rangle = \sqrt{n+1}|n+1\rangle, \quad (10)$$

$$a|n\rangle = \sqrt{n}|n-1\rangle. \quad (11)$$

It convenient to express interactions with a simple harmonic oscillator by introducing additional terms involving a and a+, and interactions between oscillators with terms such as $a_1^+a_2 + a_1a_2^+$. For now, however, we confine our attention to a single oscillator. We will assume for the purpose of discussion that an arbitrary state can be perfectly prepared, and that the state of the system can be projectively measured, but otherwise, there are no interactions with the external world, so that the system is perfectly closed.

Suppose we want to perform quantum computation with the single simple harmonic oscillator described above. The most natural choice for representation of quantum bits is the energy eigenstates $|n\rangle$. This choice allows us to perform a controlled-NOT gate in the following way. Recall that this transformation performs the mapping:

$$|00\rangle_L = |00\rangle_L \tag{12}$$

$$|01\rangle_L = |01\rangle_L \tag{13}$$

$$|10\rangle_L = |11\rangle_L \tag{14}$$

$$|11\rangle_L = |10\rangle_L \tag{15}$$

on two quantum bit states, here the subscript L is used to clearly distinguish 'logical' states in contrast to the harmonic oscillator basis states. Let us encode these two quantum bits using the mapping

$$|00\rangle_L = |0\rangle \tag{16}$$

$$|01\rangle_L = |2\rangle \tag{17}$$

$$|10\rangle_L = (|4\rangle + |1\rangle)/\sqrt{2} \tag{18}$$

$$|11\rangle_L = (|4\rangle - |1\rangle)/\sqrt{2}\ . \tag{19}$$

Now suppose that at t = 0 the system is started in a state spanned by these basis states, and we simply evolve the system forward to time $t = \pi / \hbar\omega$. This causes the energy eigenstates to undergo the transformation $|n\rangle \to \exp(-\mathrm{i}\pi a^+ a)|n\rangle = (-1)^n |n\rangle$ such that $|0\rangle$, $|2\rangle$ and $|4\rangle$ stay unchanged, but $|1\rangle \to -|1\rangle$. As a result, we obtain the desired controlled-NOT gate transformation [18].

In general, a necessary and sufficient condition for a physical system to be able to perform a unitary transform U is simply that the time evolution operator for the system, $T = \exp(-\mathrm{i}Ht)$, defined by its Hamiltonian H, has nearly the same eigenvalue spectrum as U. In the case above, the controlled-NOT gate was simple to implement because it only has eigenvalues +1 and –l, it was straightforward to arrange an encoding to obtain the same eigenvalues from the time evolution operator for the harmonic oscillator. The Hamiltonian for an oscillator could be perturbed to realize nearly any eigenvalue spectrum, and any number of quantum bits could be represented by simply mapping them into the infinite number of eigenstates of the system. This suggests that perhaps one might be able to realize an entire quantum computer in a single simple harmonic oscillator.

4 Summary

Of course, there are many problems with the above scenario. Clearly, one will not always know the eigenvalue spectrum of the unitary operator for a certain quantum computation, even though one may know how to construct the operator from elementary gates. In fact, for most problems addressed by quantum algorithms, knowledge of the eigenvalue spectrum is tantamount to knowledge of the solution.

Another obvious problem is that the technique used above does not allow one computation to be cascaded with another, because in general, cascading two unitary transforms results in a new transform with unrelated eigenvalues.

Finally, the idea of using a single harmonic oscillator to perform quantum computation is flawed because it neglects the principle of digital representation of information. A Hilbert space of 2n dimensions mapped into the state space of a single harmonic oscillator would have to allow for the possibility of states with energy $2^n \hbar\omega$. In contrast, the same Hilbert space could be obtained by using n two-level quantum systems, which has the energy of at most $n\hbar\omega$. Similar comparisons can be made between a classical dial with 2n settings, and a register of n classical bits. Quantum computation builds upon digital computation, not analog computation.

Acknowledgment. This work was supported by the Science and Technology Research Program of Beijing Municipal Education Commission (Grant No. KM201111417007), the Funding Project for Academic Human Resources Development in Institutions of Higher Learning under the Jurisdiction of Beijing Municipality (Grant No. PHR200907120), and the Innovative Talent Foundation of Beijing Union University.

References

1. Turing, A.M.: Proc. Lond. Math. Soc. 42, 230 (1936)
2. Feynman, R.P.: Int. J. Theor. Phys. 12, 467 (1982)
3. Divincenzo, D.P.: Science 270, 255 (1995)
4. Divincenzo, D.P.: Proc. R. Soc. London A 454, 261 (1998)
5. Gutzwiller, M.C.: J. Math. Phys. 12, 343 (1971)
6. Gutzwiller, M.C.: Chaos in Classical and Quantum Mechanics. Springer, New York (1990)
7. Kleppner, D., Delos, J.B.: Found. Phys. 31, 593 (2001)
8. Du, M.L., Delos, J.B.: Phys. Rev. Lett. 58, 1731 (1987)
9. Du, M.L., Delos, J.B.: Phys. Rev. A 38, 1896 (1988)
10. Du, M.L., Delos, J.B.: Phys. Rev. A 38, 1913 (1988)
11. Sadovskii, D.A., Shaw, J.A., Delos, J.B.: Phys. Rev. Lett. 75, 2120 (1995)
12. Dando, P.A., Monteiro, T.S., Owen, S.M.: Phys. Rev. Lett. 75, 2797 (1998)
13. Bohm, D.: Quantum Theory. Prentice-Hall, New York (1951)
14. Li, Q.S., Hu, X.G.: Physica Scripta 51, 417 (1995)
15. Schiff, L.I.: Quantum Mechanics. McGraw-Hill, New York (1968)
16. Lu, J., Li, Q.S., Sun, Z.: Phys. Chem. Chem. Phys (PCCP) 3, 1022 (2001)
17. Lu, J.: Physica Scripta 69, 84 (2004)
18. Lloyd, S.: J. Mod. Opt. 41, 2503 (1994)

Intrusion Detection Model Based on Set Pair Analysis Theory

Yuan Kan, Wei Jiang, and Xiaojuan Ji

General Courses Department, Military Transportation University, Tianjin 300161, China
kanyuan820@gmail.com, jwlyn2001@yahoo.com.cn, janey8284@sina.com

Abstract. Based on set pair analysis theory successfully been used in many different areas in recent years, the intrusion detection model is designed. Making use of standard data set to carry out emulate experiment, and the result of the experiment shows that applying set pair analysis theory to intrusion detection is feasible and effective.

Keywords: Intrusion Detection, Network Security, Set Pair Analysis, KDD1999.

1 Introduction

Intrusion detection, as an active security protection technology, can provide real-time guard to internal attack, external attack and misoperation, and can provide safety service to network through three-dimensional depth and multi-level defense. Thus, in information age, intrusion detection becomes a focus of security research.

As to the research on intrusion detection, one of the focuses is sorting algorithm, and frequently-used intrusion detection technology including probability statistics, neural network, support vector machine, and artificial immunity etc. Based on set pair analysis theory [1] successfully been used in many different areas in recent years, the paper raises intrusion detection model to improve the detection rate of intrusion detection and the identification of new attack.

2 Set Pair Analysis in Intrusion Detection

The basic contents of identical-discrepancy-contrary pattern recognition are: when connection degree of two sets, A and B , can be described as identical-discrepancy-contrary connection degree $\mu = a + bi + cj$, if set pair potential $a/c > 1$, A and B are two patterns with same type; if set pair potential $a/c < 1$, A and B are two patterns with opposite type; and if set pair potential $a/c = 1$, A and B are two patterns with transitional type. Especially,

Y. Zhang (Ed.): Future Communication, Computing, Control and Management, LNEE 141, pp. 35–42.
springerlink.com

if $a = 1$, A and B are completely same patterns; if $c = 1$, A and B are purely opposite patterns; and if $b = 1$, A and B are two patterns with vague and uncertain type.

Suppose identical-discrepancy-contrary connection vector are $\vec{\mu}_1 = (a_1, b_1, c_1)$ and $\vec{\mu}_2 = (a_2, b_2, c_2)$, then identical-discrepancy-contrary distance is defined as

$$\rho = \sqrt{(a_1 - a_2)^2 + (b_1 - b_2)^2 + (c_1 - c_2)^2} \tag{1}$$

This definition describes the distance between identical-discrepancy-contrary vector $\vec{\mu}_1$ and $\vec{\mu}_2$.

If $\vec{\mu}_1$ is the identical-discrepancy-contrary vector obtained from the set pair composed of A_1, the number i pattern system, and corresponding ideal system, $\vec{\mu}$ is the identical-discrepancy-contrary vector obtained from the set pair composed of sub-system B and the same ideal system, ρ_1 is the distance between $\vec{\mu}_1$ and $\vec{\mu}$, identical-discrepancy-contrary connection vectors, here i is the index, $i = 1,2,\cdots,n$, if

$$\rho_K = \min(\rho_1, \rho_2, \cdots, \rho_n) \tag{2}$$

So sub-system B is closest to pattern system A_K, thus B can be classified into pattern A_K, the equation (2) can be called the closest principle of identical-discrepancy-contrary pattern recognition.

Besides, in some problems with high precision requirement, identical-discrepancy-contrary vector can be extended from three-dimension to four-dimension, five-dimension or more dimensions, and can deep into every dimensional vector.

The basis of set pair analysis is the set pair, through related knowledge and understanding of intrusion detection technology mention above, various set pair can also be found in intrusion detection technology, such as normal access and illegal access, normal data flow and abnormal data flow, etc. All can be seen as examples of set pair, and create prerequisite for the rational application of set pair analysis method in intrusion detection. Here suppose an intrusion detection model based on set pair analysis, and its general process shown in Figure 1:

Now taking abnormal data flow formed by characteristic database and be detected data packet in the previous experiment for further explanation. The set pair composed of these two parts can be seen as two systems, so characteristic database can be seen as pattern system mentioned in the previous part and abnormal data flow can be seen

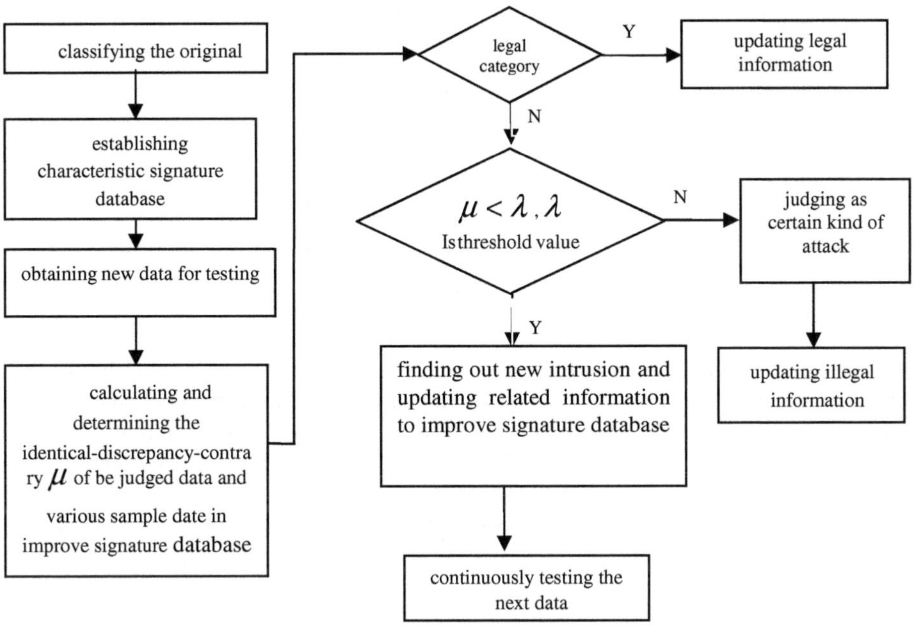

Fig. 1. The Process of Intrusion Detection Based on Set Pair Analysis

as be judged system. In the process of set pair analysis, it is very important to make sure the contrary characterization and measurement, and the common contrary concept [2] can be divided into five types, that is, reciprocal type, with or without type, positive and negative type, complementary type and imaginary-real type, among which, in domain of science and technology, reciprocal type contrary is a common and important contrary type. So taking reciprocal type contrary as starting point to form set pair of be classified data flow and data in known-class standard data set.

Suppose $count$ feature of be classified linkage record A and linkage record B in Dos type are denoted as α and β, its variation range is within [0, L], and $\alpha < \beta$. According to references [3], it is easy to know the proportion of α in contrary interval $[\frac{1}{\beta}, \beta]$ of β in number pair (α, β) is:

$$p = \frac{\alpha}{\beta - \frac{1}{\beta}} = \frac{\alpha\beta}{\beta^2 - 1} \tag{3}$$

Thus, contrary degree of α to β in interval $[\frac{1}{\beta}, \beta]$ is:

$$c = 1 - p = \frac{t^2 - st - 1}{t^2 - 1} \tag{4}$$

At the same time, identical degree of α and β is:

$$a = \frac{\alpha}{\beta} \tag{5}$$

Therefore, discrepancy degree of number pair (α, β) is:

$$b = p - a = \frac{\alpha\beta}{\beta^2 - 1} - \frac{\alpha}{\beta} = \frac{\alpha}{\beta(\beta^2 - 1)} \tag{6}$$

Thus, in set pair analysis, connection degree of number pair (α, β) can be described as:

$$\mu = a + bi + cj = \frac{s}{t} + \frac{s}{t(t^2 - 1)} i + \frac{t^2 - st - 1}{t^2 - 1} j \tag{7}$$

We also know that each data has many characteristic attribute, each data, that is, each vector in standard data set introduced in the third part has forty-one characteristic value. So, in application of set pair analysis, it can be seen as a multi-evidence recognition problem [4], the evidence mentioned here corresponds to the characteristics of data in intrusion detection.

This multi-evidence recognition problem can be described as $G = (S, E, W, D)$, where object set is $S = \{s_1, s_2, \cdots, s_m\}$, characteristic set is $E = \{e_1, e_2, \cdots, e_n\}$, characteristic weight set is $W = \{w_1, w_2, \cdots w_n\}$, recognition matrix is $D = (d_{kr})_{m \times n}$, d_{kr} is the support degree of characteristic e_r to candidate target s_k, $d_{kr} > 0$.

Suppose u_r and v_r are maximum support and minimum support of characteristic e_r in characteristic set E to object set S. Considering characteristic set $E = \{e_1, e_2, \cdots, e_n\}$ as a n-dimensional characteristic space composed of n characteristics, respective coordinates of the strongest and the weakest support point of characteristic set to object set are:

$$\alpha = (\alpha_1, \alpha_2, \cdots, \alpha_n), \text{ among which } \alpha_i = \max(d_{kr}),\ i \in [1, n] \tag{8}$$

$$\beta = (\beta_1, \beta_2, \cdots, \beta_n), \text{ among which } \beta_i = \min(d_{kr}),\ i \in [1, n] \tag{9}$$

The strongest distance of generalized Euclidean [5] is defined as:

$$d_{k\alpha} = \sqrt{\sum_{i=1}^{n} [w_i (d_{ki} - \alpha_i)]^2},\quad k \in [1, m] \tag{10}$$

It can be used to describe the distance between object S_k in object set and the strongest point in characteristic space, the smaller $d_{k\alpha}$ is, the closer S_k is to the strongest support point of characteristic space. The reciprocal of $d_{k\alpha}$ is Euclidean identical degree of object S_k and support object of characteristic set E, can be described as: $\alpha_k = 1/d_{k\alpha}$;

The weakest distance of generalized Euclidean is defined as:

$$d_{k\beta} = \sqrt{\sum_{i=1}^{n}[w_i(d_{ki} - \beta_i)]^2}, \quad k \in [1, m] \tag{11}$$

It can be used to describe the distance between object S_k in object set and the weakest point in characteristic space, the smaller $d_{k\alpha}$ is, the closer S_k is to the weakest support point of characteristic space. The reciprocal of $d_{k\alpha}$ is Euclidean contrary degree of object S_k and support object of characteristic set E, can be described as: $c_k = 1/d_{k\beta}$;

Defining Euclidean relative similarity degree in n-dimensional space composed by object S_k in characteristic set E as:

$$r_k = \frac{1/d_{k\alpha}}{1/d_{k\alpha} + 1/d_{k\beta}} = \frac{d_{k\beta}}{d_{k\alpha} + d_{k\beta}} \tag{12}$$

Analyzing equation (12), the greater value of r_k is, the closer S_k is to the best recognition result, the higher support degree of characteristic set is to S_k, which show that the greater possibility of be judged data belong to S_k attack type is; while the smaller value of r_k is, the farther S_k is to the best recognition result, the lower support degree of characteristic set is to S_k, which show that the smaller possibility of be judged data belong to S_k attack type is. Therefore, in the domain of intrusion detection, set pair analysis can be used to classify attack data, and its feasibility and availability can be proved by the following experiment.

3 Simulation Experiment and Analysis

Experimental data come from standard data set KDD1999 including normal data and intrusion data. The intrusion data totally describe twenty-two intrusion behaviors, such as back, buffer_overflow, ftp_write, teardrop, warezclient, warezmaster etc., which belong to four attack types, that is, Dos, R2L, U2R and Probe. Each link is composed of forty-one attributes marked link status and one sign attribute marked normal or abnormal. Forty-one attributes belong to four attribute set, that is, basic attribute set, content attribute set, data flow attribute set and host computer data flow attribute set.

In this experiment, only nine data flow characteristics based on time and closely related to network attack are selected. Shown in Table 1, they correspond to attributes

23-31 of the original link, and the short sequence of each characteristic value of after-extracted TCP link record forms a nine-dimensional vector. Thus, at this time, each record probably contains nine characteristics.

Table 1. Characteristic of TCP/IP Linked Record Based on Time Current Capacity

Characteristic Name	Characteristic Description
Count	In current 2s, the link time of the same host computer
Serror_rate	The percentage of SYN error in link
Rerror_rate	The percentage of REJ error in link
Same_srv_rate	The percentage of link to the same service
Diff_srv_rate	The percentage of link to the different service
Srv_count	In current 2s, the link time of the same service
Srv_serror_rate	The percentage of SYN error in link to the same service
Srv_ rerror_rate	The percentage of REJ error in link to the same service
Srv_diff_host_rate	The percentage of link to the different host computer

Being convenient for the experiment, two representative data subset, 10percent and Correct, are selected as experimental data, and for reducing data quantity, at the beginning of the experiment, only data belong to Dos and Normal types are selected.

① Using selected 10percent data as original sample database E and selected Correct data as be judged sample database, among which $E=\{E_1,E_2\}$ shows that there are only two types in known database, i.e. Normal type and Dos type, and each type has a characteristic vector, i.e. $E_i=\{N_{1i},N_{2i},\cdots,N_{9i}\}^T$, here $N_1,N_2,\cdots,N_9$ successively correspond to nine time-based current characteristic value in Table 1, $E_i\subset E$.

② Randomly selecting 9000 Normal-type records and 1000 Dos-type records from be judged sample database to form be judged set S , therefore, can form nine set pairs with characteristic vector corresponding with object of E . Before calculating the connection degree, data should be normalized.

③ As to the first record of be judged sample database, according to equation (7), successively calculating nine groups of set pair connection degree μ_1, μ_2, ..., μ_9, and comparing with threshold value λ_1, λ_2, ..., λ_9 respectively, threshold value is determined by constant adjustment in program implementation process. If $\mu_1>\lambda_1$, $\mu_2>\lambda_2$, ..., $\mu_9>\lambda_9$, thus calculating connection degree of be judged record and E_i:

$$\mu_i=\sum w_j\mu_j \tag{13}$$

Where μ_j is μ_1, μ_2, ..., μ_9 respectively, w_j is corresponding weight of μ_j and satisfying: $\sum w_j=1$ and $w_j>0$.

④ After matching all records in original sample database, suppose $\mu = \max(\mu_i)$, if $\mu < \lambda$, there is no matchable corresponding original sample; or matching successfully, according to record type of matchable original sample database, to determine whether be judged record is Normal type or Dos type.

⑤ Taking one record to repeat the procedure ② until all 10000 records in be judged sample database be matched over.

In this experiment, the ultimate detection accuracy (detection accuracy = number of correctly classified sample / total number of sample) is the only interested thing, so the effect evaluation of intrusion detection mentioned here can only account and determine total number of be judged sample of Normal type and Dos type, and will not account error reporting rate and failed reporting rate.

According to procedure above, in Windows XP system environment, using VC++6.0 to write program, simulation experiment is carried out. When selecting sample, besides selecting Dos-type and Normal-type data mentioned above, also selecting R2L-type and Probe-type data record respectively to mix with Normal-type data for programming, while U2R-type attack data, for only has more than 50 records in 10percent database set, is not necessary to treat as same type of detection. Finally, through periodic sampling method, taking all types of data record as example, observing detection accuracy of four experiments respectively, as shown in Table 2:

Table 2. Test Result

Sample Type	Detection Accuracy
Dos+Normal	98.6%
R2L+Normal	69.5%
Probe+Normal	75.7%
All Types	83.7%

In experiment, there are 494021 records in selected 10percent data subset, among which 97278 records are normal data record, 396473 records are abnormal data record, the ratio of normal data to abnormal data in selected, be judged data set is 9:1, which corresponds to the general rule of intrusion detection. From result of detection, as to detection accuracy, correct recognition rate is highest for Dos-type attack, while correct recognition rate is lowest for R2L-type attack, for usually R2L-type attack disguises as legitimate user to attack the system. Therefore, characteristic of R2L-type is similar to that of normal data packet, the correct recognition rate of the algorithm is reduced accordingly. Meanwhile, the experiment shows that the selection of threshold value has great influence on the final result of set pair analysis. Through repeated attempt and adjustment, actually, the final detection result is the highest detection accuracy getting from selected, different threshold value, and the rule of taking threshold value needs further testing and concluding. In addition, as to the inspiration of data failed to be discriminated in the experiment, set pair analysis algorithm acting on discriminating invasion classification, to some extent, shows that record with no type is different from attack data of known type; thus providing a dynamic expansion method for existing characteristic signature database.

Experiment shows that set pair analysis method can achieve similar classification result of KDD data set to that of clustering [5], decision tree [6] and K-means algorithm [7] etc, and can get better classification result. Therefore, using set pair analysis theory to intrusion detection model is feasible and effective.

4 Summary

Through understanding of characteristic of intrusion detection system and concept of set pair, the paper analyzes the possibility of applying set pair analysis to intrusion detection technology, and the result of simulation experiment shows that intrusion detection model based on set pair analysis is feasible and effective.

Actually, combining with rough set theory [8], set pair analysis can also be used in decision analysis, pattern recognition, data mining and other fields, to improve attribute reduction method [9] in complete information system. This can also inspire us that, in intrusion detection field, in the process of data preprocessing, combining the method of set pair analysis and that of rough set, keeping the classification or decision-making capacity of characteristic signature database from changing, deleting redundant or unimportant characteristic attribute, therefore, to improve anomaly detection efficiency of intrusion detection system, which is one of subsequent research of the paper.

References

1. Zhao, K.-Q.: Set Pair Analysis and Its Preliminary Application. ZK Press, HangZhou (2000)
2. Zhao, K.-Q.: Based on Set Pair Analysis Confrontation Classification, Measurement and Application. Science, Technology and Dialectics (2), 26–30 (1994)
3. Zhao, K.-Q.: Based on Set Pair Analysis of the Size Relation Characterization and Application. Non-ferrous Metals Engineering & Research 17(3), 59–62 (1996)
4. Li, Z., Zhang, B., Zhu, Y.-Z.: An Improved Set-pair Approximation and Its Application in Multi-evidence Target Recognition. Computing Technology and Automation 24(4), 80–82 (2005)
5. Li, L.-J., Ling, Y.-L., Wang, R.-C.: Numerosity Reduction Algorithm Adapting to Data Classification in IDS. Application Research of Computers 24(12), 146–148 (2007)
6. Zhao, X.-F., Ye, Z.: Intrusion Detection Model Based on Weighted Multi-Random Decision Tree. Journal of Computer Applications 27(5), 1041–1043 (2007)
7. Gu, B.-P., Xu, X.-Y., Guo, H.-Y.: Research of K-Means Algorithm Based on Particle Swarm Optimization in Network Intrusion Detection. Journal of Computer Applications 27(6), 1368–1370 (2007)
8. Chen, Z.-M., Zhou, J.-S.: Introduction to Matrix theory, pp. 54–89. BeiHang University Press, Beijing (2002)
9. Guda, S., Mishra, N., Motwani, R., et al.: Clustering Data Streams. In: Proceedings of the Annual Symposium on Foundations of Computer Science. IEEE (2000)

Research on Maximum Wind Energy Capturing of Doubly Fed Induction Wind Generator Based on Fuzzy Control Technology

Haihui Song[1,2] and Tian De[2]

[1] Shanghai Second Polytechnic University, Shanghai, China
[2] North China Electric Power University, Beijing, China
{gzshh,gz3849}@163.com

Abstract. The fuzzy controller here is used to control the doubly fed induction wind generator in order to improve the robust of the system. According to the wind changes, the optimal tip speed ratio is kept and the power output to the d maintains constant frequency. this paper designed a fuzzy controller, and established a fuzzy control rule base according to the maximum power capturing. The fuzzy controller capturs the maximum power with generator power output error and generator rotor speed error to generate reference speed, which contains the maximum tip speed ratio. The theory analysis and simulation results based on MATLAB/Simulink demonstrate this control strategy is convenient and efficient.

Keywords: Fuzzy control, DFIG, variable speed constant frequency, maximum wind energy Capturing.

1 Introduction

The variable speed constant frequency (VSCF) doubly fed induction wind generator is currently the main unit today. It is commonly used industry standard PID control system as a double-fed wind turbine controller, to achieve control of wind energy capture. Domestic and foreign experts and scholars have made a number of different control methods and strategies.Such as feedback linearization, sliding mode variable structure control method, H∞robust control methods and disturbance rejection control methods. But these studies is limited by the constraints of limited, It must first establish an effective system model, But fine deterministic system model is very complex because of the uncertainty and aerodynamic complexity of the model power electronics. We construct the fuzzy controller, and study the model of doubly fed induction wind generator based on the fuzzy control strategy. It can avoid the drawbacks of traditional methods, follow automatically changes of unit performance and wind speed. predict accurately the aerodynamic characteristics of wind turbines,Adjust the unit voltage and output power. It can not only give a constant frequency, but also enable the unit Capturing maximum wind power.

Y. Zhang (Ed.): Future Communication, Computing, Control and Management, LNEE 141, pp. 43–48.
springerlink.com

2 Double-Fed Wind Power Generation System

2.1 System Components

Double-fed wind power generation system schematic shown in Figure 1. It includes wind turbine, the gearbox, DFIG, inverter and control system. Wind acts on the rotor blades with a certain speed and angle of attack, blades rotating under the action of the rotating torque, it transforms wind energy into mechanical energy, and drives generator through the gearbox, then the generator transforms mechanical energy into electrical energy. Part of the energy goes directly to the grid through the stator,and the others delivery to the grid Through the converter from the rotor.

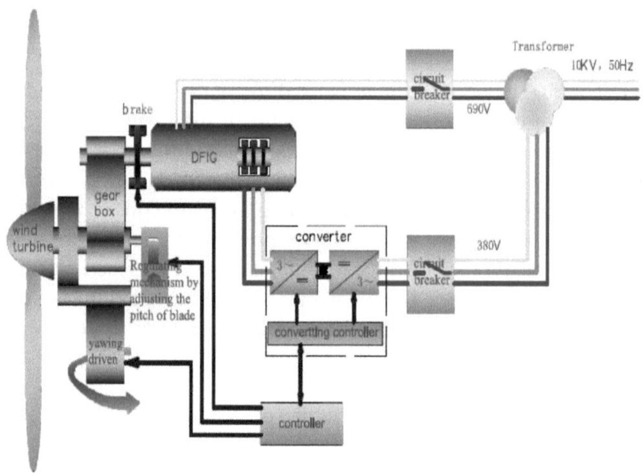

Fig. 1. Double-fed wind power generation system schematic

2.2 The Mathematical Model of Each Component

Transforming voltage, current, flux and torque in triphase into the parameters in diphase.

1) Voltage equation of stator and rotor

$$\begin{cases} u_{ds} = -R_s i_{ds} - p\psi_{ds} + \omega_1 \psi_{qs} \\ u_{qs} = -R_s i_{qs} - p\psi_{qs} + \omega_1 \psi_{ds} \\ u_{dr} = R_r i_{dr} + p\psi_{dr} - \omega_s \psi_{qr} \\ u_{qr} = R_r i_{qr} + p\psi_{qr} + \omega_s \psi_{dr} \end{cases}$$

u_{dr} , u_{qr} is the voltage component of rotor in d, q axis

u_{ds} , u_{qs} is the voltagecomponentof statorin d, q axis

i_{dr} , i_{qr} is thecurrentcomponentof rotorin d, q axis

i_{ds} , i_{qs} is the current component of stator in d, q axis

ω_s is the Angular velocity of d,q axis relative to the rotor

2) Flux equation of stator and rotor

$$\begin{cases} \psi_{ds} = L_s i_{ds} - L_m i_{dr} \\ \psi_{qs} = L_s i_{qs} - L_m i_{qr} \\ \psi_{dr} = L_r i_{dr} - L_m i_{dr} \\ \psi_{qr} = L_r i_{qr} - L_m i_{qr} \end{cases}$$

ψ_{dr} , ψ_{qr} is the flux component of rotor in d, q axis

ψ_{ds} , ψ_{qs} is the flux component of stator in d, q axis

L_m is the mutual inductance between stator and rotor in d , q axis

L_r is the equivalent self - inductance of rotor in d,q axis

L_s is the equivalent self - inductance of stator in d,q axis

3) Torque equation

$$T_e = n_p L_m (i_{ds} i_{qr} - i_{qs} i_{dr})$$

4) Equations of motion

$$T_m = T_e + \frac{J}{n_p}\frac{d\omega_m}{dt}$$

n_p is the number of pole pairs in DFIG

ω_m is the mechanical angular velocity of rotor

3 Fuzzy Controller Design

The variable speed constant frequency (VSCF) doubly fed induction wind generator is currently the main unit today.

3.1 Control Strategy

The control strategy on maximum wind energy capturing of doubly fed induction wind generator is how to get optimum tip speed ratio. We designed a fuzzy controller according to fuzzy control theory. Adjusting the anti-torque of the DFIG ,changing speed of wind turbine following wind speed to obtain the optimum tip speed ratio, tracking the best power curve, so that the wind turbine gets the highest wind energy conversion efficiency.

For a constant tip speed ratio, It as a fuzzy controller input with tip speed ratio error and its error change rate, and the output is the stator voltage variation.

3.2 Lingual Variable

Setting EP and EW as input variables of the fuzzy controller, and U as output variables.

Table 1. Table of lingual variable EP

	−6	−5	−4	−3	−2	−1	−0	+0	1	2	3	4	5	6
NB	1	0.5	0	0	0	0	0	0	0	0	0	0	0	0
NM	0	0.5	1	0.5	0	0	0	0	0	0	0	0	0	0
NS	0	0	0	0.5	1	0.5	0	0	0	0	0	0	0	0
NZ	0	0	0	0	0	0.5	1	0	0	0	0	0	0	0
PZ	0	0	0	0	0	0	0	1	0.5	0	0	0	0	0
PS	0	0	0	0	0	0	0	0	0.5	1	0.5	0	0	0
PM	0	0	0	0	0	0	0	0	0	0	0.5	1	0.5	0
PB	0	0	0	0	0	0	0	0	0	0	0	0	0.5	1

Table 2. Table of lingual variable EW

	−6	−5	−4	−3	−2	−1	0	1	2	3	4	5	6
NB	1	0.5	0	0	0	0	0	0	0	0	0	0	0
NM	0	0.5	1	0.5	0	0	0	0	0	0	0	0	0
NS	0	0	0	0.5	1	0.5	0	0	0	0	0	0	0
ZR	0	0	0	0	0	0.5	1	0.5	0	0	0	0	0
PS	0	0	0	0	0	0	0	0.5	1	0.5	0	0	0
PM	0	0	0	0	0	0	0	0	0	0.5	1	0.5	0
PB	0	0	0	0	0	0	0	0	0	0	0	0.5	1

Table 3. Fuzzy control rule

EP / U / EW	NB	NM	NS	ZR	PS	PM	PB
NB	PB	PB	PB	PB	PM	ZR	ZR
NM	PB	PB	PB	PM	PS	ZR	ZR
NS	PM	PM	PS	PS	ZR	ZR	NS
NZ	PS	PS	PS	ZR	ZR	NS	NS
PZ	PS	PS	ZR	ZR	NS	NS	NS
PS	PS	ZR	ZR	NS	NS	NM	NM
PM	ZR	ZR	NS	NM	NB	NB	NB
PB	ZR	ZR	NM	NB	NB	NB	NB

4 Simulation

4.1 The Model of DFIG

Using MATLAB/Simulink to establish a mathematical model of DFIG system, design control programs, and simulation system

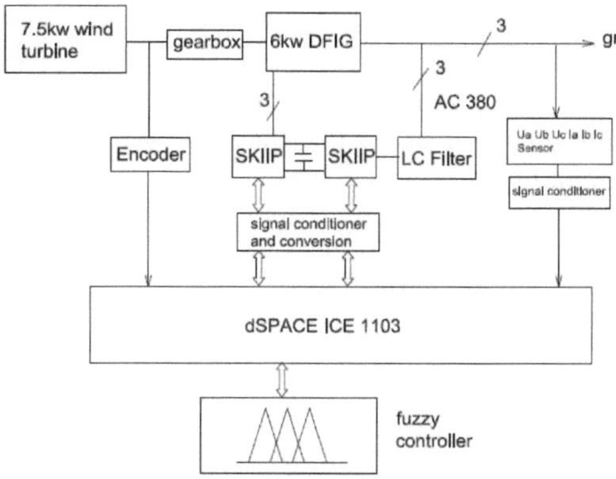

Fig. 2. Fuzzy Control System diagram

4.2 Simulation

Following is the curve after changes in wind speed. The top curve is the random wind, and middle curve is the output active power of DFIG, the bottom curve is the waveform of wind energy utilization factor.

From the above analysis, we know that, when wind speed changes, the output active power of DFIG is also to follow changes in wind speed; obtaining the optimum tip speed ratio, tracking the best power curve.

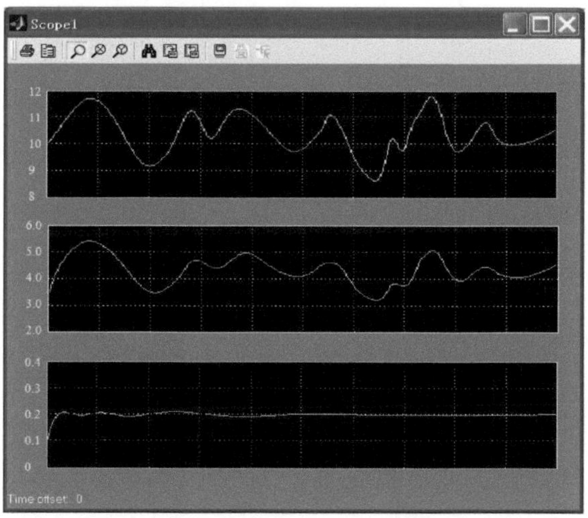

Fig. 3. Simulation waveform

5 Conclusion

The variable speed constant frequency (VSCF) doubly fed induction wind generator is currently the main unit today.

This paper designs a fuzzy controller, and establishs simulation model based on Matlab/Simulink, change parameters and simulate online. The simulation results indicates that the wind power system has good wind compression and good capacity of capturing the maximum power. It shows that the controller is convenient and effective, and this offers a good platform for researching.

Acknowledgment. Supported by Leading Academic Discipline Project of Shanghai Municipal Education Commission(No.J51801)

References

1. Tian, D., Guo, F., Liu, S., et al.: A wind-tunnel test on the entirety model of the concentrated wind energy turbine (Report NO.IIIA comp-arative test of power generation). Perpr. for the Beijing International Conference on Wind Energy, May 3-9, pp. C-76–C-85 (1995)
2. Tian, D., Huang, S., Diao, M., et al.: A Wind Tunnel Test on the Entirety Model of Concentrated Wind Energy Turbine (Report No.2 A test for hydrodynamic performance). In: Proceedings of International Agricultral Mechanization Conference, Beijing, P.R.China, April 10-13, No.088 (1995)
3. Markvart, T.: Sizing Of Hybrid Photovoltaic Wind Energy System. Solar Energy 57(4), 261–265 (1996)
4. Heier, S.: Grid integration of wind energy conversion systems. John Wiley & Sons Ltd., England (1998)
5. Hansen, H., Hansen, A., Larsen, T.J., et al.: Control design for a pitch-requlated, variable speed wind turbine. Riso National Laboratory, Demark (2005)

A Developed Switching Vector Control Method Using DSP TMS320F240

Wen Youmei

Changsha Aeronautical Vocational and Technical College, Changsha Hunan 410124

Abstract. A developed switching vector control has been realized using DSP TMS320F240. The use of such a dedicated DSP for motor control has reduced the hardware complexity and it has provided all functions of speed control. Operation of variable PI speed controller results in desired characteristics depending on the position of moving door. It is observed that there is a fast current change in accordance with the change in speed or load.

Keywords: DSP, switching control, elevator door.

1 Introduction

Permanent magnet synchronous motor (PMSM) is a popular drive element in industrial fields because of its advantages over other types of motors such as DC and induction motors[1]. These advantages include its higher energy density, faster response, higher efficiency and higher power factor. Due to absence of brushes and commentators, PMSM can operate free of maintenance and sparks with its low inertia and simple construction. PMSM has become preferable option for those applications which essentially require high power density and small size machine such as in the aerospace and robotics area.

With the rapid development of the international economy and the increment of modern buildings, the popularization rate of the elevator rises so quickly that it is becoming an indispensable tool in daily life. A reliable operation of an elevator is crucial for the passengers' safety. As an important component of elevator system, the elevator doors face passengers directly. A smooth and quick opening-and-closing motion of the elevator doors can improve the impression of the elevator greatly. This puts forward a high requirement of velocity regulation and reliability for the servo drive of the door motor.

2 Mathematical Model of Pmsm

Permanent magnet synchronous motors are usually modeled in the d-q reference frame fixed to the rotor. In this synchronous rotating reference frame, the d-axis is oriented along the permanent magnet flux. Assuming that the PMSM is of three-phased with balanced windings and no saturation, the stator voltage equation of PMSM is generally described by the following set of equations:

Y. Zhang (Ed.): Future Communication, Computing, Control and Management, LNEE 141, pp. 49–55.
springerlink.com

$$\begin{aligned} u_d &= Ri_d + P\phi_d - \omega\phi_q \\ u_d &= Ri_q + P\phi_q + \omega\phi_d \\ \phi_d &= L_d i_d + \phi_m \\ \phi_q &= L_q i_q \end{aligned} \tag{1}$$

The inverter frequency is related to the rotor speed as:

$$\omega = P\omega_r \tag{2}$$

The electromagnetic torque can be described as:

$$T_\varepsilon = \frac{3P}{2}\left[\phi_m i_q + \left(L_d - L_q\right) i_d i_q\right] \tag{3}$$

The dynamic equation of the PMSM drive is described as:

$$T_\varepsilon = T_L + B\omega_r + J\dot{\omega}_r \tag{4}$$

Therefore the nonlinear state equations are given as:

$$\begin{aligned} \dot{i}_d &= -\frac{R}{L_d} i_d + P\omega_r i_q + \frac{1}{L_d} u_d \\ \dot{i}_q &= -\frac{R}{L_q} i_q - P\omega_r i_d - \frac{P\phi_m}{L_q}\omega_r + \frac{1}{L_q} u_q \\ \dot{\omega}_r &= \frac{1}{J}\left(T_\varepsilon - T_L - B\omega_r\right) \end{aligned} \tag{5}$$

It is obvious that the above dynamic model of PMSM is highly nonlinear due to the coupling between the speed and the electrical currents. Moreover the model parameters can vary with the temperature or saturation during operation and the load torque can also change. Thus, when a high performance of PMSM drive is required, the model structure nonlinearities, parameter uncertainties and load torque variations must be considered during the controller design.

3 System Configration and Control Scheme

Fig. 1 shows the basic building block diagram of the proposed drive system. The system comprises of: (a) three phase IGBT based VSI, (b) three phase PMSM, (c) the feedback control circuits receiving sensed signals, namely, the speed signal and current signals respectively. PMSM is star connected with an isolated neutral. Sensing currents corresponding to two phases is sufficient to yield feedback signals for all three-phase currents.

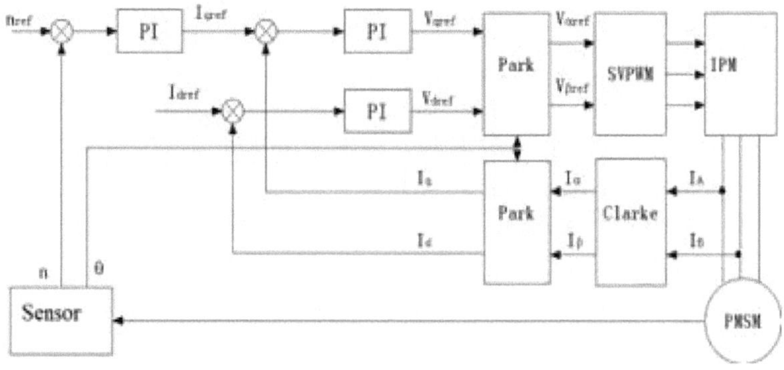

Fig. 1. Principle diagram of vector control system of PMSM

The speed of the motor is compared with its reference value and the error is processed in the PI speed controller. The output of the speed controller is considered as the reference current signal. The controller calculates the d component and the q component of the reference current signal in the synchronously rotating reference frame aligned with the rotor field. These signals are then converted into three-phase reference currents in the stationary reference frame of PMSM. These reference currents are fed into the PWM current controller which also simultaneously receives the sensed currents. For the current sensing, two of the three-phase currents are sensed using Hall Effect current sensors and the third current can be computed from the two sensed currents because the sum of all the three currents is zero for a star connected balanced motor. The current error signals are amplified as the modulating signals for the PWM current controller. A Virtual Triangular Carrier Waveform (VTCW) is generated at the required switching frequency. The point of intersection of the VTCW and the modulating signals, acts as the point of state hanging over for the resulting PWM driver signals. Six driver PWM signals emanate from the output of the PWM current controller and are then fed to the respective gate driver circuits of the VSI. The principle of detection of rotor position and speed are shown in Fig. 2 and Fig. 3 respectively.

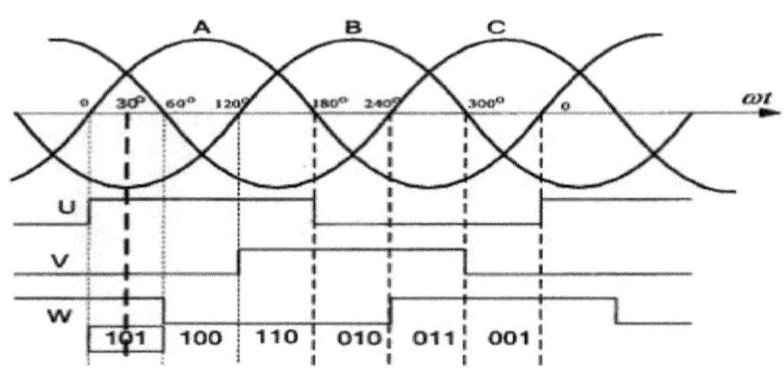

Fig. 2. Corresponding relation between the period of motor and signals of Hall sensor

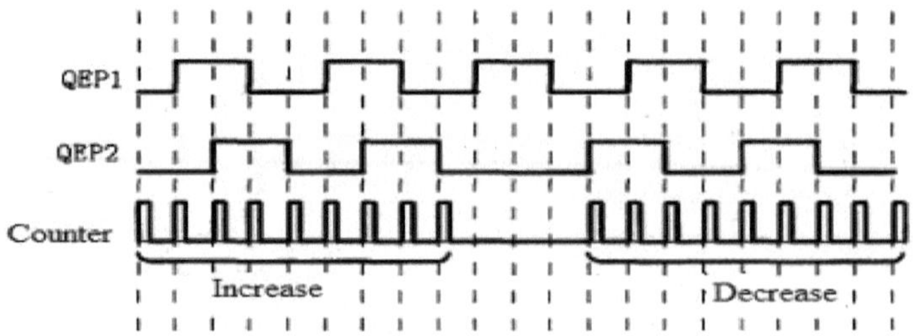

Fig. 3. Running mode of QEP (quadrature encoder pulse)

4 Hardware Implementation

Some of the major parts of the drive system, namely, the speed controller, vector controller, and PWM current controller are implemented through the control software. The external circuit hardware is developed.

The drive hardware consists of three-phase VSI, gate driver circuit, speed sensing circuits and the current sensing circuits. Four separate DC power supplies are developed which are isolated from the main circuits and each other to supply to the driver circuits. For designing the driver power supplies, a single phase supply with 220 V and 50 Hz is stepped down into four-isolated single phase signals with 24 V and 50 Hz. These signals are rectified and regulated to make necessary four DC supplies for six gate driver circuits. Each gate driver has an optical isolation to isolate the control circuit from the power circuit. Six gate signals at the output of driver chips are applied between gate and emitter of the respective IGBTs.

Sensed speed signal, reference speed signal and sensed current signals are scaled into the acceptable range of 0-5V by the respective signal conditioning circuits and then are fed to the respective ADC channels of DSP. Both sensed speed and reference speed signals are adjusted in range from -3000rpm to +3000 rpm. The hardware circuits are shown in Fig.4 and Fig. 5.

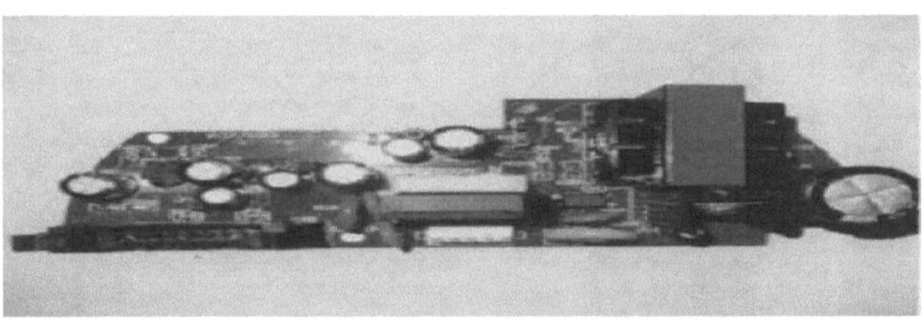

Fig. 4. PCB of switching power circuit

Fig. 5. PCB of control circuit and human-computer interaction circuit

5 Implementation Procedure

The DSP control software is executed in two modules: the initialization module and the run module. The initialization module executes once. The run module is a waiting loop interrupted by the PWM signal. When the interrupt flag is set, acknowledgement signal is generated and the corresponding Interrupt Service Routine (ISR) is served. The complete vector control algorithm with variable PI control is computed depending on the door position within the PWM ISR.

The compare unit of DSP is used to generate the PWM signal. Timer l is operated in the continuous Up-Down counting mode, which is responsible for the internal generation of the virtual triangular carrier wave. Programming code has been done to generate complementary PWM signals at a switching frequency of 10 kHz.

The developed scheme is tested for an elevator door system. Desired and test results for the opening and closing motion are respectively shown in Figs.6-9.

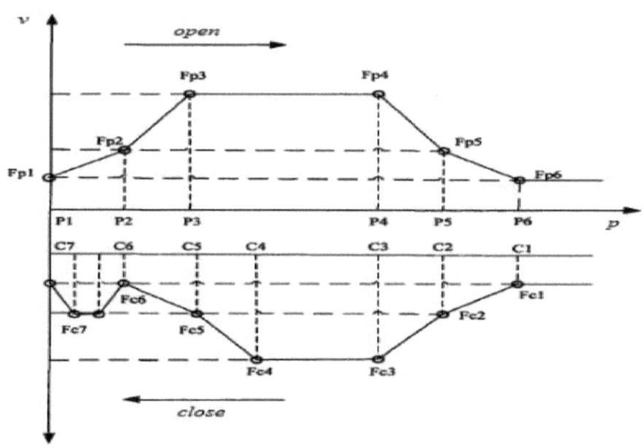

Fig. 6. Opening and closing speed of the elevator door motor

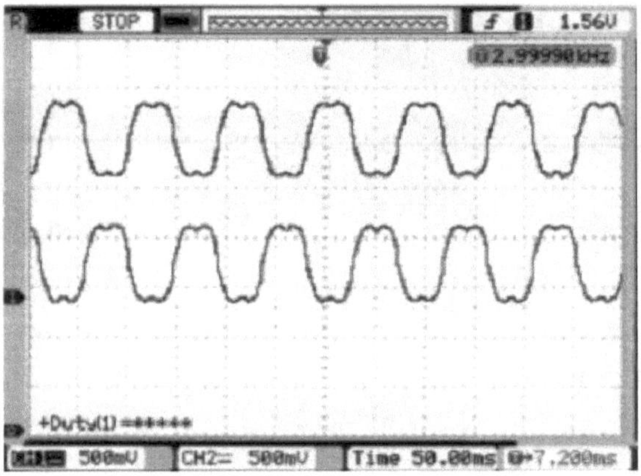

Fig. 7. SVPWM waveform output from the controller

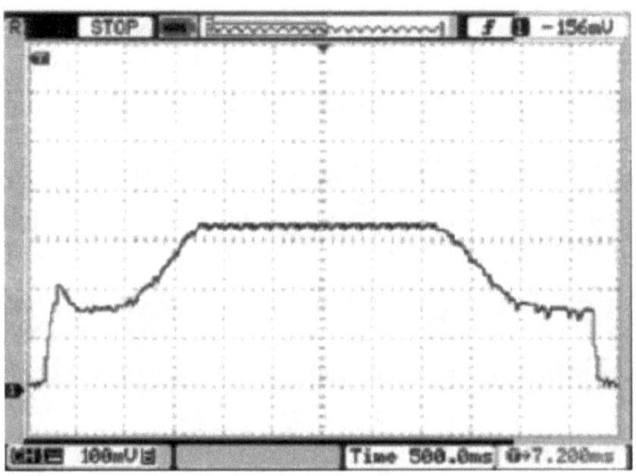

Fig. 8. Opening speed curve of the elevator door motor

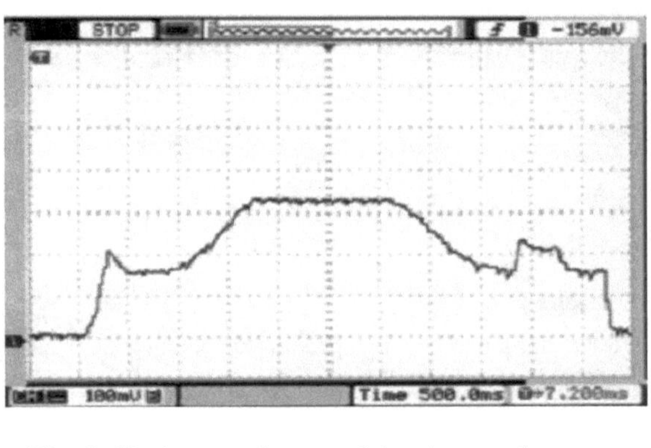

Fig. 9. Closing speed curve of the elevator door motor

6 Conclusion

The cable driven by a motor regulates the position of the doors according to the position feedback. A PMSM can deliver a high performance of the servo system. The position and velocity regulation of PMSM can be obtained by means of vector control, which can be realized in a DSP based system. In this paper, a switching control strategy is developed instead of the conventional PI control to enhance the dynamic response, such as fast tracking and precise positioning. The digitalized implementation of the control algorithm with a DSP can enhance the potentials of PMSM in such servo application.

References

1. Magota, S., Kunimatsu, S., Yamamoto, G., Fujii, T., Otsubo, R.: The Opening-and-Closing Speed Control of an Elevator Door by Frequency-Shaping ILQ Design Method. In: Proceedings of the 2006 American Control Conference, Minneapolis, Minnesota, USA, June 14-16, pp. 1960–1961 (2006)
2. Chiasson, J., Seto, D., Sun, F., Stankovic, A., Bortoff, S.: Control of two PM linear motors with a single inverter: application to elevator doors. Mechatronics 15, 95–110 (2005)
3. Singh, B., Singh, B.P., Dwivedi, S.T.: DSP based implementation of vector control scheme for permanent magnet synchronous motor drive. IETE Journal of Research 53, 153–164 (2007)

A Modeling Approach Using UML 2 for Security Protocols in Distributed System

Xiaoxiong Zhou

School of Information Science and Technology, Jiujiang University, Jiujiang, Republic of China

Abstract. We were able to use UML 2 to model the simple protocols and the Needham-Schroeder and Yahalom protocols without requiring any extensions to the language. The approach of building the framework, using progressively more complex protocols, demonstrated that UML 2 could support all of the protocols modeled without revealing any circumstances under which the language was deficient in supporting modeling within this domain.

Keywords: Security Protocols, UML 2, Distributed System.

1 Introduction

A typical distributed computer system consists of a number of different participants that range from people, companies, computers, to devices such as magnetic card readers that all must communicate through any number of different channels such as distributed computer networks, phone lines, radio, infrared or devices such as bank cards. The architectural key- stone of any secure computer system is the security protocol that provides the building blocks for ensuring secure electronic communications between the participants. While the design of secure communication protocols has improved over the years the tasks of building and validating these protocols remains inherently difficult. Failures occur due a result of unintended use, malicious attacks, incorrect logic or incorrect transition from design to code.

2 UML 2 and MDD for the Modeling of Security Protocols

Our approach was to develop a modeling technique using UML 2, without extensions, taking advantage of the new features within this version that support the definition of communication between participants through ports and protocols and provides the ability to create an executable model. The use of MDD to construct executable models provided a means to examine a mechanism to bridge the current gap that exists between the design and implementation [1]. As part of the modeling exercise, we evaluated whether or not the security protocols modeled in UML 2 were expressive enough to present the key concepts of a security protocol without losing any important details. In addition, we evaluated the use of executable models in order to determine if they were predictive in that they would reveal any non-obvious properties of the security protocols being modeled.

Y. Zhang (Ed.): Future Communication, Computing, Control and Management, LNEE 141, pp. 57–64.
springerlink.com

2.1 Design of the Security Protocol Framework

The design of the framework was partitioned based on the type of participant, where the Server was treated separately from other participants given that it is a "trusted" participant responsible for generating session keys. The remaining participants, Alice, Bob and the Adversary, shared common behaviors and so were treated as similar, with the idea that there would be a parent capsule, Participant, and the other participants would inherit the common behaviors from the parent. All capsules would be contained within the single capsule, SecureEnvironment, which would be the overall container used to automatically generate the code.

The capsule, Participant, contains the ports that allow communication between participants and the Server. The two main ports are ParticipantPort and ServerPort, which are associated to two protocol roles: ParticipantProtocol and ServerProtocol, where protocol roles specify the set of messages that can be received or sent from a port. A port plays the role of one participant in the communication relationship, so a port has an associated direction which will either be base/client or conjugate/server. Only compatible ports can be connected such that every signal in the 'out' set of one protocol role must be in the 'in' set of the other protocol role.

2.2 The Adversary

The Adversary controls the communications between all principals and can therefore obtain any message that is passing between principals. This requires that all communications between Alice, Bob and the Server must be accessible to the Adversary. In order to model this behavior within the framework, the Adversary had to have access to the messages that are sent via the ports for each of the capsules. The design decision was made to embed the capsules for Alice and Bob within the Adversary capsule through an aggregate relationship. The Adversary contains internal ports through which Alice and Bob can communicate with one another or communicate with the Server [2]. The intent of this design is that the Adversary becomes synonymous with the network, in that all communications pass through the Adversary and are routed accordingly. In Fig. 1, the capsules and relationships are described.

2.3 Communications

Ports and Protocols: Communications between all participants occurs via ports representing a protocol class. In the framework, these ports are not directly connected or wired, requiring that a participant must first establish a connection to a port and then disconnect when the port is no longer required [3]. All capsules that inherit from the Participant capsule have a port for communicating with the Server and one for communicating with another participant. Given that the Adversary contains PrincipalA and PrincipalB, the Adversary capsule contains two internal ports that PrincipalA and PrincipalB connect to when establishing communications with another participant or the Server. The Adversary has internal mechanisms to track these communications in order to ensure that messages are directed to the correct participant in the exchange. These internal mechanisms also provide the means for the Adversary to either eavesdrop or launch an attack as may be appropriate to the security protocol scenario being modeled.

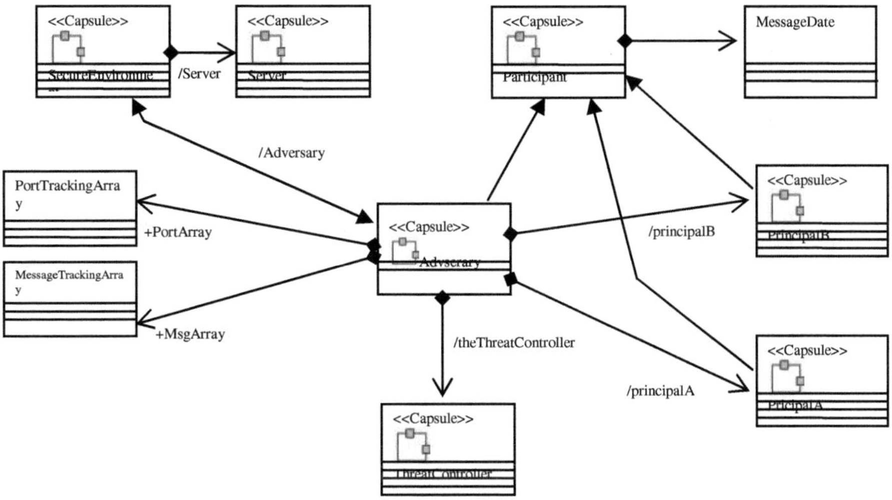

Fig. 1. Framework Structure

There are two protocols used for communications within the framework: the ServerProtocol and the ParticipantProtocol. The ServerProtocol is a binary protocol and a participant, using the services of the Server, must have a conjugate port that corresponds to the Server's public base port. Given the limited role that the Server plays within the security protocols being modeled, the protocol contains one 'in' signal, RequestKey, and one 'out' signal, SendKey. The ParticipantProtocol is different from the ServerProtocol in that it is a symmetric protocol, where the set of 'in' messages maps directly to the set of 'out' messages. This means that participants communicating with other participants through a port with the ParticipantProtocol role communicate through base ports, no conjugate port is required to define the flow of messages. This allows participants to both send and receive messages through their ports.

2.4 Modeling of Threats

The Adversary in the framework is one of the most intricate components of the model due to its omniscient role within the security protocols. It contains the ability to execute various threats during protocol runs in order to determine if one of these threats can compromise the security protocol in any way. In order to support multiple threat models the Adversary has a hierarchical state machine that decomposes the Communicating state into two possible states: Eavesdropping or Attack, as described in Fig. 2.

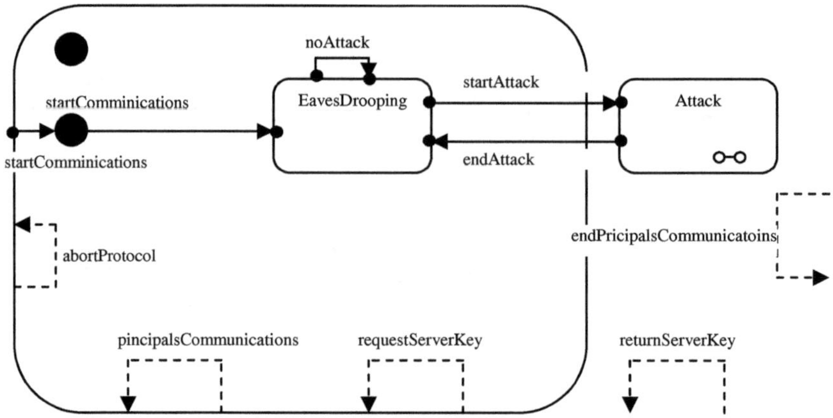

Fig. 2. Adversary Capsule Communicating State Machine

The framework can accommodate multiple attack types in order to expand the set of available threats, since an attack is simply a state within the overarching Attack state. Adding a threat simply requires the addition of a choice point for that type of attack, the addition of the attack name in the ThreatController attack list and then the definition of the specific attack state. The attacks that were modeled for the framework included the following:

- Eavesdropping - This is the most basic attack and by default the Adversary is in this passive attack state when there is no other threat being executed. In this type of attack, the Adversary can capture the information relating to the message flows between principals and any information in those messages that is not encrypted.
- Intercept - An Intercept attack involves intercepting a message from one principal to another, excluding the Server, and interjecting the Adversary's identity into the message resulting in confusion as to who is involved in the protocol and who knows the session key.

2.5 Modeling the Needham-Schroeder Protocol

In the Needham-Schroeder security protocol, nonces are used to ensure freshness in order to guard against a Replay attack. The protocol run in standard notation appears as follows:

1) $A \rightarrow S : A, B$
2) $S \rightarrow A : \{K_{PB}, B\} K_{AS}$
3) $A \rightarrow B : \{N_A, A\} K_{PB}$
4) $B \rightarrow S : B, A$
5) $S \rightarrow B : \{K_{PA}, A\} K_{BS}$
6) $B \rightarrow A : \{N_A, N_B\} K_{PA}$

7) $A \rightarrow B : \{N_B\} K_{PB}$

Within the framework it is the state machines for PrincipalA and PrincipalB that reflect the steps within the Needham- Schroeder protocol. The Behavioural state machines reflect the Protocol state machines in that the state transitions only occur based upon receiving the correct message in the correct sequence. As an example, PrincipalA initiates the protocol run by requesting the session key from the Server as in step 1 of the protocol run described above. It only transitions to the state, sendMessageToB, where it formulates and sends the message to PrincipalB, once it has received the correct response from the Server as in step 2. Figure 4 illustrates the Behavioural state machine of PrincipalA as designed for the Needham-Schroeder security protocol. PrincipalB's Behavioural state machine corresponds to that of PrincipalA in that PrincipalB only transitions to the next state when it receives the correct message from PrincipalA or the Server.

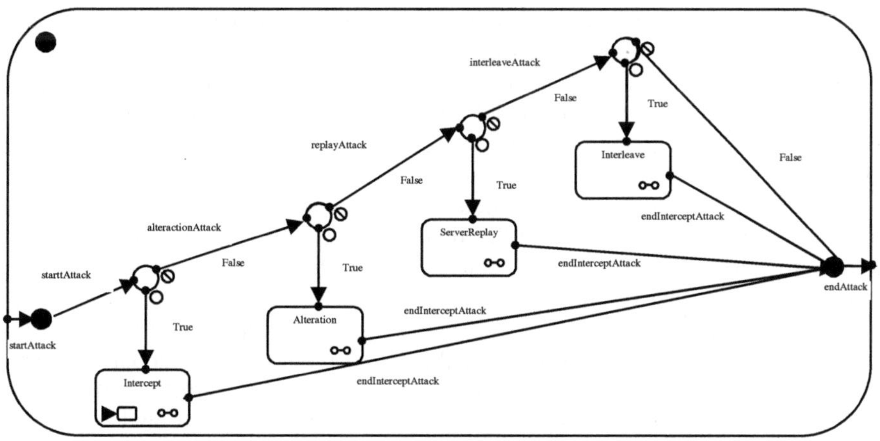

Fig. 3. Attack State Machine

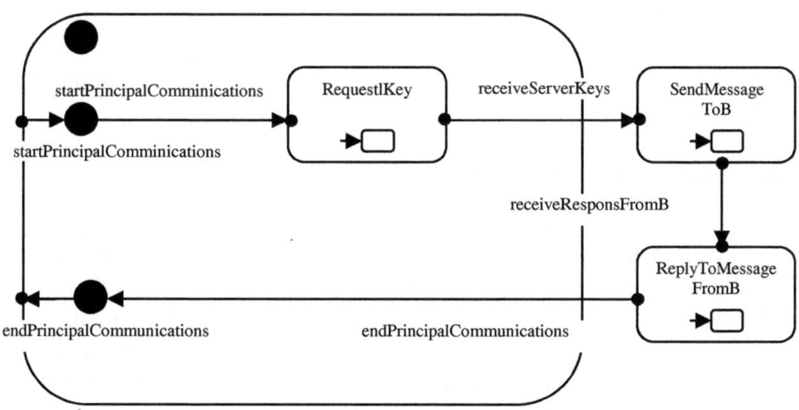

Fig. 4. PrincipalA State Machine for Needham-Schroeder

2.6 Applying Threats to the Needham-Schroeder Protocol

We applied all of the threats to the Needham-Schroeder protocol, but prior to executing the model we predicted the outcome based upon the attack type and the structure of the protocol. The only difference was in the way in which wc implemented the InterLeave attack as we designed it according to Lowe's description of the attack on the Needham-Schroeder security protocol [4]. In this attack Lowe assumes that all participants know the other principals' public key, so we can reduce the set of messages in the protocol to the following:

1) $A \rightarrow B: \{N_A, A\} K_{PB}$
2) $B \rightarrow A: \{N_A, N_B\} K_{PA}$
3) $A \rightarrow B: \{N_B\} K_{PB}$

So we had two versions of the Needham-Schroeder protocol, where we determined the threats that applied and predicted the outcomes as described in Table 1.

Validation Once the framework was constructed, we used the Needham-Schroeder Public Authentication protocol to determine if the expected behaviours and known flaws were revealed within the model. We also modeled the Yahalom protocol, a security protocol with no known flaws, in order to evaluate if the model was expressive and predictive. Initially, we applied an informal visual inspection approach to the model. Then we executed each of the models and from the execution examined the state monitor, sequence diagrams and utilized the trace capture in order to examine the artefacts to verify if the expected behaviours of the protocol are evident during execution. For all security protocol models created, we subjected them to adversarial behaviours with the intent that this would reveal any deficiencies in the design or the actual model itself. By comparing what the visual model reveals about the security protocol with actual behaviours exposed during execution, we evaluated if there are any deficiencies with our approach of using UML 2, with no extensions, and MDD for the modeling of security protocols.

Findings We were able to use UML 2 to model the simple protocols and the Needham-Schroeder and Yahalom protocols without requiring any extensions to the language. The approach of building the framework, using progressively more complex protocols, demonstrated that UML 2 could support all of the protocols modeled without revealing any circumstances under which the language was deficient in supporting modeling within this domain.

2.7 Informal Visual Inspection

The validation of the security protocols through an informal visual inspection process revealed that, while UML 2 could support modeling within this domain, the visual models did not provide any more information about the protocol than was revealed through the definition of the security protocols using the standard notation. In fact, the UML 2 visual models created using Rose-RT revealed less information about the security protocol given that the state machine transitions were identified only with a user defined label, which did not reveal anything about the underlying signal/message or its contents.

2.8 Execution of Models

In terms of analyzing and validating the security protocols, the ability to execute the model provided the greatest ability to reveal non-obvious properties of the protocol. The Rose-RT tool provides a facility to capture the sequence diagram relating to the protocol execution and this diagram can be analyzed to examine the protocol run under normal conditions and when the threats are applied. The sequence diagrams produced logged the flow of the messages and the change of states for the various participants involved in the protocol run.

Table 1. Needham-schroeder protocol: threats and expected outcomes

Attack Type	Needham-Schroeder Protocol	Needham-Schroeder Abbreviated Protocol
Intercept	Unsuccessful	Not Applicable
Alteration	Unsuccessful	Not Applicable
ServerReplay	Successful	Not Applicable
InterLeave	Not Applicable	Successful

Table 2. Security protocols and threats

Attack Type	Needham-Schroeder Protocol	Needham-Schroeder Abbreviated Protocol	Yahalom Protocol
Intercept	Attack Unsuccessful		Attack Unsuccessful
Alteration	Attack Unsuccessful		Attack Unsuccessful
ServerReplay	Attack Unsuccessful		Attack Unsuccessful
InterLeave		Attack Unsuccessful	Not Applicable

By applying a series of threats over multiple protocol runs, we were able to compare our predictions about expected outcomes with the resulting sequence diagrams. This supported our ability to determine if the implementation adhered to our protocol specification and whether any unexpected behaviours were revealed. The advantage of having a set of potential threats available to apply to the model, is that we can then test the model under various conditions and observe the behaviours of the implementation of that security protocol. In our research, we developed a limited set of threats but this provided a basis for the testing of our models under adverse conditions even when the threat was not expected to be effective against a particular protocol.

Table 2 illustrates the set of threats available within this version of the framework and the vulnerabilities of those security protocols that were implemented. Note that in the case of the Needham-Schroeder abbreviated protocol we only applied two of the threat types: Eavesdropping and the InterLeave attack.

3 Conclusion

While a number of approaches have been developed to address the correct modeling and implementation of security protocols, there is still a significant semantic gap between the design of security protocols in the context of software engineering and the transition from design to implementation. Our approach is to use UML 2 without extensions to support the design, construction and verification of security protocols. The use of UML 2 and Model Driven Development (MDD) to create executable models provides the means to ensure that design intent is directly translated into code and to validate the implementation as the design, as expressed by the model, can be automatically compiled and built into executable code. Within our framework, once the executable code has been created, the protocol can be executed under normal conditions as well as under different threat scenarios.

References

1. Jürjens, J.: Formal Development and Verification of Security-Critical Systems with UML. In: AVoCS 2010 - Workshop on Automated Verification of Critical Systems, Oxford (2010)
2. Devanbu, P.T., Stubblebine, S.: Software engineering for security: A roadmap. In: ICES 2009 Special Volume on the Future of Software Engineering (2009)
3. Jürjens, J.: UMLsec: Extending UML for Secure Systems Development. In: Jézéquel, J.-M., Hussmann, H., Cook, S. (eds.) UML 2002. LNCS, vol. 2460, pp. 412–425. Springer, Heidelberg (2002)
4. Debbabi, M., Mejri, M.: Towards the correctness of security protocols. Electronic Notes in Theoretical Computer Science, vol. 83 (2004)

The Edge-Detection Algorithm Based on Enlarged Block-Range

Wang Weiqing

Department of Information Management, Southwest University Rongchang Campus, Chongqing 402460, China

Abstract. Because the contrasts of different regions in an image are different, as it detecting the edge, it is very difficult to choose different detection thresholds in different image areas. So the edge which detected by the traditional algorithm is more fuzzy, its edge-line thickness is uneven, and there is more burr and fracture in the edge line. Firstly, the algorithm divides the image into same size of $n \times n$ pixel blocks. To search for the maximum value b and the minimum value a in the pixel block, and obtained the range $[a,b]$ of the pixel block. Then unify the range $[a,b]$ of the pixel block to the total range $[c,d]$ of the image. Finally, use the traditional detection operator to detect the image edge. The Algorithm can choose different detection thresholds in different image areas. The experimental results show that the detection edge detected by the algorithm is more clear and exquisite than that detected by the classical algorithm. And its edge-line thickness is even, and there is lesser burr and fracture in the edge line.

Keywords: Burr, Edge Detection, Image segmentation, Operator, Pixel Block.

1 Introduction

Edge detection is the very important foundation in the arrears of image segmentation, region-form extraction and measurement, regional recognition of target areas and so on. It is also the primary research of medical image processing in extracting a variety of feature information of diseases and pathological diagnosis, and in identifying diseased tissue or organs [1]. In engineering applications, edge detection plays a very important position [2]. Currently, the research of edge detection is relatively hot, and its major researches focus on: Edge detection based on wavelet function [3-4], edge detection algorithm based on texture [5-6], Improvements of the traditional algorithms and its applications [7-8]. However, these methods are more effective for edge detecting in specific image, and the thinning degree of its detected edge is not enough, the edge is fuzzier, its burr and fracture is more. This was mainly due to the different regions in an image have different contrasts, when detects the edge it is very difficult to choose different edge detection thresholds for different areas. In this paper, based on the traditional edge operators, it magnifies different pixel blocks to different degrees, further enhance the contrast of the edge value between both sides, to unify the range of different areas. It could choose different edge-detection thresholds for different contrast regions, so that it could detect the edges whose thinning degree are better.

Y. Zhang (Ed.): Future Communication, Computing, Control and Management, LNEE 141, pp. 65–72.
springerlink.com

2 Algorithm Theory

2.1 Edge-Detection Operator

Classical edge detection is based on the original image, and study the gray of each pixel in a neighborhood region. Using the variation regularity of the first order or second order directional derivative detect the edge. The common edge-detection methods are: Roberts operator, Sobel operators, Prewitt operator and Canny edge detector and so on. Roberts operator is as follow:

Roberts operator: If we seek the cross-partial derivatives along the direction in figure 1, then it got the crossover edge operator which was proposed by Roberts in 1963. The biggest advantage of this method is its calculating quantity is very small, and its calculating speed is very fast. However, because this method used the even template, that is shown in figure 1, the gradient magnitude that was found at the point(x, y) is actually the value at the cross point, thus it results that the gradient magnitude at the point (x, y) in the image point was offset half a pixel.

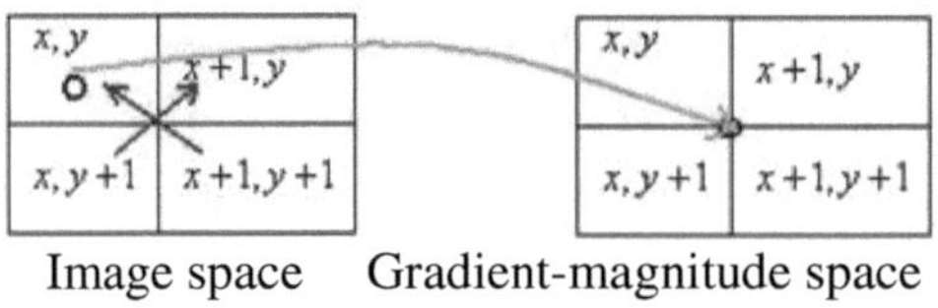

Fig. 1. Roberts partial derivative operator

1	0
0	-1

0	1
-1	0

Fig. 2. Roberts operator

Specific methods are shown as follows:

Assume the gradient magnitude of image $f(x, y)$ is: $G(x, y) = |G_x| + |G_y|$

The Roberts operator is the 2×2 operator template. Two convolution cores forms the Roberts operator, it is shown as figure 2. Every point in the image are convoluted with this two cores. The calculating result of the template is as follows:

$$G_x = 1 * f(x, y) + 0 * f(x+1, y) + 0 * f(x, y+1) + (-1) * f(x+1, y+1) = f(x, y) - f(x+1, y+1)$$

$$G_x = 0 * f(x, y) + 1 * f(x+1, y) + (-1) * f(x, y+1) + 0 * f(x+1, y+1) = f(x+1, y) - f(x, y+1)$$

$$G(x, y) = |G_x| + |G_y| = |f(x, y) - f(x+1, y+1)| + |f(x+1, y) - f(x, y+1)|$$

If $G(x, y)$ is greater than a certain threshold, we believe that the point (x, y) is an edge point.

Sobel operator: Sobel operator is a first-order differential operator, it uses the gradient value in a neighborhood region to calculate the gradient of the pixel, and then use a certain threshold to choose the pixel. It is given by:

$$s = (dx^2 + dy^2)^{1/2} \tag{6}$$

The Roberts operator is the 3×3 operator template. Two convolution core dx and dy form the Sobel operator, it is shown as figure 3. One core has the largest response to the vertical edge, and another core has the largest response to the horizontal edge. The larger value of two convolutions is the output value at the point. The calculating result is an edge-magnitude image.

2.2 The Amplification Principle of a Pixel Block Range

From the above principle of edge detection, if the $G(x, y)$ of a pixels (the Roberts operator is as an example, the other operators are similar to it) is greater than a threshold, we think that point is an edge point. But to a medical image, the $G(x, y)$ in different parts of the edge has larger difference, it is shown as a, b and c in figure 3. If the threshold is chosen too small, the edge that has larger contrast is thicker. (such as the region b); If the threshold is chosen too large, the edge that has smaller contrast can not be detected out(such as the region c). Therefore, it is very difficult to choose the threshold. For some complex images, there is even no appropriate threshold can be chosen. Therefore, we have to choose different threshold in different regions in the same image. So the image processing becomes very complex. By careful analysis of the images we found, the pixel value ranges in different regions of the same image are quite different, but there is small pixel-values difference within the region of the image. Therefore, we can magnify the pixel range of different region to the same range of the image, thus, it can select the same threshold for different regions in an image. The amplification principle of range is as follows:

Step one: Zoning: An image is divided into the same size of $n \times n$ pixel blocks, and each block is a processing region of the image. Suppose value of each pixel within a $n \times n$ pixel block is $f(i, j)$, where $i \in [1, n]$ and $j \in [1, n]$;

Step two: Seek the range within a pixel block: In the $n \times n$ pixel block, assume $\max(x_1, x_2, ..., x_m)$ is the maximum value in the m numerals $(x_1, x_2, ..., x_m)$, $\min(x_1, x_2, ..., x_m)$ is the minimum value in the m numerals $(x_1, x_2, ..., x_m)$, when $a = \min(f(1,1), f(1,2), ..., f(n,n))$ and $b = \max(f(1,1), f(1,2), ..., f(n,n))$, then $a \le f(i, j) \le b$ ($i \in [1, n]$, $j \in [1, n]$), that is $f(i, j) \in [a, b]$. Therefore, the range of the $n \times n$ pixel block is $[a, b]$.

Step three: Unify the threshold of all pixel blocks: Assume the range value which need to be unified in each $n \times n$ pixel blocks is $[c, d]$, then the amplification factor of the range value in the $n \times n$ pixel block is: $k = (d - c)/(b - a)$, and the value of each pixel within the pixel block is: $f(i, j) = f(i, j) * k - a * k + c$,where $i \in [1, n]$ and $j \in [1, n]$.

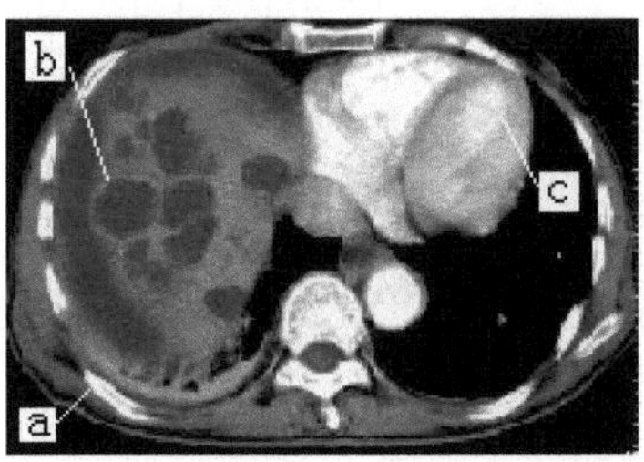

Fig. 3. Value differences of $G(x, y)$

3 Algorithm Design

Firstly, the image is divided into pixel blocks of $n \times n$, and seek its maximum value b and minimum value a within the block, then unify the pixel range value within $[c,d]$, finally, within each image block use edge-detection operator to detect image edge. Thus, the edge-detection algorithm based on the enlarged block-range is as follows(the selected language is VC# .net):

```
for j←0 to Height – 1 step 15
    for i←0 to Width – 1 step 15
        if (i + 16 < Width and j + 16 <Height)
            then   call kuai(i, j, 16, 16, 300, box1
```

Where the core of the algorithm is function kuai, it is the edge detection process that is used to calculated each pixel block, the algorithm is as follows:

```
        kuai(int left, int top, int m, int n, int a, Bitmap box1)// pixel block of•a is the
threshold
    for j←top to top+n // get the information
        for i←left to left + m
            juxing[i-left,j-top]←box1.GetPixel(i, j).R
            if (min > juxing[i - left, j - top])
                then min ← juxing[i - left, j - top]
            if (max < juxing[i - left, j - top])
                then   max ← juxing[i - left, j - top]
    if (min !← max)
        then   k ← (int)(1000 / (max - min));
               k1 ← min * k;
               for   j← 0 to   n// zoom data
                   for i←0 to m
                       juxing[i, j]← juxing[i, j] * k - k1
               for j← 0 to n - 1// handle edge points
                   for i←0 to m - 1
                       c1 ← juxing[i, j]
                       c2 ← juxing[i + 1, j + 1]
                       c3 ← juxing[i + 1, j]
                       c4 ← juxing[i, j + 1]
                       rx ← c1 - c2
                       ry ← c3 - c4
                       rr ← Math.Abs(rx) + Math.Abs(ry)
                       if (rr > a)
```

```
                                    then rr ← 0
                                          SetPixel(left + i, top + j,
Color.FromArgb(rr, 255, rr))
        else
                                       SetPixel(left + i, top + j,
Color.FromArgb(255, 255, 255))
```

4 Experiments and Comparative Analysis of the Algorithm

The experimental environment is: CPU is AMD Athlon (tm) 64 X2 Dual Core Processor 3600+, its frequency is 2GHz; the memory is 1G; Operation system is Microsoft Windows XP Professional Service Pack 3; the programming language is Visual C#.net 2005. The experimental results are as follows:

Figure 4 for a classical image, it is the comparative results that were tested by the traditional Roberts operator and the edge-detection algorithm based on enlarged block-range. Figure 4a is the original image, and figure 4b is the result that is detected by the traditional Roberts edge detection operator, and figure4c is the result that is detected by the edge-detection algorithm based on enlarged block-range. It can be found from the figure, the edge detected by the algorithm in this article is more clearly and exquisite, the thickness of the edge line thickness is uniformity, and there are fewer burrs and fractures in the edge line. However, the edge detected by the classical edge detection algorithm is blurrier, the thickness of the edge line thickness is not uniformity, and there is more burrs and fractures in the edge line.

a Original image b Roberts operator c Threshold is 200 and pixel block is 16 * 16

Fig. 4. The result detected by the algorithm in this article and Roberts operator

When the pixel block size is 8*8, the results in figure 5 are detected by the edge-detection algorithm based on enlarged block-range with different thresholds. Figure 5a, figure 5b, figure5c, figure 5d, figure 5e and figure 5f are the detection when their thresholds are respectively 200,300,400,500 and 600. Experiments show that if the threshold is too small then the detected edges are much fuzzier, and if the threshold is too large then there are more fractures in the detected edge. Therefore, the thresholds in different the size of the pixel blocks need to be appropriate.

a Original image b Threshold is 200 c Threshold is 300
d Threshold is 400 e Threshold is 500 f Threshold is 600

Fig. 5. The size of pixel block is8*8

When the threshold is 300, the results in figure 6 are detected by the edge-detection algorithm based on enlarged block-range with different size of pixel blocks. Figure 6a, figure 6b, figure 6c, figure 6d, figure 6e and figure 6f are the detection when their pixel blocks are respectively 4*4,8*8,12*12,16*16 and 20*20. Experiments show that the size of pixel blocks in different images need to be appropriate, if the size of block is too small then there are too many detected edges in the image, the detected image is much fuzzier, and if the size of block is too large then detected edges is thin, but there are some missed edges and some fractures in the image. Therefore, when the threshold in an image has been chosen the size of its pixel block should be chosen properly.

a Original image b pixel block is 4*4 c pixel block is 8*8
d pixel block is 12*12 e pixel block is 16*16 f pixel block is 20*20

Fig. 6. Threshold is 300

When the threshold is 300 and the pixel block size is16*16, the results in figure 7 are detected by the edge-detection algorithm based on enlarged block-range with different images. Figure 7a is the original images, and figure 7b is the corresponding detection results. Experiments show that there are different edge-detection effects in different images. There are better edge-detection results in the images whose edge contrast is more obvious.

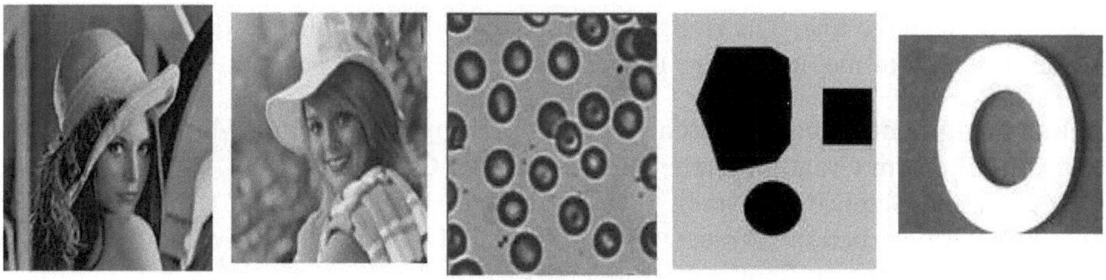

a Original image

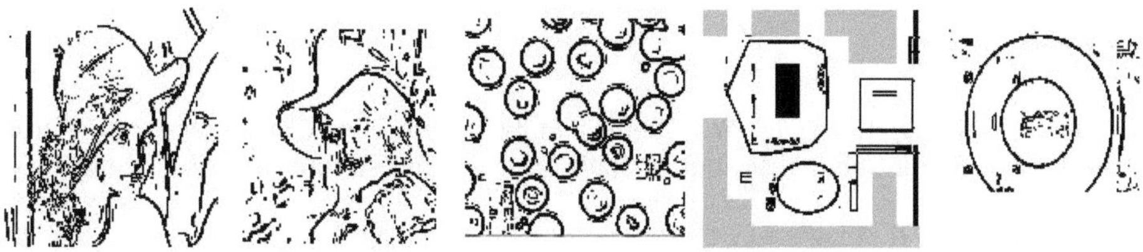

b results detected by the algorithm in this article

Fig. 7. Size of pixel block is16*16 and threshold is 300

These experiments show that: Different images, we should choose the appropriate threshold and pixel block for different image. The edge-detection effect which use the algorithm in this article to detect is more obvious, and the edge-detection result have significantly improved than that detected by the traditional operator, there is good connectivity and fewer fractures in its detection results, its edge line is very uniform, and there are fewer burrs in them.

5 Conclusion

Edge detection is the very important foundation in the research arrears of image segmentation, and it is also the firs step of research. In engineering applications, edge detection plays a very important position. Currently, the research of edge detection has been very mature. Because the contrasts of different regions in an image are different, as it detecting the edge, it is very difficult to choose different detection thresholds in different image areas. So the edge which detected by the traditional algorithm is more fuzzy, its edge-line thickness is uneven, and there is more burr and fracture in the edge line. Firstly, the algorithm divides the image into same size of $n \times n$ pixel blocks. To search for the maximum value b and the minimum value a in the pixel block, and obtained the range $[a,b]$ of the pixel block. Then unify the range $[a,b]$ of the pixel block to the total range $[c,d]$ of the image. Finally, use the traditional detection operator to detect the image edge. The Algorithm can choose different detection thresholds in different image areas. The experimental results show that the detection edge detected by the algorithm is more clear and exquisite than that detected by the classical algorithm. And its edge-line thickness is even, and there is lesser burr and fracture in the edge line. The research of this algorithm established the foundation for further research of medical image processing.

Acknowledgment. This research is supported by Chongqing Municipal Higher Education Reform Grant Fundamental Research of Chongqing(09-3-181), Chongqing Education and Science "Eleventh Five-Year Plan" funded projects (08-GJ-044),Key Discipline Construction Project of Southwest University, cultivating discipline construction projects of Southwest University.

References

1. Yuan, X., Wang, L., Geng, W., Hu, H.: Fully Autom atic Liver Segm-entation from Multi-slice Spiral Com puted Tom ographic Images. Chinese Journal of Electronics 18(4), 677–680 (2009)
2. Yang, N., Chen, H., Li, Z., Yao, C.: A New Algorithm of the Circle Detection in A Complex Background Image. Journal of Beijing Jiaotong University 34(2), 67–70 (2010)
3. Yun, Y., Dai, S.-G., Mu, P.-A.: Algorithm for automobile headlamp image edge detection based on wavelet transform. J. University of Shanghai for Science and Technology 31(6), 593–596 (2009)
4. Zeng, L., Guo, H.-Y., Ma, R.: Edge detection of industrial CT 3D image based on wavelet locating and Facet model. Optics and Precision Engineerin. 18(2), 443–450 (2010)
5. Xiao, C.-X., Chu, Y., Zhang, Q.: Texture Image Segmentation Using Level Set Function Evolved by Gaussian Mixture Model. Chinese Journal of Computers 33(7), 1295–1304 (2010)
6. Wang, S., Xia, Y., Jiao, L.C.: Mean Shift Based Adaptive Texture Image Segmentation Method. Journal of Software 21(6), 1451–1460 (2010)
7. Li, J., Xue, Q.: Canny edge detector and its application to analyze the brazed joint interface. China Welding 1(18), 63–66 (2009)
8. Shiu, G.T., Yuan, Y., Yang, S.: Image segmentation of small defects in precision weldments with complex structure. China Welding 16(1), 52–56 (2007)

The Image Correction Algorithm Based on Combined Transformation

Wang Weiqing

Department of Information Management, Southwest University Rongchang Campus, Chongqing 402460, China

Abstract. Currently, there are many image correction algorithms in researching. However, due to bad selection of the transformation parameters, the correction effect is not very good in these algorithms. In this algorithm, according to the transformation features of an image, we can correct an image by using the combined transformation of translation transformation, rotation transformation and scaling transformation. Experimental results show that there are more noise points in this correction algorithm, but the corrected image is much better. Its results can be further processed for the research of image fusion. It laid the groundwork for future image processing.

Keywords: Correction, algorithm, matrix, image.

1 Introduction

Image correction plays an important role in digital image processing [1-2]. When we input medical image data through the scanner or digital camera, it can not avoid causing the image changes such as its ratio, position and angle, and these changes will result in that the adjacent data slices can not be inosculated. Obviously, such data can not be directly applied to the subsequent processing. Therefore, we must correct these images. Through a series of geometric transformation, we can get regular volume data. For relatively simple image warping, it seems that using panning, zooming, rotating and other basic transformation is feasible. However, in more cases, the space transformation is uncertain, so it can not be simply completed by the basic transform and its complex transformation. At this time, we can specify the control points, or use a polynomial transformation or use controlling grid interpolation transformation to achieve the image corrections. At present, there many researches about the image correction [3-5]. Among them, the commonly methods about image correction are perspective transformation and control point transformation [6]. The perspective transform method is mainly based on the principle of camera imaging to correct the image [7]; and the control point transformation method can adapt the general image correction, its main principle is based on the correspondence between control points, and using the bilinear interpolation method to correct the image. However, the results by using both of the methods have much deformation, and it is not satisfy the expected the correction results. Therefore, we propose a correction algorithm based on combined transformation.

Y. Zhang (Ed.): Future Communication, Computing, Control and Management, LNEE 141, pp. 73–81.
springerlink.com

2 The Algorithm Theory

The basic idea of the combined transformation is based on the principle of transformation to correct image. In the two images that are the original image before correction and the corrected image, to select the four corresponding control points (i_1, j_1), (i_2, j_2), (i_3, j_3), (i_4, j_4) 及 (i_1', j_1'), (i_2', j_2'), (i_3', j_3'), (i_4', j_4'). As shown in figure 1, after a combination of the following four-step transformation, the four control points $1,2,3,4$ became the corrected control points $1',2',3',4'$.

2.1 Translation Transformation

Shift the original image control point 1 to the corrected control point $1'$. Then the translation transform coefficient is:

$$h = i_1' - i_1 \qquad k = j_1' - j_1 \tag{1}$$

Then the translation transformation matrix A_1 of (1) is:

$$A_1 = \begin{bmatrix} 1 & 0 & 0 \\ 0 & 1 & 0 \\ i_1' - i_1 & j_1' - j_1 & 1 \end{bmatrix} \tag{2}$$

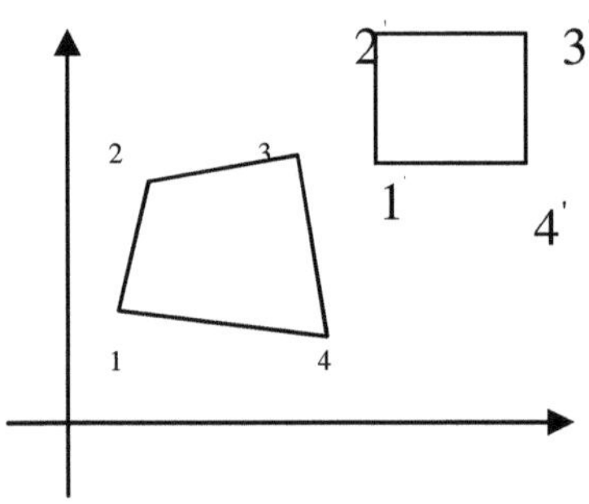

Fig. 1. Combined transformation

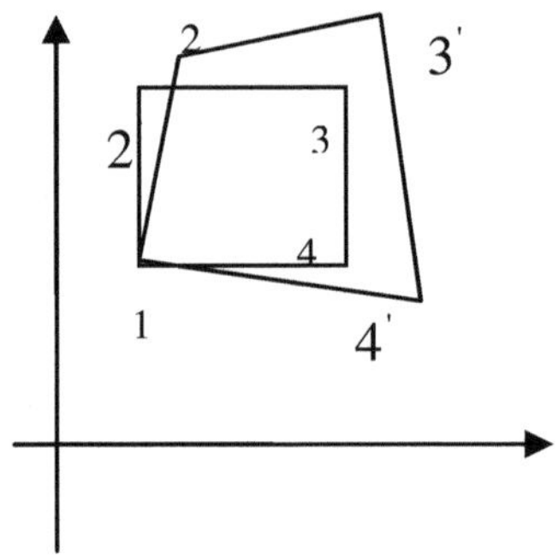

Fig. 2. Translation transformation

2.2 Rotation Transformation

The translated image is rotated angle θ_1, and we provides for the angle is positive when it rotates counterclockwise. So makes the three points $1,3,3'$ be in a straight line. As shown in figure 3, the three sides of $\Delta 133'$ are respectively:

$$s_{13} = \sqrt{(i_3 - i_1)^2 + (j_3 - j_1)^2} \tag{3}$$

$$s_{13'} = \sqrt{(i_{3'} - i_1)^2 + (j_{3'} - j_1)^2} \tag{4}$$

$$s_{33'} = \sqrt{(i_{3'} - i_3)^2 + (j_{3'} - j_3)^2} \tag{5}$$

Form the cosine theorem, we can get:

$$\theta_1 = \arccos\left(\frac{s_{13}^2 + s_{13'}^2 - s_{33'}^2}{2 s_{13} s_{13'}}\right) \tag{6}$$

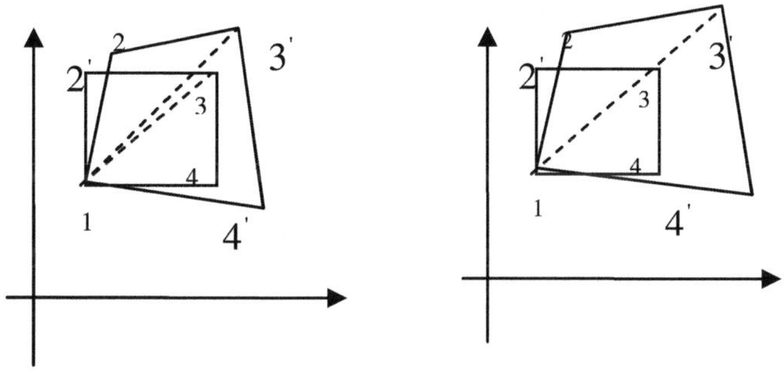

Fig. 3. Calculation of the rotation angle

Because the angle range of arc cosine is $[0, \pi]$, in order to determine the sign of rotation angle θ_1, we first determine the point $3'$ in figure 3 is on the left or right of the line 13. If it is on the left then θ_1 is positive, otherwise θ_1 is negative. Then the equation of the line 13 is:

$$\frac{j - j_1}{i - i_1} = \frac{j_3 - j_1}{i_3 - i_1} \tag{7}$$

Thus we can conclude:

$$j = \frac{j_3 - j_1}{i_3 - i_1} * (i - i_1) + j_1 \tag{8}$$

Put the abscissa of the point $3'$ in formula (8), then:

$$j = \frac{j_3 - j_1}{i_3 - i_1} * (i_{3'} - i_1) + j_1 \tag{9}$$

If $j > j_{3'}$ then we can conclude that point $3'$ is on the left of the line 13, and θ_1 is positive. Otherwise θ_1 is negative.

After rotating the angle θ_1, the diagonal lines of the two control rectangles will overlap together. In order to the scaling transformation, we need to rotate the image again, thus the diagonal lines of the two control rectangles will be in parallel with the horizontal axis. So the rotation angle is: $\theta_2 = -45^{\circ}$. As shown in figure 4.

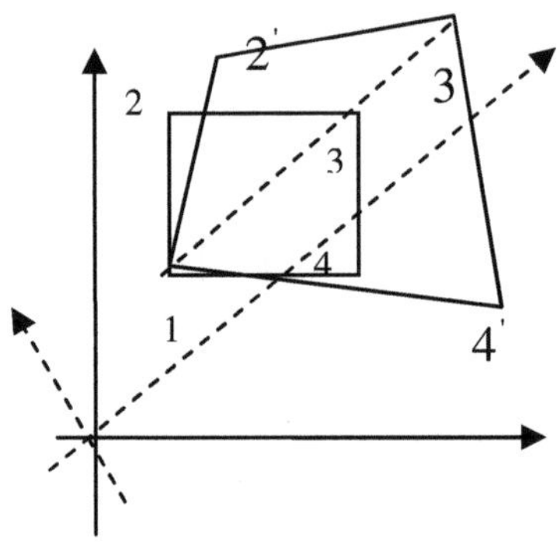

Fig. 4. Rotate angle θ_2

Therefore, the total rotation angle is as:

$$\theta = \theta_1 - \theta_2 \tag{10}$$

We provides for the angle is positive when it rotates counterclockwise. Thus the rotation transformation matrix A_2 is:

$$A_2 = \begin{bmatrix} \cos\theta & \sin\theta & 0 \\ -\sin\theta & \cos\theta & 0 \\ 0 & 0 & 1 \end{bmatrix} \tag{11}$$

2.3 Scaling Transformation

After rotating the image, the proportion of two diagonal lines of the control points is still inconsistent. So we will make it consistent by scaling transformation. As shown in figure 4, the distance between control points were:

$$s_{13} = \sqrt{(i_3 - i_1)^2 + (j_3 - j_1)^2} \tag{12}$$

$$s_{13'} = \sqrt{(i_{3'} - i_1)^2 + (j_{3'} - j_1)^2} \tag{13}$$

$$s_{24} = \sqrt{(i_4 - i_2)^2 + (j_4 - j_2)^2} \tag{14}$$

$$s_{2'4'} = \sqrt{(i_{4'} - i_{2'})^2 + (j_{4'} - j_{2'})^2} \tag{15}$$

Thus the vertical and horizontal ratios are:

$$s_x = \frac{s_{13'}}{s_{13}} \tag{16}$$

$$s_y = \frac{s_{2'4'}}{s_{24}} \tag{17}$$

So the scaling matrix A_3 can be obtained:

$$A_3 = \begin{bmatrix} s_x & 0 & 0 \\ 0 & s_y & 0 \\ 0 & 0 & 1 \end{bmatrix} \tag{18}$$

2.4 Rotation Reverse Transformation

After scaling transformation, the proportions between the corrected image and the need image have been consistent. In order to let the diagonal lines of the two control rectangles be in parallel with the horizontal axis, we have rotated angle is: $\theta_2 = -45^\circ$. Therefore, the image must be reversely rotated. Thus the edge of the image rectangle of the control points will be in parallel to the axes. It is shown as figure. The reverse rotation angle is: $\theta = 45°$. Therefore, the rotation transformation matrix A_4 is:

$$A_4 = \begin{bmatrix} \cos\theta & \sin\theta & 0 \\ -\sin\theta & \cos\theta & 0 \\ 0 & 0 & 1 \end{bmatrix} = \begin{bmatrix} \cos 45° & \sin 45° & 0 \\ -\sin 45° & \cos 45° & 0 \\ 0 & 0 & 1 \end{bmatrix} = \begin{bmatrix} \frac{\sqrt{2}}{2} & \frac{\sqrt{2}}{2} & 0 \\ -\frac{\sqrt{2}}{2} & \frac{\sqrt{2}}{2} & 0 \\ 0 & 0 & 1 \end{bmatrix} \tag{19}$$

2.5 Combined Transformation Matrix

From the above, the combined transformation matrix A is:

$$A = A_1 A_2 A_3 A_4 \tag{20}$$

So the final combined transformation is:

$$(i', j', 1) = (i, j, 1)A = (i, j, 1)A_1 A_2 A_3 A_4 \tag{21}$$

3 The Design of the Combined Transform Algorithm

According to the above analysis, in the two images that are the original image before correction and the corrected image, to select the four corresponding control points (i_1, j_1), (i_2, j_2), (i_3, j_3), (i_4, j_4) 及 (i_1', j_1'), (i_2', j_2'), (i_3', j_3'), (i_4', j_4'). If We use (21) to correct the image, then the correction algorithm is as following:

```
1.   hx ← i11 - i1
2.   ky ← j11 - j1
3.   i1 ← i1 + hx
4.   j1 ← j1 + ky
5.   i2 ← i2 + hx
6.   j2 ← j2 + ky
7.   i3 ← i3 + hx
8.   j3 ← j3 + ky
9.   i4 ← i4 + hx
10.  j4 ← j4 + ky
11.  s13 ←Sqrt(((i3 - i1) * (i3 - i1) + (j3 - j1) * (j3 - j1)))
12.  s131 ←Sqrt((i13 - i1) * (i13 - i1) + (j13 - j1) * (j13 - j1))
13.  s1131 ←Sqrt((i13 - i11) * (i13 - i11) + (j13 - j11) * (j13 - j11))
14.  s33 ←Sqrt((i13 - i3) * (i13 - i3) + (j13 - j3) * (j13 - j3))
15.  s24 ←Sqrt((i4 - i2) * (i4 - i2) + (j4 - j2) * (j4 - j2))
16.  s2141 ←Sqrt((i14 - i12) * (i14 - i12) + (j14 - j12) * (j14 - j12))
17.  jiao ←Acos((s13*s13+s131*s131-s33*s33)/(2*s13*s131))  // To
calculate the rotation angle
18.  zf ← (j3 - j1) / (i3 - i1) * (i13 - i1) + j11-j13 //To determine the angle sign
19.  if (zf < 0)// If the angle is negative then add the negative sign
         then  jiao ← (-1) * jiao
20.  jiao ← jiao -3.1415926 / 4// To minus 45-degrees
21.  cx ← Width / 2//Look for the midpoint of the image
22.  cy ← Height / 2
23.  for  i ← 0 to Width
         x0 ← i
         for  j ← 0 to Height
             y0 ← j
             x2 ← x0 + hx// Translation transformation
             y2 ← y0 +ky// Translation transformation
             x1← (x2 - cx)* Cos(jiao) - (y2 - cy) Sin(jiao) + cx// Rotate
```

the image

```
y1←(x2 - cx) * Sin(jiao) + (y2 - cy) * Cos(jiao) +cy
x2 ← (x1 * s1131 / s13);// Scaling transformation
y2 ← (y1 * s2141 / s24)
jiao1 ← 3.1415926 / 4// Reversal rotation angle
x1 ←(x2 - cx) *Cos(jiao1)-(y2 - cy) *Sin(jiao1) +cx// Rotate
```

the image

```
y1 ← (x2 -cx)* Sin(jiao1)+(y2 -cy) *Cos(jiao1)+cy
if (x1 >= 0 and x1 < Width and y1 >= 0 and y1 < Height)
    then    SetPixel(x1, y1, GetPixel(i, j))
```

4 The Experiment and Comparative Analysis of the Algorithms

To verify the correctness and effectiveness of the filtering algorithm, we select this experimental environment: CPU is AMD Athlon (tm) 64 X2 Dual Core Processor 3600+, its frequency is 2GHz; the memory is 1G; Operation system is Microsoft Windows XP Professional Service Pack 3; the programming language is Visual C#.net 2005. We select a gynecology image to do the experiment. The experimental results are as follows:

Figure 5 shows the experimental results by the scaling transformation. Where image a is the original image needs to be corrected, and the expecting result is to scale the original image with ratio 90%. Image b is the results by combined transformation correction algorithm. Image c is the results by perspective transformation correction algorithm.

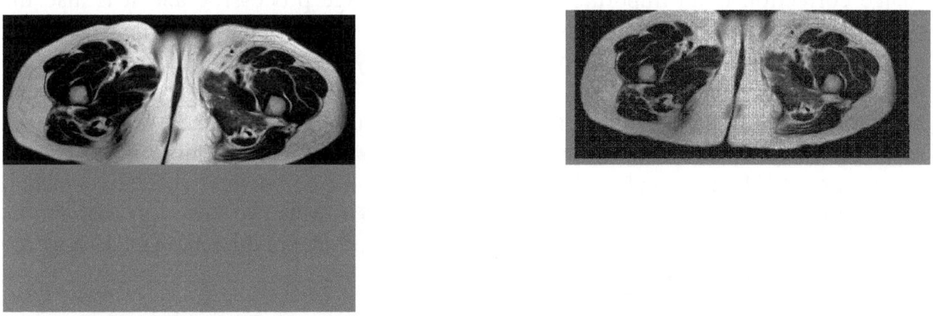

a. The original image b. The results by combined transformation c. The results by perspective transformation

Fig. 5. Correct the image with the scaling transformation

Figure 6 shows the experimental results by the rotation transformation. Where image a is the original image needs to be corrected, and the expecting result is to rotate the original image with 30 degrees. Image b is the results by combined transformation correction algorithm. Image c is the results by perspective transformation correction algorithm.

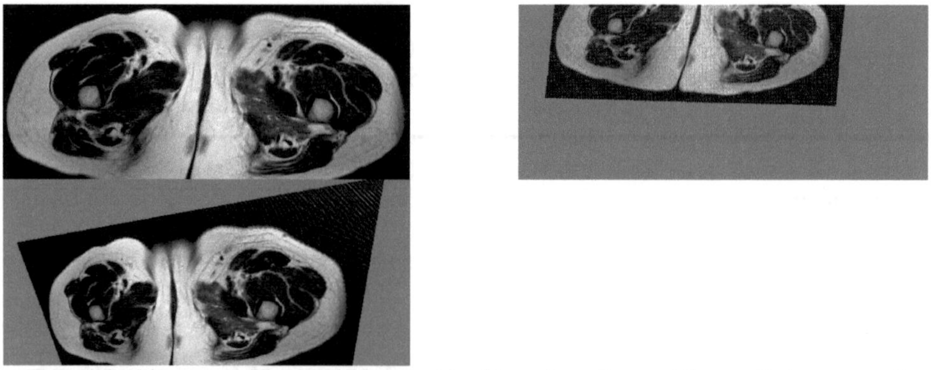

a. The original image b. The results by combined transformation c.The results by perspective transformation

Fig. 6. Correct the image with the rotation transformation

From figure 5 and figure 6, it shows that there are more noises in the image corrected by the combined transformation algorithm, but there is relatively small deformation in it, and the correct results are very satisfactory. On the contrary, there are smaller noises in the image corrected by the perspective transformation algorithm, but there is relatively more deformation in it, and the correct results are not satisfactory.

5 Conclusion

Image correction is an important foundation for image processing and it is also the preliminary work of image processing. It has a very important role in the medical image analysis and other engineering applications. Currently, the research of image correction has been mature. However, the image correction effect is not very good for the selection of the transformation parameters is difficult. The corrected image can not meet the further processing of image fusion. Although there are more noises in the image corrected by the combined transformation algorithm, but there is relatively small deformation in the corrected image and the correct results are very satisfactory. Its results can be further processed for the research of image fusion. And it laid the groundwork for future image processing.

References

1. Cui, H., Chen, J., Wang, D.: Study and Implementation of Distortion Correction and Mosaics of Fisheye Images. Computer Engineering 33(10), 190–192 (2007)
2. Zhu, Y., He, Y.-H., Li, P., Gao, Y.-J., Shao, Y.-H., Ma, H.: A New Method to Remove Dithering in Optical Coherence Tomography without Information Loss. Laser & Infrared 37(3), 288–291 (2007)
3. Li, Q., Fu, Z., Liu, Q.: An Effective Skew Image Correction Method. Computer Engineering 32(11), 194–196 (2006)

4. Yu, D.-B., Su, Z.-W., Yan, K.H.: A New Type of Machine Vision Systems with Algorithm for Image Correction. Laser & Infrared 38(11), 1173–1176 (2008)
5. Su, Z., Wang, J., Huang, M., et al.: A machine vision system with an irregular imaging function. In: The 5th International Conference on Image and Signal Processing and Analysis, ISPA 2007, pp. 458–463. IEEE (2007)
6. Hu, D.-H., Wang, H., Ai, J., Zhang, L., Zhang, S.-L., et al.: Comparisons of two kinds of image rectification algorithms. Computer Engineering and Application 45(13), 191–193 (2009)
7. Zheng, D., Ge, W., Zhang, D., Ge, W.: Application of Improved BP Algorithm Based on Numerical Optimization to Fault Diagnosis of Rotating Machinery. Journal of Transduction Technology 18(3), 510–513 (2005)

Designing Network Immunization Strategies Based on Graph Partitioning Methods

Changxi Niu, Lemin Li, and Du Xu

The Key Laboratory of Broadband Optical Fiber Transmission and Communication Networks, University of Electronic Science and Technology of China (UESTC), Chengdu, China
chxniu@gmail.com, lml@uestc.edu.cn, xudu@uestc.edu.cn

Abstract. Designing effective immunization strategies is very important to fight against epidemic attacks such as worms and viruses. In this paper, we design two immunization strategies based on graph partitioning methods. We compare the proposed strategies in various simulation scenarios with the widely used High Degree First (HDF) immunization strategy. We find the proposed immunization strategies based on graph partitioning methods are better than the HDF immunization strategy in all of the simulation scenarios.

Keywords: Epidemic, Immunization, Graph Partitioning.

1 Introduction

Epidemic attack such as worms and viruses [1] is one of the most serious security problems in networks. Designing effective network immunization strategy is an important way to fight against them [2]. Immunization resources are expensive and limited. The goal of network immunization is to pick out a set of nodes in the network to immunize so that the network is infected by the epidemic as little as possible.

The Internet topology has the power law characteristic [3]. The highly connected nodes (hubs) play important roles in the epidemic spreading dynamics. The High Degree First (HDF) strategy targeting immunization resources towards the hubs in the network has been widely used [2]. All nodes in the network are sorted based on their degrees and the most connected one is picked out to be immunized first.

We propose to use the graph partitioning methods to design immunization strategies. The intuition is that if we can find a set of nodes to partition the network into a set of isolated segments, the epidemic can only infect the segment in which the epidemic starts. In such a way, the epidemic attack is contained locally.

Epidemic spreading in the Internet is usually modeled using the SIS (Susceptible-Infected-Susceptible) model [4]. In the SIS model, nodes in the network can be in the susceptible state or the infected state. At each time step, every susceptible node is infected with probability β if it is connected to the infected nodes. In the meanwhile, each infected node is cured and goes back to the susceptible state with probability δ. In this paper, we use a nonlinear dynamical systems (NLDS) approach [5] which accurately models SIS type epidemic propagation in arbitrary network topologies.

Y. Zhang (Ed.): Future Communication, Computing, Control and Management, LNEE 141, pp. 83–88.
springerlink.com

Internet topology modeling work can be done in either the Autonomous Systems (AS) level or the router level. In this paper, we use AS level Internet topology. In a comparison work among the many AS models [6], Positive Feedback Preference (PFP) [7] is the most successful of all in accurately modeling the AS level Internet topology so far. Besides model topology, we also use a snapshot of the Internet AS data downloaded from the DIMES project [8] which is a distributed scientific research project, aimed at studying the structure of the Internet.

2 The Proposed Immunization Strategies

The First Immunization Strategy. The first proposed immunization strategy aims at partitioning the network so that the largest segment of the immunized network contains as few nodes as possible. When the epidemic starts spreading, it can only infect the segment in which the seed epidemic starts, and its damage is minimized.

However, this partition problem is NP-complete [9]. In order to implement this immunization strategy, we find a heuristic algorithm aimed at solving the Key Player problem (KPP) [10] can be used here.

The first immunization strategy is as follows:

1. Select k nodes randomly in the network to initialize a set S
2. Set F = fit as the size of the largest segment after deleting the nodes in the set S
3. For each node u in S and each node v not in S
   a. ΔF = improvement in fit if u and v were swapped
4. Select pair with largest DELTAF
   a. If $\Delta F == 0$ then terminate
   b. Else, swap pair with greatest improvement in fit and set $F = F - \Delta F$
5. Go to step 3

Return S as the set of nodes picked out by the first immunization strategy.

The Second Immunization Strategy. The above implementation of the first immunization strategy uses a kind of greedy optimization algorithm which is very costly. Here, we design another immunization strategy.

The second proposed immunization strategy uses a combination of the well known graph partitioning software package METIS [11] and a heuristic algorithm for the vertex cover problem [9].

1. Use the METIS to partition the network into some almost equally sized segments and get the edges connecting different segments.

2. Compute the vertex cover of the sub graph induced by the edges connecting different segments. We use a simple heuristic algorithm to do the vertex cover computation: first it picks the node of the highest degree of the sub graph induced by the edges connecting different segments as a node in the vertex cover and then deletes it and its edges. Then iteratively, it picks the node of the highest degree of the sub graph after deleting the already picked out nodes and deletes it and its edges. The algorithm stops when all the edges are deleted.

3. Rank the nodes in the vertex cover according to how many different segments it connects to in the partitioned network after using METIS. The node which connects the largest numbers of different segments ranks the top.

When we are asked to immunize k nodes, we pick the top k ranked nodes. In this way, we aim to disconnect the different network segment as much as possible.

3 Simulation Results

In this section, we perform extensive simulations to compare our proposed immunization strategies with the widely used HDF immunization strategy in various simulation scenarios.

We use three network topologies to do the simulations. We regard all of the networks as undirected. First, we use the tool provided by the author of PFP [7] to generate two AS level network topologies of different sizes. We generate 20 instances of 1000-node PFP model network topologies and 20 instances of 5000-node PFP model network topologies. Epidemic starts from 5‰ nodes randomly chosen from the networks except the immunization nodes we pick out. On each of the 20 instances, the epidemic spreading simulations are carried out 100 times, each time with differently randomly chosen epidemic starting nodes and the final infected node numbers are averaged over the 100 simulation runs. Then we average the results over all of the 20 instances. Here we only report the final average results. The snapshot of the Internet AS data downloaded from The DIMES project contains 19049 nodes. The SIS epidemic parameters are $\beta = 0.2$, $\delta = 0.2$.

We use the kmetis program in the METIS package to partition the network into some parts. We choose how many parts to partition based on how many nodes we want every segment contains at most. In the following we partition the 1000-node, 5000-node network into 50 parts and partition the 19049-node into 400 parts.

In the figures, we use KP to represent the first proposed immunization strategy and use VC to represent the second proposed immunization strategy. The running time of the first proposed strategy is unacceptable in the 5000-node networks. For the 5000-node PFP network topology, we only present the second proposed strategy to compare with the HDF strategy.

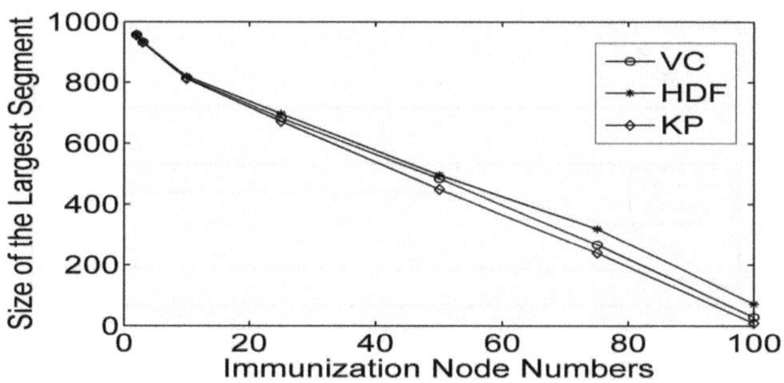

Fig. 1. Size of the largest segment after immunizing 2 to 100 nodes in PFP 1000-node model

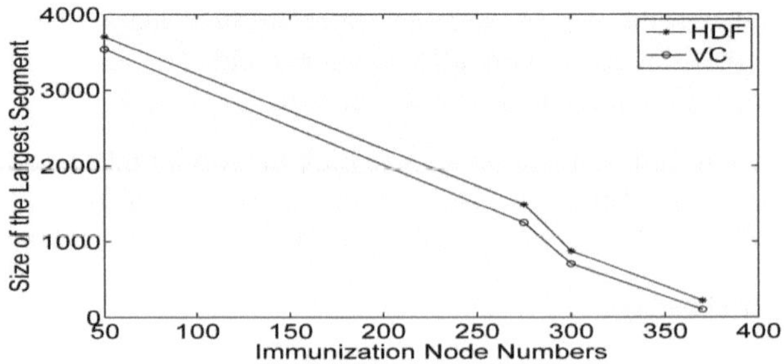

Fig. 2. Size of the largest segment after immunizing a set of nodes in the PFP 5000-node model

Fig. 1 and Fig. 2 plots the node numbers of the largest segment after immunizing a set of k nodes in the 1000-node PFP model and 5000-node PFP model network topology using the HDF strategy and our strategies separately. We can see from Fig. 1 and Fig. 2 that our immunization strategies can always get a smaller size of the largest segment.

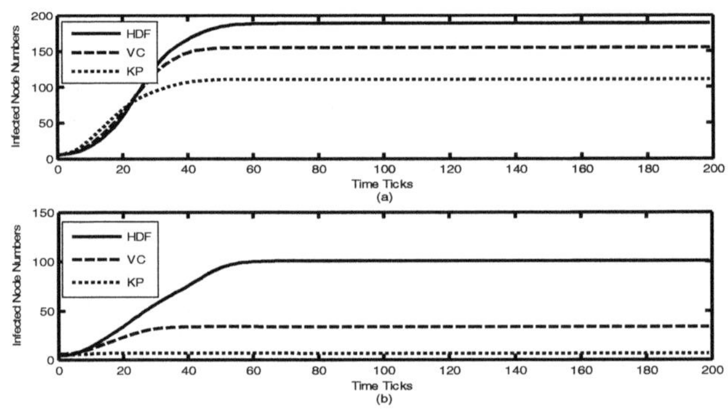

Fig. 3. Infected node numbers vs. Time ticks after immunizing 7% (a) and 10% (b) of all the nodes in the PFP 1000-node model.

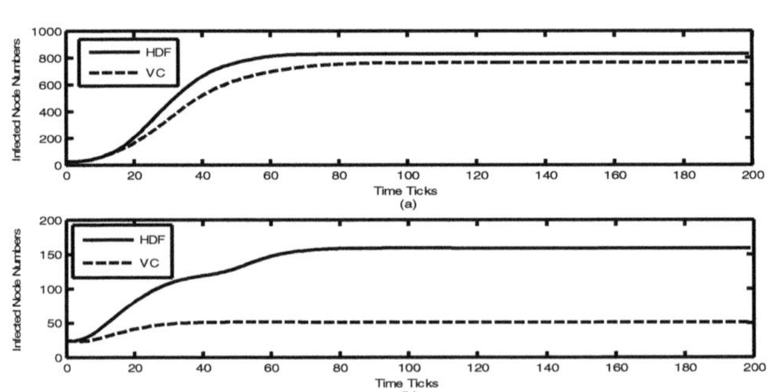

Fig. 4. Infected node numbers vs. Time ticks after immunizing 275 (a) and 370 (b) nodes in the PFP 5000-node model.

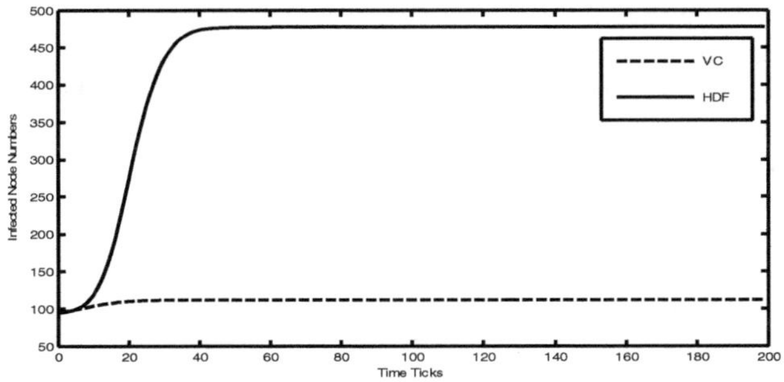

Fig. 5. Infected node numbers vs. Time ticks after immunizing 850 nodes in the 19049-node network.

Then we present in Fig. 3 and Fig. 4 the immunization effects in the 1000-node PFP model and 5000-node PFP model network topology using the HDF strategy and our strategies separately. Fig. 3 plots the infected node numbers when immunizing 7% and 10% of all the nodes. Fig. 4 plots the infected node numbers when immunizing 275 and 370 nodes. The simulation runs 200 time ticks when the infected node numbers go to steady. From Fig. 3 and Fig. 4, we can see that the proposed immunization strategies perform better than the HDF immunization strategy in fighting against the epidemic.

Finally in Fig. 5, we present the immunization effects of fighting against the epidemic when immunizing 850 nodes in the 19049-node network of the snapshot of the Internet AS data. We can see that the second proposed strategy performs nearly 4 times better than the HDF strategy in containing the epidemic.

4 Summary

In this paper, we design two network immunization strategies based on graph partitioning methods to fight against epidemic spreading in networks. We find the proposed immunization strategies are better than the High Degree First (HDF) immunization strategy in all of the simulation scenarios.

Acknowledgement. This work was supported by the State Key Development Program of Basic Research of China (973 Program, 2007CB307104 of 2007CB307100).

References

1. Staniford, S., Paxson, V., Weaver, N.: How to own the internet in your spare time. In: Proc. 11th Symp. USENIX Security, pp. 149–167 (2002)
2. Pastor-Satorras, R., Vespignani, A.: Immunization of complex networks. Phys. Rev. E 65(3), 036104 (2002)
3. Faloutsos, M., Faloutsos, P., Faloutsos, C.: On power-law relationships of the internet topology. In: SIGCOMM 1999, pp. 251–262 (1999)
4. Pastor-Satorras, R., Vespignani, A.: Epidemic spreading in scale-free networks. Phys. Rev. Lett. 86(14), 3200–3203 (2001)

5. Chakrabarti, D., Wang, Y., Wang, C., Leskovec, J., Faloutsos, C.: Epidemic thresholds in real networks. ACM Trans. Inf. Syst. Secur. 10(4), 1–26 (2008)
6. Haddadi, H., Fay, D., Jamakovic, A., Maennel, O., Moore, A., Mortier, R., Rio, M., Uhlig, S.: Beyond node degree: evaluating as topology models, University of Cambridge, Computer Laboratory, Tech. Rep. UCAM-CL-TR-725 (2008)
7. Zhou, S., Mondragon, R.J.: Accurately modeling the internet topology. Phys. Rev. E 70(6), 66108 (2004)
8. http://www.netdimes.org/
9. Garey, M., Johnson, D.: Computers and intractability. Freeman, San Francisco (1979)
10. Borgatti, S.: The key player problem. SSRN eLibrary (2002)
11. Karypis, G., Kumar, V.: A fast and high quality multilevel scheme for partitioning irregular graphs. SIAM Journal on Scientific Computing 20(1), 359 (1999)

Researches into Semiorthogonality Quality of Multivariant Wavelet Packages with Short Support

Yongcai Hu

Department of Fundamentals, Henan Polytechnic Institute, Nanyang 473009, P.R. China
nysslt88@126.com

Abstract. In this article, the notion of semiorthogonal nonseparable multivariant wavelet packages is introduced. A novel algorithm for constructing them is presented by virtue of iteration method. The semiorthogonality properties of multivariant wavelet packages associated with composite dilation matrix, are investigated, by means of functional analysis, time-frequency analysis method, and operator theory. Two biorthogonality formulas concerning these wavelet packas are presented.

Keywords: Semiorthogonality, multivariant wavelet packages, refinable functions, wavelet transfer, filter bank theory, Riesz bases, symmetry integral transform.

1 Introduction

At present, image interpolation algorithms based on wavelet transform are mainly based on multiresolution analysis of the wavelet. Raditionally, short-time Fourier Transform and Gabor Transform were used for harmonic studies of nonstationary power system waveforms which are basically Fourier Transform-based methods. To overcome the limitations in these existing methods,wavelet transform based algorithm has been developed to estimate the limitations the frequency and time information of a harmonic signal.The rise of wavelet analysis in applied mathematics is due to its applications and the flexibility. Multiwavelets can simultaneously possess many desired properties such as short support, semiorthogonality, symmetry, and vanishing moments, which a single wavelet cannot possess simultaneously. This suggests that multiwavelet systems can provide perfect reconstruction, good performance at the boundaries (symmetry), and high approximation order (vanishing moments). Already they have led to exciting applications in image processing [1]signal analysis [2], and fractals [3] and so on. Vector-valued wavelets are a class of special multiwavelets. Chen [4] introduced the notion of orthogonal vector-valued wavelets [5], Every frame(or Bessel sequence) determines an analysis operator, the range of which is important for a lumber of pplications.prefiltering is usually required for discrete multiwavelet transforms [6] but not necessary for discrete vector-valued transforms. Wavelet packets, owing to their nice characteristics, have been widely applied to signal processing [7], code theory, image compression, solving integral equation and so on. Coifman and Meyer firstly introduced the notion of univariate orthogonal wavelet packets. Yang [5] constructed a-scale orthogonal multiwavelet packets which were

Y. Zhang (Ed.): Future Communication, Computing, Control and Management, LNEE 141, pp. 89–97.
springerlink.com

more flexible in applications. It is known that the majority of information is multi-dimensional information. Shen [6] introduced multivariate orthogonal wavelets which may be used in a wider field. Thus, it is necessary to generalize the concept of multivariate wavelet packets to the case of multivariate vector-valued wavelets. The goal of this paper is to give the definition and the construction of biorthogonal vector-valued wavelet packages and construct several new Riesz bases of space $L^2(R^u, C^v)$. In order to enhance the application of wavelet analysis in the signal processing, and to improve the accuracy of signal processing as far as possible, predecessors have done a large number of practices and explorations, resulting in a lot of signal processing methods based on wavelet transform.

2 Symbols and Fundaments on Vector-Valued Function Space

We begin with some notations. Set $Z_+ = \{0\} \cup N$, s, $u, v \in N$ and $s, u, v \geq 2, Z^u = \{(z_1, z_2, \cdots, z_u):$ $z_r \in Z, r = 1, 2, \cdots, u\}$, $Z_+^u = \{\{(z_1, z_2, \cdots, z_4)::z_r \in z_+, r = 1, 2, \cdots, u\}$. Suppose that P is an $u \times u$ matrix whose all entries are integers and all eigenvalues is large than one in modulus. The absolute value of the determinant of matrix P is denoted by a, i.e., $|\det(P)| = a$, There exist a elements $\lambda_0, \lambda_1, \cdots, \lambda_{a-1}$ in Z_+^u by algebra theory such that $Z^u = \bigcup_{\lambda \in \Gamma_0}(\lambda + PZ^u)$; $(\lambda_1 + PZ^u) \cap (\lambda_2 + PZ^u) = \varnothing$, where $\Gamma_0 = \{\lambda_0, \lambda_1, \cdots, \lambda_{a-1}\}$ denotes the set of all different representative elements in the quotient group $Z^u/(PZ^u)$ and λ_1, λ_2 denote two arbitrary distinct elements in Γ_0. Set $\lambda_0 = \underline{0}$, where $\underline{0}$ is the null of Z_+^u. Let $\Gamma = \Gamma_0 - \{\underline{0}\}$ and Γ, Γ_0 be two index sets. By $L^2(R^u, C^v)$, we denote the aggregate of all vector- valued functions $G(x)$, i.e., $L^2(R^u, C^v) := \{G(x) = (g_1(x), g_2(x), \cdots g_v(x))^T : g_\iota(x) \in L^2(R^u), \iota = 1, 2, \cdots v\}$, where T means the transpose of a vector. Video images and digital films are examples of vector-valued functions where $g_l(x)$ in the above $G(x)$, denotes the pixel on the l th column at the point x. For $G(x) \in L^2(R^u, C^v)$, $\|\hbar\|_2$ denotes the norm of vector-valued function $G(x)$, , i.e.,

$$\|G\|_2 =: (\sum_{l=1}^{v} \int_{R^u} | g_l(x) |^2 \, dx)^{1/2}$$

In the below * means the transpose and the complex conjugate, and its integration is defined to be

$$\int_{R^u} G(x)dx := (\int_{R^u} g_1(x)dx, \int_{R^u} g_2(x)\,dx, \cdots\cdots, \int_{R^u} g_v(x)dx)^T.$$

The Fourier transform of $\hbar(x)$, is defined as $\hat{\hbar}(\gamma) := \int_{R^u} \hbar(x) \cdot e^{-ix\cdot\gamma} dx$, where $x \cdot \gamma$ denotes the inner product of real vectors x and γ. For $\lambda, \hbar \in L^2(R^u, C^v)$, their symbol in ner product is defined by

$$[\lambda(\cdot), \hbar(\cdot)] := \int_{R^u} \lambda(x) \hbar(x)^* dx, \tag{1}$$

For any subsets W, $W_1, W_2 \subset R^u$, denoting $PW = \{Px : x \in W\}$, $4W = \{4x : x \in W\}$, $W_1 + W_2 = \{x_1 + x_2 :$

$x_1 \in W_1, x_2 \in W_2\}$, $W_1 - W_2 = \{x_1 = \{x_1 - x_2 : \ x_1 \in W_1,\ x_2 \in W_2\}$

Definition 1. We say that a pair of vector-valued functions $G(x), \widetilde{G}(x) \in L^2(R^u, C^v)$ are biorthogonal ones, if their translations satisfy

$$[G(\cdot), \widetilde{G}(\cdot - n)] = \delta_{\underline{0},n} I_v, \qquad n \in Z^u, \tag{2}$$

where I_v denotes the $v \times v$ indentity matrix and $\delta_{\underline{0},n}$ is the Kronecker symbol.

A set of vector-valued functions $\{G_n\ (x)\}_{n \in Z^s} \subset U \subset L^2(R^u, C^v)$ is called a Riesz basis of U if it satisfies

(i) for $\lambda(x) \in U$, there exists a unique $v \times v$ matrix sequence $\{M_k\}_{k \in Z^u} \in \ell^2(Z^u)^{v\times v}$ such that

$$\lambda(x) = \sum_{k \in Z^u} M_k G_k(x), \quad x \in R^u, \tag{3}$$

where $\ell^2(Z^u)^{v\times v} = \{M : Z^u \to C^{u\times u},\ \| M \|_2 = \sum_{\iota,s=1}^{v} \sum_{k \in Z^u} | m_{\iota,s}(k) |^2)^{1/2} < +\infty\}$,

(ii) there exist two constants

$0 < L_1 \le L_2 < +\infty$ such that, for any matrix sequence $\{Q_n\}_{n \in Z^u}$, the following equality follows,i.e.,

$$C_1 \|\{Q_n\}\|_* \le \| \sum_{n \in Z^u} Q_n \Phi_n(x) \| \le C_2 \|\{Q_n\}\|_*, \tag{4}$$

where $\|\{Q_n\}\|_*$ denotes the norm of the sequence $\{Q_n\}_{n \in Z^u}$, and choose $\|\{Q_n\}\|_* = (\sum_{n \in Z^u} (\|Q_n\|_F)^2)^{1/2}$ where $\| Q_n \|_F$, for every $n \in Z^u$, is the Frobenius norm of a matrix Q_n.

In what follows, we introduce the notion of vector- valued multiresolution analysis and give the definition of semiorthogonal vector-valued wavelets of space $L^2(R^u, C^v)$

Definition 2. A vector-valued multiresolution analysis of the space $L^2(R^u, C^v)$ is a nested sequence of closed subspaces $\{Y_\ell\}_{\ell\in Z}$ such that (i) $Y_\ell \subset Y_{\ell+1}, \forall \ell \in Z$; (ii) $\bigcap_{\ell\in Z} Y_\ell = \{0\}$ and $\bigcup_{\ell\in Z} Y_\ell$ is dense in $L^2(R^u, C^v)$, where 0 denotes an zero vector of space R^v; (iii) $G(x)\in Y_\ell \Leftrightarrow G(Px)\in Y_{\ell+1}, \forall \ell \in Z$; (iv) there exists $H(x)\in Y_0$, called a vector-valued multivariant scaling function, such that its translates $H_n(x) := H(x-n), \quad n\in Z^s$ forms a Riesz basis of subspace Y_0.

Since $H(x)\in Y_0 \subset Y_1$, by Definition 2 and (4) there exists a finitely supported sequence of constant $v\times v$ matrice $\{M_k\}_{k\in Z^u} \in \ell^2(Z^u)^{v\times v}$ such that

$$H(x) = \sum_{k\in Z^u} M_k H(Px-k). \tag{5}$$

Equation (6) is called a refinement equation. Define

$$\mathcal{M}(\gamma) = \frac{1}{a}\sum_{k\in Z^4} M_k \cdot \exp\{-ik\cdot\gamma\}, \quad \gamma\in R^u. \tag{6}$$

where $\mathcal{M}(\gamma)$, which is a $2\pi Z^u$ periodic function, is called a symbol of $H(x)$. Thus, (6) becomes

$$\widehat{H}(P\gamma) = \mathcal{M}(\gamma)\widehat{H}(\gamma), \quad \gamma\in R^u. \tag{7}$$

Let $X_{j,} j\in Z$ be the direct complementary subspace of Y_j in Y_{j+1}. Assume that there exist $a-1$ vector-valued functions $\psi_\mu(x)\in L^2(R^u, C^v), \mu\in\Gamma$ such that their translations and dilations form a Riesz basis of $X_{j,}$ i.e.,

$$X_j = \overline{(span\{\Psi_\mu(P^j\cdot -n): n\in Z^u, \mu\in\Gamma\})}, \quad j\in Z. \tag{8}$$

Since $\Psi_\mu(x)\in X_0 \subset Y_1, \quad \mu\in\Gamma$, there exist $a-1$ finitesupported sequences of constant $v\times v$ matrice $\{B_n^{(\mu)}\}_{n\in Z^4}$ such that

$$\Psi_\mu(x) = \sum_{n\in Z^u} B_n^{(\mu)} F(Px-n), \quad \mu\in\Gamma. \tag{9}$$

By implementing the Fourier transform for the both sides of (9), we have

$$\hat{\Psi}_\mu(P\gamma) = \mathcal{B}^{(\mu)}(\gamma)\hat{\Phi}(\gamma), \quad \gamma\in R^u, \quad \mu\in\Gamma. \tag{10}$$

where $$\mathcal{B}^{(\lambda)}(\gamma) = \frac{1}{a}\sum_{n\in Z^u} B_n^{(\lambda)} \cdot \exp\{-in\cdot\gamma\}, \ \lambda\in\Gamma. \tag{11}$$

If $H(x), \widetilde{H}(x) \in L^2(R^u, C^v)$ are a pair of biorthogonal vector-valued scaling functions, then

$$[H(\cdot), \widetilde{H}(\cdot - n)] = \delta_{\underline{0},n} I_v, \quad n \in Z^u. \tag{12}$$

We say that $\Psi_\mu(x), \tilde{\Psi}_\mu(x) \in L^2(R^u, C^v),\ \mu \in \Gamma$ are pairs of biorthogonal vector-valued wavelets associated with a pairof biorthogonal vector-valued scaling functions $H(x)$ and $\widetilde{H}(x)$, if the family $\{\Psi_\mu(x-n), n \in Z^u, \mu \in \Gamma\}$ is a Riesz basis of subspace X_0, and

$$[H(\cdot), \tilde{\Psi}_\mu(\cdot - n)] = 0, \quad \mu \in \Gamma, \quad n \in Z^4. \tag{13}$$

$$[\widetilde{H}(\cdot), \Psi_\mu(\cdot - n)] = 0, \quad \mu \in \Gamma, \quad n \in Z^4. \tag{14}$$

$$X_j^{(\mu)} = \overline{Span\{\Psi_\mu(P^j \cdot - n) : n \in Z^u\}}, \ \mu \in \Gamma, j \in Z. \tag{15}$$

Similar to (5) and (9), there exist 256 finite supported sequences of $v \times v$ constant matrice $\{\widetilde{M}_k\}_{k \in Z^u}$ and $\{\tilde{B}_k^{(\lambda)}\}_{k \in Z^u}$, $\mu \in \Gamma$ such that $\widetilde{H}(x)$ and $\tilde{\Psi}_\lambda(x), \lambda \in \Gamma$ satisfy the refinement equations:

$$\widetilde{H}(x) = \sum_{k \in Z^u} \widetilde{M}_k \widetilde{H}(Px - k), \tag{16}$$

$$\tilde{\Psi}_\lambda(x) = \sum_{k \in Z^u} \tilde{B}_k^{(\lambda)} \widetilde{H}(Px - k), \ \lambda \in \Gamma. \tag{17}$$

3 The Traits of Vector-Valued Multivariant Wavelet Packages

Denoting by $G_{\underline{0}}(x) = H(x),\ G_\lambda(x) = \tilde{\Psi}_\lambda(x),\ \widetilde{H}_{\underline{0}}(x) = \widetilde{H}(x)$, $\tilde{G}_\lambda(x) = \tilde{\Psi}_\lambda(x),\ Q_k^{(\underline{0})} = M_k,\ Q_k^{(\lambda)} = B_k^{(\lambda)}$,

$\tilde{Q}_k^{(\underline{0})} = \widetilde{M}_k\ \tilde{Q}_k^{(\lambda)} = \tilde{B}_k^{(\lambda)},\ \lambda \in \Gamma,\ k \in Z^u$, order $P = 4I_v, u = 4$. For any $\alpha \in Z_+^4$ and the given vector-valued semiorthogonal scaling functions $G_{\underline{0}}(x)$ and $\tilde{G}_{\underline{0}}(x)$, iteratively define, respectively,

$$G_\alpha(x) = G_{4\sigma+\mu}(x) = \sum_{k \in Z^4} Q_k^{(\mu)} G_\sigma(4x - k), \tag{18}$$

$$\tilde{G}_\alpha(x) = \tilde{G}_{4\sigma+\mu}(x) = \sum_{k \in Z^4} \tilde{Q}_k^{(\mu)} \tilde{G}_\sigma(4x - k). \tag{19}$$

where $\mu \in \Gamma_0, \sigma \in Z_+^4$ is the unique element such that $\alpha = 4\sigma + \mu$, $\mu \in \Gamma_0$ follows.

Lemma 1[4]. Let $F(x), \tilde{F}(x) \in L^2(R^4, C^v)$. Then they are biorthogonal ones if and only if

$$\sum_{k \in Z^4} \hat{F}(\gamma + 2k\pi)\hat{\tilde{F}}(\gamma + 2k\pi)^* = I_v. \tag{20}$$

Definition 3. We say that two families of vector-valued functions $\{G_{4\sigma+\mu}(x), \sigma \in Z_+^4, \mu \in \Gamma_0\}$ and $\{\tilde{G}_{4\sigma+\mu}(x),\ \sigma \in Z_+^4,\ \mu \in \Gamma_0\}$ are vector-valued wavelet packages with respect to a pair of semiorthogonal vector-valued scaling functions $G_{\underline{0}}(x)$ and $\tilde{G}_{\underline{0}}(x)$, respectively, where $G_{4\sigma+\mu}(x)$ and $\tilde{G}_{4\sigma+\mu}(x)$ are given by (18) and (19), respectively.

Applying the Fourier transform for the both sides of (18) and (19) yields, respectively,

$$\hat{G}_{4\sigma+\mu}(\gamma) = Q^{(\mu)}(\gamma/4)\hat{G}_{\sigma}(\gamma/4), \quad \mu \in \Gamma_0, \tag{21}$$

$$\hat{\tilde{G}}_{4\sigma+\mu}(4\gamma) = Q^{(\mu)}(\gamma)\hat{\tilde{G}}_{\sigma}(\gamma), \quad \mu \in \Gamma_0, \tag{22}$$

$$\mathcal{Q}^{(\mu)}(\gamma) = \frac{1}{4^4}\sum_{k \in Z^4} Q_k^{(\mu)} \cdot \exp\{-ik \cdot \gamma\}, \quad \mu \in \Gamma_0 \tag{23}$$

$$\tilde{\mathcal{Q}}^{(\mu)}(\gamma) = \frac{1}{4^4}\sum_{k \in Z^4} \tilde{Q}_k^{(\mu)} \cdot \exp\{-ik \cdot \gamma\}, \mu \in \Gamma_0. \tag{24}$$

We are now in a position to characterize the semiorthogonality traits of the wavelet packages.

Lemma 2[6]. Assume that $G_\mu(x),\ \tilde{G}_\mu(x) \in L^2(R^4, C^v)$, $\mu \in \Gamma$ are pairs of semiorthogonal vector-valued wavelets associated with a pair of semiorthogonal scaling functions $G_{\underline{0}}(x)$ and $\tilde{G}_{\underline{0}}(x)$. Then, for $\mu, \nu \in \Gamma_0$, we have

$$\sum_{\rho \in \Gamma_0} \mathcal{Q}^{(\mu)}((\gamma + 2\rho\pi)/4)\tilde{\mathcal{Q}}^{(\nu)}((\gamma + 2\rho\pi)/4)^* = \delta_{\mu,\nu} I_v. \tag{25}$$

Lemma 3[6]. Suppose $\{G_\alpha(x),\ \alpha \in Z_+^4\}$ and $\{\tilde{G}_\alpha(x), \alpha \in Z_+^4\}$ are wavelet packets with respect to a pair of biorthogonal vector-valued functions $G_{\underline{0}}(x)$ and $\tilde{G}_{\underline{0}}(x)$. Then, for $\alpha \in Z_+^4$, we have

$$[G_\alpha(\cdot), \tilde{G}_\alpha(\cdot - k)] = \delta_{\underline{0},k} I_v,\ k \in Z^4. \tag{26}$$

Theorem 1. Assume that $\{G_\beta(x),\beta\in Z_+^4\}$ and $\{\tilde{G}_\beta(x),\beta\in Z_+^4\}$ are vector-valued wavelet packages with respect to a pair of semiorthogonal vector-valued functions $G_{\underline{0}}(x)$ and $\tilde{G}_{\underline{0}}(x)$, respectively. Then, for $\beta\in Z_+^4,\mu,\nu\in\Gamma_0$, we have

$$[G_{4\beta+\mu}(\cdot),\tilde{G}_{4\beta+\nu}(\cdot-k)]=\delta_{\underline{0},k}\delta_{\mu,\nu}I_v,\ \ k\in Z^4. \tag{27}$$

Proof. Since the spaces

R^4 has the following partition: $R^4=\bigcup_{u\in Z^4}([0,2\pi]^4+2u\pi)$, and $([0,2\pi]^4+2\pi u_1)\bigcap([0,2\pi]^4+2\pi u_2)=\emptyset$ where $u_1\neq u_2,u_1,u_2\in Z^s$,then by Lemma 1, we have

$$[G_{4\beta+\mu}(\cdot),\tilde{G}_{4\beta+\nu}(\cdot-k)]=\frac{1}{(2\pi)^4}\int_{R^4}\hat{G}_{4\beta+\mu}(\gamma)\hat{\tilde{G}}_{4\beta+\nu}(\gamma)^*\cdot\exp\{ik\cdot\gamma\}d\gamma$$

$$=\frac{1}{(2\pi)^4}\int_{R^s}\mathcal{Q}^{(\mu)}(\gamma/4)\hat{G}_\beta(\gamma/4)\hat{\tilde{G}}_\beta(\gamma/4)^*\tilde{\mathcal{Q}}^{(\nu)}(\gamma/4)^*e^{ik\cdot\gamma}d\gamma$$

$$=(\frac{2}{\pi})^4\sum_{k\in Z^4}\int_{[0,2\pi]^4+2k\pi}\mathcal{Q}^{(\mu)}(\gamma)\hat{G}_\beta(\gamma)\hat{\tilde{G}}_\beta(\gamma)^*\cdot\tilde{\mathcal{Q}}^{(\nu)}(\gamma)^*e^{i4k\cdot\gamma}d\gamma$$

$$=\frac{1}{(2\pi)^4}\int_{[0,8\pi]^4}\mathcal{Q}^{(\mu)}(\gamma/4)\tilde{\mathcal{Q}}^{(\nu)}(\gamma/4)^*e^{ik\cdot\gamma}d\gamma$$

$$=\frac{1}{(2\pi)^4}\int_{[0,2\pi]^4}\delta_{\mu,\nu}I_ue^{i4\cdot\gamma}d\gamma=\delta_{\underline{0},k}\delta_{\mu,\nu}I_v.$$

This completes the proof of Theorem 1.

Theorem 2. If $\{G_\alpha(x),\alpha\in Z_+^4\}$ and $\{\tilde{G}_\alpha(x),\alpha\in Z_+^4\}$ are vector-valued wavelet packages with respect to semiorthogonal vector-valued scaling functions $G_0(x)$ and $\tilde{G}_0(x)$, then for any $\alpha,\sigma\in Z_+^4$, we have

$$[G_\alpha(\cdot),\tilde{G}_\sigma(\cdot-k)]=\delta_{\alpha,\sigma}\delta_{\underline{0},k}I_v,\ \ k\in Z^4. \tag{28}$$

Proof. When $\alpha=\sigma$,(28) follows by Lemma 3. as $\alpha\neq\sigma$ and $\alpha,\sigma\in\Gamma_0$, it follows from Lemma 4 that (28) holds, too. Assuming that α is not equal to β, as well as at least one of $\{\alpha,\sigma\}$ doesn't belong to Γ_0, we rewrite α,σ as $\alpha=4\alpha_1+\rho_1$, $\sigma=4\sigma_1+\mu_1$, where $\rho_1,\mu_1\in\Gamma_0$. Case 1. If $\alpha_1=\sigma_1$, then $\rho_1\neq\mu_1$. (28) follows by (24), (25) and Lemma 1 and Lemma 2,i.e.,

$$[G_{\alpha}(\cdot),\tilde{G}_{\sigma}(\cdot-k)] = \frac{1}{(2\pi)^4}\int_{R^4}\hat{G}_{4\alpha_1+\rho_1}(\gamma)\hat{\tilde{G}}_{4\sigma_1+\mu_1}(\gamma)^*\cdot\exp\{ik\cdot\gamma\}d\gamma$$

$$=\frac{1}{(2\pi)^4}\int_{[0,8\pi]^4}\mathcal{Q}^{(\rho_1)}(\gamma/4)\{\sum_{u\in Z^s}\hat{G}_{\alpha_1}(\gamma/4+2u\pi)$$

$$\cdot\hat{\tilde{G}}_{\alpha_1}(\gamma/4+2u\pi)^*\}\tilde{\mathcal{Q}}^{(\mu_1)}(\gamma/4)^*\cdot e^{ik\cdot\gamma}d\gamma$$

$$=\frac{1}{(2\pi)^4}\int_{[0,2\pi]^4}\delta_{\rho_1,\mu_1}I_v\cdot\exp\{ik\cdot\gamma\}d\gamma=O.$$

Case 2 If $\alpha_1\neq\sigma_1$, order $\alpha_1=4\alpha_2+\rho_2$, $\sigma_1=4\sigma_2+\mu_2$, where $\alpha_2,\sigma_2\in Z_+^s$, and $\rho_2,\mu_2\in\Gamma_0$. Provided that $\alpha_2=\sigma_2$, then $\rho_2\neq\mu_2$. Similar to Case 1, (28) can be established. When $\alpha_2\neq\sigma_2$,we order $\alpha_2=4\alpha_3+\rho_3$, $\sigma_2=4\sigma_3+\mu_3$, where $\alpha_3,\sigma_3\in Z_+^4$, $\rho_3,\mu_3\in\Gamma_0$. Thus, after taking finite steps (denoted by κ), we obtain $\alpha_\kappa\in\Gamma_0$, and $\rho_\kappa,\mu_\kappa\in\Gamma_0$. If $\alpha_\kappa=\sigma_\kappa$, then $\rho_\kappa\neq\mu_\kappa$. Similar to the Case 1, (28) follows. If $\alpha_\kappa\neq\sigma_\kappa$, then it gets from (12)-(15):

$$[G_{\alpha_\kappa}(\cdot),\tilde{G}_{\sigma_\kappa}(\cdot-k)]=O \Leftrightarrow \sum_{u\in Z^4}\hat{G}_{\alpha_\kappa}(\gamma+2u\pi)\hat{\tilde{G}}_{\beta_\kappa}(\gamma+2u\pi)^*=O,\ k\in Z^4.$$

Furthermore, we obtain

$$16\pi^4[G_{\alpha}(\cdot),\tilde{G}_{\sigma}(\cdot-k)]=\int_{R^4}\hat{G}_{\alpha}(\gamma)\hat{\tilde{G}}_{\sigma_1}(\gamma)^*\cdot e^{ik\cdot\gamma}d\gamma$$

$$=\int_{R^4}\hat{G}_{4\alpha_1+\lambda_1}(\gamma)\hat{\tilde{G}}_{4\beta_1+\mu_1}(\gamma)^*\cdot\exp\{ik\cdot\gamma\}d\gamma=\cdots\cdots\cdots$$

$$=\int_{[0,2\cdot4^\kappa]^4}\{\prod_{l=1}^{\kappa}\mathcal{Q}^{(\rho_l)}(\frac{\gamma}{4^l})\}\{\sum_{u\in Z^4}\hat{G}_{\alpha_\kappa}(\frac{\gamma}{4^l}+2u\pi)$$

$$\cdot\hat{\tilde{G}}_{\sigma_\kappa}(\frac{\gamma}{4^l}+2u\pi)^*\}\{\prod_{l=1}^{\kappa}\tilde{\mathcal{Q}}^{(\mu_l)}(\frac{\gamma}{4^l})\}^*\cdot e^{ik\cdot\gamma}d\gamma$$

$$=\int_{([0,2\cdot4^\kappa\pi]^4}\{\prod_{l=1}^{\kappa}\mathcal{Q}^{(\rho_l)}(\frac{\gamma}{4^l})\}\cdot O\cdot\{\prod_{l=1}^{\kappa}\tilde{\mathcal{Q}}^{(\mu_l)}(\frac{\gamma}{4^l})\}^*\cdot\exp\{-ik\cdot\gamma\}d\gamma=O.$$

Therefore, for any $\alpha,\sigma\in Z_+^4$, result (28) is established.

Theorem 3[7]. Let $\phi(x),\tilde{\phi}(x),\hbar_\iota(x)$ and $\tilde{\hbar}_\iota(x)$, $\iota\in J$ be functions in $L^2(R^2)$. Assume that conditions in Theorem 1 are satisfied. Then, for any function $f(x)\in L^2(R^2)$, and any integer n, we have

$$\sum_{u\in Z^2}\langle f,\tilde{\phi}_{n,u}\rangle\phi_{n,u}(x)=\sum_{\iota=1}^{7}\sum_{s=-\infty}^{n-1}\sum_{u\in Z^2}\langle f,\tilde{\hbar}_{\iota:s,u}\rangle\hbar_{\iota:s,u}(x)\,. \tag{29}$$

4 Conclusion

Three semiorthogonal ity formulas of biorthogonal vector-valued four-dimensional wavelet packages are obstablished. The direct decomposition of space $L^2(R^u, C^v)$ is proposed by constructing a series of subspaces of the wavelet packages.

References

1. Zhang, N., Wu, X.: Lossless Compression of Color Mosaic Images. IEEE Trans. Image Processing 15(16), 1379–1388 (2006)
2. Telesca, L., et al.: Multiresolution wavelet analysis of earthquakes. Chaos, Solitons & Fractals 22(3), 741–748 (2004)
3. Iovane, G., Giordano, P.: Wavelet and multiresolution analysis: Nature of ε^{∞} Cantorian space-time. Chaos. Solitons & Frac-tals 32(4), 896–910 (2007)
4. Shen, Z.: Nontensor product wavelet packets in $L_2(R^3)$. SIAM Math. Anal. 26(4), 1061–1074 (1995)
5. Yang, S., et al.: Construction of biorthogonal multiwavelets. Math. Anal. Appl. (2002); Weber, E.:Orthogonal frames of translates. J. Appl. Comput. Harmon. Anal. 17, 69-90 (2004)
6. Chen, Q., Shi, Z.: Biorthogonal multiple vector-valued multivariate wavelet packets associated with a dilation matrix. Chaos. Solitons & Fractals 35, 323–332 (2008)
7. Chen, Q., Shi, Z., Cao, H., Shi, Z.: Construction and decomposition of biorthogonal vector-valu-ed wavelets with compact support. Chaos, Solitons & Fractals 42(5), 2765–2778 (2009)

The Research of a Pair of Canonical Frames Yielded by Finitely Supported Refinable Functions

Hailin Gao

Department of Fundamentals, Henan Polytechnic Institute, Nanyang 473009, China
sxxa11xauat@126.com

Abstract. Wavelet analysis is a particular time-or space-scale representation of signals which has found a wide range of applications in physics, signal processing and applied Mathematics in the last few years. We show that there exist wavelet frame generated by two functions which have good dual wavelet frames, but for which the canonical dual wavelet frame does not consist of wavelets, according to scaling functions.

Keywords: Dual pseudoframes, bivariate, orthogonal pseudoframes, Banach frames, frame operator,wavelet frame, the canonical wavelet frame, scaling functions, composite matrix, time-frequency analysis method.

1 Introduction

Recently, wavelet tight frames have attracted more and more attention, just because they have nice time-frequency localization property, shift-invariances, and more design freedom. Wavelet tight frames have been widely used in denoising and image processing. Tight frames generalize orthonormal systems. They preserve the unitary property or the relevant analysis ans synthesis operator. Frames are intermingled with exciting applications to physics, to engineering and to science in general. Frames didn't start in isolation, and even now in its maturity, surprising and deep connections to other areas continue to enrich the subject. The subjects are well explained, and they are all amenable to the kind of numerical methods where wavelet algorithms excel. Wavelet analysis is a particular time-or space-scale representation of signals which has found a wide range of applications in physics, signal processing and applied Mathematics in the last few years. The rise of frame theory in pplied mathematics is due to the flexibility and redundancy of frames. The structured frames are much easier to be constructed than structured orthonormal bases. The notion of frames was introduced by Duffin and Schaeffer[1] and popularized greatly by the work of Daubechies and her coauthors[2,3]. After this ground breaking work, the theory of frames began to be more widely studied both in theory and in applications[4-7], such as signal processing, image processing, data compression and sampling theory. The introduction for biorthogonal wavelet packets attributes to Cohen and Daubechies Every frame determines a frame operator which is invertible and also yields the standard dual frame of the original frame. The notion of Frame multiresolution analysis as described by [5] generalizes the notion of multiresolution analysis by allowing non-exact affine frames. However, subspaces at different resolutions in a FMRA are still generated by a frame formed by

Y. Zhang (Ed.): Future Communication, Computing, Control and Management, LNEE 141, pp. 99–107.
springerlink.com

translates and dilates of a single function. This article is motivated from the observation that standard methods in sampling theory provide examples of multirsolution structure which are not FMRAs. Inspired by [5] and [7], we introduce the notion of a generalized multiresolution structure (GMRS) of $L^2(R)$ generated by several functions of inte ger translates in space $L^2(R)$. We have that the GMRS has a pyramid decompoposition scheme and obtain a frame-like decomposition based on such a GMRS. It also lead to new constructions of affine frames of $L^2(R)$.

2 The Traits of Frames and Frame Operator

First, we introduce some notations and definitions. Let Ω be a separable Hilbert space. We recall that a sequence $\{\eta_\iota : \iota \in Z\} \subseteq \Omega$ is a frame for H if there exist positive real numbers C, D such that

$$\forall \lambda \in \Omega, \qquad C\,\|\lambda\|^2 \le \sum_\iota |\langle \lambda, \eta_\iota \rangle|^2 \le D\,\|\lambda\|^2 \tag{1}$$

A sequence $\{\eta_\iota\} \subseteq \Omega$ is a Bessel sequence if (only) the upper inequality of (1) holds. If only forall $\lambda \in X \subset U$, the upper inequality of (1) holds, the sequence $\{\eta_\iota\} \subseteq \Omega$ is a Bessel sequence with respect to (w.r.t.) Ω. If $\{\eta_\iota\}$ is a frame, there exist a dual frame $\{\eta_\iota^*\}$ such that

$$\forall\, g \in \Omega, \quad g = \sum_\iota \langle g, \eta_\iota \rangle \eta_\iota^* = \sum_\iota \langle g, \eta_\iota^* \rangle \eta_\iota . \tag{2}$$

where C and D are called the lower frame bound and upper bound, respectively. In particular, when $C = D = 1,$ we say that $\{\xi_v\}_{v\in\Gamma}$ is a (nor- malized) tight frame in X. The frame operator $S : X \to X$, which is associated with a frame $\{\xi_v\}_{v\in\Gamma}$, is defined to be

$$S\varphi = \sum_{v\in\Gamma} \langle \varphi, \xi_v \rangle \xi_v, \qquad \varphi \in X \tag{3}$$

Obviously, $\{\xi_v\}_{v\in\Gamma}$ is a frame in X with lower frame bound C and upper frame bound D if and only if S is well defined and $CI \le S \le DI$, where I denotes the identity operator on X. A sequence $\{f_\iota\} \subseteq X$ is a Bessel sequence if (only) the upper inequality of (1) follows. If only for all $\Upsilon \in \Lambda \subset X$, the upper inequality of (1) holds, the sequence $\{f_\iota\} \subseteq X$ is a Bessel sequence with respect to (w.r.t.) the subspace Λ. Let $\{\xi_v\}_{v\in\Gamma}$ be a frame in a Hilbert space X with lower frame bound

C and upper bound D. Then its frame operator S is invertible and $C^{-1}I \le S^{-1} \le D^{-1}I$, that is to say, S^{-1} is also a frame operator. Moreover, we have the following result.

Theorem 1 Let $\{\xi_v\}_{v\in\Gamma}$ be a frame in X. Then the frame operators $S: X \to X$, and S^{-1}: $X \to X$ are self-adjoint operators, respectively.

Proof. For any $h, g \in X$, we have $Sh = \sum_{v\in\Gamma}\langle h, \xi_v\rangle \xi_v$, $Sg = \sum_{v\in\Gamma}\langle g, \xi_v\rangle \xi_v$. Then,

$$\langle Sh, g\rangle = \left\langle \sum_{v\in\Gamma}\langle h, \xi_v\rangle \xi_v, g\right\rangle = \sum_{v\in\Gamma}\langle h, \xi_v\rangle\langle \xi_v, g\rangle$$

$$= \sum_{v\in\Gamma}\overline{\langle g, \xi_v\rangle}\langle h, \xi_v\rangle = < h, \sum_{v\in\Gamma}\langle g, \xi_v\rangle \xi_v >= \langle h, Sg\rangle$$

Furthermore,

$$\langle S^{-1}h, g\rangle = \langle S^{-1}h, SS^{-1}g\rangle = \langle SS^{-1}h, S^{-1}g\rangle = \langle h, S^{-1}g\rangle .$$

Hence, the frame operators S and S^{-1} are all self-adjoint operators.

Note that $\{S^{-1}\xi_v\}_{v\in\Gamma}$ is a dual frame of the frame $\{\xi_v\}_{v\in\Gamma}$ since

$$h = SS^{-1}h = \sum_{v\in\Gamma}\langle S^{-1}h, \xi_v\rangle \xi_v = \sum_{v\in\Gamma}\langle h, S^{-1}\xi_v\rangle \xi_v, \quad \forall h \in X; \tag{4}$$

$$\langle h, g\rangle = \left\langle \sum_{v\in\Gamma}\langle h, S^{-1}\xi_v\rangle \xi_v, g\right\rangle = \sum_{v\in\Gamma}\langle h, S^{-1}\xi_v\rangle\langle \xi_v, g\rangle \quad \forall h, g \in X \tag{5}$$

The family $\{S^{-1}\xi_v\}_{v\in\Gamma}$ is said to be the canonical dual frame of the frame $\{\xi_v\}_{v\in\Gamma}$ and (4) is the canonical representation of any h in X using the frame $\{\xi_v\}_{v\in\Gamma}$ in X .Generally speaking, for a given frame $\{\xi_v\}_{v\in\Gamma}$ in X, there are many dual frames other than its standard dual frame. However, the standard dual frame $\{S^{-1}\xi_v\}_{v\in\Gamma}$ enjoys the following optical property(see[1]•. For any element $h \in X$,

$$\sum_{v\in\Gamma}\left|\langle h, S^{-1}\xi_v\rangle\right|^2 \le \sum_{v\in\Gamma}|c_v|^2, \quad \text{whenever} \quad h = \sum_{v\in\Gamma} c_v\xi_v, \ \{c_v\}_{v\in\Gamma} \in \ell_2(Z).$$

In other words, the representation in (4) using the canonical dual frame has the smallest norm of the frame coefficients to represent a given element in X.

It is well known that frames with certain structures, such frames of translates, Gabor frames, and wavelet frames, are very useful in both theory and application. The interested reader can consult[1,2,5] and references therein on the theory of frames. In the following, we will recall two nice families of frames and their canonical dual frames. A frame of translates is derived from two functions ψ_1, ψ_2 in $L^2(R)$ by integer translates. That is, $\{\psi_l(\cdot - k), k \in Z, l = 1,2\}$ is a frame in $L^2(R)$ defined as the closed linear span of $\psi_l(\cdot - k), k \in Z, l = 1,2$. Then it is well known that its standard dual frame is given by $\{[S^{-1}\psi_l](\cdot - k) : k \in Z, l = 1,2,3\}$; like the original frame, it is generated by the integer translates of two functions since. $S[h(\cdot - k)] = [Sh](\cdot - k)$ for all $k \in Z$ and $h \in L^2(R)$ .A Gabor frame is generated from twofunctions in $L^2(R)$ by modulates and integer shifts.In other words, a Gabor frame is a frame in $L^2(R)$ given by the set $\{e^{2i\pi jb}\psi_l = 1,2(\cdot - ka) : j,\ k \in Z, l = 1,2\}$, where a and b are real positive numbers. A direct calculation shows that $S[\{e^{i2\pi jb} f(\cdot - ka)] = \ e^{i2\pi jb}[Sf](\cdot - ka)$ for all $j,k \in Z$ and $f \in L^2(R)$. Therefore, the standard dual frame of a Gabor frame generated by $\{\psi_1,\ \psi_2\}$ is also yielded by modulating and shifting the two functions $\{S^{-1}\psi_1, S^{-1}\psi_2\}$. We are about to discuss the standard dual frame of a wavelet frame. Throughout this paper, we shall use the following notation

$$h_{l:j,k} = 2^j h_l(4^j \cdot - k), \quad j,k \in Z, \text{ where } h_l \in L^2(R), \quad l = 1,2,3.$$

A wavelet frame is generated from several functions in $L^2(R)$ by dilates and integer translates. We say that $\{\psi_1, \psi_2, \psi_3\}$ generates a wavelet frame in $L^2(R)$ if $\psi_{l:j,k} : j,k \in Z,\ l = 1,2,3\}$ is a frame in $L^2(R)$. We say that $\{\psi_1, \psi_2\}$ and $\{\tilde{\psi}_1, \tilde{\psi}_2\}$ generates a pair of dual wavelet frame if each of them generates a wavelet frame and

$$\forall h,\ g \in L^2(R), \ \langle h, g \rangle = \sum_{l=1}^{3} \sum_{j \in Z} \sum_{k \in Z} \langle h, \tilde{\psi}_{l:j,k} \rangle \langle \psi_{l:j,k}, g \rangle, \qquad (6)$$

3 The Concept of Bivariate Generalized Multiresolution Structure

Let r be a positive integer, and $\Lambda = \{1, 2, \cdots, r\}$ be a finite index set. We consider the case of multiple generators, which yield multiple pseudoframes for subspaces of $L^2(R^2)$. In what follows, we consider the case of generators, which yield affine pseudoframes of integer grid translates for subspaces of $L^2(R^2)$. Let $\{T_v\phi_\iota\}$ and $\{T_v\widetilde{\phi}_\iota\}$ $(\iota\in\Lambda,\ v\in Z^2)$ be two sequences in $L^2(R^2)$. Let Ω be a closed subspace of $L^2(R^2)$. We say that $\{T_v\phi_\iota\}$ forms a pseudoframe for the subspace Ω with respect to (w.r.t.) $\{T_v\widetilde{\phi}_\iota\}$ $(\iota\in\Lambda,\ v\in Z^2)$ if

$$\forall\, \hbar(x)\in\Omega, \qquad \hbar(x)=\sum_{\iota\in\Lambda}\sum_{v\in Z^2}\langle \hbar, T_v\phi_\iota\rangle T_v\tilde{\phi}_\iota(x). \tag{7}$$

It is important to note that ϕ_ι and $\tilde{\phi}_\iota$ need not be contained in Ω.Consequently,the positions of $T_v\phi_\iota$ and $T_v\tilde{\phi}_\iota$ are not generally commutable [5], i.e., there exists $\Gamma(x)\in\Omega$ such that

$$\sum_{\iota\in\Lambda}\sum_{v\in Z^2}\langle \hbar, T_v\phi_\iota\rangle T_v\tilde{\phi}_\iota(x)\neq\sum_{\iota\in\Lambda}\sum_{v\in Z^2}\langle \hbar, T_v\tilde{\phi}_\iota\rangle T_v\phi_\iota(x)=\hbar(x). \tag{8}$$

Definition 1. We say that a Generalized multiresolution structure (GMS) $\{V_n,\phi_\iota(x),\tilde{\phi}_\iota(x)\}_{n\in Z,\iota\in\Lambda}$ of $L^2(R^2)$ is a sequence of closed linear subspaces $\{V_n\}_{n\in Z}$ of $L^2(R^2)$ and 2r elements $\phi_\iota(x),\tilde{\phi}_\iota(x)$ $\in L^2(R^2)$ such that (i) $V_n\subset V_{n+1}$, $n\in Z$; (ii) $\bigcap_{n\in Z}V_n=\{0\}$; $\bigcup_{n\in Z}V_n$ is dense in $L^2(R^2)$; (iii) $g(x)\in V_n$ if and only if $g(2x)\in V_{n+1}$ $\forall n\in Z$ implies $T_v g(x)\in V_0$, for $v\in Z$; (v) $\{T_v\phi_\iota(x),\ \iota\in\Lambda, v\in Z^2\}$ forms an affine pseudoframe for V_0 with respect to $\{T_v\tilde{\phi}_\iota(x),\ \iota\in\Lambda, v\in Z^2\}$.

A necessary and sufficient condition for the construction of an affine pseudoframe for PaleyWiener subspaces is presented as follows.

Proposition 1[6]. Let $f(x)\in L^2(R^2)$ satisfy $|\widehat{f}|$ a.e. on a connected neighbourhood of 0 in $[-\frac{1}{2},\frac{1}{2})^2$, and $|\widehat{f}|=0$ a.e. otherwise. Define $\Xi\equiv\{\omega\in R^2: |\widehat{f}(\omega)|\geq C>0\}$, and $V_0=PW_\Xi=\{\phi\in L^2(R):$ $\text{supp}(\hat{\phi})\subseteq\Lambda\}$. Then for $\widetilde{f}\in L^2(R^2)$, $\{T_v f, v\in Z^2\}$ is an affine pseudoframe for V_0 with respect to $\{T_v\widetilde{f}, v\in Z^2\}$ if and only if

$$\overline{\hat{f}(\omega)}\,\hat{\widetilde{f}}(\omega)\chi_\Xi(\omega)=\chi_\Xi(\omega) \quad \text{a. e.}, \tag{9}$$

where χ_Λ is the characteristic function on Λ. Moreover, if $\widetilde{f}(\omega)$ is the above conditions then $\{T_v\phi, v\in Z^2\}$ and $\{T_v\widetilde{\phi}, v\in Z^2\}$ are a pair of commutative affine pseudoframes for V_0, i.e., $\forall\, \hbar\in V_0$,

$$\hbar(x)=\sum_{k\in Z^2}\langle \hbar, T_k\widetilde{f}\rangle T_k f(x)=\sum_{k\in Z^2}\langle \hbar, T_k f\rangle T_k\widetilde{f}(x). \qquad (10)$$

The filter banks associated with a GMS are presented as follows. Define filter functions $B_0(\omega)$ and $\widetilde{B}_0(\omega)$ by the relaton $B_0(\omega)=\sum_{v\in Z} b_0(v)e^{-2\pi i\omega}$ and $\widetilde{B}_0(\omega)=\sum_{v\in Z}\widetilde{b}_0(v)e^{-2\pi i\omega}$ of the sequences $b_0=\{b_0(v)\}$ and $\widetilde{b}_0=\{\widetilde{b}_0(v)\}$, respectively, wherever the sum is defined. Let $\{b_0(v)\}$ be such that $B_0(0)=2$ and $B_0(\omega)\neq 0$ in a neighborhoood of 0. Assume also that $|B_0(\omega)|\leq 2$. Then there exists $f(x)\in L^2(R)$ (see ref.[8]) such that

$$f(x)=2\sum_{v\in Z^2} b_0(v) f(2x-v). \qquad (11)$$

Similarly, there exists a scaling relationship for $\widetilde{f}(x)$ under the same conditions as that of b_0 for a sequence $\widetilde{b}_0=\{\widetilde{b}_0(v)\}$, i.e.,

$$\widetilde{f}(x)=2\sum_{v\in Z^2}\widetilde{b}_0(v)\widetilde{f}(2x-v) \qquad (12)$$

4 The Canonical Dual Frame of a Wavelet Frame

In this part, we shall discuss the standard dual frame of a wavelet frame with two generators in general. Then we present the main results. Define the dilation and translation operators on $L^2(R)$ (see[6]) as follows:

$$D^j f := 2^j f(4^j\cdot) \quad \text{and } T_k f := f(\cdot-k), \quad j,k\in Z, \quad f\in L^2(R).$$

Suppose that $\{\psi_1,\psi_2\}$ generates a wavelet frame in $L_2(R)$. Let $\boldsymbol{F}$ denote its frame operator defined as follows: (for $h\in L^2(R)$)

$$Sh=\sum_{l=1}^{3}\sum_{j,k\in Z}\langle h,\psi_{l,j,k}\rangle\psi_{l,j,k}. \qquad (13)$$

It is evident to observe that $SD^j=D^jS$ for all $j\in Z$ and $D^jT_k=T_{4-jk}D^j$. We define the period $P(\{\psi_1,\psi_2\})\in N\cup\{0\}$ of the wavelet frame generated by $\{\psi_1,\psi_2\}$ as follows:

$$\langle P(\{\psi_1,\psi_2\})\rangle := \{k\in Z : S^{-1}T_{k+n}\psi_l = T_k S^{-1}T_n\psi_l, \forall n\in Z, l=1,2\} \quad (14)$$

where $\langle x\rangle$ denotes the additive group generated by x. Let W stand for the closed linear span of $\psi_l(\cdot - k)$, $k\in Z, l=1,2$, and $\tilde{W} := S^{-1}W$. Then (8) implie

$$\langle P(\{\psi_1,\psi_2\})\rangle = \{k\in Z : S^{-1}T_k f = T_k S^{-1} f, \forall f\in W\}$$
$$= \{k\in Z :\ T_k S g = f T_k g,\ \forall g\in \tilde{W}\}.$$

when $\{\psi_1,\psi_2\}$ generates a wavelet frame in $L^2(R)$ then the set $\{\psi_{l:j,k} : l=1,2,3$ and $k=0,1,\cdots,\ 4^J-1\}$ also generates the same wavelet frame for any nonnegative integer J since $[\psi_{l:J,k}]_{j,m} = \psi_{l:j+J,k+4^J m}, l=1,2,3$.

Theorem 2. Suppose that $\{\psi_1,\psi_2,\psi_3\}$ generates a wavelet frame in $L^2(R)$. For any nonnegative integer J, the following statements are equivalent:(i) There exist $2\cdot 4^J$ functions $\tilde{\psi}_1,\tilde{\psi}_2,\tilde{\psi}_{2\cdot 3^J}$ such that they generate the canonical dual frame of the wavelet frame $\{\psi_{l:j,k} : j,k\in Z$ and $l=1,2\}$; (ii) $\{F^{-1}[\psi_{l:Jk}] : k=0,1,\cdots 4^J-1$ and $l=1,2\}$ and $\{\psi_{l:j,k} : k=0,1,\cdots 4^J-1$ and $l=1,2\}$ generate a pair of dual wavelet frame in $L^2(R)$; (iii) $P(\{\psi_1,\psi_2,\psi_3\}) | 4^J$, where $P(\{\psi_1,\psi_2\})$ defined in (8). Moreover, when $\{\psi_1,\psi_2\}$ generates a wavelet frame which is a Riesz basis, (i)-(ii) are also equivalent to (iv) $V_J(\{\psi_1,\psi_2\})$ is shift-invariant where $V_J(\{\psi_1,\ \psi_2\})$ is the closed linear span of $\psi_{l:j,k} : k\in Z,\ \ j<J$ and $l=1,2,3.$

Proof. If (a) is true, by setting $F^{-1}\ [\psi_{l:J,k-1}] = \tilde{\psi}_{k+4^J(l-1)}, l=1,2,3$ and $k=1,2,\cdots,4^J$, we obtain statement (b) is established. The converse obviously holds. Since

$$S[S^{-1}\psi_{l:J,k}]_{j,m} = SD^j T_m S^{-1} D^J T_k \psi_l = D^j S(T_m S^{-1} T_{3^{-J}k} D^J \psi_l)$$
$$= D^j T_{m+4^{-J}k} D^J \psi_l = \psi_{l:j+J,k+4^J m},$$

we note that (b) is equivalent to(9) below,

$$[S^{-1}\psi_{l;J,k}]_{j,m} = S^{-1}\psi_{l:j+J,k+3^J m},\quad \forall j,k,m\in Z. \quad (9)$$

In particular, taking $j=-J$ in equation (9), we have

$$S^{-1}T_{k+4^J m}\psi_l = S^{-1}\psi_{l:0,k+4^J m} = [S^{-1}\psi_{l:J,k}]_{-J,m} = D^{-J}T_m S^{-1} D^J T_k \psi_l$$
$$= D^{-J}T_m D^J S^{-1} T_k \psi_l = D^{-J} D^J T_{4^J m} S^{-1} T_k \psi_l = T_{4^J m} S^{-1} T_k \psi_l,\quad l=1,2,3.$$

Therefore, $P(\psi_1,\psi_2)\,|\,3^J$ Conversely, if $P(\psi_1,\psi_2)\,|\,4^J$ then

$$S^{-1}\psi_{l:j+J,k+4^J m} = D^{j+J}S^{-1}T_{k+4^J m}\psi_l = D^{j+J}T_{3^J m}S^{-1}T_k\psi_l$$
$$= D^jT_mD^JS^{-1}T_k\psi_l = D^jT_mS^{-1}D^JT_k\psi_l = S^{-1}[\psi_{l:J,k}]_{j,m.}$$

Hence, (9) holds and (b) $\Leftrightarrow$ (c).

When $\{\psi_1,\psi_2\}$ generates a wavelet frame in $L^2(R)$ which is also Riesz basis, it is obvious that (a) $\Leftrightarrow$ (d). To prove the converse, define $\phi_{l:j,k,}$ as the orthogonal projection of $\psi_{l:j,k}$ onto the space which is the closed linear span of $\{\psi_{l':j',k'}: l'=1,2,\ j'k'\in Z\}/\{\psi_{l:j,k}\}$. Define a function

$$\tilde{\psi}_{l:j,k} = (\psi_{l,j,k,} - \phi_{l:,j,k})/c_{l:,j,k,} \text{ where } c_{l:,j,k} := \left\langle \psi_{l:,j,k} - \phi_{l:,j,k}, \psi_{l:,j,k}\right\rangle \neq 0.$$

When (d) holds, It is easy to verify that $\{\tilde{\psi}_{l:,j,k,} : k=0,\cdots,4^J-1, l=1,2,3\}$ generate its canonical dual frame.

In general, it is difficult to compute $P(\{\psi_1,\psi_2\})$ because computing the space $\tilde{W}$ and the inverse S involve inverting an infinite matrix. The situation would be much simpler if we require that S commute with both dilates and integer translates in the whole space $L^2(R)$ as in case of frames of translates. In order to state our result, the Fourier transform of an integrable function $f\in L^1(R)$ is defined by

$$\hat{f}(\omega) = \int_R f(x)\exp\{-ix\omega\}dx,\ \ \omega\in R, \tag{10}$$

which can be naturally extended to functions in $L^2(R)$.

5 Summary

In the paper, we prove that the frame operator and its inverse are all self-adjoint operators. We prove that the canonical dual wavelet frame cannot be generated by the translates and dilates of a single function.

References

1. Duffin, R.J., Schaeffer, A.C.: A class of nonharmonic Fourier series. Trans. Amer. Math. Soc. 72, 341–366 (1952)
2. Daubechies, I., Grossmann, A., Meyer, Y.: Painless nonorthogonal expansions. J. Math. Phys. 27(5), 1271–1283
3. Ron, A., Shen, Z.W.: Affine systems in $L_2(R^d)$ (II) Dual systems. Fourier Anal. Appl. 4, 617–637 (1997)

4. Chen, Q., Shi, Z., Cao, H.: The characterization of a class of subspace pseudoframes with arbitrary real number translations. Chaos, Solitons & Fractals 42(5), 2696–2706 (2009)
5. Weber, E.: Orthogonal frames of translates. J. Appl. Comput. Harmon. Anal. 17, 69–90 (2004)
6. Daubechies, I., Han, B.: Pairs of dual wavelet frames from any two refinable functions. J. Constr. Appro. 20, 325–352 (2004)
7. Daubechies, I., Han, B.: The canonical dual. frame of a wavelet frame. J. Appl. Comput. Harmon. Anal. 12, 269–285 (2002)
8. Bownik, M.: A Characterization of Affine Dual Frames in $L_2(R^n)$. J. Appl. Comput. Harmon. Anal. 8, 203–221 (2000)
9. Christensen, O., Eldar, Y.C.: Oblique dual frames and shift-invariant spaces. J. Appl. Comput. Harmon. Anal. 17, 48–68 (2004)

The Study on Private Investment Fund Risk Assessment Using Analytic Hierarchy Process

Li Xin

Economics and Management School, Wuhan University, Wuhan China 430072
10485842@qq.com

Abstract. In this paper, we established system risk assessment and evaluation of the private provision of fund risk level of each of the evaluation index, the comprehensive evaluation method using the hierarchical fuzzy comprehensive evaluation. We provide a quantitative funds research in China the risk of private. It not only helps investors the right decision, and make the government implement policies, so as to ensure the stability of the financial system.

Keywords: Private Offered Fund, Risk Assessment, Analytic Hierarchy Process.

1 Introduction

Provide funds from private experience "stead" and "the valet Fried" in 1993 to 1994. After ten years, it develops through the high speed is flat. Rapid development stage is between 1996 and 2001. According to estimation, the amount of private funds in China provide is about 95 to 110 billion years of 1996 yuan, the rapid growth of 760 years of 880 billion yuan. The scaled is 600 yuan to 700 billion yuan between 2002 years later. Part of the reason is the development of private provides large sums of money for the requirements of the Chinese market. In the next few decades, the Chinese have a rapid development of the economy. Two securities market and private institutions will need fresh capital growth. Another part of the answer may be due to the high return the attraction of the private provide the funds. However, private provision of fund has been in the underground state before 2004. Nearly two years, the private provision of fund recently due to start warming up fund laws change this will directly cause the legalization of it. At the same time, it is strong equity investment after the full implementation of the development of the joint stock system reform. Private funding are divided into private equity securities funds and private equity fund in China. The former listed securities market is active, and the latter mainly is the investment to a private company. And to provide funds, supervision of public funds to provide more private capital loose flexible way to run, it can bring greater profit. Many private investment of funds to provide more than 150%, some even back at 200% to 300%, and this is a research institution named chao Yang yong xu released private funding 2009 annual ranking. High. Back to often represent the High risk. First of all, private funding risk comes from its own characteristic. The supervision of the leak, disclosure of information, such as less could lead to credit risk, moral hazard. Long time investment will also lead to the liquidity risk. Second, from private funding from the Chinese law, start is a short various mechanisms and not perfects some specific risk; improve the

Y. Zhang (Ed.): Future Communication, Computing, Control and Management, LNEE 141, pp. 109–115.
springerlink.com

level of risk above products. Private provides the research funds in China belong to the initial stage. Domestic scholars mostly concentrated in private funds, the paper discuss the significance of provide legal supervision and standard operation problem. Research also didn't do thorough evaluation system risk. Even scholars mainly used qualitative, quantitative analysis method is almost this aspect.

2 Selection of Index System

This study analyzes the existing main risk characteristics on the basis of development, provide funds in private. Choice of the indicators suggests that are as follows:

(1) Market risk. It means that the market price of the asset to adverse change investors loss dropped. It includes the interest rate risk, foreign exchange risk, commodity price risk, stock price risk. Market risk inevitable risk belongs to the macro environment.

(2) Credit risk. It means the money without the loss caused by, including market credit risk, enterprise credit risk and personal credit risk.

(3) The liquidity risk. It shows that, the loss of uncertainty risk fund assets of the realization of the uncertainty, as investors leave quantity and time. Go from the currency risk and the trade, the public will ban the exit of risk and financial leverage blotted out.

(4) Legal risk. The risk of illegal position in China's current risk capital source, illegal stock price manipulation of risk and risk of illegal the internal structure of the company are included.

(5) Business risk. Business risk is a loss caused by the potential business process and failed to perform. It includes the operation risk, trade risk, technology risk, and client compensation risk, capital loss risk, etc. The system of risk assessment of private offered fund in China is showed in table 1.

Table 1. The system of risk assessment of private offered fund in china

Objective layer	Criteria layer	Index layer
Risk of private offered fund in China	Business risk B_1	Operational risk C_{11}
		Transaction risk C_{12}
		Technical risk C_{13}
		Customer compensatory risk C_{14}
	Market risk B_2	Assets loss risk C_{15}
		Interest rate risk C_{21}
		Foreign exchange rate risk C_{22}
	Credit risk B_3	Stock price risk C_{23}
		Commodity price risk C_{24}
	Liquidity risk B_4	Market credit risk C_{31}
		Enterprise credit risk

Table 1. *(continued)*

	Legal risk B_5 Moral risk B_6	C_{32} Individual credit risk C_{33} Exit difficult C_{41} High financial leverage risk C_{42} Existence form risk C_{51} Capital source risk C_{52} Stock price manipulation risk C_{53} Internal structure riskC_{54} Information disclosure risk C_{61} Capital abuse risk C_{62} Excessive investment riskC_{63} Decreased service efficiency rate C_{64}

3 Methodology and Materials

We calculate the weight of each index used the analytic hierarchy process (AHP) method. Then the risk level can be obtained, the index of using the fuzzy comprehensive evaluation method. Data from the questionnaire survey, the experts. Specific steps are listed below:

Analytic Hierarchy Process (AHP) method

Building judgment matrix . The element aij in the judgment matrix means the score ratios given by experts using 1 to 9 scale methods. The score is explained by the relative importance of the index.

$$A = \begin{bmatrix} a_{11} & a_{12} & \cdots & a_{1n} \\ a_{21} & a_{22} & \cdots & a_{2n} \\ \cdots & \cdots & \cdots & \cdots \\ a_{n1} & a_{n2} & \cdots & a_{nn} \end{bmatrix}$$

Computing maximum eigenvalue and weights

The calculation formula is listed below:

$$M_i = \prod_{k=1}^{n} a_{ij}$$

$$\overline{w}_i = \sqrt[n]{M_i}$$

$$w_i = \frac{\overline{w}_i}{\sum_{i=1}^{n} \overline{w}_i} (i = 1, 2, \ldots, n)$$

$$\lambda_{\max} = \sum_{i=1}^{n} \frac{(Aw)_i}{nw_i}$$

Consistency inspection

$$CI = \frac{\lambda_{\max} - n}{n - 1}$$

$$CR = \frac{CI}{RI}$$

RI is a random index. The result is acceptable when CR< .0.1.

Fuzzy comprehensive evaluation. The evaluation factors set $U = \{u_1, u_2, \ldots, u_n\}$, means the index of system. The evaluation set $V = \{v_1, v_2, \ldots, v_n\}$ means the risk level we ranked.

- The element r_{ij} in the matrix equal to the ratio that the number of experts who choose the level v_j on factor u_i to the whole number of experts. The judgment matrix is below:

$$R = \begin{bmatrix} r_{11} & r_{12} & \cdots & r_{1n} \\ r_{21} & r_{22} & \cdots & r_{2n} \\ \cdots & \cdots & \cdots & \cdots \\ r_{m1} & r_{m2} & \cdots & r_{mn} \end{bmatrix}$$

- The result of fuzzy comprehensive evaluation:
  The first layer: $B_i = W_i * R_i$
  The second layer: $B = W * (B_1, B_2, \ldots, B_n)$
  Score is given to each level we ranked: $V = (v_1, v_2, v_3, v_4, v_5) = (5, 4, 3, 2, 1)$.

4 Results and Discussion

Model analysis

The result of AHP method The weight obtained by AHP method is showed in table 2.

Table 2. The evaluation of system and weight obtained by ahp

Index	Weight on the objective layer	Weight on the criteria layer
B1	0.2017	--
C11	0.0502	0.2489
C12	0.0457	0.2268
C13	0.0210	0.1042
C14	0.0266	0.1317
C15	0.0582	0.2884
B2	0.0786	---
C21	0.0170	0.2163
C22	0.0076	0.0961
C23	0.0229	0.2909
C24	0.0312	0.3968
B3	0.2282	---
C31	0.0863	0.3780
C32	0.1031	0.4520
C33	0.0388	0.1701
B4	0.2414	---
C41	0.1636	0.6779
C42	0.0778	0.3221
B5	0.1345	---
C51	0.0169	0.1259
C52	0.0509	0.3781
C53	0.0446	0.3314
C54	0.0221	0.1645
B6	0.1157	---
C61	0.0339	0.2930
C62	0.0408	0.3528
C63	0.0261	0.2260
C64	0.0148	0.1283

The result of fuzzy comprehensive evaluation

a) The result of first layer

The result of first layer is showed in table 3.

Table 3. Result of fuzzy comprehensive evaluation of first layer

Index	High	Relatively high	Moderate	Relatively low	Low	Score
B1	0.0277	0.2307	0.4028	0.3388	0.0000	2.9473
B2	0.0000	0.3831	0.2711	0.1015	0.2445	2.7932
B3	0.1193	0.1090	0.3607	0.1300	0.0000	3.5746
B4	0.0753	0.3296	0.4444	0.1506	0.0000	3.3293
B5	0.0183	0.3661	0.2816	0.3156	0.0183	3.0502
B6	0.0326	0.2291	0.4996	0.2389	0.0000	3.0560

b) The result of second layer

$$B = (0.0572 \quad 0.3253 \quad 0.3878 \quad 0.2124 \quad 0.0217)$$

The analysis of result Weight business risk, credit risk, liquidity risk, are beyond 0.2, the most important factors in the class B. Is said management risk, credit risk, liquidity risk is vital to the private provision of constituent fund. At the same time, the index to exit the difficulties and enterprise credit risk is greater than 0.1, C level of the importance of standing in the 1 st 2 nd places respectively. They are the most important indexes in the C level.

Scores of credit risk and liquidity risk is of the highest two factors to evaluate a grade 3.5746 and 3.3293, respectively. We can safely draw the conclusion that to credit risk, liquidity risk standing in a private fund system to provide the commanding heights of the established in China. Through the second grade evaluation score, comprehensive risk fund system in China is 3.1973; the private provision is a very high level

5 Discussion and Conclusions

In short, it is a clear result can be found from the above discussion.

The liquidity risk is one of the important links in China private fund system to provide. Leave time and number of profits have great influence on the funds. In adversity, to succeed can't loss to the minimum degree, in contrast, can increase the investment from difficult losses. According to the results of vague overall value and provide liquidity risk fund private is on the high side. At first, provide regular fund is private than the public funds to provide. The manager always investors signed the contract from withdrawing money to assure that prevent them from stable, long-term interests of the fund investment projects. In addition, private funding not circulation and generally do not enter the market can reduce their cash flow. Finally, still not exit mechanism is in China.

(1) Credit risk is a consequence in the private provision of fund risk to the establishment of the system in China. On one hand, private funding is by the lack of financial guarantee laws and regulations. As long as the fund management company default deciding factor, investors are faced with huge losses. On the other hand, the credit is the main standard, factors when investors choose a institutions. A poor reputation will lead to difficulties in raising money, the result in the UN-going concern. According to the result of fuzzy comprehensive value, and credit risk is China's private provide higher than normal current fund system. First of all, the credit mechanism is not effective in our country, investors on the basis of signing a contract with a fund management company's trust. There are few legal ways for them, to protect and restore their interest's agencies cannot revered agreement. In addition, the incomplete information private provide Fund Company is open to the public for investors is fully aware of the difficult situation. They require the company to business and performance through private channels or interpretation of conversation. Last but not least, private provision of Fund Company in our country has great number, but scale is small, the good and bad are intermingled credit.

(2) Private provision of funds is a relatively flexible financing method of the capital and improve the utilization ratio and effective use of funds still. At the same time, private funding lower the threshold of the funds provided more than public, can ease

the burden of financing of small and medium enterprises. The experience of other countries should be used for reference guide our control risk and regulate the development of private provide funds. This is a very important financial industry and the healthy development of the national economy.

References

1. Wallsten, S.J.: The small business innovation research program: encouraging technological innovation and commercialization in small firms? Rand Journal of Economics 31, 82–100 (2000)
2. Mason, C.: Public policy support for the informal venture capitalmarket in Europe: a critical review. International Small Business Journal 27, 536–556 (2009)
3. Cochrane John, H.: The risk and return of venture capital. Journal of Financial Economics 75, 3–52 (2005)
4. Jääskeläinen, M., Maula, M., Murray, G.: Profit distribution and compensation structures in publicly and privately funded hybrid venture capital funds. Research Policy 36(7), 913–929 (2007)
5. Benz, M.: Entrepreneurship as a non-profit-seeking activity. International Entrepreneurship and Management Journal 5, 23–44 (2009)
6. Lerner, J.: The government as venture capitalist: the long-run impact of the SBIR program. Journal of Business 72(3), 285–318 (1999)
7. Lerner, J., Watson, B.: The public venture capital challenge: the australian case. Venture Capital 10(1), 1–20 (2007)
8. Leleux, B., Surlemont, B.: Public versus private venture capital: seeding or crowding out? A pan-European analysis. Journal of Business Venturing 18, 81–104 (2003)
9. Lerner, J.: When bureacrats meet entrepreneurs: The design of effective "Public Venture Capital" programmes. The Economic Journal 112(2), F73–F84 (2002)
10. Cumming, D.J., Macintosh, J.G.: Crowding out private equity: Canadian evidence. Journal of Usiness Venturing 21(3), 569–609 (2006)

An Improved Background Modeling Method for Target Detection*

Cheng Mingzhi[1] and Gao Junxiang[2,*]

[1] College of Information & Mechanical Engineering, Beijing Institute of Graphic Communication, Beijing, 102600, China
[2] School of Science, Huazhong Agricultural University, Wuhan 430070, China
Gao200@gmail.com

Abstract. To promote the efficiency of moving-object extraction using Gaussian Mixture Model, an improved method was proposed for intelligent video surveillance system. Based on the fact that the majority of the background pixels remain steady, these pixels were detected according to certain criteria before the procedures of matching and updating GMM parameters, and then the frequency of matching and updating was decreased. As a result, the computation redundancy could be reduced sharply. A couple of criteria to estimate the stable and unstable status as well as the flow chart were presented to illustrate the proposed method. Experimental results of two video sequences show that the speed of the improved method increase to 7.2 and 5.4 times, respectively, indicating that the efficiency is promoted significantly while keeping approximately equal quality.

Keywords: Background modeling, moving-target detection, Gaussian mixture model.

1 Introduction

Background modeling for moving-object extraction is a fundamental and crucial problem in intelligent video surveillance system. Accordingly, researchers have proposed quite a few adaptive modeling algorithms for dynamic background [1-3]. Among these algorithms, Gaussian Mixture Model (GMM) method presented by Stauffer and Grimson can represent the complex distribution of each pixel, though it became a popular technique for background modeling in various applications [4], and some measures have been taken in literatures [5-8] to enhance its performance. Unfortunately, conventional Gaussian mixture model suffers from the heavy computational complexity. As a result, the algorithm is difficult to be used if special hardware support is absence.

To cope with this problem, we present an improved moving object detection method on the basis of Gaussian mixture model. Considering the majority of the background

* This work is supported by Scientific Research Common Program of Beijing Municipal Commission of Education (Grant No. KM201010015003).

* Corresponding author.

Y. Zhang (Ed.): Future Communication, Computing, Control and Management, LNEE 141, pp. 117–123.
springerlink.com

pixels remain steady in video surveillance applications, the approach identified these pixels according to certain criteria and decreased the frequency of matching and updating GMM parameters. In this phase, two criteria are introduced in this paper to judge whether a pixel keeps steady or not. The improved method can reduce the computation redundancy sharply while keeping approximately equal quality.

2 Background Modeling Using Gaussian Mixture Model

The pixel-based Gaussian mixture model of background is a statistical model that supposes the data originates from a weighted combination of several Gaussian functions. If K Gaussian functions are used to describe a pixel, the observation of the given pixel will be in one of the K states at one time. The probability that a pixel has observation value X_t at time t is approximated by

$$P(X_t) = \sum_{i=1}^{K} \omega_{i,t} * \eta(X_t, \mu_{i,t}, \Sigma_{i,t}) \tag{1}$$

where $\omega_{i,t}$ is the weight estimation of the ith Gaussian distribution in the mixture model at time t, $\mu_{i,t}$ and $\Sigma_{i,t}$ are the mean value and covariance matrix of the ith Gaussian distribution in the mixture model at time t, and η is a Gaussian probability density function represent by

$$\eta(X_t, \mu, \Sigma) = \frac{1}{(2\pi)^{n/2} \left|\Sigma_{i,t}\right|^{1/2}} e^{-\frac{1}{2}(X_t - \mu_{i,t})^T \Sigma_{i,t}^{-1} (X_t - \mu_{i,t})} \tag{2}$$

where n is the dimension of vector X_t. Suppose that components of X_t are independent and have the same variance, the covariance matrix can be computed according to equation (3)

$$\Sigma_{i,t} = \sigma_{i,t}^2 I \tag{3}$$

The parameters of the Gaussian distribution $\omega_{i,t}$, $\mu_{i,t}$ and $\Sigma_{i,t}$ are updated with new frames. A retrieved pixel value is compared with all the distributions of the mixture assigned to that pixel to find out whether there is a match. If the pixel value is within 2.5 times standard deviations of one Gaussian component, a match is found. If a pixel value does not match any one of the K components, the component with the smallest probability is replaced by a new component. The new component is initialized with a mean value equal to the current pixel value, and a variance equal to a predetermined constant.

For both matching and not matching components, the updating for the weights is done according to the following equation.

$$\omega_{i,t} = (1-\alpha)\omega_{i,t-1} + \alpha D_{i,t} \tag{4}$$

where $D_{i,t} = 1$ for the matching distribution and $D_{i,t} = 0$ for the others. As a result, the matching distribution weight will relatively increase. In this equation, α denotes the learning parameter, and it is usually determined experimentally.

For a matching distribution, the parameters $\mu_{i,t}$ and $\Sigma_{i,t}$ are adjusted by the following formulas.

$$\mu_{i,t} = (1-\rho)\mu_{i,t-1} + \rho X_{i,t} \tag{5}$$

$$\sigma^2_{i,t} = (1-\rho)\sigma^2_{i,t-1} + \rho(X_{i,t} - \mu_{i,t})^{\mathrm{T}}(X_{i,t} - \mu_{i,t}) \tag{6}$$

where ρ represents the second learning parameter, and it is computed by equation $\rho = \alpha\eta(X_t \mid \mu_{i,t-1}, \sigma_{i,t-1})$.

The K Gaussian components are ordered by the ratio value of ω/σ, consequently, the first B components are chosen as the background model, where B is estimated according to:

$$B = \arg\min_b\left(\sum_{i=1}^{b} \omega_i > T\right) \tag{7}$$

where T is a measure of the minimum portion of the data that should be accounted for by the background, and the pixels match any one of the B components are classified as background pixels.

3 Improved Gaussian Mixture Mode

The improved method proposed in this paper is based on the fact that the majority of the background pixels keep steady; therefore these pixels may enter stable status. For the pixels entering this status, the GMM parameters hardly change before and after updating. Thus we can define two criteria, and they are used to judge a pixel entering and exiting stable status, respectively. After a new observation value is obtained, mixture model matching and updating are performed if the pixel is not in stable status. On the contrary, if the pixel is in stable status and the exiting criterion does not hold, the frequency of matching and updating will be decreased. The detailed steps are described in the following sections.

(1) Define a variable status for every pixel; the pixel is either in stable status Sstab or unstable status Sunstab,. If a pixel represents background pixel for a relatively long time, then the pixel is in stable status. On the other hand, if a pixel represents foreground pixel or unsteady background pixel, then the pixel is in unstable status. In Gaussian mixture model, stable status means one distribution dominate the mixture.

(2) The status of a pixel alternate between these two statuses, which need two criteria Cstab and Cunstab. The former is used to judge whether a pixel enters stable status, while the latter is used to judge whether a pixel exits stable status. The two criteria are formulated as follows.

$$C_{stab} \equiv \begin{cases} i = 1 & i \text{ is the matching distribution of previous data} \\ \omega_{1,t} > T \\ \omega_{1,t} - \omega_{1,t-1} < \delta \end{cases} \tag{8}$$

$$C_{unstab} \equiv (\left|X_t - \mu_{1,t-1}\right| > \lambda \sigma_{1,t-1}) \tag{9}$$

where the meaning of T is same as equation (7), δ is the threshold, and it can be determined experimentally (for example: $\delta = 0.01\alpha \sim 0.1\alpha$). In formula (8), it is obvious that the weight of the first distribution is far larger than else, and there is no much difference between the weight of the first distribution at time t and time t-1. In formula (9), to avoid frequent status alternation of a pixel, parameter λ should be a little larger than β (for example: $\beta=2.5$, $\lambda=3$).

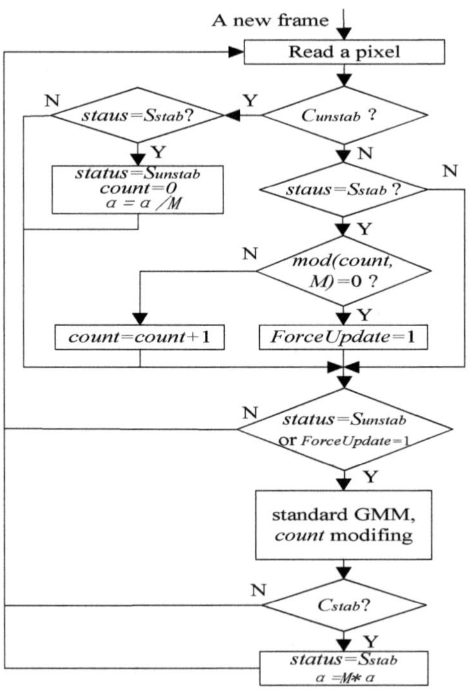

Fig. 1. The flow chart of the proposed method

(3) After a new observation value is obtained, the new status of the pixel and various parameters are determined as the following description.

a) If the pixel is in unstable status, i.e., status= Sunstab, mixture model matching and updating are performed according to standard GMM method.

b) If the pixel is in stable status, and the criterion Cunstab holds, then assign Sunstab to status.

c) If the pixel is in stable status, and the criterion Cunstab does not hold, the frequency of matching and updating will be decreased, i.e., matching and updating are done every M frames. Suppose matching and updating counter is a variable count, then the force matching and updating flag ForceUpdate set 1 if Mod(count, M)=0.

d) In stable status, the updating frequency decrease M times lead to learning parameter decrease M times. To compensate this reduction, learning parameter α should increase M times.

The flow chart of improved GMM algorithm is shown in Fig. 1.

4 Experimental Results

We test the proposed method on different surveillance sequences. An indoor sequence (Hall) and an outdoor one (Road) are extracted to demonstrate the performance of the algorithm. The parameters of the algorithm were set experimentally, i.e., $T = 0.8$, $\alpha = 0.02$, $\beta = 2.5$, $\lambda = 3$, $M = 20$, $\delta = 0.1\alpha$.

Fig. 2 demonstrates the efficiency promotion of the proposed algorithm comparing with standard GMM. We can see that the time consumption of improved GMM method are reduced sharply, in other word, the efficiency of processing sequences increase 7.2 times and 5.4 times, respectively.

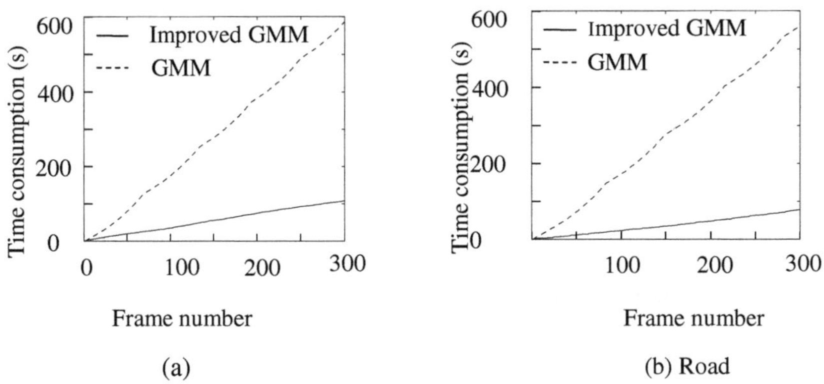

Fig. 2. Comparison of time consumption between standard GMM and Improved GMM

The increase of method efficiency should not damage the quality of moving object extraction. To compare the quality of the images processed by standard GMM and improved GMM methods, Fig. 3 and Fig.4 show the results of the two methods. It is can be seen that there is no obvious difference between Fig. 3-b and Fig. 3-c, as well as Fig. 4-b and Fig. 4-c.

(a) Three frames of the sequence

(b) Processing results of standard GMM method

(c) Processing results of Improved GMM method

Fig. 3. Quality comparison of processing results (sequence: Hall)

(a) Three frames of the sequence

(b) Processing results of standard GMM method

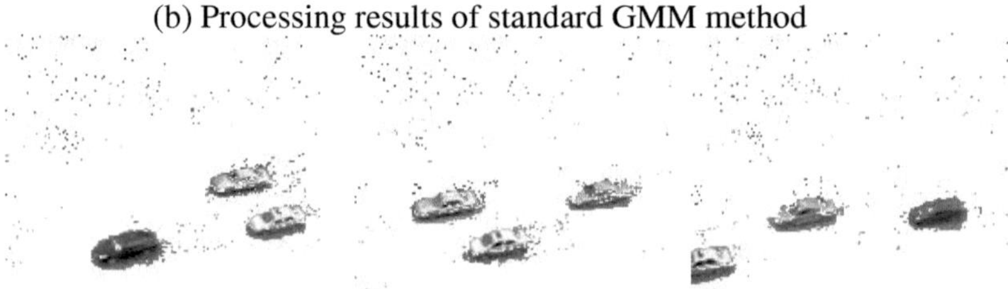

(c) Processing results of Improved GMM method

Fig. 4. Quality comparison of processing results (sequence: Road)

5 Conclusion

In this paper we present a fast algorithm of moving-object extraction based on Gaussian mixture mode. Because the majority of the background pixels remain steady in video surveillance applications, the proposed approach identify these pixels according to certain criteria, and decrease the frequency of matching and updating GMM parameters. Experimental results of two video sequences show that the speed of the improved method increases sharply, while keeping approximately equal quality. The advantage above-mentioned make the method more practicable for visual surveillance system.

References

1. Colombari, A., Fusiello, A.: Patch-based Background Iinitialization in Heavily Cluttered Video. IEEE Transactions on Image and Processing 19(4), 926–933 (2010)
2. Lin, H.H., Liu, T.L., Chuang, J.H.: Learning a Scene Background Model via Classification. IEEE Transactions on Image and Processing 57(5), 1641–1654 (2009)
3. Prokofiev, A.V.: Fast Background Modeling Algorithm Based on Stability Intervals. Journal of Automation and Information Sciences 40(6), 72–79 (2008)
4. Stauffer, C., Grimson, W.E.L.: Adaptive Background Mixture Models for Real-time Tracking. In: Proceedings of the IEEE Computer Society Conference on CVPR, pp. 246–252 (1999)
5. Chen, Z.J., Yu, H.B.: Moving Vessels Detection Based on Background Subtraction. Journal of System Simulation 19(18), 4316–4319 (2007)
6. Hu, J.S., Su, T.M.: Robust Background Subtraction with Shadow and Highlight Removal for Indoor Surveillance. EURASIP Journal on Advances in Signal Processing 10(1), 1–14 (2007)
7. Yang, G.L., Kong, L.F.: Approach of Constructing Background Model Based on Image Blocks. Robots 29(1), 29–34 (2007)
8. Tang, Z.: Fast Background Subtraction and Shadow Elimination using Improved Gaussian Mixture Model. In: Proceedings of the 6th International Workshop on HAVE, pp. 38–41 (2007)

The Intelligent Management System of the Buses on Busy Roads

Gu Wenyang, Chen Huanlin, Zhang Jiajian, and Wu Xiang

School of Science, Hohai University
gwy.hhu@163.com

Abstract. In order to solve the current problem of crowded buses, develop the citizens' consciousness of "waiting in line, taking with civility", improve the operating efficiency of buses, form good traffic order, build a green, environment-friendly and new mode of bus operation, we have discussed and researched an intelligent operation and management system of buses. We hope that through our efforts, we can solve the problem of crowded buses in Nanjing on the basis of certain feasibility.

Keywords: Buses, crowded buses, intelligent management system.

1 Research Background

Urban public transportation is an important component of city integrated transport system. It is the main body of urban passenger transport. The development of urban modernization exacerbates the conflict between the rapid increase of the urban passenger capacity and insufficient carrying capacity of city's public transport system.

At present, China's transportation system has gradually become complete, especially the short-distance public transport system, which has become quite perfect because of the needs of construction and people's livelihood of each city.

However, due to too large passenger flow, buses of some lines (specially in suburban districts) are very crowded. "It is hard to squeeze into the bus, but there is no place to stand". This situation has become quite common. People's travel is still not very convenient.

"Difficult in getting on, vexed on bus" has become a big problem restricting citizens to go out.

Take Nanjing for example: as long as No. 101 bus (to and fro Dongshan, master stop, in Zhonghua Gate) and No. 105 bus (to and fro Jiangning Campus of Hohai University and Zhonghua Gate) arrive at the stop, people rush to the bus door and crowd around the door.

If two buses arrive at Zhonghua Gate stop at the same time, the crowded phenomenon is more "spectacular". This greatly affects the city's appearance and the building of Nanjing into a civilized city.

At the same time, there are a series of problems brought by the difficulty in taking buses. First, the stops and scheduled buses where passengers are numerous become the places of unlawful actions.

Y. Zhang (Ed.): Future Communication, Computing, Control and Management, LNEE 141, pp. 125–130.
springerlink.com

They often steal other's property in the chaotic situation when passengers crowdedly get on a bus. They also steal in a crowded bus and cause unnecessary losses to passengers. The consequences are very serious in many cases. To find out its cause, the inefficient urban public transport system can hardly absolve itself from the blame.

Second, crowded buses are usually packed with people at the several earlier stops. Even the standees nearly have no place to place their feet and people waiting at the latter stops can do nothing but watch helplessly as the buses rush before their eyes.

Over time, in order to be able to take the bus, people have to walk several stops backward and wait at the stop nearer the starting stop in the hope of vying for a place on the bus. The result of this will inevitably be that some stops will perform practically no function. It can't reach the expectancy at the beginning of urban planning. The city's public resources cannot play a role and at the same time, it will be more inconvenient for residents to go out.

Therefore, the Nanjing Institute of Traffic Planning proposed and implemented a series of policies and measures, such as: making scientific and reasonable planning and arrangement for bus lines to optimize the bus network; gradually expanding the scale of public transport, scientifically increasing the number of buses, raising the departure frequency, etc. Moreover, setting up long wooden fences for passengers to wait in line at the bus stops（such as the railway stop） where the passenger flow volume is very large.

But, as a whole, it has little effect---not all bus stops have enough places to set up the long wooden fences. And setting up wooden fences can not fundamentally change the situation of different utilization ratio of different sites. So, this needs more effective and reasonable methods to manage the city public transport system.

2 Research Purpose

We design this project first for the sake of serving the recent innovative thinking training contest. We hope that by virtue of this project and the opportunity the contest offers, we can open our mind, design daringly, exercise our ability of innovative thinking, and cultivate our rigorous and earnest style of scientific exploration. An ancient proverb says: learning without thought is useless.

Most of the time, the lack of innovation is not because of lack of innovation ability, but lack of motivation and goal of innovation. In this innovative thinking training contest, we decide to see this project as goal and direction of our thinking and we will unswervingly go ahead in this direction, whether we are successful or not. We will innovate our thinking, carefully design, attentively research and constantly make improvements. We believe that we will achieve something finally.

In addition, considering from a longer-term stage, we hope our research achievements can be put into practice finally and contribute pygmy efforts to the improvement of busy and redundant city public transport system. As students of Jiangning Campus of Hohai University, we fully realize how difficult it is to squeeze into a bus. Putting ourselves in the place of others, we can understand the bitter of the citizens in Nanjing and other cities. We have the ambition to build a more reasonable and more humanized urban public transport system to make the way people go out safer, more convenient, comfortable and quicker, thus making the city life more harmonious and beautiful.

Finally, if our project could be designed successfully in the end and could be put into production and use, we hope this can be the starting point for us to establish a business.

3 Operation Scheme

The specific operation scheme of the intelligent management system is as follows:

(1) Put "ticket machine" at the stops where the passenger volume is very large. According to the principle of "first coming, first served", people who come first waiting at the stop get the ticket whose order has already been scheduled (nothing to do with the stops). The tickets are produced by the "ticket machine". The tickets note the bus number, the starting, terminal stops and the bar code.

Passengers can toss coins or scan their bus cards for the ticket charges, so as to reduce human involvement to the greatest extent. The "ticket machine" swaps data with the buses by the local area network or the wireless to realize the dynamic allocation of the tickets in different time and stops.

The ticket price is concerned with the bus numbers, starting and terminal stops. It is generally managed by the "ticket machine" (preventing the ticket price from human intervention) and the tickets are one-to-one correspondence with the bus number. This realized "one person, one ticket, one bus" and rational allocation of resources. Once a bus is full, the ticketing machine will stop selling tickets for this bus as soon as possible, in order to prevent overloading behavior.

In order to operate normally, a series of software data must be written into the "ticket machine" in advance by programming. The related hardware data (such as ticket price, the corresponding bus number, etc.) are fixed in the "ticket machine" by program.

(2)The tickets produced by the "ticket machine" needs a corresponding ticket reader. Set a ticket reader near the front door of the bus and install the electronic display screen. Besides, the bus door must be turned into an electronic door, which is controlled by the ticket reader.

According to the number indicated on the electronic display screen, in another words, the order, passengers can correspondingly scan the tickets by their ticket order in their hands. Then the electronic door opens and the passengers can get on at this moment. If the ticket order in hand do not correspond with the number showed on the display screen, the ticket scanning is invalid and the door of course can not be opened. Because the ticket reader can only correspondingly read the ticket according to the number displayed on the electronic display screen, the passenger who gets the order first has the priority to take the bus.

In this way, the order of getting on bus completely depends on the order ticket in the passenger's hands. And the order of the order ticket depends on the order's priority of the passenger coming to the stop. Thus, it corresponds with the principle of "first coming, first served". And this can avoid the situation of the passengers swarming to the front door of the bus as soon as the bus stops. It can make the passenger consciously queue up and orderly getting on. It is because of the one-to-one corresponding between the ticket and the bus number, so passenger can only get on one bus by one ticket. We think it is very necessary for the order of getting on the bus when more buses arriving at the stop.

In case some passengers who bought the ticket in advance, but can not get on the bus for special reasons and passenger stagnation is caused, we make the correlative stipulation with the ticket reading time(such as: 5 seconds). If the ticket reader does not check within the regulated time, it will automatically leap over and present the next number to check next ticket. However, the passenger who bought the ticket and can not get on the bus for special reasons can refresh this ticket at the "ticket machine". If there are seats which are not taken, the refreshed ticket order will be ranged after the present order. If all the seats are taken on the bus, the refreshed ticket order will be ranged after the order of the next bus number. In this way, we can make the measures of reasonable distribution for tickets fully realize.

(3) In order to solve the problem of over crowding in the bus, the intelligent management system must scientifically and reasonably control the number of passenger getting on the bus. Set a display screen both inside and outside of the bus to count the passengers of the bus. When there is no one in the bus, it notes 0 and the number on the screen add 1 with a passenger getting on. As long as the number on the screen reaches the maximum loading number, the "ticketing reader" stops selling the order ticket corresponding to the bus number immediately. In this way, as long as the bus hasn't been fully "filled", passengers can smoothly scan their tickets and take the bus. Once the bus seats are fully taken, the passengers have to wait for the next bus.

Besides, in order to make the operation of the bus more standardized, humanized, environmental-friendly and modernized, we make the following adjustment of the bus:

(1)Install a trumpet for the steering wheel accessory of each bus. This will facilitate the driver to inform passengers in a timely and correct way, strengthening contact between the driver and passengers;

(2)Put several boxes of scented water in the bus to replace the original petrol fume with the faint scent and make the passengers feel the comfort of the bus;

(3)Put a trash barrel in the rear door accessory of the bus and encourage passengers to put the rubbish into the barrel consciously to keep a clean environment inside the bus;

(4)Post more publicity slogans related to environmental protection and health inside the bus and less profitable advertisements to raise environmental awareness among people.

As the intelligent management system is mainly used to solve the problem of over-crowding, it will reduce the speed at which passengers get on the bus instead of generating a good effect if it is still used at bus stops where the passenger flow has been very low or in a period of average demand. Therefore, the buses should be installed with two kinds of mode: usual bus operating mode and intelligent management system mode, for drivers to adjust and choose. In the rush hours, drivers can switch to intelligent management system mode; otherwise they can adjust to the usual mode. Thus, the operating efficiency of buses can be greatly improved.

After the intelligent management system is launched, the way citizens take buses will undergo drastic changes. The following is the specific flow citizens take buses:

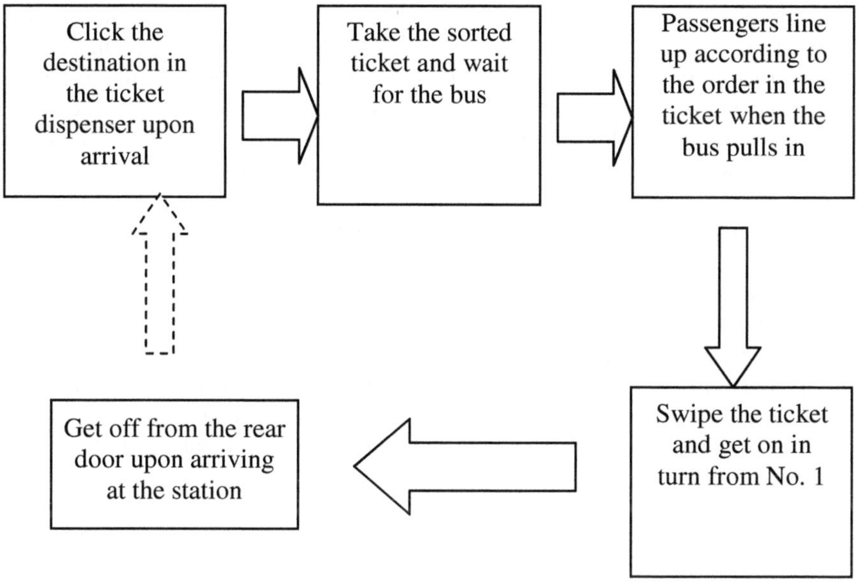

4 System Maintenance and Management

Each new thing or method will be confronted with certain resistance when it just appears due to people's strangeness or dependence on the old thing. In order to prevent such situation, the intelligent management system is forcibly enforced after it is just put into use, and personnel are arranged to guide passengers to use at each bus stop. The queue ticket machine should be equipped with brushing card (bus card) booking system apart from accepting cash to be convenient and efficient. As for the money in the queue machine at each bus stop, specially-assigned persons are sent to take it out after the buses stop operation in the evening. The total sum can easily be calculated according to the number of sold tickets recorded in the electronic system. Checking these two numbers can effectively monitor the personnel taking the money. In addition, vigorous and effective publicity is needed to call for citizens to use this system in an appropriate way.

From the above introduction, you can know that our system is interlocked. The failure of queue machine at any bus stop will exert a huge impact on the operation of whole system and cause passengers great inconvenience. To reduce the probability of such risk, we need to properly install several "queue ticket machines" for each stop. Moreover, we will arrange professional personnel to regularly maintain and update the machines to ensure the long-term and normal operation of this system.

5 In-depth Exploration

To improve the efficiency of our system, provide better service for passengers and make citizens more satisfied with public transport, we need to explore and popularize this system in an in-depth way. For example, we need to develop an efficient prioritization scheme for bus dispatching. To this end, we can know the previous situation of passenger flow distribution and get the on-off passenger flow data of related bus stops (including rush-hour, period of average demand and low usage period) through field investigation, visiting bus company or collecting on-line information. Then we can analyze the data with relevant knowledge of mathematical modeling and obtain a reasonable, efficient and highly satisfying bus departure timetable according to the number of people of full load.

When to depart or how many buses depart can be executed in accordance with the data on the timetable. Of course, dissimilar problems should be approached differently. It is inevitable that buses can be late or delayed. The timetable can be referred to but can not be completely executed to the letter. The management of departure time and number of departing buses, together with our intelligent dispatching, can greatly reduce the empty-loaded rate and consumption of energy like fuel to indirectly increase the business benefit of the bus company, effectively avoid traffic jam, improve the traffic condition and air quality of the roads throughout the city and truly realize the goal of serving passengers.

References

1. Chen, X., Li, S.: Targets, Effect and Measures of Priority of Urban Public Transport. China Municipal Engineering (June 2001)
2. Wang, X., Yao, Z., Song, R.: Dispatching Management of Operation of Urban Public Transport. China Railroad Press, Beijing (2001)
3. Su, Y., Yan, K., Han, H.: Research of Operational Characteristic Model of Public Transport. Journal of Chongqing Jiaotong University 20(1) (March 2001)
4. Liu, C., Xu, X.: Priority and Control Informatization of Public Transport. Henan Scientific and Technological Publishing House (June 2002)

Car Speed Measurement Based on Feature Point Tracking in Traffic Video

Shi-Song Zhu[1], Min Gu[1], and Toshio Koga[2]

[1] School of computer science and technology, Henan Polytechnic University, Jiaozuo, Henan, China
zss@hpu.edu.cn, gm_lyq@163.com
[2] Graduate School of Science and Engineering, Yamagata University, Yonezawa-shi Yamagata, Japan
koga@yz.yamagata-u.ac.jp

Abstract. Moving car information in traffic video signals contains plenty of important information on road safety. In this paper, we propose an algorithm that extracts and tracks Feature Points of cars, and analyze their trajectories obtained by the tracking. The information is indispensable for measuring moving car speed, counting the number of cars, monitoring distance between two cars, classifying running directions, and monitoring congestion situation. Utilizing Harris operator, Feature Points such as edges and corners are extracted, followed by block-matching to track the Feature Points in successive video frames. Many cars can be tracked at the same time automatically since the information is obtained from video sequences. As an example, this paper shows how car speed can be measured.

Keywords: Background updating, image processing, traffic video, vehicles detection.

1 Introduction

In recent years, as the number of cars increases more and more rapidly, road safety is increasingly of a public concern. To address this concern, research on monitoring how cars are running on the roads is mandatory. Possible outcomes of this kind of research will help enhance traffic efficiency by early warning congestion ahead, letting cars slow down, solving various problems such as environmental pollution, etc. One of the outcomes may be to detect the speed of running cars from the video signals. With image processing, rigorous carefulness is not needed in installing a video camera. Besides the speed measurement, various information of traffic situation can be presented [1].

In this paper, we propose a method in which extraction of the moving cars along with formation and tracking of the car areas is carried out using only the video signals. Feature Points in the car area are extracted by using the Harris operator. Stable detection of Feature Points is realized by application of block-matching algorithm, producing good tracking results. Speed of the cars can be detected with considerably high accuracy investigating trajectories of the Feature Points in the car area.

Y. Zhang (Ed.): Future Communication, Computing, Control and Management, LNEE 141, pp. 131–139.
springerlink.com

2 Overview of the System

In a basic system, picture of running cars is taken by a fixed video camera on one side of road where the camera is placed 4m high and installed so that its viewing direction approximately parallel to the traffic flow. The video signal is processed by our proposing algorithm. As a result, information of running cars can be monitored, e.g. the number of cars, moving speeds, distance between two cars, etc. In this paper, measured is speed of outgoing cars passing through the fixed camera.

3 Extraction of Moving Objects

For measuring car speed, it is important that car areas can be extracted as exactly as possible. Typically, there are two methods to get moving cars from traffic video stream. The first method is Subtraction from Background technique used to separate car areas from the background. However, the background is vulnerable to change in lighting and weather conditions, although the background is calculated as an average of all image sequence in [2][3]. There is another method which may be called Frame Subtraction technique. This is a technique considerably independent of lighting and weather conditions. We use this method in our experiment.

To extract only the moving cars from the dynamic scene, changed parts showing the moving objects in the image is detected after binarizing the difference using three successive frames Frame(t-1), Frame(t) and Frame(t+1).

Letting I(x,y,t) represents intensity of Frame(t), difference between Frame(t) and Frame(t+n) is defined as follows,

$$e(x,y,t+n) = \left| I(x,y,t+n) - I(x,y,t) \right|. \tag{1}$$

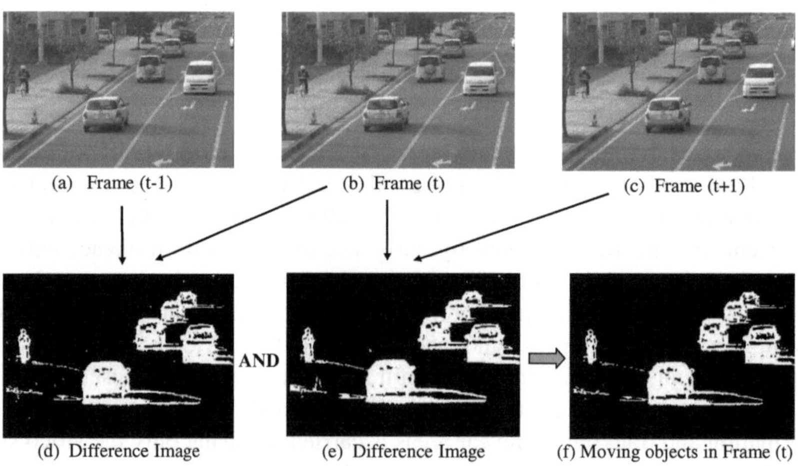

Fig. 1. Binary difference images

Two difference images are produced, one between Frame (t) and Frame(t-1), and the other between Frame (t) and Frame(t+1). The two difference images are binarized and shown in Fig.1 (d) and (e). Then, the two binarized differences are calculated with logical 'AND' operation and shown with white dots as in Fig.1 (f), where the moving objects in Frame (t) are extracted.

4 Formation of Car Areas

If two cars or more are mixed up in a single car area, it is very difficult to obtain accurate information about the cars, e.g. when counting the number of cars. A separation technique for measuring the number of cars running one after another is proposed, which is based on edge information on the cars [4]. However, we propose a technique for separating the overlapped area by using the chain code. Two typical patterns of the overlapped cars are exemplified in Fig.2.

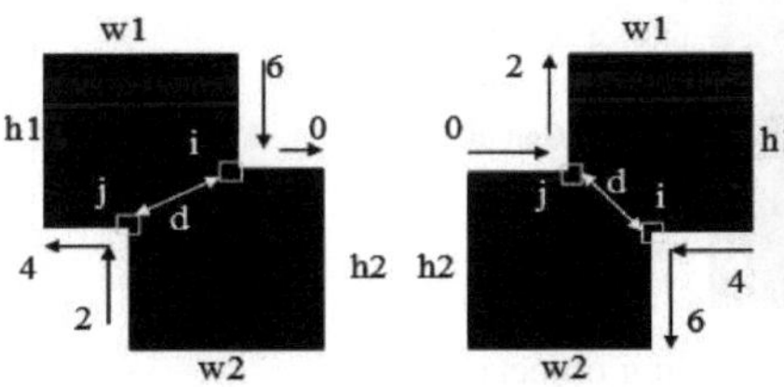

Fig. 2. Typical patterns for overlapped cars

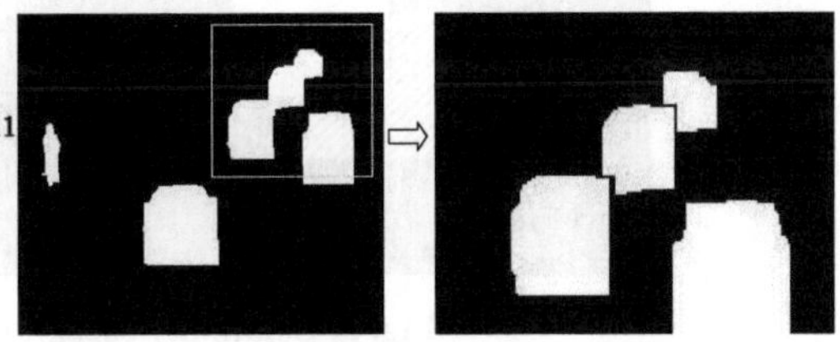

Fig. 3. Objects separation

By raster scanning, a point is found at first in a closed area. Then, take this point as a start point, check the chain code along contours in a clockwise direction. Using the chain code output, it is easy to look for an intersection or a corner at right angle, such as i and j, as shown in Fig.2, where the corner is found when there is code number difference of two (2) between two successive codes. If the distance "d" between i and j is smaller than at least either width or height of the car area, overlapped area can be separated with a line connecting i and j. For better representation, an appropriate boundary is created as shown in Fig.3.

5 Car Tracking

A window is provided on the road in PC display beforehand as a starting point for measuring the speed. Cars only passing the window position are measured. Fig.4(a) shows the window position (black) in the image. It shows that outgoing cars passing the window can be easily selected. The moving objects such as bicycles and incoming cars are discarded as shown in Fig.4(b).

(a) Window for one-way traffic (b) Outgoing cars

Fig. 4. Window and extracted cars

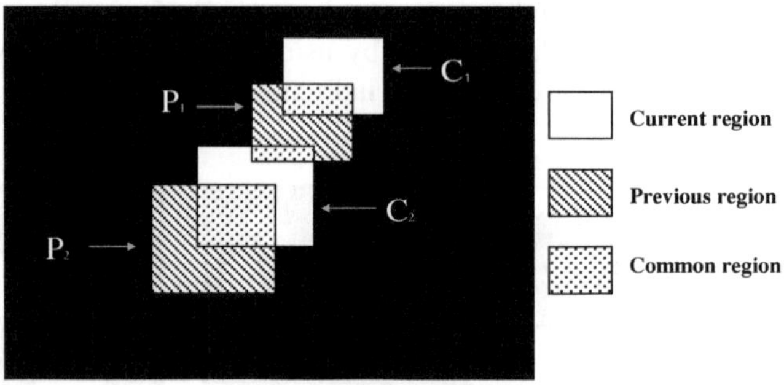

Fig. 5. Continuity check for moving objects

For tracking, a car area in the previous frame is designated by Pi (i =1, 2, ...) and that in the current frame by Cj (j=1, 2, ...) as shown in Fig.5, representing two cars running with a close distance. The current region named C2 has two common regions representing two different cars in the previous frame. But the area P2 is considered to be the same area, because there are a larger number of pixels in common regions with P2. Typically, S is calculated in Eq. 2, where S shows an overlapped area relative to the number of pixels in Pi or Cj. When S becomes maximum between Pi and Cj, these two areas are determined as in continuity i.e. the same object. The area identified to be the same is tracked successively [5].

$$S_{P \to C} = \frac{A(P_i \cap C_j)}{A(P_i)}, \quad S_{C \to P} = \frac{A(P_i \cap C_j)}{A(C_j)}. \tag{2}$$

6 Feature Point Extraction and Tracking

To detect speed of moving cars, running distance and the consumed time are measured with a reference placed in the extracted car area. However, measurement error usually occurs because the detected car area shape and/or size might be different from an actual size of the car, frame by frame. Extraction of accurate and stable Feature Point is very important to detect more accurate speed of the cars. Extraction of the Feature Point is described in many papers, such as [6][7]. The Harris operator is a technique for deriving the Feature Point based on the correlation of image signals, where the correlation output value grows for the Feature Point existing in the edges and corners [8].The Feature Point is extracted using the Harris operator in this paper.

The Feature Point is extracted only in the extracted car area. The Feature Point for each object sometime differs frame by frame in the actual dynamic scene. To detect the car speed with higher accuracy, the extracted Feature Point should be stable in the sense that it is found at the same position in the corresponding object even if the object is in motion. In this paper, we use the Block Matching Algorithm (BMA) to track the Feature Point in the car area. Block Matching Algorithms has been used in practice since 1981 [9].

Among several feature points calculated by the Harris Operator, one with the largest value is chosen as Feature Point. The Feature Point is extracted for each car in an initial frame in the measurement. After defining a block of 9*9 pixels, place the Feature Point at its center, Block Matching Algorithm is applied to find the position or displacement of the Feature Point in the next frame, where the displacement shows highest correlation. Thus, the new Feature Point is repeatedly found in successive frames to produce a trajectory representing a car.

As an example, Fig.6 shows trajectories of the Feature Point as tracking results for two different cars in the same video sequence. Fig.6(a) shows that the trajectory is almost a straight line. It suggests that first car is tracked very well. Fig.6(b) shows the result is not necessarily a straight line, which suggests the result is somewhat erroneous. Improvement is made in the next section.

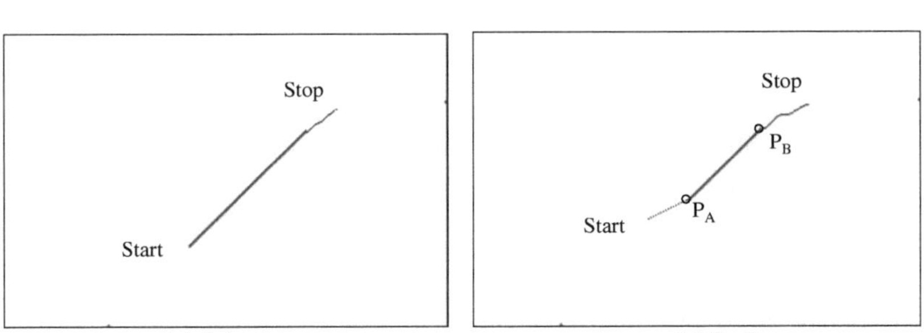

(a) 1st car (b) 2nd car

Fig. 6. Trajectories for Feature Points

7 Speed Detection Result

Here, let us use three methods for detecting speed of cars. In the first method, the speed is calculated from a specified distance and time consumed in passing through it. The number of pixels in the specified distance is known in advance, hence an equivalent distance per pixel can be calculated. The time is equal to the number of frames multiplied by 33 msec. In the second method, length of a reliable part of the trajectory is extracted for the calculation. Both methods above make use of average distance as the equivalent distance per pixel. In the third method, the equivalent distance per pixel is computed so as to closely approximate actual distance.

7.1 Method 1: Based on Whole Length of Trajectory

The Feature Point in the car area is used for detecting speed after the car has passed the window and until it reaches end point. The number of pixels between the start and the end point of the trajectory is counted from the coordinates of the trajectory. After multiplying the distance per pixel by the number of pixels in the trajectory, speed can be calculated, since both the distance and time are already known.

The speed measurement results for the two cars are as follows. For the 1st car, there are 74 frames consumed, with Start frame number 74 and End frame number 148. The distance is 32m. Therefore, speed is calculated to be 41.6 km/h. For the 2nd car, there are 72 frames consumed, with Start frame number 129 and End frame number 201. The distance is 25m. Therefore, speed is calculated to be 37.5 km/h.

7.2 Method 2: Based on Reliable Part of Trajectory

If the trajectory is not a straight line from the start to the end point, the result should be more or less in error. So, we only used the straight part of each trajectory to measure the speed of the two running cars, since the straightness is equal to high reliability.

The calculated result of the experiment is as follows. For the 1st car, the entire trajectory is determined as straight since fitting a straight line onto the trajectory produced an error within one pixel difference. Therefore, the result is the same as Method 1. For the 2nd car, there are 52 frames consumed, with start frame number 147 and End frame number 199. The distance of the straight part is 18.6m. Therefore, speed is calculated to be 38.6 km/h.

7.3 Method 3: Improvement on Method 2

By Method 2, the accuracy of calculated car speed has been improved. But these still is room for further improvement. When calculating distance from the length of the straightness, the average distance per pixel is used. However, Fig.7 shows the distance per pixel at a point near start point on the image is different from that at the far side, and the latter is larger.

So, it is possible to calculate more accurate distance per pixel at any point in the trajectory. The ratio of the width to the height of a car in the image is constant everywhere. The change in the distance per pixel along the trajectory is linear as seen in Fig.8.

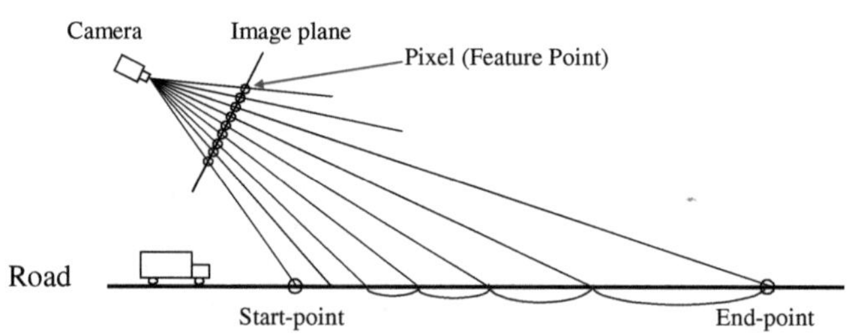

Fig. 7. Distance per pixel

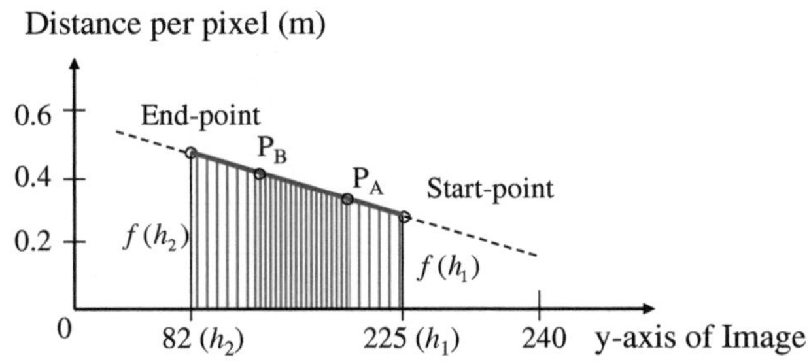

Fig. 8. Equivalent distance per pixel along a trajectory

The width of lane is 3.17m, corresponding to 142 pixels at the start-point while 64 pixels at the end-point. Therefore, we can figure out that one pixel at the start-point corresponds to 0.21m, while at the end-point to 0.46m.

As a consequence, a linear equation is obtained as follows,

$$f(h) = -0.00175\,h + 0.5983\ . \tag{3}$$

Distance D between any two points along the trajectory, for example, from PA to PB (the y-axis value of PA is 134, and PB is 84), can be computed by the following equation,

$$D = \int_{P_A}^{P_B} (-0.00175\,h + 0.5983)\,dh\,. \tag{4}$$

The computed results of the experiment are as follows. For the 2nd car, there are 52 frames consumed, with start frame number 147 and End frame number 199. The distance (D) of the straight part is 20m. Therefore, speed is calculated to be 41.5 km/h.

7.4 Error Analysis

Speed of the moving cars is measured for the same dynamic scene by these methods. The results obtained in the measurements are summarized in Table 1. For the 1st car, all the these methods show almost the same result, which is very accurate in comparison to the actual speed. For the 2nd car, a considerable detection error of -2.8km/h (6.9%) by Method 1 is reduced to -1.7km/h (4.2%) by applying Method 2. This is mainly thanks to the use of straight part i.e. reliable part along a trajectory.

Method 1 gives fairly good results, while Method 2 shows apparent improvement from Method 1 for the 2nd car. This result shows that the straightness of trajectories can produce higher accuracy [10].

Method 3 improves accuracy in the definition of distance per pixel corresponding to a position in a picture, while Method 2 makes use of average distance per pixel regardless of position. The same result is obtained among the three methods for the 1st car, while Method 2 and 3 show improvement over Method 1 for the 2nd car, although they produce similar error amount, -1.7km/h and +1.2km/h, respectively. Basically, Method 3 can measure the speed with higher accuracy than Method 2, even if the difference is small.

Table 1. Detected Car Speed (km/h)

		1st car	2nd car
actual speed		40.5	40.3
Method 1	detected speed	41.6	37.5
	error	1.1	-2.8
Method 2	detected speed	41.6	38.6
	error	1.1	-1.7
Method 3	detected speed	41.5	41.5
	error	1.0	+1.2

8 Conclusion

In this paper, we have proposed a speed measurement method by tracking the Feature Point of more than one car in video signals at the same time. The effectiveness of our tracking algorithm using the Feature Point for cars is shown in the experiment and tells us that speed measurement based on video signal processing is considerably usable in traffic control applications.

Extraction and tracking of Feature Point in a more accurate and stable manner are of a dominant issue. In addition, after improving this issue, our algorithm will find many other applications such as monitoring of the numbers of cars, distance between two cars, running direction of individual cars, etc.

References

1. Ozawa, S.: Image Processing for Intelligent Transport Systems. IEICE Trans. Inf. & Syst. E82-D(3), 629–636 (1999)
2. Cathey, F.W., Dailey, D.J.: A novel technique to dynamically measure vehicle speed using uncalibrated roadway cameras. In: Proceedings of IEEE Intelligent Vehicles Symposium, Las Vegas, Nev, USA, pp. 777–782 (June 2005)
3. Atkociunas, E., Blake, R., Juozapavicius, A., Kazimianec, M.: Image Processing in Road Traffic Analysis. Nonlinear analysis: Modelling and Control 10(4), 315–332 (2005)
4. Kuboyama, H., Ozawa, S.: Measurement of heavy traffic in a tunnel from image sequences. Systems and Computers in Japan 35(13), 52–62 (2004)
5. Gupte, S., Masoud, O., Martin, R.F.K., Papanikolopoulos, N.P.: Detection and Classification of Vehicles. IEEE Transactions on Intelligent Transportation Systems 3(1), 37–47 (2002)
6. Tommasini, T., Fusiello, A., Trucco, E., Roberto, V.: Making good features track better. In: Proc. IEEE Conf. on Computer Vision and Pattern Recognition, pp. 178–183 (1998)
7. Zitova, B., Flusser, J., Kautsky, J., Peters, G.: Feature point detection in multiframe images. Technical report, Czech Pattern Recognition Workshop, 104 (February 2000)

8. Harris, C., Stephens, M.: A combined corner and edge detector. In: Proc. 4th Alvey Vision Conf., Manchester, U.K., pp. 147–151 (August 1988)
9. Koga, T., Linuma, K., Hirano, A., Lijima, Y., Ishiguro, T.: Motion-compensated interframe coding for video conferencing. In: Proc. Nat. Telecommun. Conf., New Orleans (LA), pp. G5.3.1–G5.3.5 (1981)
10. Zhu, S., Koga, T.: Feature Point Tracking for Car Speed Measurement. In: IEEE Asia Pacific Conf. on Circuits and Systems (APCCAS 2006), Singapore, pp. 1144–1147 (2006)

Adaptive Edge-Detection Implementation for Video Processing Based on DM642

Yixuan Wu[1] and Zhiwei Tang[2]

[1] The Third Research Institute of Ministry of Public Security, Shanghai, China
wuyixuan876@163.com
[2] The Third Research Institute of Ministry of Public Security, Shanghai, China
99chaoyang@163.com

Abstract. This article introduces the implementation of an adaptive edge-detection filter on a DSP (dm642) using a combination of hardware and software components, which can detect textural features of digital image, makes it non-linear enhancement and suppresses the noise to a certain extent. As a benchmark, preliminary results are presented for this system and evaluated with respect to the different edge detector. And how to use the advantages of the DM642 processor to improve the efficiency of edge-detection is mainly discussed in this article. In order to improve the performance of edge-detection algorithm on DM642, a series of rapid image processing algorithms are optimized and proposed. The emphasis of the experiments is put on the feasibility of distributed high-performance processing from both hardware and software aspects, which may be easily applied to other larger scale or more hard real-time intelligent information processing. Experiments show that it can detect edges with high accuracy, more details and fine distinction.

Keywords: Edge-detection, video, dm642, adaptive.

1 Introduction

Edge-detection algorithm is one of the most important image processing in intelligent video surveillance. Edges are boundaries between different textures which also can be defined as discontinuities in image intensity from one pixel to another. The edges of a given image are always important characteristics that offer an indication for a higher frequency. Consequently, edge detection with one image is not only helpful to achieve image segmentation, data compression; but also applicable in well matching, such as image reconstruction. Now, there have been a number of solutions proposed to extract edges, such as Roberts operators, Sobel operators, Prewitt, Laplacian operators and Fractional differential which is non-integer order differential as well, attracts more attention compared with integer-order differential algorithms currently. Historically, image processing such as Edge-detection algorithm has been a relatively difficult function to implement optimally in hardware, leading many software designers to use digital signal processors in soft implementations. Unfortunately, because of the function's computationally intensive nature, such an approach typically requires powerful digital signal processors within the system to support the processing requirements.

Y. Zhang (Ed.): Future Communication, Computing, Control and Management, LNEE 141, pp. 141–147.
springerlink.com

2 Edge Detection Background

Edge detection is a fundamental tool used in most image processing applications to obtain information from the frames as a precursor step to feature extraction and object segmentation. This process detects outlines of an object and boundaries between objects and the background in the image. An edge-detection filter can also be used to improve the appearance of blurred or anti-aliased video streams. The basic edge-detection operator is a matrix-area gradient operation that determines the level of variance between different pixels. The edge-detection operator is calculated by forming a matrix centered on a pixel chosen as the center of the matrix area. If the value of this matrix area is above a given threshold, then the middle pixel is classified as an edge. Examples of gradient-based edge detectors are Roberts, Prewitt, and Sobel operators. All the gradient-based algorithms have kernel operators that calculate the strength of the slope in directions which are orthogonal to each other, commonly vertical and horizontal. Later, the contributions of the different components of the slopes are combined to give the total value of the edge strength.

3 Hardware Implementation of System

These operations typically require very high computation power. Standard definition NTSC video is digitized at 720x576 or full D1 resolution at 30 frames per second, which results in a 30MHz pixel rate. The TMS320DM642 device is the highest-performance fixed-point DSP generation in the C6000 DSP platform. With performance of up to 4800 million instructions per second (MIPS) at a clock rate of 600 MHz, the DM642 device offers cost-effective solutions to high-performance DSP programming challenges. The DM642 DSP possesses the operational flexibility of high-speed controllers and the numerical capability of array processors.

The system is designed to carry out the algorithm described in Fig.2 shows its structure which is made up of DSP, A/D, SDRAM, FLASH, D/A, internet interface, etc. The DSP device is DM642, which are the highest-performance fixed-point DSP. The DM642 device is based on the second-generation high-performance, advanced VLIW architecture developed by TI, making these DSPs an excellent choice for digital media applications. With performance of up to 5760 million instructions per second at a clock rate of 720 MHz, the DM642 device offers cost-effective solutions to high-performance DSP programming challenges.

The A/D device is TVP5146 with an enhanced video input processor, which consists of a two-channel analog preprocessing circuit including source selection, anti-aliasing filter and ADC, a digital multi-standard decoder, a brightness/contrast/saturation control circuit. It consists of TWO high speed, 10-bit, video A/D converters with complementary inputs, a standard TTL input interface and a high impedance, analog output current source. video must conform to BT656 protocol to satisfy A/D decoding. When the target is working, the A/D device will convert analog video to digital video data. Meanwhile, the blank and synchronization signal will be separated from video. The DSP device will store video data into SDRAM in terms of the timing of the blank and synchronization signal.

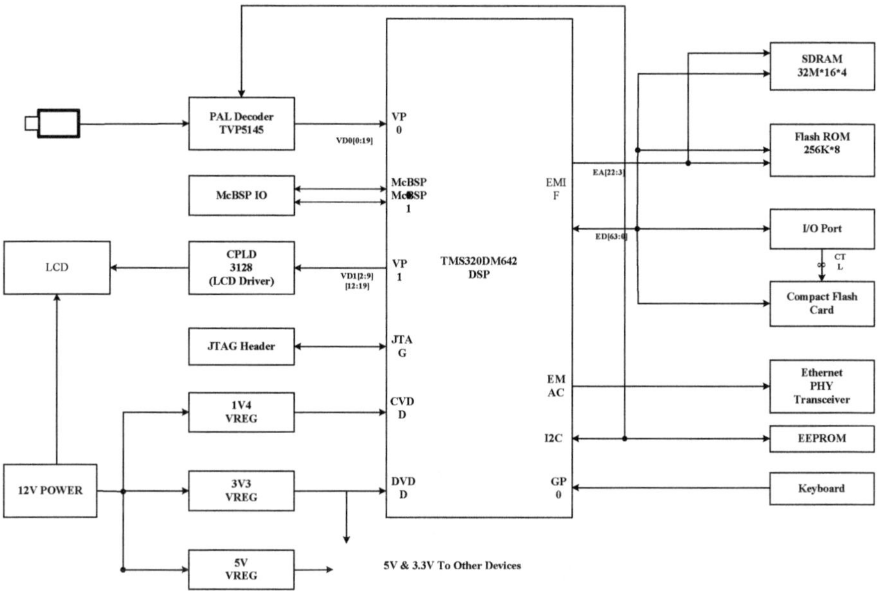

Fig. 1. Structure of the hardware system

4 The Optimization Methods for DSPs

The algorithm was designed and the code implementing it was written in consideration of its designated platform, hence a DSP-friendly algorithm was chosen and implemented. The optimization methods in the flow and the memory usage of the programs should be modified based on the platform-dependent considerations. The main accelerator of the embedded systems is the direct memory access (DMA) unit which can share the loading of DSPs and can perform the parallel processing for moving and tracking the data. The DM642 platforms have 64 and 6 independent channels DMA, respectively, and the bandwidth of each one is up to 2.4 GB per second from the TI technical reports. The enhanced DMA (EDMA) supported by TI provides the efficient and convenient usage, and its definitions and the operations. Both the addresses of the source and the destination have to be set, and EDMA will move the data directly without the help of DSP. The memory reduction methods by using EDMA efficiently and optimizing the program for DASS are described below.

The DSP BIOS is used to allocate the required resources, and the system performance can be greatly increased by accessing the high-speed small-size internal memory (IM) and the low-speed large-size external memory (EM) well. The bottleneck of the real-time image processing of DASS is the bandwidth of accessing the memory. Fig. 4 shows the three-level hierarchy architecture of DASS, and each block works at different operating frequencies. The fast cache memory in level 1 is separated into two parts, one is for the level 1 program, L1P, and the other is for

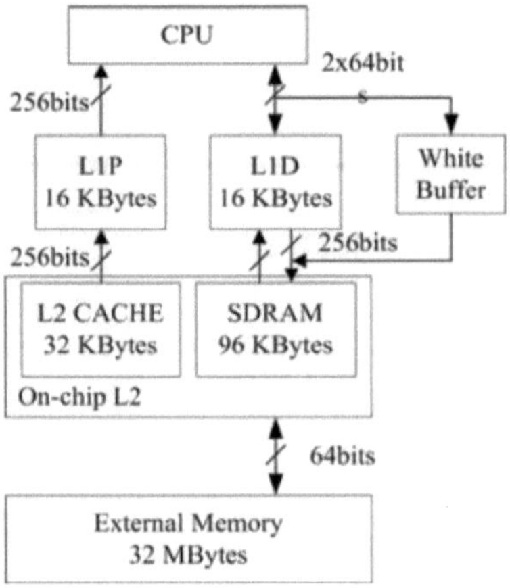

Fig. 2. Three-level hierarchy architecture

caching the image data, L1D. Since L1D is too small to store the original image for processing, efficient data fetching algorithm is adopted in DASS to fully utilize the IM to increase the speed. L1P and L1D are much faster than EM, and through the EDMA operations, the waiting time of DSP can be reduced.

Therefore, a suitable memory allocation which can enormously increase the performance for DASS is proposed. The memory map of DASS is shown in Fig. 5, and the limited size of IM, 256 KB, can be used. However, the system also needs IM to run applications and store some statistic data, such as system buffers, vectors, heaps and stacks.

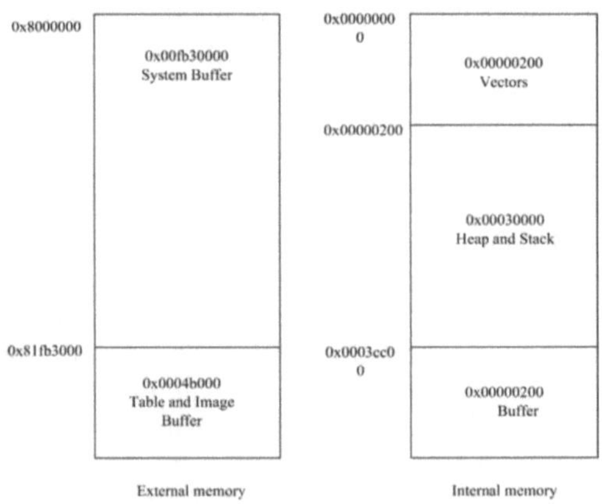

Fig. 3. Memory architecture

Therefore, the algorithms should be modified to reduce its IM usage, so only two lines of the captured image are required after testing and scheduling. In DASS, the width of a frame is 352 pixels, so 704 bytes are necessary for cache usage. Furthermore, ping-pong buffer technology is adopted to increase the overall speed, and these buffers are allocatedin a block of IM. Moreover, EM can be pre-defined by using DSP/BIOS at the beginning of the program, and can reduce the system overhead. Based on the very long instruction words (VLIW) properties provided by TI, some redundant data memory could be reduced. In the overall algorithm flow, the upper and lower bound of the variables can be estimated, and appropriate declarations of the data type can be defined. Using the skill has two advantages, one is to decrease the memory usage, and the other is to make the instructions operate fully based on VLIW characteristics to increase the performance.

5 Software Implementation of System

Traditional development flows in the DSP industry have involved validating a C model for correctness on a host PC or Unix workstation and then painstakingly porting that C code to hand coded DSP assembly language. This is both time consuming and error prone. This process tends to encounter difficulties that can arise from maintaining the code over several projects. The recommended code development flow involves utilizing the DM642 code generation tools to aid in optimization rather than forcing the programmer to code by hand in assembly. These advantages allow the compiler to do all the laborious work of instruction selection, parallelizing, pipelining, and register allocation. This allows the programmer the ability to focus on getting the product to market quickly. These features simplify the maintenance of the code, as everything resides in a C framework that is simple to maintain, support, and upgrade.

The recommended code development flow for the DM642 involves the phases described below. The tutorial section of the Programmer Guide focuses on phases 1-3. These phases will instruct the programmer when to go to the tuning stage of phase 3. What is learned is the importance of giving the compiler enough information to fully maximize its potential. An added advantage is that this compiler provides direct feedback on the entire programmer high MIPS areas (loops). Based on this feedback, there are some very simple steps the programmer can take to pass complete and better information to the compiler. This allows for a quicker start in maximizing compiler performance. You can achieve the best performance from your DM642 code if you follow this code development flow when you are writing and debugging your code:

Because most of the millions of instructions per second (MIPS) in DSP applications occur in tight loops, it is important for the DM642 code generation tools to make maximal use of all the hardware resources in important loops. Fortunately, loops inherently have more parallelism than non-looping code because there are multiple iterations of the same code executing with limited dependencies between each iteration. Through a technique called software pipelining, the DM642 code generation tools use the multiple resources of the VelociTI architecture efficiently and obtain very high performance. This section shows the code development flow recommended to achieve the highest performance on loops and provides a feedback list that can be used to optimize loops with references to more detailed documentation.

DSP-oriented coding style was used in order to help the compiler in producing a highly optimized code. Branches were avoided, if possible, or carefully written. If any apriori knowledge on a branch is available, it was used for writing the branch

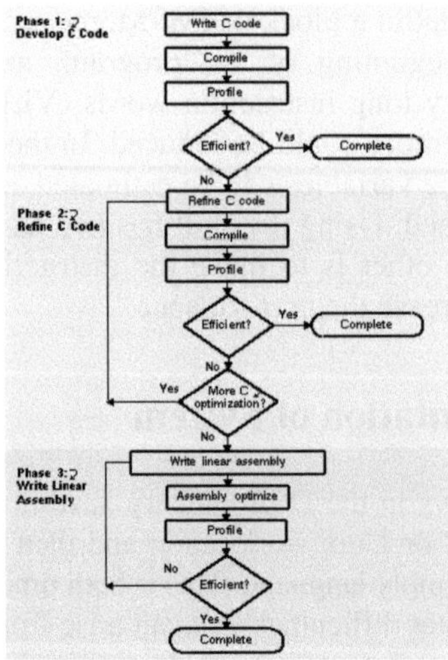

Fig. 4. Code development flow

efficiently. Divisions were performed by shift operations with numbers which are a power of 2. Frequently used function values, such as Cos and Sin, were pre-calculated and saved in tables of constants. The code was organized and written for good data locality in order to allow good cache utilization. Hence, different code snippets that use the same memory areas were moved together as long as the algorithm consistency remained intact. Small size data types were chosen so that memory usage is minimized and better parallelism is achieved. Additional parallelism was achieved by loop unrolling. #pragma directives were used in order to allow better compiler optimizations. Video input/output raises many performance issues in transferring the video data in real-time. In order to deal with these performance issues, the EDMA (Enhanced Direct Memory Access) mechanism of the DM642 chip was used, hence allowing faster I/O through video peripherals without any CPU computation directly involved.

6 Summary

Field testing the system with the DM642 board and a standard video camera from the experimental results, we obtained the improved effect of differential operates, which is thin edge and clear and obvious besides allowing edge enhancement, demonstrated in Figure 6.

This paper discussed the implementation of the edge-detection algorithm on DM642. We presented preliminary results, which show the capability of supporting adaptive edge detection for real-time image processing. Further improvement in the edge-detection research area has resulted in a broad range of evaluation techniques, which include using automatic scale selection, neural networks and probabilistic models. While these advanced algorithms provide better results, a common challenge among these techniques is the increasing computational cost as the complexity of the algorithm increases. DSPs are good alternatives, which can be used to reduce these computationally-intensive.

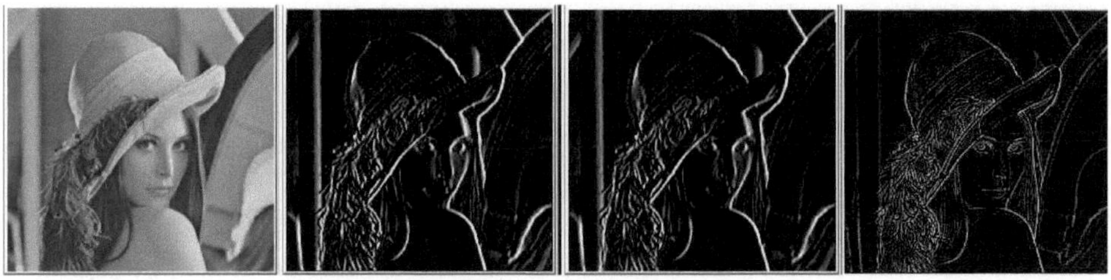

Fig. 5. Experimental results example

Acknowledgment. This paper is supported and financed by the Key Research Project of Ministry of Public Security (Grant No. 2009ZDYGGSS025).

References

1. Stiller, C., Konrad, J.: Estimating Motion in Image Sequences. IEEE Signal Processing 16(7), 70–91 (1999)
2. Pu, Y., Wang, W.: The fractional differential mask of digital image and its numerical implementation algorithm. Acta Autom Sin. 33(11), 1128–1135 (2007) (in Chinese)
3. Koch, C., Ellis, T.J., Georgiadis, A.: Real-time Occupant classification in High Dynamic Range Environments. IEEE Intelligent Vehicle Symposium 18(2), 284–291 (2002)
4. Owell, J., Remagino, P., Jones, G.A.: From Connected Components to Object Sequences. In: Proc. 1st. IEEE International Workshop on Performance Evaluation of Moving and Tracking and Surveillance, Grenoble, France, March 31, pp. 72–79 (2000)
5. Gonzalez, R.C., Woods, R.E.: Image segmentation, Digital Image Processing, 2nd edn., pp. 578–579. Prentice Hall, Inc., New Jersey (2002)
6. Nche, C.F., Parish, D.J., Phillips, I.W., Powell, W.H.: A New Architecture for Surveillance Video Networks. International Journal of Communication Systems 9, 133–142 (1996)
7. Wu, B.F., Chen, Y.H., Peng, H.Y., Chen, C.J.: A real-time vision-based safety assist system. In: Proceedings of IEEE International Conference on Systems, Man and Cybernetics, pp. 2994–2999 (2008)
8. Senior, S., Pankanti, A., Hampapur, L., Brown, Y.-L., Tian, A.: Ekin, Blinkering surveillance: Enabling Video Privacy Through Computer Vision, IBM Technical Report, Vol: RC22886 (2003)

Dynamic Modeling, Simulation and Experiment of the Delta Robot

Wang Pan Feng, Zhang Li Min, and Zhao Xue Man

School of Mechanical Engineering, Tianjin University, 92#, Weijin Road, Nankai District, Tianjin, China
panfengwang@tju.edu.cn, tju.zlmwan@yahoo.com.cn, zhaoxueman@gmail.com

Abstract. The analysis of dynamic characterization combined with dynamic modeling, simulation and experiment of the Delta robot is presented. The substructure displacement method, Ansys software and LMS dynamic testing equipment are respectively used for dynamic model, simulation and experiment. Meanwhile, the dynamic characterization data has been obtained in typical configurations. The results show that it is consistent in theoretical analysis and simulation and experiment test, so the validity of the analysis approach is verified.

Keywords: Dynamic modeling, simulation, delta robot.

1 Introduction

Delta robot with low movement inertia and high accelerations have been widely applied in packaging, electronics and many other light industries for handling and assembly. To achieve high acceleration lightweight materials such as carbon fiber has been adopted to make the active proximal link and distal link of the Delta robot, which will lead to the structural flexibility causing unwanted structural vibration during operation and reduce the dynamic accuracy.

The analysis of dynamic characterization is important prerequisite and basis for improving the dynamic accuracy of the robot. The analysis approaches includes theoretical analysis, Finite element simulation and experiment test. The separated analysis approach above-mentioned has been extensively studied.

The flexible model is the basis for theoretical analysis. Various approaches included the flexible multi-body dynamics(FMD)[1-2]、the linear elasto- dynamics(KED)[3-4]、substructure method[5] have been widely used. FMD can obtain accurate model, but the modeling and solving is very complex and mostly be used for geometric nonlinearity. On the contrary, linear elasto-dynamics approach ignored the coupling between rigid motion and flexible motion, which leads to a simplified linear dynamic model, so it has often been used in the design and dynamic analysis of the mechanism with flexible links. The Finite element simulation is usually carried out with Abaqus, Ansys or other software. Long[6]、Portman[7] created the Ansys model with typical configuration of the Gough-Stewart manipulator. Huang[8] considered the flexible of joint and base, and created the Ansys model of parallel robot with parallelogram chain. Moreover, the

Y. Zhang (Ed.): Future Communication, Computing, Control and Management, LNEE 141, pp. 149–156.
springerlink.com

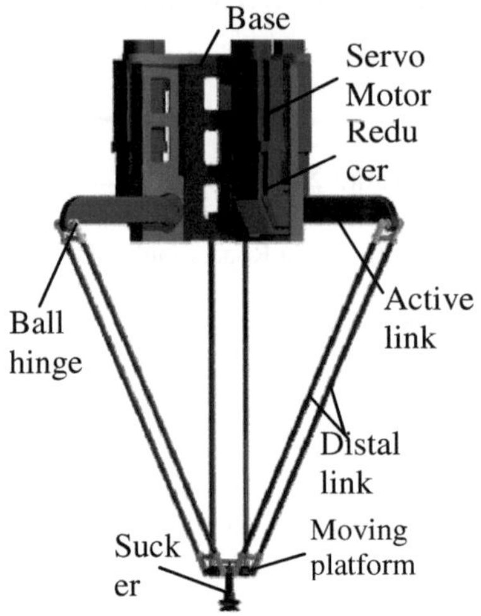

Fig. 1. 3D model of the Delta robot

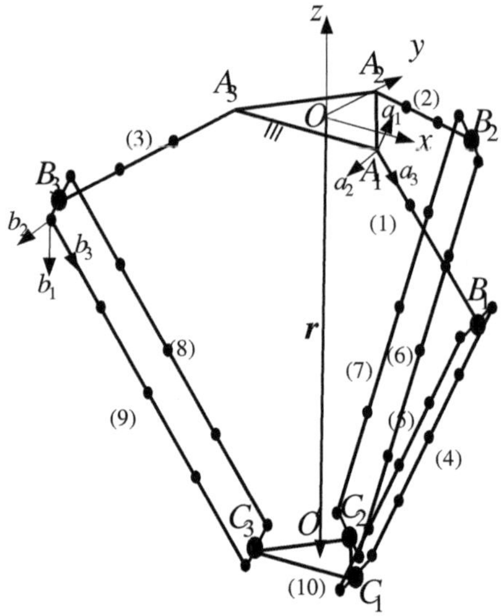

Fig. 2. Schematic diagram of the Delta robot

experiment test needs to implement for verifying the effectiveness of the theoretical analysis and finite element analysis. Tokhi M. O.[9] completed the comparative analysis between theoretical and experiment for the single flexible manipulator, Wang X. Y.[10] presented multiple experiment modal identification approach for testing the 3-DOF planner linkage in different configurations. However, it is noted that the approach integrated the three above-mentioned analysis is rare for referring.

In view of this, this paper present an approach integrated the theoretical, simulation and experiment analysis for the Delta robot. First, the dynamic model is established by substructure displacement method, and then the finite element model is provided by Ansys software. At last, the experiment test of the prototype and test data are presented.

2 Theoretical Analysis

Fig. 1 shows a 3D solid model of the Delta robot, which is composed of a base, a movable platform and three identical limbs. The moving platform and active proximal links are made of aluminum alloy and the distal links are made of carbon fiber for light-weight design.

By means of the substructure approach, the system can be divided into 10 components, as shown in Fig. 2, (1), (2), (3) are the components of active proximal links, (4), (5), (6), (7),(8), (9) are the components of distal links and (10) is component of the moving platform. In the formulation of the dynamic model, the following assumptions are made: moving platform and the hinge are treated as rigid body; elastic deformation satisfies the assumptions of small displacement, real motion is regarded as the superposition of rigid body motion and elastic displacement, and rigid body motion is known.

2.1 Modeling of the Proximal Link

As shown in Fig. 2, the active proximal link is regarded as a homogeneous cantilever beam. The mass of the active proximal link is equivalent to four lumped masses on two endpoints and two nodes, the joint connected between the active proximal link and distal link is equivalent to the lumped mass added at the end of the active proximal link. We omit the subscript for the time being. The vibration equation along the axis a_1、a_2 and a_3 in the frame $A_i - a_1a_2a_3$ can be expressed as follows

$$\boldsymbol{M}_{ak}\ddot{\boldsymbol{x}}_{ak} + \boldsymbol{K}_{ak}\boldsymbol{x}_{ak} = \boldsymbol{Q}_{ak}, \quad k = 1,2,3 \tag{1}$$

where $\boldsymbol{x}_{ak}$, $\boldsymbol{M}_{ak}$, $\boldsymbol{K}_{ak}$, $\boldsymbol{Q}_{ak}$ represents the elastic displacement matrices, mass matrices, stiffness matrices and inertia force and interface force matrices along $a_k\,(k=1,2,3)$ direction.By use of the analysis of structural mechanics, the dynamic model along $a_k\,(k=1,2,3)$ direction of the active proximal link is given

$$\boldsymbol{M}_a\ddot{\boldsymbol{x}}_a + \boldsymbol{K}_a\boldsymbol{x}_a = \boldsymbol{Q}_a \tag{2}$$

where
$\boldsymbol{M}_a = \boldsymbol{T}_a^{\mathrm{T}} diag\{\boldsymbol{M}_{a1}, \boldsymbol{M}_{a2}, \boldsymbol{M}_{a3}\}\boldsymbol{T}_a$, $\boldsymbol{K}_a = \boldsymbol{T}_a^{\mathrm{T}} diag\{\boldsymbol{K}_{a1}, \boldsymbol{K}_{a2}, \boldsymbol{K}_{a3}\}\boldsymbol{T}_a$, $\boldsymbol{Q}_a = \boldsymbol{T}_a^{\mathrm{T}} diag\{\boldsymbol{Q}_{a1}, \boldsymbol{Q}_{a2}, \boldsymbol{Q}_{a3}\}\boldsymbol{T}_a$ includes the inertia force of the active proximal link and interface force, $\boldsymbol{T}_a$ is rotation matrix.

2.2 Modeling of the Distal Link

As shown in Fig. 2, modeling using similar procedures with the active proximal link, the dynamic model along $b_k\,(k=1,2,3)$ direction of the active proximal link is given

$$\boldsymbol{M}_b\ddot{\boldsymbol{x}}_b + \boldsymbol{K}_b\boldsymbol{x}_b = \boldsymbol{Q}_b \tag{3}$$

where $\boldsymbol{M}_b$, $\boldsymbol{K}_b$, $\boldsymbol{Q}_b$ is the whole mass, stiffness and external force matrix of the distal link.

2.3 Modeling of the Moving Platform

Based on the assumptions mentioned above, the moving platform is considered as a rigid body, which bears the elastic displacement due to the influence of the flexible links. The modeling of the moving platform can be derived using the Alembert principle and ignoring the high-order items.

$$\begin{bmatrix} m_p \boldsymbol{E}_3 & 0 \\ 0 & \boldsymbol{I}_p \end{bmatrix} \ddot{\boldsymbol{x}}_p = \boldsymbol{Q}_p \tag{4}$$

where $\boldsymbol{x}_p$, m_p , $\boldsymbol{I}_p$, represents the generalized coordinates of 6-dimensional vibration, mass and inertia matrix of the moving platform, $\boldsymbol{E}_3$ a unit matrix of order 3, $\boldsymbol{Q}_p$ represents the inertia force and external force.

2.4 Modeling of the System

The modeling of the system is constructed using the interface constraints of the sub-structure, based on the dynamic model of each sub-structure. The deformation compatibility condition between point C (represented as $\boldsymbol{x}_{cij5}$) of the jth distal linkof the ith chain, and the connection point (represented as $\boldsymbol{x}_p$) of the moving platform is represented as follows

$$\boldsymbol{T}_{bi} \boldsymbol{x}_{cij5} = \begin{bmatrix} \boldsymbol{E}_3 & -\tilde{\boldsymbol{a}}_{ij} \end{bmatrix} \boldsymbol{x}_p \tag{5}$$

where $\boldsymbol{E}_3$ is a unit matrix of order 3, $\tilde{\boldsymbol{a}}_{ij}$ represents the vector from the center of the platform to the center of the spherical joint, $\boldsymbol{T}_{bi}$ represents the rotation matrix from the frame $B_i - b_1 b_2 b_3$ to the based frame $O - xyz$. The deformation compatibility condition between point B (represented as $\boldsymbol{x}_{cij1}$) of the distal link j of the chain i and the connection point (represented as $\boldsymbol{x}_{zi3}$) of the active proximal link is represented as follows

$$\boldsymbol{T}_{bi} \boldsymbol{x}_{cij1} = \boldsymbol{T}_{ai} \boldsymbol{x}_{zi3} \tag{6}$$

where $\boldsymbol{T}_{ai}$ represents the rotation matrix from the frame $A_i - a_1 a_2 a_3$ to the based frame $O - xyz$. From Eq. (5) and (6), the relationship between the coordinates $\boldsymbol{u}$ of substructure and the coordinates $\boldsymbol{U}$ of the system can be represented

$$\boldsymbol{u} = \boldsymbol{B}\boldsymbol{U} \tag{7}$$

Substituting the Eq.(7) and imposing the boundary constraints, all substructure dynamic model equations can be assembled into the system dynamic model equation

$$\boldsymbol{M}\ddot{\boldsymbol{U}} + \boldsymbol{K}\boldsymbol{U} = \boldsymbol{Q} \tag{8}$$

where $\boldsymbol{M} = \boldsymbol{B}^{\mathrm{T}} \boldsymbol{M}' \boldsymbol{B}$ is the mass matrix of the system, $\boldsymbol{K} = \boldsymbol{B}^{\mathrm{T}} \boldsymbol{K}' \boldsymbol{B}$ is the stiffness matrix of the system, $\boldsymbol{Q} = \boldsymbol{B}^{\mathrm{T}} \boldsymbol{Q}' \boldsymbol{B}$ is the generalized inertia force and external force.

3 Finite Element Simulations

Ansys software has been used for creating the finite element model in this paper. Figure 3 is the simulation model of Solidworks and Ansys, in which the appropriate simplification about chamfer, fillet and hinge is made. The contact types are defined in accordance with kinematic pair of robot, the tetrahedral and hexahedral solid element is employed, the fix support is added in driving terminal.

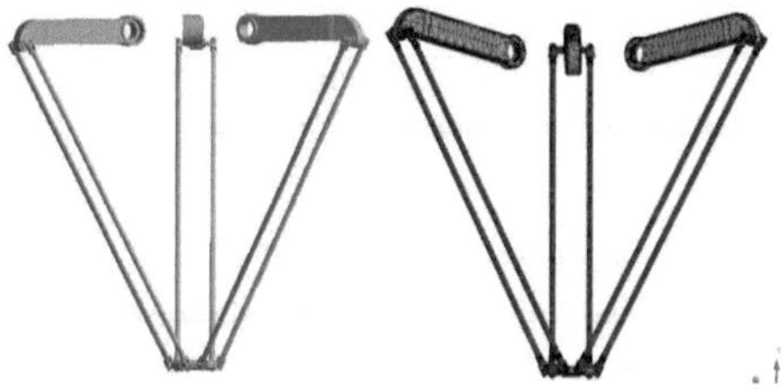

Fig. 3. Simulation model of Solidworks and Ansys

Fig. 4. Testing of the Delta robot

4 Experiment Test

LMS dynamic test equipment is employed in experiment test. In addition, PCB086C02 force hammer, B&K acceleration transducer, PC and so on. Fig. 4 show the test prototype and the test photo, which connected to the earth through the solid frame, similar to that of the earth admittance to zero. According to structural characteristics of the system, the transient excitation of the hammer incentives is used.

5 Results

Without losing generality, the dimensional and inertia parameters of the Delta robot are shown in Table 1 and Table 2. D and h represent the diameter and height of the workspace; H represents the vertical distance between the motor axis and the up-plane of the workspace; r_a and r_b represent the radius of the base and the moving platform; l_1 , H_a , W_a and t_a represent the length, height, width and thickness of the active proximal link, l_2 , D_d and t_d represent the length, outside diameter and thickness of the distal link. The material of the active proximal link is aluminum alloy with density ρ_a , moment of inertia I_{a_1} , I_{a_2} and elastic modulus E_a . The material of the distal link is carbon-fiber with density ρ_d , moment of inertia

I_c and elastic modulus E_d . The material of rigid elements, which include hook joint and moving platform, is structure steel and aluminum alloy, respectively. M_m , I_{mx} , I_{my} and I_{mz} are the mass and moment of inertia of the moving platform, M_e is the mass of hook joint.

Table 1. Dimensional Parameters [mm]

l_1	l_2	H	r_a	r_b	D	h
350	950	690	200	50	1100	250

Table 2. The Inertia Parameters

ρ_a (kg/m^3)	E_a (Pa)	I_{a_1} (kg·m^2)	I_{a_2} (kg·m^2)	ρ_d (kg/m^3)	I_c (kg·m^2)
2770	7.1×10^{10}	3.38×10^{-7}	1.79×10^{-7}	1700	1.167×10^{-9}
E_d (Pa)	M_m (kg)	I_{mx} (kg·m^2)	I_{my} (kg·m^2)	I_{mz} (kg·m^2)	M_e (kg)
1.16×10^{11}	0.35	1.312×10^{-3}	1.387×10^{-3}	2.014×10^{-3}	0.1

Table 3. Lower Frequency Comparison [Hz]

		First order frequency	Second order frequency	Third order frequency		First order frequency	Second order frequency	Third order frequency
	Theoretical result	62.61	62.61	63.71		53.20	61.31	63.70
A	Analysis result	63.06	63.09	64.48	*B*	50.29	62.59	64.49
	Experiment result	38.82	40.46	73.36		22.37	37.88	58.55

The above results show the effectiveness of the dynamic model mentioned before and the consistency of theoretical and simulation. One to three-order natural frequencies are almost the same and the maximum error is less than 3%. Due to space limitations, Fig. 5 (a) and (b) show that one order shape comparison between theoretical and simulation by Ansys at point A $(0,0,0)$ and B $(0,500\text{ mm},-250\text{ mm})$. The figures show the shape computed by theoretical and software Ansys is very consistent. Figure 6 shows that the FRF in x, y and z direction at point A and B.

However, it is noticed that the value of theoretical analysis and experiment values vary widely, so we should re-examine the real connection between the active proximal link and the output shaft. That is the servo motor connects the reducer, then reducer gearbox output rotor connects with the active proximal link. As the effect of the stiffness of the internal structure of the motor rotor and reducer, it will lead big error between theoretical value and experiment assuming the active proximal link fixed connected with the motor. Therefore, we added the stiffness $k_e = 2.5\times10^4$ N/m to the drive terminal of the active proximal link. As showed in Table 4, we found that the theoretical and experiment have similar results, thus we can determined the stiffness of the reducer has a great influence for dynamic characteristic of the system.

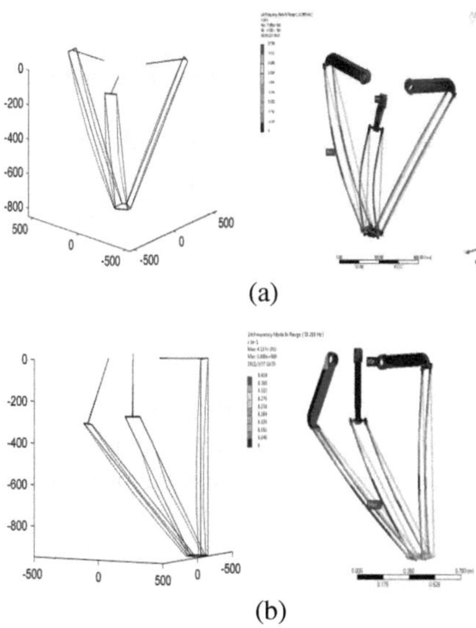

Fig. 5. One order shape comparison

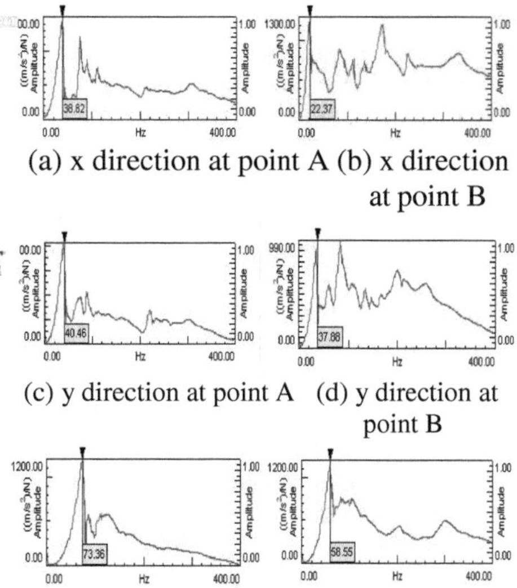

(a) x direction at point A (b) x direction at point B

(c) y direction at point A (d) y direction at point B

(e) z direction at point A (f) z direction at point B

Fig. 6. FRF in different direction

Table 4. Lower Frequency Comparison

	First frequency /Hz	Second frequency /Hz	Third frequency /Hz
Theoretical value of point *A*	39.68	39.68	72.16
Experimential value of point *A*	38.82	40.46	73.36
Theoretical value of point *B*	30.57	37.48	60.84
Experimential value of point *B*	22.37	37.88	58.55

6 Conclusions

This paper presents an approach integrated the theoretical analysis, simulation and experiment aimed to investigate the dynamic characterization of the Delta robot. The conclusions are drawn as follows:

(1) A linear dynamic model of the Delta robot with the flexible links is proposed, and the experiment test platform is built. The effectiveness of the theoretical model is verified by the simulation and experiment results.

(2) The results of dynamic characteristics show that the stiffness of the reducer has a great influence for dynamic characteristic of the system.

References

1. Shabana, A.A.: Flexible multibody dynamics: review of past and recent developments. Journal of Multibody System Dynamics 1, 189–222 (1997)
2. Wu, H.T., Mani, N.K., Ashrafiuon, H.: Selection of modal basis for flexible bodies of mechanical systems. Mechanism and Machine Theory 30, 471–489 (1995)
3. Liao, C.Y., Sung, C.K.: An elastodynamic analysis and control of flexible linkages using piezoceramic sensors and actuators. ASME Journal of Mechanical Design 115, 658–665 (1993)
4. Fattah, A.J., Misra, A.K.: Dynamics of a 3-DOF spatial parallel manipulator with flexible links. In: Proceeding of the 1995 IEEE International Conference on Robotics and Automation, Nagoya, Japan, pp. 627–632 (1995)
5. Wang, X., Mills, J.K.: Dynamic modeling of a flexible-link planar parallel platform using a substructuring approach. Mechanism and Machine Theory 41, 671–687 (2006)
6. Portman, V.T., Sandler, B.-Z., Zahavi, E.: Rigid 6×6 parallel platform for precision 3-D micromanipulation theory and design application. IEEE Transactions on Robotics and Automation 16, 629–643 (2000)
7. Long, C.S., Snyman, J.A., Groenwold, A.A.: Optimal structural design of a planar parallel platform for machining. Applied Mathematical Modelling 27, 581–609 (2003)
8. Huang, T., Mei, J.P., Zhao, X.Y., et al.: Stiffness estimation of a tripod-based parallel kinematic machine. In: Proceedings of the IEEE International Conference on Robotics & Automation, vol. 4, pp. 3280–3285 (2001)
9. Tokhi, M.O., Mohamed, Z., Amin, S.H.M., et al.: Dynamic characterization of a flexible manipulator system, theory and experiments. Proceedings of TENCON 3, 167–172 (2000)
10. Wang, X.Y., Mills, J.K.: Experimental modal identification of configuration-dependent vibration using smart material transducers with applicaiton to a planar parallel robot. In: Proceeding of IEEE International Conference on Robotics and Biomimetics, pp. 234–239 (2005)

From CAD to Search, Rebuild the Industrial Chain of Architecture by IT

Jing Zhang[1], Jia Lang[1], Mingjiu Yu[1,*], and Lei Qiu[2]

[1] Architecture and Environment Institute of Sichuan University, Chengdu, China
zj-11@21cn.com, 715071735@qq.com,
348630521@qq.com
[2] Business Institute of Guangzhou University, Guangzhou, China
tears@pai8.com

Abstract. After the introduction of CAD technology in the 80s of the 20th century, the Information Technology (IT) is influencing the upper stream and the down stream of the architectural industry at an unprecedented speed, in which the architectural design is regarded as the core section. This paper will retrospect and summarize the process of the IT's restructuring the architectural design and the whole industrial chain, including the former two stages which have happened and the third stage which is emerging. Furthermore, this paper will give some effective suggestions on the development direction of the third stage, discuss how it is being put into practice, and present the conceptions of the object database of all industries, which takes the algorithm for the Object Rank as its essence and its application mode of searching in order to solve the issues such as evaluation and sequence of searching involved with all the stakeholders, enterprises and products in the domain of architecture and facilitate users to make the best choice to meet their needs quickly and effectively when facing mass data. In addition, this methodology can also be applied for other similar industries.

Keywords: Architecture, IT, Search, Objects Rank.

1 Introduction

The introduction of IT actually has restructured the architectural design and its relevant industrial chain thoroughly. This process has been persisting for more than 20 years, and will continue to develop at a higher acceleration. The whole process can be divided into three stages of theme, e.g. "The tools", "The communication" and the "Selection". Attention should be drawn into the fact that the theme content of every former stage does not disappear at the next stage, but continue to develop with a less outstanding position.

2 The First Stage: Tools

In the 80s of the 20th century, the milestone for the introduction of IT to the architectural industry can be defined as the widespread use of PCs, which made the application of the

* Corresponding author.

Y. Zhang (Ed.): Future Communication, Computing, Control and Management, LNEE 141, pp. 157–163.
springerlink.com

CAD (Computer Aided Design) technology possible. However, the slogan of the day was overly simplistic: "Throw off the drawing board!". It is evident that the so called "Computer Aided Design" was nothing but "Computer Aided Drawing". In terms of nature, the "walls" in the computerized graphics files still were merely made up of lines, same as these on the hand painted draft. And the CAD technology of the day could not help architects to actualize view stimulation, calculation and revision for the design plan.

The "3D"(Three Dimensions) is another widely accepted computer design concept, which was introduced into the architectural industry through the design software, 3DS of Autodesk, a tool for architectural impression drawing in the 90s of the 20th century. Thus, computers had been preliminarily endowed with the function of "Aided Design". Personally, as the first batch of explorers who tentatively attempted to use cartoon software as 3DS to carry out architectural presentation and cartoon, I experienced the whole evolution process of architectural design software, from 3DS 3.0, 4.0 and 3DS MAX to LightScape with Radiosity lighting technology. During this process, IT is a very powerful tool which can simulate the forms, material quality and luminous beam of the real world more and more vividly. And in the late 90s of the 20th century, whether it be the architectural impression drawing, cartoon or movie simulation production, all looked very genuine. As the down stream of the architectural design industry, the industry of real estate also has utilized the 3D technology to display its products extensively.

"VR"(Virtual Reality) was the next technology on which people put great expectation. Due to maturity of the real-time rendering technology and break-neck speed of development of hardware of the day, VR expression, similar to large-scale 3D game scenario, once became a popular trend. It is really a historical breakthrough that users could enter into the virtual architecture for a free walk, other than following the camera routes given by the cartoonists. With the explosive development of the Internet, VR was even applied on webpage display and the VRML (Virtual Reality Modeling Language) was the mainstream language for the application of the VR technology.

"BIM"(Building Information Modeling) became the new generation IT tool in the new millennium which integrated aforesaid three mainstream technologies essentially. In the operational environment of BIM, architectural virtual modeling is no longer made up of lines as well as that in CAD, nor simple shapes with material quality in the operational environment of 3D technology, but "genuine objects" with physical attributes. Thus, architectural virtual modeling is no longer the private goods of architects but a powerful tool which can be extensively applied on the entire process of the industrial chain: from plot survey, planning and design, materials producing, building construction...to real estate developing, advertisement marketing, and property management. Only one modeling could meet the needs of the entire process, which is the complete replication of architecture of real world in the virtual space.

Another IT tool, "GIS"(Geographic Information System) is influencing architectural industry gradually in recent years. The typical application is Google Earth which has greatly reshaped the domain of architecture since Google became the mammoth of the Internet. All architectures are no longer a cluster of data existing in a graphics library of a certain design company solely, but virtual objects erecting on a virtual globe with longitude and latitude in accordance with its real site within a true environment where it will be built. This software has an extreme goal that each architecture on this planet will have its own virtual substitute.

As a powerful tool utilized in the domain of architecture, IT still has a long way to go. The coming of other stages would never weaken, but strengthen its attribute of tool.

3 The Second Stage: Communication

In the era of CAD, floppy disk was ever the most effective tool for architects to share their architectural drawings. Besides this, face to face discussion as a traditional communication way may be another choice. Undoubtedly, face to face discussion, post mail and telefacsimile with low resolution were regarded as the mainstreams of the day. It seems, most people even had not perceived it necessary to improve this situation.

Driven by technologies, Internet as a great breakthrough in the domain of communication reshaped human's life and work. "E-mail" is the first extensively used tool of the Internet with which the user can conduct a long-distance communication through attaching files in CAD and 3D formats to it. "IM"(Instant Messaging) comes after "E-mail", ICQ is regarded as its most popular application, with which users can easily communicate with their counterparts without delay or waiting.

"News Group" is a repository usually within the Usenet system, for messages posted from many users in different locations. The term may be confusing to some, because it is usually a discussion group. Newsgroups are technically distinct from, but functionally similar to, discussion forums on the World Wide Web. Newsreader software is used to read newsgroups. Bulletin board system (BBS) is its advanced form which realized the large-size group communication. BBS is a computer system running software that allows users to connect and log in to the system using a terminal program. Once logged in, a user can perform functions such as uploading and downloading software and data, reading news and bulletins, and exchanging messages with other users, either through electronic mail or in public message boards. BBS's marriage with the Internet opened a new era for the global communication within architectural industry. It is worth mentioning that the ABBS based in China, with a history of over 13 years, is the biggest architectural forum around the world, on which more than 2m architects and relevant professionals are discussing about their topics. On the other side, another sort of BBS, home owner's forum (e.g. www.focus.cn) supplies a communication platform for the ultimate consumers of architectural products.

"SNS" (Social Networking Services) (e.g. Facebook and Twitter) became the focus of the Internet later, with which thousands applications have been developed. However most of these applications are merely games. Nowadays many developers (including ourselves) are planning to develop professional applications and integrate them with the "SNS".

Under the background of communication, all tools on the aforesaid first stage have a trend of networking, which facilitate more intensive networking cooperation. Architects' communication would not only be limited to languages, texts, drawings and cartoons, but can organize themselves as teams in which every team member can be engaged into a specific architectural design.

"LBS" (Location Based Service) became the new focus around one year ago. Every communicator can be located to a real location precisely on the earth by the "GPS"(Global Positioning System) that endows the communication itself a meaning of localization. Four Square is the most popular LBS style website which help to extend the communication from computer screen to cell phone and tablet in a real time and a real address. For the architectural industry based on real addresses, this is a great leap.

4 The Third Stage: Selection

Under the background of communication, every section of the industrial chain is becoming more transparent and being operated faster than ever. However, serious information asymmetry still exists between the up stream and down stream. Especially in the period of the Internet, we are experiencing a process from information scarcity to serious information overload (duplicated information and spam).

How to solve this issue? "Search Engine" is the first solution. With PR(Page Rank) function, where a website's relevance is determined by the number of pages, and the importance of those pages, that linked back to the original site, Google can sort and classify page information, with which people can select the information that they need from the vast information ocean to improve their work efficiency.

For architectural industry, the objects, from professionals (Architects, engineer, interior designer, cartoonist...), companies (developers, governments, building materials manufacturers, furniture producers, construction company) to products (building materials, furniture, commercial residential building and architecture) have individual web pages or websites existing online. Even some of them have their own web pages or websites, their actual rankings and statuses are not equal to the ranking of their web pages. In other words, a conclusion cannot be drawn that the more links the web page of a certain product has the higher the ranking of the web page is, in turn the higher the quality of the product is. On the other side, it is simplistic and hasty to merely use a key word to search these objects because you cannot get what you expect at all. Tentatively to say, traditional searching engines are out of date within a certain industry.

The team with which I work is trying to solve this cutting edge technological issue by establishing a specialized search engine integrating objective data with subjective grading, which can be applied on a specific industry or sector. The methods of specific practices are set out below.

1. Duplicating all objects of the industrial chain to the virtual world
All companies, products and professionals of the up and down stream of the architectural industrial Chain would be categorized into an industrial database, and real geographic location of every object has been indicated on the Google Earth. Simultaneously the BIM models of these products are being built up through property owners and designers. In addition, reality sense of users can be strengthened by applying following technologies: VRML and flash cartoons which are equipped with 360° pan-shot and mouse-followed visual angles.

2. Marking every object with a score

There are two categories of scores:

One category is objective scores similar to Page Rank, which is defined in terms of the physical behavior, quality, qualification acquired in real world, reputation online, etc. from the objects. These data could be obtained from the Internet by program. Additionally objects also can submit their information of themselves.

The other category is generated from the subjective score of industry insiders and users. The quality of each product is subdivided into several sub-items which are evaluated by industry insiders and users. Weight of each sub-item can be generated by their voting as well.

The objects also can be sorted to two categories:

One group is the object that can be evaluated directly such as commercial residential building, fitments, furniture, home appliances, building materials, architectures which are not commercial residential building and architectures.

The other group is the object that is only evaluated indirectly by the scores of the objects evaluated directly, such as developers, the design company, the material company, product's brands, etc.

3. How to make subjective scoring unbiased?

The fairness of the subjective scoring would influence the validity of the final ranking of the object directly. Thus, only the middle 60% of the entire scoring are regarded as valid under algorithm of selective preference. Data deviated too far away would not be admissible. Furthermore, an overwhelming restrictive mechanism works in this way: the unfair scoring would influence the score for the values themselves. Besides these, for the objects that have not yet attracted enough values, their scoring would be withhold, because only adequate values could deliver a relatively objective total score.

4. The scoring of all objects are intertwined and interacted.

For example, in case a commercial residential building is evaluated for a high score, the rankings of its Design Company, developer, Construction Company, material company would be escalated accordingly. If subdivided further, in case its sub-item "quality of structures" wins a high score, the construction company of it also would acquire a remarkably higher score accordingly. By the same rule, a higher score for the "space design" would benefit the design company and the tagged principal conceptual designers. Vice versa, a certain designer whose works win high scores would benefit the company for which he is working.

Under this recycle mechanism, when the amounts of the evaluated score have been accumulated for a certain level, the benchmarks of objects in the virtual world would be closer to their counterparts in the real world as much as possible. Every one of us may not know or be aware these benchmarks currently.

Using algorithm, the intertwined and interacted scoring is a relatively scientific method to achieve the fairness for all scoring. Our Object Rank is similar to the algorithm of Page Rank. Under the algorithm of Page Rank, every web page would be assigned with a PR value firstly, then the set of key words Q = <key1,key2,…,keyn> which is generated after users search string segmentations would be processed by the indexer of the searching engine to produce a matched set of web pages PageSet=<pi,pk,pm,pn,…>, next step is to sequence the web pages in the <pi,pk,pm,pn,…> in accordance with PR value(from the highest to the lowest), in the end the K-web pages with high rankings would be sent back to users. The algorithm of Object Rank works in a almost same way, but it focuses on obtaining the information and direct evaluated scoring of the object online by searching to generate the OR values of the object, and these OR values would be processed by the indexer of database to produce a matched set of objects, ObjectSet=<oi,ok,om,on,…>, next step is to sequence the objects in the <oi,ok,om,on,…> in accordance with OR value(from the highest to the lowest), finally the K-objects with high rankings would be sent back to users.

$$ObjectRank(A) = (1 - C) + C\sum_{P \subset A} \frac{ObjectRank(p)}{|P|} \tag{1}$$

Thus, it can be seen that the Object Rank of a certain object is a summation of the Object Rank value of all other objects relevant to this object dividing the number of their respective objects. A represents the all objects set related to object p, and |P| represents the related objects' number of object p.

Power method also can be used to calculate this value, it is transformed to solve the value of lim Anx, within which the matrix is

$$A = CP^t + (1 - C) \times ee^T / m \tag{2}$$

In this formula, eT is the full-line 1 of the n-dimension, and m is the number of all the objects. This linkage relationship can be demonstrated by matrix P. If the object i has relationship with the object j, pij = 1, if not, pij = 0. If the total number of objects is N, this objects-related matrix is a matrix of one row and N-columns.

$$p = \begin{bmatrix} 0 & 1 & 1 \\ 0 & 0 & 1 \\ 1 & 1 & 0 \end{bmatrix} \tag{3}$$

5. A Searching Engine of Users

Based on aforesaid theories, the virtual world database of architectural industry is similar to the Internet itself, which is a database available for searching engine. However one thing should be made clear is such searching is different from Google which searches by using key words in terms of two substantive aspects, first is it supply so many options, let us take the commercial residential building as an example, tens of searching options are set for it. The real needs of users only can be clearly reflected by adequate options, and only one key word is quite not enough. Secondly, weight design is integrated into searching. Users can set their own preferences within location, environment, layout, style, appreciation potential, developers' reputation, construction quality and architectural materials. For example, location could be valued for 30%, layout for 25%, environment for 20%...thus everyone can set up him/herself a weight system, with which a ranking of searching objects that belongs to them could be acquired in turn. If the score of every user's personalized searching is UserScore,

$$\text{UserScore} = \sum_{p=1}^{n} (Sp \times Kp\%) = \text{S1 k1\%+S2 k2\%+S3 k3\%......+Sn Kn\%} \tag{4}$$

In which （k1+k2+k3......+kn） = 100, in the case where the summation of the personalized items given by users is not equal to 100, the value would be increased or decreased to 100 by program.

6. Real time online communication

When browsing the homepage of the real estate, users can conduct a direct dialogue of words, audio or video with principal conceptual designers, interior designers or salesmen of developers by IM, avoiding unnecessary procedures, intermediary agents or barriers.

7. An environment for cooperation

For every stakeholder in this industry, especially personnel, this would be a huge cooperation platform. For example every furniture producer could upload real BIM model of its product to web page and all designers could utilize these virtual product as a plug-in in the software installed in their own computer. Once designers are satisfied with this product and home owner also click to approve, the web page would send purchasing information to furniture manufacturers automatically. Few days later, physical furniture would be arranged in a specific room in a certain building as what has been illustrated in the space of design, after being delivered by logistics chain. Design, communication, selection of materials and electronic commerce could be integrated into one platform to improve the operating efficiency of whole industrial chain.

5 Conclusion

Whole application has entered into its beta testing stage (Please refer to the website of the searching engine: www.pinfang.com). We are dedicating to replicate whole industrial chain to the virtual world and hope the process for each object to obtain its own attributes and clear quality score would be a great leap of restructuring by IT in architectural industry.

References

1. Ma, H.: Architectural Forum Highlights, vol. 8, pp. 112–120. China Building Industry Press
2. Kang, Y.: Architect full 3D Manual, vol. 12, pp. 50–61. Beijing Hope Electronic Press (2000)

Research on Development and Strategy for Hebei Wushu Schools

Shiru Hao

Department of Physical Education, Baoding College, Baoding Hebei 071000, China
kjiansuoz@163.com

Abstract. Hebei Wushu industry have lots of problem like: disorderly uncontrolled development on amount and scale, Inappropriate resource configuration and wasting, disorderly management and unfair competition. Inadequate commercial operation, low grade, un-enjoyment matches; Shortage on equipment development, variety simple and single. Backward in detection methods and technology. No standardized, Intensive, systemic development. No brand effect and lack of competitiveness. All these problem restricting the development of Hebei Wushu industry. This article made a analysis and give strategy to the solution of those problems.

Keywords: Hebei Province, Wushu School, Wushu Industrial.

1 Introduction

As the primary period of Wushu industry, Wushu School shows its huge business opportunities it contains. After several decades development, over 12000 Wushu school had been established. As for Hebei, there are 146 Wushu school that most of them are the family-style. But such kinds of business model can not satisfy the requirement of market economy, what's worse is that it also become the obstacle of the development for the scale and collectivization. And the unconventionally development is also imperfect and abnormal, We can see it from it's disordered development and unfair competition. So a strategy to solve these problem is imperative. And good for resuscitate the Wushu industry.

2 Object and Method

Questionnaires with details had been finished from 50 registered Wushu school. 20 of the schools have more than 300 students while the other 30 are not. Method been used during this researching are: literature material study, visit investigation, questionnaire investigation, mathematical statistics. 50 questionnaires been sent, and the data retrieve rate is 100% with 50 of them been sent back. Besides, we made questionnaires on education and management of Wushu school, 200 been sent with 187 be retrieved, retrieve rate is 93.5%, logical analysis been used on those questionnaires. EXCEL, SPSS been used in data analysis.

Y. Zhang (Ed.): Future Communication, Computing, Control and Management, LNEE 141, pp. 165–171.
springerlink.com

3 Result and Analysis

3.1 Current Situation of Hebei Wushu Schools

3.1.1 Way and Level for Running Wushu School

After investigation, all 50 Wushu are private with 46 of them are running by individual, it takes 92% of all, the other 4 schools are joint venture which take 4% of all. This situation shows that most Wushu school in Hebei province are private, even the joint venture ones are running by contract with individual. As for education level, 42 of the school have integration of primary and middle school, which take 84% of all. 6 of them have have an integration of high school and vocational school, take 12% of all; 1 is primary school, take 2% of all; There's also one of the school have junior college education, but attached to Hebei Teachers' College, students attend self-taught examination, those who pass 3 years and 4 year examination will respectively gain a national recognition junior college and bachelor degree. All above show that those Wushu school are low level, some of them even make trade on diploma.

3.1.2 Stadium and Facility

Sufficient and complete facility is an essential condition for education in Wushu school. As the survey shows in table 1, 10 of the schools have their own indoor and outdoor stadiums and facilities, but can only basically meet current requirements; 20 of them have some simple, poor facilities, the other 20 even have to rent stadium and facilities. None of them have high level stadium, classroom, audio-visual classroom, mult-media classroom, language lab, reading room, computer room and laboratory room. This situation shows that the Wushu school running condition for Hebei is not consummate, basic constructions and facilities need a improvement.

Table 1. Facility of Hebei Wushu school

Wushu School	account	rate
With consummate facility	10	20%
With simple&basic facilities	20	40%
Renting facilities	20	40%

3.1.3 Teacher Resource

Teacher and trainer is the mainstay of education and training, also make an important affection on the quality of teaching, thus make it a prerequisite for the development of Wushu school. For teacher resource detail see table 2 and 3.

Table 2. Teachers diploma statistics

Diploma	amount	rate
Master	0	——
Bachelor	48	14.7%
College	135	41.4%
Technical	103	31.6%
High School	40	12.3%

Table 3. Employment situation

Employment situation	amount	rate
Full time	42	21.2%
Part time	156	78.8%

Table 2 and 3 shows that qualification of most teachers are still at a low level, part time staff take a high percentage. Among those who have College degree or above, most of them are culture class teacher with a lot of retired people. specialized teachers, are mainly graduates from their own schools. Lack of professional dedication and sense of responsibility is a universal phenomenon since most of the teachers are part time employee. As for the specialized teachers, their deficiency on culture makes a obstacle in the improvement of teaching method. Thus make the quality of education in these schools can't be satisfied.

3.1.4 Administration

School to government department and school internal management consist the management condition.

Examine and approval is what government management take charge for. Some of them are approved by provincial department, some are by city department, there are also approved by country level department. For the details of department, there are Sports Commission, Education Commission, Mass Sports Department, Wushu Association(subordinates of Sports Commission), Education Technical Secondary department, Social Resource Education Department; some even have relate to Public Security Bureau, Industry and Commercial Bureau, Civil Affair Bureau, Cultural Affair Bureau, Personal Bureau etc. About the daily management process, most of the departments do is just making some formalist inspection. Mult-approval and mult-management make all departments only pay attention to the economic benefit rather than management. And all of these are the reason of disorder in this industry.

School internal management includes: education management and life management. For the life management, close type, centralized living, is the only way Wushu schools choose, which means this is feasible. But the education worrying all because only 5 of them following the national syllabus, program and examination standards, as well as a perfect responsibilities and management regulations. The others do not have these conditions or imperfect on this. With the disorder in education management and noneffective measures, students who graduate from these schools can not meet the social requirement, and this is also disadvantageous for the development of Wushu Schools.

3.1.5 Advertisement

Advertisement, as an important approach to raise the popularity and attract more new students, always being highly valued. But the method and enforcement are various. Schools with small scale usually choose television, letters, and recruit brochure as their means of propaganda. investment on this is small due to the incompetent financial strength they have. As for those school which have strong financial support, they will make full use of television, Internet, feature show, picture album etc. But all the advertisement have a same problem on approval progress and content authenticity, some even use advertisement to vilify others. Figure 1 shows the disorder situation on advertisement.

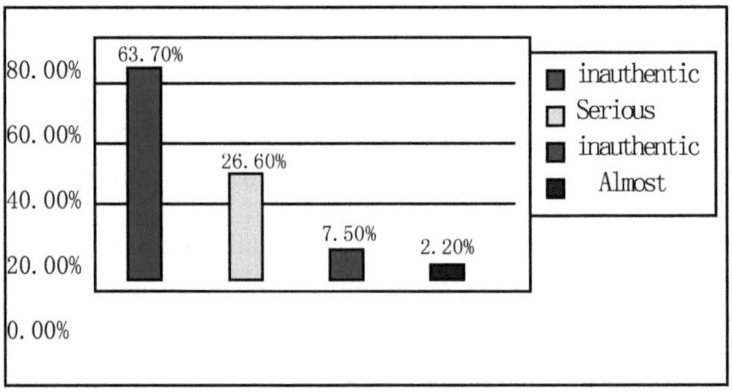

Fig. 1. Content authenticity of advertisement n=182

3.1.6 Competition

Competition can well reflect the education quality. And the result can mirror the level of education and training. 45 of the 50 schools usually attend variety competitions, take 90% of all. Especially in routine and Sanshou competitions. 5 small scale school which take 10% of the schools rarely attend competitions, because of the shortage of entry fee. As for the institution, most of the school reach an agreement on reformation. The primary reason is about the fairness. First of all is that all school should have the same chance to attend the national and provincial competition, qualification depend on equal competition among all students from all schools rather than only choose students from some large scale schools. On the other hand, some school will introduce high level competitor to attend the competition instead of their students. Identity authenticity need a strict scrutiny. Also, the business operation still need an improvement or it can not arouse the enthusiasm of all. At last, there's a gap and many difference between traditional Wushu and the national competitive Wushu competition. Traditional Wushu is lagging behind.

3.2 Problem in Development

After the survey and analysis we found that low-level education, too much attention on propaganda with neglect on school construction; emphasize economic benefit, charging and skill education but neglect education quality, management and culture study; irrational institution; big amount with unbalanced location and development are the main problems for Wushu school development.

3.3 Strategy for Development

3.3.1 Leapfrog Development, High Level, July-Gradation School

How to achieve the modernization for Wushu, the long-history sport is a big task for a long time. As a primary and important part of Wushu industry, Wushu schools should try to accomplish modern development, try to run high level, mult-gradation schools. But the schools in Hebei province are still at the initial stage of development with "mill type" management, unsubstantial foundation, emphasize short-term benefit, depends on their own funds strategies. These strategies only works for primitive accumulation of capital. For modernization, Wushu schools need more fund to set up national standard facilities, hire teachers who proficient in both economic, Wushu industry and education. Fund's necessary, too for shortage on funds restrict the development of Wushu school development and operation. But, what we need to do to solve the problem of fund insufficient?

First of all, carry out merger integration, make a rational distribution,rearrange social resource to make it shared to all. Cooperation with enterprises, absorb social capital, make a large scale, collectivize development, set up high level educational facilities. Make sure it meet the national standard and the need for a long-term development. Next, build a high level Wushu school which can lead the direction of a well-organized development, then the Wushu pre-school, primary school, middle school and college. Let the variety Wushu schools revive Hebei Wushu education. The third one is about the competition. Establish Wushu competition clubs, attend more professional competition, training in practising. At last, make industrial development based on Wushu school, integrate education, competition and industrial development, improve market and input-output cycle system, make a good support for Wushu school development. Otherwise, the conservative will become a massive roadblock for Wushu school development.

3.3.2 Improve the Level of Teacher and School-Running Ability

《China Education Reformation and Development synopsis》 indicated that the hope for the future of our nation is education, while teachers play a vital role in it. So teachers with good political accomplishment, reasonable structure design is very important. So each Wushu school should try their best to set up a high-quality team of teacher with reasonable structure design to improve their ability to run a Wushu school.

3.3.3 Reformation and Professionalism on Competition

The current market of competition is still a depressed one with small scale, inequity, less audience. So the reformation on institution is urgent. First, set up a high quality umpire team with good sense of responsibility and professional skill, improve umpire team management, make the competition equal to all. Second, establish an active competition system with powerful commercial operation. Take national fighting sports operation as an example, also combine market-oriented mode from "China Wushu & Sanda Tournament " to develop Hebei Wushu competition market. Every

Wushu school should establish competition club, carry out registration institution, create student archive, make fair competition among schools. Separate the organizer and competitor, change the irrational situation that one could be organizer, umpire and competitor. completely eradicate cheating. Thirdly, modify current rules of competition, make it simple and easy to understand, learn experience from competitive Wushu. Try to finalize the reformation of tradition Wushu to make it reasonable, fair and distinctive. Forth, improve competition level and its enjoyment to meet the audience's requirement. Also, it can help in income promotion by make the gate receipt, advertisement, television broadcast right profit. Meanwhile, better economic benefit, better propaganda and social benefit.

3.3.4 Regulating School Management and Laws

Only by regulating school management and laws can we make sure the development of steady and healthy. First thing we need to do in regulating is confirm the manager of Wushu schools by set up relevant legal documents. Author suggest that "Education bureau" is the best choice. Second is about the approval process. Give the only right on approval to provincial education department. Next, "Hebei Education bureau" should set up a series of policy and institution, especially evaluation system. Make a comprehensive and accurate evaluate among Wushu schools. Grading the schools and their tuition by the evaluation result. To those who make malign competition, must be stroke and treated seriously. For the ones who can not meet the requirement of evaluation system, let them make a rectification with time limited, or banned. Evaluation should become a regular part in Wushu school management. And make gradation according to the result, upgrade the good ones while demote the unqualified ones. As for the advertisement, make strict investigation and management on it, set up report and supervise system.To those who make advertisement without approval or cheating in it, should be treated with serious punishment.

4 Conclusion

The disorder and large scale development in Hebei Wushu school result in unbalanced distribution and development, resource wasting; low-level education with conservative attitude, unclear education mission, too much attention on advertisement but ignore school construction and facility improvement, highly value economic benefit, skill education, but neglect education quality and culture study, Backward business model etc. All above have become the internal burden on Wushu school development. No clear approval and management department, high value in charging with inattentive management, no efficient measure to disordered institution, irregular competition institution and insufficient on students' employment opportunity, are the external factor which restrict the development of Wushu school.

5 Suggestion

Free our mind, change operation principle, set up market-oriented system, achieve a leaf in the development. Build a high level Wushu school which can lead the direction of a well-organized development.

Carry out merger integration, make a rational distribution, rearrange social resource and share to all. Cooperation with enterprises, absorb social capital, make a large scale, collectivize development, combine eduction, competition and industrial development, set up beneficial input-output system, and a active, lively Wushu school structure.

References

1. Zhen, X.: Situation and strategy for Nongovernmental Wushu School. Journal of Shanghai Sports College 12(6), 47–50 (1996)
2. Jiang, B., Zeng, Y., Chen, X., et al.: Discussion On Strategy For Wushu School Development. Sports Software Science Achievement Complication (7), 78–81 (2003)
3. The State Council China Education Reformation And Development Synopsis China Development, 3rd edn. (1993)
4. Yi, T.: Modern Capitalism in Wushu Industry. 21 Century Economy (April 1, 2002)

Stock Prediction with Directed Cross-Correlation Network

Hua Chen and Qiming Sun

School of Economics and Management Beijing University of Posts and Telecommunication, Beijing, 100876, P.R. China
hua_0110@hotmail.com, sunqm@bupt.edu.cn

Abstract. In this paper, the price trading data for every minute from Shanghai A-share stock market has been studied, and the directed complex network model of Chinese stock market has been constructed. By computing cross-correlation between each pair of stocks, it showed that directed complex networks exist in Chinese stock market. Furthermore, using the directed complex network model and the cross-correlation value between different stocks, the moving trends of stock price and the range of it has been predicted. We find that the prediction results are consistent with the real exchange results for most stocks in each trading day except for the first 10 minutes after opening and 8 minutes before the close. After the prediction value is rounded, the average accuracy rate of price prediction with 640 Shanghai A-share stocks can reach to 83%. All of this can be a good guide for stock investors of decision making.

Keywords: Stock market, complex network, directed cross-correlation, stock price prediction.

1 Introduction

From the perspective of investors, the stock market is a high-risk market. The reason is that the stock market is a so complex system with a lot of uncertainty factors, and too many factors can affect the stock price fluctuation, so it is too hard to make the stock price prediction. Especially, most investors think that it is very difficult to judge the moving trends of stock price and how much the ups and downs along it. Therefore, it is not easy to decide when is the right time to buy or to sell.

Actually, stock market is affected by many factors such as macro economical, political events, psychology of investors, policies of firms, etc[1]. In addition, since the stock market is essentially complicated, nonlinear, dynamic, nonparametric, chaotic [2]. Therefore, the stock price prediction is becoming a challenging task.

However, there are still many researchers have predicted the trends, index and price of the stocks with a series methods in recent years [3-16]. The prediction methods include nonlinear time series model [3,4], artificial neural networks [5-7], decision trees [8,9], genetic algorithms [10,11], Markov models [12] support vector set [13-15] and so on.

In this paper, we study 640 stocks and the time-dependent cross-correlation between different stock returns in Chinese A-share stock market, and compute the

Y. Zhang (Ed.): Future Communication, Computing, Control and Management, LNEE 141, pp. 173–179.
springerlink.com

time-dependent cross-correlation coefficient with each pair of stocks. The directed complex network model in Chinese stock market is constructed. With the value of time-dependent cross-correlation coefficient and the model of the directed complex network, the stock price of moving trend and the strength will be predicted.

2 Directed Time-Dependent Cross-Correlation Complex Network

In the nature and society, from the internet to microbiology, there are so many complex systems just like stock market. A great deal of information is included in every complex system, and the information is often not easy to deal with. Since 1998, after the small-world network model was presented by Watts [17], more and more complex systems in the world were studied with some different complex networks [18].

Stock market is a complex system including a lot of varying information, so it offers another new field to use complex network method [19]. We suppose that the nodes in the network are companies or companies' stocks; the edges between nodes are the cross-correlations of companies or stocks; the edge coefficient is the strength of association between the companies or stocks. The sum of all edge coefficient connected one node is larger, and the node will have greater influence in the network.

Without considering the direction of impact between any two nodes in the network, the network is an undirected network, otherwise it is a directed network. For example, there are two nodes in the network, if it is an undirected network, the cross-correlation between two nodes A and B, the intensity of it can be obtained, but we cannot know the influence resulting from them is from node A to node B or from node B to node A, or the impact in which direction is greater. However, if the network is a directed network, all the questions can have clear answers. Furthermore, from the perspective of financial investment, risk management and portfolio optimization, the directed network is more important than undirected network in stock market, because the directed network model can help investors to obtain more specific useful information, and make more reasonable decision, even reduce the losses.

After the first network model [20] in stock market appeared, there are so many papers based on stock price correlations were presented [21-26]. L. Kullmann et al. [21] studied the clustering of companies in the Dow Jones Industrial Average and the Standard & Poor's 500 by different methods based on the q-state Potts model. Using the correlation matrix of stock returns, Onnela et al. [22-24] constructed dynamic asset tree of the Standard & Poor's 500 indexes with daily closure prices of 116 stocks. They showed how the tree evolves over time and how the tree length shrinks during a stock market crisis. But asset tree is also more fragile in structure, so they constructed dynamic asset graph [25] based on the split-adjusted daily closure prices for a total of N=477 stocks traded at the New York Stock Exchange Market. At the same time, they make the comparison between asset tree and the asset graph [26].

However, up to now, there is no more work have actively concerned about the directed network in stock market. In the following, we will construct the directed complex network model in stock market.

Variable definition: t-a time moment; $A(t, j)$-time and price matrix of Shanghai A-share stock market; $P(t)$-trading price of one stock at time moment t; $R^{s}_{\Delta t}(t)$-price returns of stock *a* at time moment *t* based on Δt time intervals; $Q_{\Delta t}^{ab}(\tau)$-the impact cross-correlation coefficient from stock *a* to stock *b* after time shift τ.

In this paper, we investigate every minute transaction price closing data traded in Shanghai A-share stock market. From 07-March-2011 to 01-Aprial-2011, except for stocks trading halt during this time and the ST stocks which the highs and lows are limited to 5 percent but not 10 percent, at last, 640 stocks with every minute closing price trading data has been studied. Corresponding with the clear time t of each trading day, every stock has 4800 price and time trading data P(t). Therefore, the time and price data of Shanghai A-share stock market can be described as a matrix A(t, j) with 4800 lines and 640 columns.

During the time interval Δt, stock price returns of a stock at time *t* is: $R(t) = \ln[p(t)] - \ln[p(t - \Delta t)]$, the size of time interval Δt is divided in minutes as $\Delta t = 1m, 2m, 3m \ldots$, it is reasonable in efficient market theory. Calculated with the price trading data matrix *A*(*t*, *j*) in Shanghai A-share stock market, the matrix *R* of time price returns with 4799 lines and 640 columns can be obtained.

For any two stocks *a* and *b*, after computing stock returns of transaction price data with Δt time intervals, in the whole statistic time, the impact cross-correlation coefficient from stock *a* to stock *b* after time shift τ computed as:

$$Q_{\Delta t}^{ab}(\tau) = \frac{\langle R_{\Delta t}^{a}(t) R_{\Delta t}^{b}(t+\tau)\rangle - \langle R_{\Delta t}^{a}(t)\rangle\langle R_{\Delta t}^{b}(t+\tau)\rangle}{\sqrt{\langle [R_{\Delta t}^{a}(t) - \langle R_{\Delta t}^{a}(t)\rangle]^2\rangle\langle [R_{\Delta t}^{b}(t+\tau) - \langle R_{\Delta t}^{b}(t+\tau)\rangle]^2\rangle}}$$

Where $\langle \cdots \rangle$ means a temporal average over a certain period, i.e., $\langle R_{\Delta t}^{a}(t)\rangle = \frac{1}{n}\sum_{t=1}^{n} R_{\Delta t}^{a}(t)$,

$\langle R_{\Delta t}^{a}(t) R_{\Delta t}^{b}(t+\tau)\rangle = \frac{1}{n}\sum_{k=0}^{\frac{n}{\Delta t}-\tau} R_{\Delta t}^{a}(k\Delta t) R_{\Delta t}^{b}(k\Delta t+\tau)$, $n = 4799$. Then, the cross-correlation coefficient of each two stocks in 640 Shanghai A-share stocks is computed with this formula, so the cross-correlation coefficient matrix can be obtained.

The cross-correlation coefficient $Q_{\Delta t}^{ab}(\tau)$ for any two stocks *a* and *b* is between -1 and 1. If $0 < Q_{\Delta t}^{ab}(\tau) < 1$, stock *a* and stock *b* has the positive cross-correlation, it means that the price fluctuation of stock *a* pulls the price changing of stock *b* with the same direction. If $-1 < Q_{\Delta t}^{ab}(\tau) < 0$, stock *a* and stock *b* has the negative cross-correlation, it means that the pulling effect reversed, the price changing of stock *a* pulls the price changing of stock *b* with the opposite direction. Stock *a* and stock *b* are not related while $Q_{\Delta t}^{ab}(\tau) = 0$, there is no influence existed.

In the cross-correlation coefficient matrix, if $|Q_{\Delta t}^{ab}(\tau)| < |Q_{\Delta t}^{ba}(\tau)|$, we think that there is a directed edge from *b* to *a* in directed network; if

$|Q_{\Delta t}^{ab}(\tau)| > |Q_{\Delta t}^{ba}(\tau)|$, there is a directed edge from *a* to *b*; if $|Q_{\Delta t}^{ab}(\tau)| = |Q_{\Delta t}^{ba}(\tau)|$, then the impact between two stocks *a* and *b* is equivalent, there is no directed influence in their cross-correlation, and no directed edge connected. According to this method, the directed network model of stock market has been constructed.

In order to offer useful network graphs for investors; meanwhile, considering the noisy impact cross-correlation between stocks, we suppose that there is a certain threshold value δ, if $Q_{\Delta t}^{ab}(\tau) < \delta$ and $Q_{\Delta t}^{ba}(\tau) < \delta$ •then let $Q_{\Delta t}^{ab}(\tau)$ and $Q_{\Delta t}^{ba}(\tau)$ equal 0, so there is no edge connected between two nodes *a* and *b*, it also means that the cross-correlation between stocks *a* and *b* has no more useful information for stock investment, it can be ignored.

21 stocks are randomly selected from Shanghai A-share stock market, let δ=0.14, an example of directed network graph is presented as fig.1.

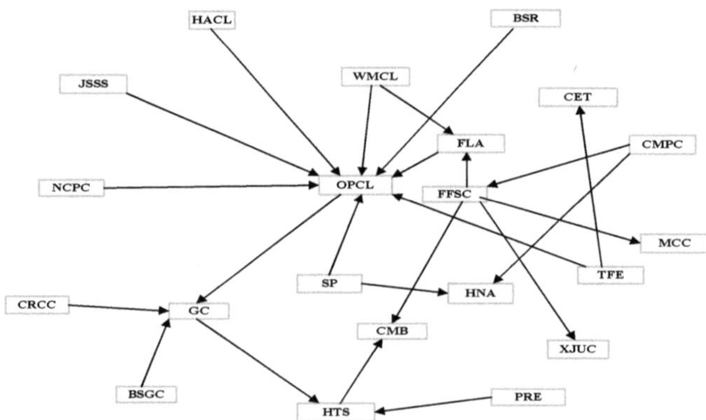

Fig. 1. An example of the directed network connecting 21 stocks used to represent the pulling effect between stocks in Chinese stock market

3 Stock Price Prediction

There is a phenomenon in the stock market: a number of stocks with good profit are so active that they will impact the price fluctuation of other stocks. By constructing the directed network model and calculating the cross-correlation coefficient for each two stocks in Shanghai A-share stock market, we find that directed cross-correlation relationship existed. It means that the price fluctuation of one stock is affected by other stocks' price fluctuation. It also means that the price fluctuation in the same direction has different time shift.

Assume that there is a stock b, the price changing of it is affected by other N stocks, or the price fluctuation of stock b lagged behind other N stocks, in addition, in the whole statistic time, they have the same direction in price fluctuation. Therefore, the price changing of stock b can be predicted by the price changing trend and range of other N stocks.

In the cross-correlation coefficient matrix, the stocks which have the cross-correlation with stock *b* can be obtained, the cross-correlation coefficient value $Q_{\Delta t}^{ib}(\tau)$ between them is also known from the matrix, where *i* is any one stock in the market. The value of cross-correlation coefficient is corresponding with the cross-correlation strength between stocks. We choose a threshold value δ, and $-1 < \delta < 1$. If $Q_{\Delta t}^{ib}(\tau) \leq \delta$, price changing of stock *i* has no influence to stock *b*, so it is excepted for price prediction of stock *b*. If $Q_{\Delta t}^{ib}(\tau) > \delta$, the price changing of stock *i* impact the price changing of stock *b*, then price changing trend and range of stock *b* can be predicted with cross-correlation coefficient $Q_{\Delta t}^{ib}(\tau)$ between stock *i* and stock *b*. It is assumed that there are *N* stocks which have the obvious influence to stock *b* price changing, and the range of price changing for every stock is μ^i in the forecasting time, then the prediction price of stock *b* is $\sum_{i=1}^{N} Q_{\Delta t}^{ib}(\tau) \times \mu^i$.

During those *N* stocks, although they have the same direction in price fluctuation, different stock has different time shift, and how long it will be lagged is also not known. Therefore, in our experiment, the data is analyzed with different time shift τ.

To describe the price prediction of stocks clearly, an example of it is showed as fig.2.

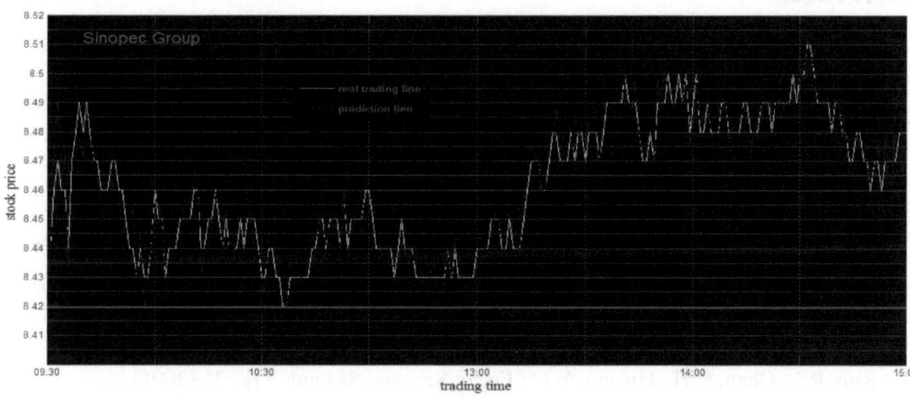

Fig. 2. The comparing graph with real trading data line and price prediction data line of Sinopec Group on 16-March-2011

Compare with the prediction result and real exchange data, it is known that the trend of price changing is same, but there is a deviation existed in the range of price changing. For the reason, the value of price prediction computed and saved as floating type in the experiment, but in Shanghai A-share stock market, the smallest trading price is 0.01 Yuan, so after rounding the prediction value, the distance of deviation will be reduced. For example, after rounding the price prediction value of Sinopec Group on 16-March-2011, the prediction accuracy rate has been improved from 67 percent to 86 percent.

4 Conclusion

In this paper, we have studied 640 stocks with one minute trading data and the cross-correlation between different stock returns in Shanghai A-share stock market. The cross-correlation coefficient matrix has been obtained by calculating cross-correlation coefficient for each pair of stocks, and the directed complex network model has also been constructed. In order to explain the directed network model, an example of directed network graph is given with 21 stocks chosen randomly from Shanghai A-share stock market. It means that the directed network existed in Chinese stock market.

With the impact cross-correlation coefficient matrix and the directed network model, the strength and the trend of price changing have been predicted. By comparing with the prediction value and the real trading price data, we find that most stocks have the same fluctuation trend between prediction result and real trading data except for the two time intervals as 10 minutes after opening and 8 minutes before closing in every trading day. After the prediction value is rounded by 0.01yuan, removing two abnormal time intervals, the average accuracy rate of price prediction for 640 stocks can reach to 83 percent. It can be seen that the stock price prediction method is feasible and effective. This will provide useful information for investors to make right investment decisions.

References

1. Tan, T.Z., Quek, C., Ng, G.S.: Computational Intelligence 23, 236 (2007)
2. Abu-Mostafa, Y.S., Atiya, A.F.: Applied Intelligence 6, 205 (1996)
3. Mills, T.C.: Journal of Economic Surveys 5, 215 (1991)
4. Clements, M.P., Franses, P.H., Swanson, N.R.: International Journal of Forecasting 20, 169 (2004)
5. Baba, N., Kozaki, M.: In: Proc. of IJCNN, vol. 1, p. 371 (1992)
6. Hawley, D.D., Johnson, J.D., Raina, D.: Financial Analysts Journal 46, 63 (1990)
7. Schoneburg, E.: Neurocomputing 2, 17 (1990)
8. Sorensen, E.H., Miller, K.L., Ooi, C.K.: The Journal of Portfolio Management 27, 42 (2000)
9. Wang, J.L., Chan, S.H.: Expert Systems with Applications 30, 605 (2006)
10. Kim, K., Han, I.: Expert Systems with Applications 19, 125 (2000)
11. Kuo, R.J., Chen, C.H., Hwang, Y.C.: Fuzzy Sets and Systems 118, 21 (2001)
12. Hassan, M.R., Nath, B.: In: Proc. of ISDA, p. 192 (2005)
13. Muller, K.R., Smola, A.J., Ratsch, G., Scholkopf, B., Kohlmorgen, J., Vapnik, V.: In: Proc. of ICANN, p. 999 (1997)
14. Trafalis, T.B., Ince, H.: In: Proc. of IJCNN, vol. 6, p. 348 (2000)
15. Ni, L.P., Ni, Z.W., Gao, Y.Z.: Expert Systems with Applications 38, 5569 (2011)
16. Yeh, C.Y., Huang, C.W., Lee, S.J.: Expert Systems with Applications 38, 2177 (2011)
17. Watts, D., Strogatz, S.: Nature 393, 440 (1998)
18. Albert, R., Barabasi, A.L.: Review of Modern Physics 74, 47 (2002)
19. Caldarelli, G., Battiston, S., Garlaschelli, D., Catanzaro, M.: Complex Networks 650, 399 (2004)

20. Mantegna, R.N.: The European Physical Journal B 11, 193 (1999)
21. Kullmann, L., Kertesz, J., Mantegna, R.N.: Physica A 287, 412 (2000)
22. Onnela, J.P., Chakraborti, A., Kaski, K., Kertesz, J.: The European Physical Journal B 30, 285 (2002)
23. Onnela, J.P., Chakraborti, A., Kaski, K., Kertesz, J.: Physica A 324, 247 (2003)
24. Onnela, J.P., Chakraborti, A., Kaski, K., Kertesz, J., Kanto, A.: Physical Review E 68, 056110 (2003)
25. Onnela, J.P., Chakraborti, A., Kaski, K., Kertesz, J., Kanto, A.: Physica Scripta T 106, 48 (2003)
26. Onnela, J.P., Kaski, K., Kertesz, J.: The European Physical Journal B 38, 353 (2004)

Solution for Transforming Web Application to Multi-tenant SaaS Application

Yangpeng Zhu[1,2], Jing Zhang[1,3], and Junhuai Li[1]

[1] School of Computer Science and Engineering, Xi'an University of Technology, Xi'an 710048, China
LiJunHuai@xaut.edu.cn
[2] School of Economic and Management, Xi' an Shiyou University, Xi'an 710065, China
zyp_hello@126.com
[3] State Key Laboratory for Manufacturing Systems Engineering (Xi'an Jiaotong University), Xi'an 710048, China
ZhangJing@xaut.edu.cn

Abstract. Software as a service is wildly used for its merit of low cost, free maintain and extensibility. How to transform a well developed typical web application to a cloud based multi-tenant SaaS application is a problem for many software vendors. In this paper we provided an example typical web application and described the process and some needed factors for transforming it to SaaS application and detail showed the step for achieving the goal in three different type database model. The result showed that the transformation was feasible and efficient and the solution would have some references for transforming web application to SaaS application.

Keywords: Software as a service, Multi-tenant, Web application, Cloud computing, Customization.

1 Introduction

As a new software service model, Software as a Service (SaaS) provider provides the entire network infrastructure and software updates in Cloud for users. Users need not buy software and hardware and hire IT professional workers. They can use the SaaS system by internet and pay for selected services according to their network flow, data store capacity and computing power [1]. The growing number of users and elastic cloud infrastructure bring down the total costs of software provider which make it popular for medium and small enterprise to use SaaS system [2]. So, the software providers who have a well development web application want to meet the market requirement to transforming his old application to a SaaS application. How to transform a web application to a SaaS quickly, efficiently and meet multi enterprises to use is a challenge problem. Now, many researches are focusing on SaaS for its creation, security, multi-tenant architecture, scalability and configurability, but few researches focus on how to convert a well developed typical web application to a SaaS application.

In [2] the authors provide a high-level description of the architecture of a SaaS application, and present four lever-maturity model to measure the configurable, efficient and scalable for SaaS application. They also discuss three type database models adapting to different SaaS application condition. In [3] the authors present three

Y. Zhang (Ed.): Future Communication, Computing, Control and Management, LNEE 141, pp. 181–188.
springerlink.com

distinct approaches for creating multi-tenant data architecture which including separate databases, shared database separate schemas and shared database shared schemas. They also discuss database security and implement SaaS extensibility by means of preallocated fields, name-value pairs or custom columns. In [4] the authors address the problem that each individual customer has different requirements for the same SaaS application. For personalizing the tenants' business process, they provide a set of variability points which can be modified by customers and transform variability descriptors into WS-BPEL process model which can be used to guide users to personalize their SaaS application. In [5] the authors present a modeling and verification framework for user business process customization. In this model, they divide the model into four layers: business processes layer, services layer, tables layer and table fields layer. In each layer user can implement customization for himself. In [6] the authors present a kind multi-tenancy SaaS electronic contract management application and describe several novel methods used in metadata, security and shared services, as well as customization and tenant extensions to support multi-tenancy SaaS application. In [7] the authors discuss the quality of service (QoS) in SaaS and present a specific multi-tenant architecture-MDSA to evaluate the application performance. In [8] the authors propose a data combination privacy preservation mechanism for SaaS which is used to describe the data combination privacy requirement and fragments the SaaS data to different data chunks. In this manner, although the tenant's data is in insecurity place, the SaaS administrator can't reveal the tenant's data. In [9] the authors propose a cloud computing architecture focusing SaaS, which address the issues of security, performance and load management in SaaS. In [10] the authors present a level 4 healthcare industry SaaS application which offers metadata configuration, superior ability to scale and handle large volumes of traffic and transactions. In [11] the authors present architecture for enterprise SaaS application which combines asymmetric and symmetric cryptography to protect the users' privacy. All above researches are focusing on SaaS for multi-tenant architecture, security, configurability, extensibility, QoS, or performance monitor, but few researches focus on transforming web application into SaaS application. In this paper, we have an example to transform a well used web restaurant management system to a cloud based, multi-tenant SaaS application [12]. We describe the process of the transform detail and talk about the problems we met.

The paper is organized as follows. Section 1 describes the difference between web application and SaaS application, while Section 2 presents new SaaS architecture and proposes some required functions to administrate the SaaS application. Section 3 shows the detail process to transform the restaurant application to SaaS application in three different types of database model, and introduces the conversion result and effect. Finally, Section 4 provides concluding remarks and outlines directions for future work.

2 Web VS SaaS

The difference between the web application to SaaS application mainly is multi-tenant character. SaaS is not just a multi-user application, but a multi-tenant application. Each customer organization is a tenant in a SaaS application. [10]. In SaaS model, lots of tenants access the system together, and each tenant has no relation to others. Tenants can create their needed users for himself and assign right to his users. SaaS application in architecture is as show in Fig. 1.

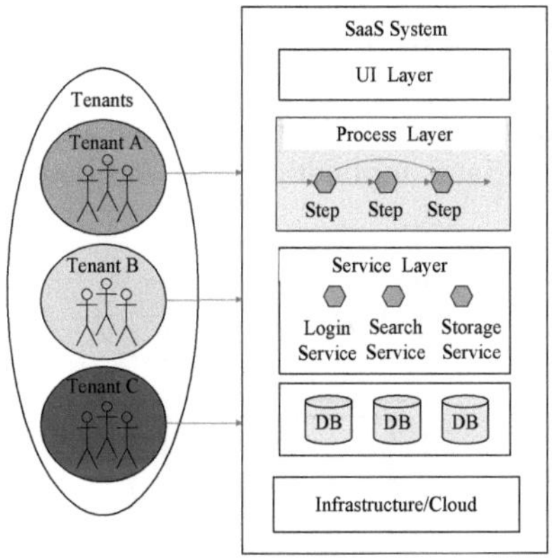

Fig. 1. SaaS application architecture

UI layer is responsible for presenting information to user, and process layer is for business processes and tenant can personalize his business process or arrange the service at this layer. Service layer focuses on providing services for an atom function and database layer is for connecting and storing user data. Cloud layer is charged for providing infrastructure such as computing and storage capacity according to users' need.

SaaS model has some differences to traditional software as follow:

A) Support Multi Tenant

The main difference between SaaS application and traditional web software is whether it supports multi-tenant. With increasing number of users, Cloud automatic assigns computing and storage capacity to different tenants according to users' needs, which reduces the total cost of ownership of SaaS application provider.

B) Software License

In SaaS model, users can lease selected service instead of buying it, and can give up it at any time he does not want to use it. This reduces the risk of user investment.

C) Reduce the User's Investment

In traditional model, user needs to buy hardware such as Router, Hub and Servers. Even more, development a custom-built software is either a hight risk investment. If the software can't meet the user's needs, developer needs to change the software or even to redesign it which increases the user and developer's investment risk. But in SaaS model, user needn't buy any hardware and can try to use the software first, if it can meet the user's demand, user pays just for rent.

D) Easy to Implement

In SaaS model, user needs not install software in his own computer, he can easily access the application through internet.

E) Convenient to Update

In traditional model, users need frequently update application in his computer. In SaaS model, software provider updates the application in the server and users need not do any more.

3 Preparing for Transformation

The new SaaS application combines the restaurant application and some additional management functions for SaaS application. The architecture is shown in Fig 2.

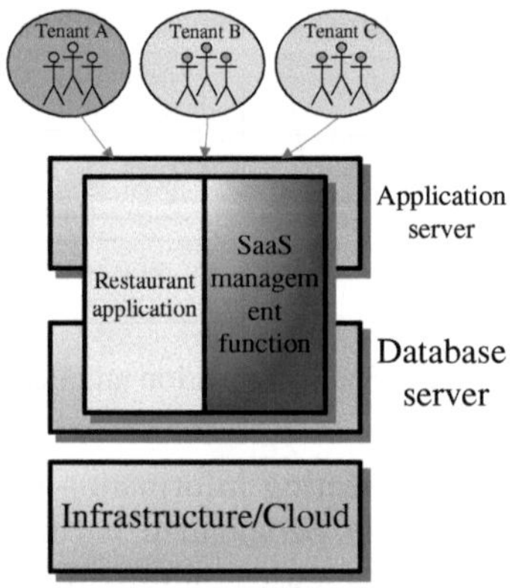

Fig. 2. New SaaS architecture

The web restaurant application has functions such as ordering, select food, billing, food management, business querying and statistics [12] which is well used in many restaurants. The SaaS management functions include security, tenant management, resources assigning and monitoring, etc. For transformation the restaurant application to SaaS application, we must combine the old restaurant application with the SaaS management functions and reform some old application. To make the saas application work well, the SaaS management functions should include functions as shown follow:

Tenant register function, old restaurant application users get their accounts by registering themselves or obtain accounts from system administrator. In SaaS application, tenant registers its enterprise information self-help and assigns roles and permission for his users.

(1) Tenant management function, SaaS administrator can check the tenant's register information and manage tenant status such as trial user, official user, suspend user or lacking fees user.
(2) Lease fees management function, In SaaS application, tenant needn't buy the software and who can pay lease fees by month or year. SaaS application can automatic remind tenant paying for fees according to their paid time.
(3) Lease fees settings function, in this function, SaaS application can set charge standard by user selected functions, lease time, user number, network bandwidth, computing power and storage capacity.

(4) Tenant template settings function, for implementation tenant's personalized user interface, SaaS application provides templates for each tenant in which tenant can specify their own login pages, logo, font, color scheme and unique URL.
(5) Log Management function: log management function can record user operation trace and is convenient for tenant administrator to trace errors. If the user in the system made some operations and caused important data lost, the user may suspect that these errors are caused by SaaS system or other users. User log is a good way to record user's operations and data changed history which ensures that user's behavior can not be faked, be destroyed, and deniable.
(6) Resource assign and monitor, SaaS application can assign and monitor resources in Cloud platform according to users' indeed requirement which can reduce the user's cost and save energy.

4 Transformation Process

In multi tenant SaaS system development, database is commonly designed for three models: separate database, separate database independent schema and shared database shared schema which have different methods to meet the SaaS application requirement [3] [13].

4.1 Separate Database

The seperate database model is shown as Fig 3.

Each tenant's data stores in an independent database and each tenant shares the public application. It is a simplest storage way and is convenient for user to backup and restore their own data. This model adapt to users who need high security but not focus on cost, such as bank, hospital etc.

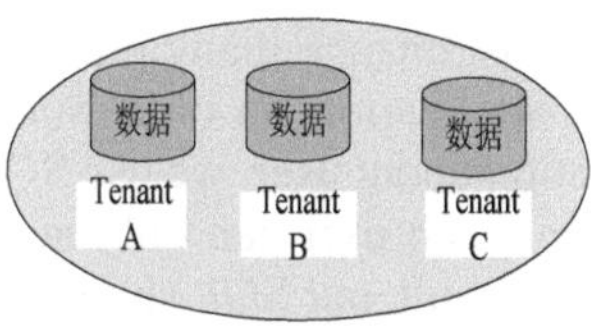

Fig. 3. Separate database

For transforming to independent database model from the restaurant application, we needn't change the old database any more including tables and columns. In this database model, we needn't change the old program except having some update in user login function. We use old database as Meta database and add it to new SaaS application. The transformation process is as follows:

(1) tenant registers in SaaS application and get a unique tenantID;
(2) The application copies the meta database as the tenant's database and names the new database with the tenantID;

(3) Change the database connection method, So, the SaaS application can connect to different tenant database according to the login tenantID;
(4) Change old login program make user login to right database.

4.2 Shared Database and Independent Schema

The Shared Databases Separate Schemas database model is shown as Fig 4.

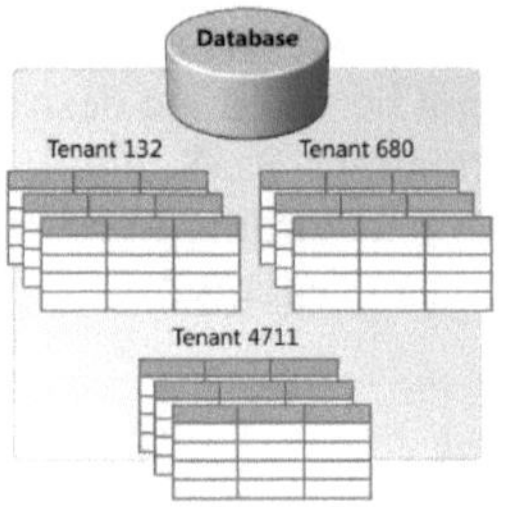

Fig. 4. Shared Databases Separate Schemas

All tenants store their data in one database, but each tenant has dependent schema, that means every tenant have a dependent set of table. In this type of database model, old restaurant database needn't change any more, but program have a big updates. The process is as follow:

(1) Tenant registers in saas application and get a unique tenantID;
(2) The SaaS application read the old database as meta database and create a set of new tables for the new registered tenant with name of the tenantID plus the corresponding old table name. For example, the tenantID is "001" and old table name is "user", then the new table name is "001user". All table belonging to the tenant whose new name is "001" plus corresponding old tables name.
(3) In this model, we must change many places in old restaurant application to transform to SaaS application. In old restaurant application, if we want to see all the information about user, we use SQL like "select * from user". But in SaaS application, each user can only access his enterprise's data. If we want to get the same result, our SQL should be "select * from tenantIDuser". The "tenantIDuser" is the table name which belongs to the new tenant generated in step 2.
(4) Certainly, the login program is must be update to adapt user login.

In this model, all SQL must be update in old application so that the users are restricted to access their own data and this would be a heavy workload. If Tomcat is used as web server in java application, we can solve this problem easily. Filter is a program which runs priority to related servlet or JSP on the web server. Filer can attach to one or more servlet or JSP and check the request for resources. Filter can use modified request information to access resources or can access resources first and modify response information before send it to user. In our example, we can use Filer and configure web.xml file to achieve our goal without modifying any other program. Certainly, in .NET we can also solve the problem by using Filter and configuring web.config file.

4.3 Shared Database and Shared Schema

The Shared Databases and shared Schemas database model is shown as Fig 6.

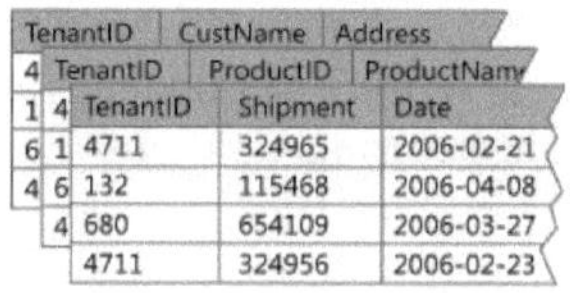

Fig. 5. Shared database shared schemas

In this database model, all tenants store data in a same database and schema. We add a column tenantID to each relation table to distinguish tenant to others. This model has the highest data sharing and lowest data isolation and has lowest hardware and maintenance cost. Because of no modifying in old database except adding a Column tenantID, we can use Filter technology to retrieve each tenant's data as the model above and need to modify little code in login program.

Now, we finish the transformation and we can see that we modify the old restaurant in some aspects as follow:

(1) Add SaaS management functions such as security, charge, tenant management, resources assign and monitor etc;
(2) Database change in three types;
(3) Modifying login functions;
(4) Using Filer technology to guarantee user can access his own data.

Table 1. The comparison among the three types transformation

Database type	complex	cost	update file numbers	update line number	customization	scalibility
type 1	low	high	10	268	easy	easy
type 2	hi	media	15	352	easy	easy
type 3	hi	low	20	376	hard	hard

The comparison among the separate database, separate database independent schema and shared database shared schema transformation is shown as Table 1.

5 Conclusion

In this paper, we provided a solution to transform a web restaurnat to a multi-tenant SaaS application and talked aout the processed in three different database model. The result showed that the transformed SaaS is well adapt to deal with the multi-tenant requirment and the tansformation was feasible and efficient.

In our future work, we plan to investigate how to turn our solution in a automation transformation and can deal with different progran language and database and pay more attention to out solution on scalability and configurability.

Acknowledgments. This research is supported by the 2007 Nationality 863 Project "Software Product Line based on Workflow" under grant number 2007AA010305, by State Key Laboratory for Manufacturing Systems Engineering Open Project under grant number sklms2011011.

References

1. Ebert, C.: A Brief History of Software Technology. IEEE Software 25(6), 22–25 (2008)
2. Chong, F., Carraro, G.: Architecture Strategies for Catching the Long Tail (April 2006), http://msdn.microsoft.com/en-us/library/aa479069.aspx
3. Chong, F., Carraro, G., Wolter, R.: Microsoft Corporation. Multi-Tenant Data Architecture (June 2006), http://msdn.microsoft.com/en-us/library/aa479086.aspx
4. Mietzner, R., Leymann, F.: Generation of BPEL Customization Processes for SaaS Applications from Variability Descriptors. In: IEEE International Conference, Honolulu, HI, July 7-11, vol. 2, pp. 359–366 (2008)
5. Li, H., Shi, Y., Li, Q.: A Multi-granularity Customization Relationship Model for SaaS. In: Proc. IEEE Symp. 2009 International Conference on Web Information Systems and Mining, Shan dong, China, November 7-8, p. 611 (2009)
6. Thomas, K., Thao, N., Linh, L.: A Software as a Service with Multi-tenancy Support for an Electronic Contract Management Application. In: SCC 2008, pp. 179–186. IEEE Computer Society (2008)
7. Lin, H., Han, Y.: Performance Management for Multi-Tenant Web Application. Chinese Journal of Computers 33(10), 1881–1895 (2010)
8. Zhang, K., Li, Q., Shi, Y.: Research on Data Combination Privacy Preservation Mechanism for SaaS. Chinese Journal of Computers 33(11), 2044–2054 (2010)
9. Pervez, Z., Lee, S., Lee, Y.-K.: Muti-Tenant, Secure, Load Disseminated SaaS Architecture. In: 12th IEEE Advanced Communication Technology (ICACT), February 7-10, pp. 214–219 (2010)
10. Hudli, A.V., Shivaradhya, B., Hudli, R.V.: Level-4 SaaS applications for healthcare industry. In: Proceedings of the 2nd Bangalore Annual Compute Conference on 2nd Bangalore Annual Compute Conference, COMPUTE 2009, Bangalore, India, January 9-10 (2009)
11. Karabulut, Y., Nassi, I.: Secure Enterprise Services Consumption for SaaS Technology Platforms. In: Proceedings of the 2009 IEEE International Conference on Data Engineering, ICDE 2009, March 29-April 02, pp. 1749–1756. IEEE Computer Society, Washington, DC (2009)
12. http://www.xa001.net/saas/
13. Aulbach, S., Grust, T., Jacobs, D., Kemper, A., Rittinger, J.: Multi-Tenant Databases for Software as a Service: Schema-Mapping Techniques. In: SIGMOD 2008, Vancouver, BC, Canada, June 9-12, pp. 1195–1206 (2008)

Risk Acceptance Criteria of Oil and Gas Pipelines Projects in Construction Period

Wu Wenping and Feng Xia

Department of Civil Engineering and Architecture, Xi'an University of Technology, Xi'an 710048, China
wwp@mail.xaut.edu.cn, fengxia862@163.com

Abstract. The transportation of oil and gas in pipelines is a valuable and essential link for oil and gas industry. However, due to the complexity of the oil and gas pipeline projects, the safety and environment issues during construction and operation are matters of great concern that should be treated carefully. Therefore, oil and gas pipeline projects must have effective risk management and risk acceptance criteria as well. Currently, indicators used as the correct factor of the quantitatively calculation for the value of acceptable risk is lacking. In order to obtain the quantitative value of risk acceptability for oil and gas pipeline projects, we have summarized the acceptability of risk development status in China and further proposed basic principles of risk acceptance index models. Risk acceptance criteria of oil and gas pipeline projects can thus be determined using the models proposed. Overall, this study provides an important basis for the risk assessment in oil and gas pipeline projects during construction period.

Keywords: Oil and gas pipelines, Risk acceptance criteria, Individual risk, Economic risk.

1 Introduction

The demand for oil and gas will continue to increase. Many countries, including China, are undertaking large-scale oil and gas pipelines constructions. However, there are many uncertainties during the construction process since the characteristics of oil and gas pipeline projects itself. The uncertainties, therefore, may result in many kinds of accidents, which is a potential risk for safety, environment issues, and economic loss as well. Risk management and control are urgent needed for carrying out suck projects to improving the construction quality of oil and gas pipeline project, guarantee the safety of pipeline construction workers and reduce construction risk. The risk management for enterprises in China has just started, and the criteria that are using mainly adopted from developed countries. Given the specific conditions in China, many of the standards need to be amended to match the rapid development. Because of the relatively underdeveloped of the technology in China, risk management faces many difficulties. Thus, it is necessary to develop a method of risk acceptance criteria for oil and gas pipeline projects, improve the basis for frisk analysis and evaluation.

Y. Zhang (Ed.): Future Communication, Computing, Control and Management, LNEE 141, pp. 189–195.
springerlink.com

2 Perspectives of Risk Acceptance Criteria

Risk acceptance criteria are tools that are used to evaluate and control risks. Although such criteria have been used for many years in different sectors of applications, their rationale and use are still being discussed [1]. The risk acceptance criteria in oil and gas industry are widely used to provide reference for risk analysis and develop risk mitigation measures, and more. Risk acceptance criteria are described in both quantitative and qualitative methods according to the type of risk presentations. Risk management standard should reflect the value of the public and the affordability of a disaster, also need to consider the affordability of community. Standards are so stringent that society cannot afford the high economic costs, will hamper social economic development. So determination of risk criteria is a process of finding the optimal solution essentially.

Different industries have different acceptable levels of risk, different scale of the project and different decision-makers towards risk attitude. Currently, determine the risk acceptance criteria generally follow two basic principles: (1) as low as reasonably possible (ALARP); (2) as low as reasonably achievable (ALARA; Fig. 1) [2-9]. Furthermore, it needs to be determined by quantifiable criteria.

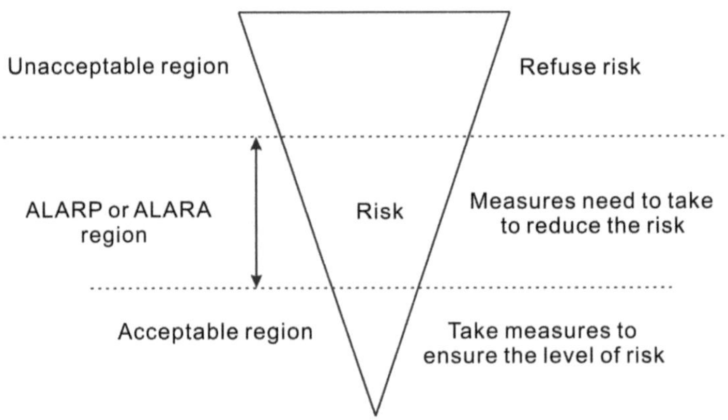

Fig. 1. The degree of risk and ALARP/ALARA guidelines

3 Models of Risk Acceptance Criteria

The impact factors of risk for oil and gas pipelines include (1) individual risk and social risk, (2) economy; economic loss related to pipeline accidents, including the cost of potential accidents, shutdown and delay production and operations, etc., and (3) environment; gas leakage cause loss of life and property and environmental damage, etc. Some of these factors can be used to measure quantitative indicators while some can only be used to describe qualitatively. In order to develop oil and gas pipelines risk acceptance criteria, it is critical to consider not only casualties, property loss and damage of pipeline system, but also the environmental pollution and the potential impact on human health and other important factors. Because many factors associated are difficult to measure accurately. Therefore, determining the risk acceptance criteria in oil and gas pipeline projects is a complex work.

Life Loss Model. Individual Risk (IR) used to determine particular time and place for a personal injury probability. Jonkman et al. [10] investigated a variety of accidents that caused a number of deaths by statistical analysis. They found that a certain period, in the same type of activity the individual risk of death remained stable. In addition to the technological progress that is likely to lead to slightly lower the risk, the risk of death had consistency and stability in a certain period and scope. Individual acceptable risk can be obtained from the above discussed statistics of the risk of death in various activities. It needs to be addressed that the Netherlands Water Control Technical Advisory Committee (TAW) accorded the degree of willingness individual participates in various activities. It is suggested that strong will of the activities (hiking adventure) to the passive activity of poor will (with hazardous facility location), located acceptable risk standards. The equation (1) is used for calculation:

$$IR < \beta 10^{-4} / a \tag{1}$$

in which β is the willingness coefficient. The change is based on the difference between individual participation and obtains benefits levels; Fig. 2 shows β values on the activities of several representatives [10].

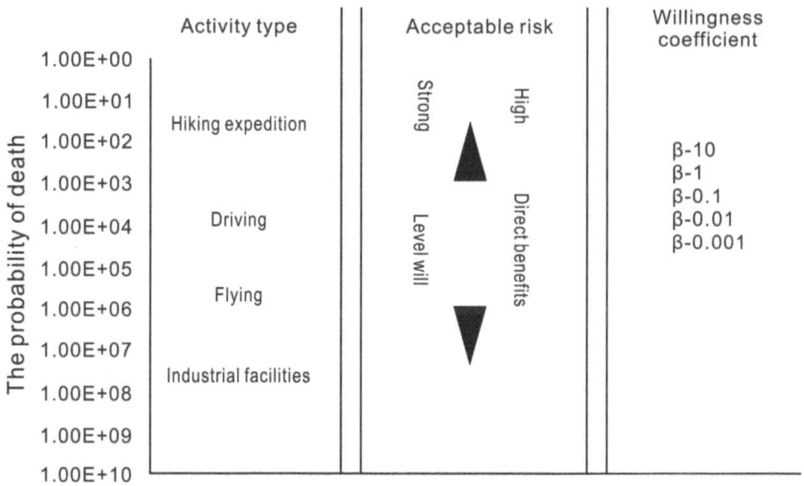

Fig. 2. Will coefficient of several representative activities

Property Risk Model. Risk of property loss used to measure total economic risk of the actual objects. For the purposes, oil and gas pipelines, risk of property loss are not factors for a particular spill directly, but the entire pipeline system.

Acceptability criteria of property loss is usually expressed by P-L curves (log-log plot) [8], (see Fig. 3). P is the level of risk or beyond the risk level, each incident on the pipeline due to the cumulative probability of loss, damage, delay; L refers to the accident-related loss. Property loss should include the loss of the structure, parts, and large maintenance costs, delay time, lost production, rehabilitation costs, and decreased reliability, etc. The area above the P-L curve is unacceptable risk area; most risks that located in the acceptable risk area cannot be assessed.

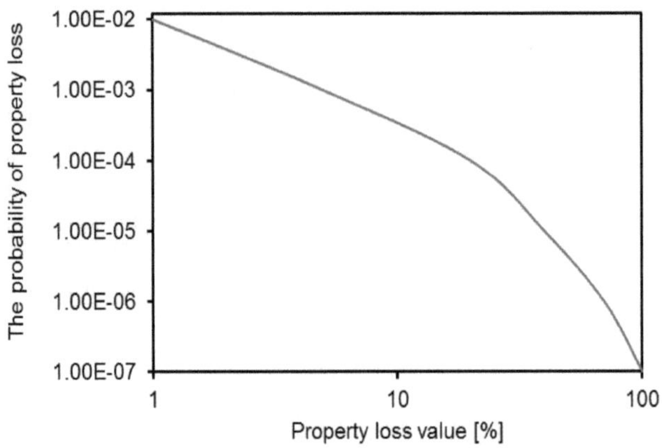

Fig. 3. Risk acceptability criteria of property P-L curve

10^{-6} Probability: Serious consequences of the accident, and without prior warning, is mainly the probability of life loss, such acceptable criteria should be less than10^{-6}; 10^{-5} Probability: Serious consequences of the accident, but has obvious signs in advance. Probability of property loss is 10^{-6} or between10^{-5}-10^{-6}; 10^{-4} Probability: Serious consequences of the accident, failure of structural elements, staff was able to escape. Property loss value is about 50% of the value of property; 10^{-3} Probability: Consequences of the accident is not serious; Extent of damage to property is certainly less than 50%, Property loss value is about 10% of the value of property; 10^{-2} Probability: Property loss is very small, for example, property loss value is 1% of the property value.

Environmental Risk Model. Environmental risk is different from the economic risk and individual risk, due to a variety of engineering activities that are exposed in a certain environment. Therefore, activities are likely to impact the environment. In actual application, in order to determine the acceptability criteria of environmental risk of a gas pipeline, it should be integrated in the acceptability criteria of three aspects of life loss, property damage and environmental damage, and get a comprehensive acceptable criteria.

Integrated Risk Acceptance Criteria Model. Integrated risk acceptance criteria not only facilitate the management for environmental risk analysis of pipelines, but also the pipeline integrity maintenance provides the basis for decision.

Purpose. To ensure safe operation of oil and gas pipeline projects, and to reduce or avoid all unnecessary risk of accidents, according to the national gas pipeline safety standards "norms" and other related content, and China current operating conditions of oil and gas pipelines develop the integrated risk acceptance criteria. In view of the possibility of the incident closely related the probability of its occurrence (Table 1, Table 2) and current precautions, prevention, detection and control measures. If the control measures in place, and in good condition, then the likelihood of the event will be greatly reduced.

Table 1. Probability of the risk classification table

Symbol	Coefficient	Quantitative analysis of the range of risk occurrence probability	Qualitative description of probability
P1	1	>1	Often
P2	2	$1\text{-}10^{-1}$	Sometimes
P3	3	$10^{-1}\text{-}10^{-2}$	Accidental
P4	4	$10^{-2}\text{-}10^{-4}$	Seldom
P5	5	$10^{-4}\text{-}10^{-6}$	Very low
P6	6	$<10^{-6}$	Rare

Table 2. The severity level of risk table

Symbol	Coefficient	Quantitative description of risk severity	Qualitative description of risk severity
S1	1	Disastrous	Damage this section of the pipeline or protection system in this paragraph, number of deaths
S2	2	Fatal	Most damage to pipes or protection systems, someone died
S3	3	Serious	Some damage to pipes or protection systems, and personnel injured
S4	4	Mild	Slight damage to pipes or protection systems, personnel sustained minor injuries

Coefficient Analysis of Risk Acceptance Criteria. Risk acceptance criteria posed by the matrix of risk acceptance (Table 3), which was used to determine the acceptable risk areas. Acceptable coefficient AC:

$$AC = Pi \times Si \tag{2}$$

Widely acceptable area $AC \geqq 12$, risk is acceptable within this region, and do not need to take action of risk control. Reasonable and feasible area $4 < AC < 12$, this region should be considered the benefit of risk acceptance and further reduce the feasibility of risk, then compare the risks and benefits. If the benefits outweigh the risks, the risk is acceptable; the contrary is not the case. Any risk should be reduced to the lowest level possible. Does not allow area $AC \leqq 4$, if risk within this region cannot be reduced, the view about that is not allowed.

Table 3. Gas pipeline risk assessment criteria

	Severity	Mild (S4)	Serious (S3)	Fatal (S2)	Disastrous (S1)
Probability		Slight damage to pipes or protection systems, personnel sustained minor injuries	Some damage to pipes or protection systems, and personnel injured	Most damage to pipes or protection systems, someone died	damage this section of the pipeline or protection system in this paragraph, number of deaths
Often (P1)	>1	4 Not	3 Not	2 Not	1 Not
Sometimes (P2)	$1\sim10^{-1}$	8	6	4 Not	2 Not
Accidental (P3)	$10^{-1}\sim10^{-2}$	12	9	6	3 Not
Seldom (P4)	$10^{-2}\sim10^{-4}$	16 can	12 can	8	4 Not
Very low (P5)	$10^{-4}\sim10^{-6}$	20 can	15 can	10	5
Rare (P6)	$<10^{-6}$	24 can	18 can	12 can	6

Note: 1. risk level = Probability × Severity 2. "can" indicates an acceptable "Not" says the unacceptable. Unlabeled says need to be further weigh the risks and benefits.

4 Summary

Risk assessment in oil and gas pipeline projects during construction period should be used to prevent exceeding the maximum acceptable risk, to reduce risk to as low as possible to a reasonable level. Three of the most important things that are assessment that can be done to reduce the risk, including risk identification and estimation, risk assessment and decision making, risk management and control, and the core issue of basis is the need for a scientific and reasonable risk acceptance criteria. In front of various risk assessment methods have their advantages and disadvantages. Therefore, the risk assessment process of oil and gas pipeline should be scientific, systematic, applicability, economic and other aspects into account, carefully choosing evaluation method, order to achieve the requirements of evaluation purposes. Finally, hope that through this research of the Chinese oil and gas pipelines to facilitate and promote for the risk acceptance criteria study.

References

1. Dai, L.S., Zheng, H.L., Cheng, W.Y., Zhang, H.B.: Risk Acceptability Criterion and Index Model for Oil and Gas Pipelines. Oil & Gas Storage and Transportation 28(8), 1–4 (2009)
2. Hu, Q.F., Huang, H.W.: Study on the Modeling of Risk Acceptance Criteria for Tunnel and Underground Engineering. Chinese Journal of Underground Space and Engineering 2(1), 543–550 (2006)

3. Wang, D.Q.: Flashover Fire Risk Analysis of Gas Pipeline Failure. Natural Gas and Oil (1999)
4. Chen, L.J.: Investment Project Risk Analysis. Mechanical Industry Press, Beijing (2004)
5. Faber, M.H., Stewart, M.G.: Risk Assessment for Civil Engineering Facilities: Critical Overview and Discussion. Reliability Engineering and System Safety (80), 173–184 (2003)
6. Wilson, R., Grouch, E.A.C.: Risk Assessment and Comparison: an Introduction. Science 236, 267–270 (1987)
7. Health and Safety Executive, Reducing Risks, Protecting People: HSE's Decision Making Process (2001)
8. PetroChina Pipeline Branch. Pipeline construction project risk management guidebook. Petroleum Industry Press, Beijing (2009)
9. Tao, L.B., Li, Y.S.: Theory and practice of risk analysis. Tongji University Press (2007)
10. Zhou, Y.: Risk acceptance criteria of long large tunnels in construction period. M.S. Thesis, Nankai University (2007)

The SWOT Analysis on the Development of Renewable Energy Constructions in Xi'an

Wu Wenping and Liang Xuelan

Department of Civil Engineering and Architecture, Xi 'an University of Technology, Xi 'an, 710048, China
wwp@mail.xaut.edu.cn, pingxingyuan@163.com

Abstract. Energy is the basic material guarantee of human life and production and is the foundation to ensure continuing growth of national economy in the modern society. The energy shortage problem will be more and more prominent as the society keeps developing. In this paper, the development of the renewable energy constructions in Xi'an is analyzed with SWOT method, The results of the analyze will illustrate the advantages and disadvantages of the existing renewable energy constructions in Xi'an as well as possible , opportunities and threats in this field, which can act as a reference for its future development in Xi'an.

Keywords: SWOT, Renewable energy construction.

1 The Current Situation of Development of Renewable Energy Constructions in Xi'an

Xi'an is located in the middle of shaanxi basin that in the hinterland of the Yellow River basin in China, It is between 107°40' to 109°49'east longitude and 33°39' to 34°45' north latitude, It belongs to the warm temperate zone's half wet monsoon climate zone, The typical climates in such region are characterized by moderate rainfall, four distinct seasons, an average annual sunshine time of 2065.6 hours and an average annual total radiation value for 4771 MJ/m2·a, Generally the years highest temperature is around 40 degrees Celsius and the lowest temperature is around -8 degrees Celsius, The frost-free period lasts 219 to 233 days on average.

Since the reform and opening, with the rapid growth of China's economic, the level of people's life is improved gradually, but also causes the issue of environmental pollution and energy shortage. Especially in recent years,with the rapid growth of China's economic, with the acceleration of the economic development and urbanization process, as well as the emergence of new constructions, energy consumptions is on a sharp rising trend, The contradictions between the development of social economy and the pollution of the energy consumptions has became increasingly sharp. To solve the problem, The State encourages industries to use renewable energy sources instead, and in response to such calling, the industries are now developing renewable energy construction actively.

Y. Zhang (Ed.): Future Communication, Computing, Control and Management, LNEE 141, pp. 197–205.
springerlink.com

At present, the applied renewable energy sources within various architectures in Xi'an are mainly solar energy, Shallow energy, Biomass energy, and so on. Those renewable energies are mainly used to solve the need of building heating, hot water supply and life lightening. During the "eleventh five-year plan", Xi'an closely followed the general guidelines of developing a resource-conserving, environmentally-friendly society, and further extended the renewable energy constructions to a larger district. So far, we have accumulatively completed more than twenty demonstration projects of the renewable energy, including seven National demonstrations, six Provincial demonstrations, and ten City-level demonstrations. The total area using solar energy technology is about 467917.26 square meters, and the accumulative total area using ground source heat pump technology is about 295100 square meters.

2 The Prospect of the Development of Renewable Energy Construction in Xi'an

2.1 Feasibility

The solar energy and ground source heat pump technology applications are flexible, safe reliable, and space saving. The solar energy and ground source heat pump technology can be used for new project, expansion or reconstruction projects, the actual constructions can be done in stages. Solar energy equipment can be located at the top of buildings or in peripheral structures, so as to conserve space. Ground source heat pump units have three types: one is monotype units and can be hoisted within the top, without taking up any indoor space; the second type is a large screw machine, and should be installed in a special room; the third type is something in between, that is, units generally scattered inside the building instead of taking up special room space. But this type of machines does need to take some indoor space. Overall, ground source heat pump for central air conditioning unit covers an area of about a third of the area needed for traditional air conditioning systems.

2.2 Superiority

At present, there are generally two ways of utilizing solar energy. One is to transform solar energy into heat energy, and the other is to transform solar energy into electrical energy. Solar energy, being inexhaustible, clean and secure, is the ideal renewable energy type. No matter considering from the amount of existing resources, the universality of the distribution, the cleanness of energy, and the reliability of the technique, solar energy has more advantages than other renewable energy types.

The applications of the shallow energy mainly utilized the ground source heat pump technology, when the systems provide energy for a building, 70% of the energy comes from the soil, and 30% comes from electrical energy. This 30% electrical energy is mainly used for the compressor and air conditioning system, In general, the ground source heat pump systems save more than two-thirds of electric power than heating by electric boiler, and more than half of energy than the fuel boiler. The characteristics of high efficiency and energy saving of ground source heat pump system determines its low operating cost, and save 30% to 70% energy than any other various heating and refrigeration equipment. Typical service life for such systems is more than 50 years, and also depreciation cost and maintenance costs are much lower than the conventional air conditioning.

As to the entire society, the solar technology as well as the ground source heat pump technology, once being applied, can not only play a certain role in environmental protection, but also reduce coal and electricity consumption, and the emissions of CO_2 to the atmosphere, which is in consistence with the requirement of low carbon life advocated by the society.

3 The SWOT Analysis of the Development of Renewable Energy Constructions in Xi'an

3.1 Analysis Theory of the SWOT

SWOT is the strategic method for analyzing the opportunities and threats existing in the external environment, and the Strengths and Weakness existing in the internal environment. Among them S refers to Internal Strengths, W refers to Internal Weakness, O refers to opportunities exist in the external environment, and T refers to Threats exist in the external environment.

Table 1. The SWOT analysis results table of Xi'an energy saving building

Strengths	Weakness
Rich in natural resources Mature technology Some successful demonstration projects Enormous development potential The guidance of foreign market	Lack of the talents Incomplete management system Competition market has not been formed Lack of operational level policy National financial subsidies are reduced Low enthusiasm is put into the related subject
Opportunities	**Threats**
The twelfth Five-Year Plan target of energy and emission reduction goals The rapid development of China's economy The transformation of people's notion on life The influence of the power market	The influence of the foreign market The development of renewable energy building is slow

Now the SWOT analysis is used in energy conservations of constructions industry in Xi'an to analyze and compare. The results are shown in the table above.

3.2 Strengths Analysis

(1) Rich in natural resources. Xian renewable energy resources include geothermal resources, solar energy resources and water resources. Being classified as belonging to the Ⅲ resource region or region with an average amount of natural resources, solar energy resources is in the medium level. This means Xi'an has more natural resources than some of the southern and eastern cities in China. The climate in Xi'an has the characteristics that it is warm in spring, hot and rainy in summer, mild in autumn and dry in winter. Under such conditions, the total amount of usable solar energy in a heating period can be up to 1352 MJ/m2•a. So Xi'an has the suitable weather conditions for implementations of renewable energy constructions in a reasonable scale.

Xi'an has a comparatively well developed geothermal geological structure and also a great amount of geothermal resources. These facts can provide excellent geological conditions for the development and utilization of shallow geothermal energy. Geothermal development and utilization of resources has been gradually developed during recent years with Chang'an - Lintong fault, soap River fault, Chana River fault and the Weihe River zone composing the borders of the construction region of an area of 466km2. The total reserves of underground water with 5.37 × 108m3 have been proven already. By the end of 1997, more than 63 geothermal wells have been completed with an average annually exploitation of about 250 × 104m3, these water resources can provide the conditions necessary for using water source heat pump. If we use ground source heat pump, in addition to the use of existing water resources, we can also make use of the water in urban sewage treatment plant after being processed. Sewage treatment plant can provide a large volume of water and guarantee constant quality of the water, with a year-round temperature of 13 to 25 degrees Celsius and other characteristics of the urban heat well. There are 14 sewage treatment plants in Xi'an,urban sewage centralized treatment rate of 85% or more, the capacity of city sewage treatment reach up to 1,166,000 tons / day, this is a valuable and abundant hot water resources.

(2) Mature technology. Our country, base on the development experience in foreign countries, has analyzed the actual situation in domestic area, and has done active researches in energy saving technology, thus making renewable energy construction application technology to be more and more mature. The application of solar photovoltaic panels and ground source heat pump technology in large scale, have confirmed this. In China, the solar energy water heater has developed from simple cyclical heating systems to central heating systems, centralization and decentralization integrated heating system, the water cycle heating system, and building heating function such as complex "comprehensive utilization of solar energy system". Then development extended from using light and heat to using solar thermal and photovoltaic energies.

(3)Some successful demonstration projects. From the time when Xi'an first started to develop energy-saving architects, in completed constructions of renewable energy demonstration projects, the accumulated areas being occupied by solar energy constructions have reached 467917.26 m2, The applications of ground source heat pump technology has covered a total area of about 295100 m2. Among these projects are some highly successful demonstrations such as export processing zone with 1 MW photovoltaic (pv) grid construction projects, and the "Dongshang—Forest movement third phase" underground water source heat pump air conditioning system energy saving heat demonstration projects. These successful demonstration projects to construction provide experience for each subject: investors think in energy-saving building, considering the central financial subsidies and these price is higher than traditional commodity house, their profits is also higher than traditional building; The owner in the use of energy-saving building, can use the solar or ground source heat pump all year, and to reduce the cost of operation, but also improve the indoor environment within; The government take the demonstration project to propaganda ,encouraging people actively use renewable energy construction.

(4) Enormous development potential. China's total amount of energy consumption in buildings is rising year by year, increasing from 10% in total national energy consumption at the end of the 1970s all the way to 27% in total national energy consumption during recent years. Compared with some developed countries, the consumption of energies during construction operation processes in our country is 2-3 times more. In recent years, because of the rapid development of economic, the speeding up of industrialization and urbanization, and the reform of the city being vigorously promoted in Xi'an, the demands for new buildings has been increasing. Among these buildings, most are the traditional heating cooling building. The proportion taken up by renewable energy building is very small. This ensures possible further development of renewable energy building.

(5) The guidance of foreign market. So far, 75% of the world's ground source heat pump installations exist in North America, in which in the application technology, the equipment, construction machines, buried tube material and back to cycle fluid, packing and auxiliary software has advanced equipment and research. European developed countries on the other hand mainly focus on the researches of the solar energy application. In recent years, developed countries in Europe have established solar heating plant supplying hot water and indoor heating for residential district living. Their utilization of solar energy is much more effective than that in China. Foreign technology in ground source heat pump and solar technology, market development mode can all be guidance for our use of renewable energy in new buildings.

3.3 Weakness Analysis

(1) Lack of the talents. The energy conservation of constructions is a new technology. It needs a large number of professionals and technical personnel, specializing in energy saving design, construction to operation and even the development of related industries, However, Xi'an is in shortage of this kind talents.

(2) Incomplete management system. In 2009, the energy conservation of constructions and material management institution was established by the urban and rural construction committee in the city of Xi'an. In 2011, the leadership team was set up, with the leader being the vice mayor of Xi'an. These show that the city government of Xi'an pays tremendous attentions to the work of the energy conservation of constructions. But since the organization is established preliminarily, the working content is still in the stage of trials and errors, and the policy measures of the energy conservation of constructions are not perfect. In addition, the supervision and administration in the building process of renewable energy construction is not strict. These will affect the development of renewable energy constructions.

(3) Competition market has not been formed. Market competition is the basic characteristics of the market economy. In the conditions of market economy, the enterprises consider their respective interests in order to get a better production conditions and more market resources. Through the competitions, enterprises achieve the survival of the fittest, and then realize the optimizing allocate-ion of resources. The development of energy conservation of constructions in the city of Xi'an up to now still stays in the stage of demonstration project. The lack of enough investors and customers makes it impossible to form a competitive market.

(4) Lack of operational level policy. Since the implementation of national energy conservation of constructions, many macroeconomic policy, industry standards, laws and regulations, and so on that are related to renewable energy constructions and energy conservation of constructions were introduced such as "The implementation opinions to promote the applications of renewable energy in the construction"(2006), "Standard for Energy Efficiency Inspection of Heating Residential Buildings"(JGJ132-2001), "Renewable energy law"(2009). These policies on the development of China's renewable energy construction play an improving role, but only in the macroscopic aspect. No operational policy that can be executed on local exists. This circumstance causes the blank of operability policy, and hindered the development of renewable energy construction.

(5) National financial subsidies are reduced. The State implements financial subsidies for the enterprises that adopt renewable energy construction actively. In the beginning of "eleventh five-year plan", the State adopted the "721" subsidies, which regulates that the central financial subsidies take up by 70%, the local government subsidies take up by 20%, and the construction unit investment take up10%. Under such policy, the construction units shall bear 30% of the cost. Along with the development of renewable energy construction, the central financial subsidies are reduced. In July 2009, the "The notice of the city demonstration implementation plan of the application of renewable energy construction (CJ[2009]305)" is issued by the ministry of finance. Stated in such notice, energy-saving building project become city demonstration. If the local government wants to declare the renewable energy construction demonstration city, no less than 3 million m2 of renewable energy construction must be added every year. The central finance will give subsidies, when this goal is reached. And even after achieving such subsidies , the central government subsidies are still no more than 10%. The reduction of national financial subsidies also influences the investor to develop renewable energy construction.

(6) Low enthusiasm is put into the related subject. Most of existing users use the central heating cooling meter age mechanism. The charges is calculated based on the area in this mechanism, whether or not the user take the energy saving measures have no influence on their own interests. So the users have no desire for the energy saving, and the demand of energy saving market is weakened greatly. In addition, due to the short development time of renewable energy construction, weak propaganda, and the hysteresis of public understanding, the enthusiasm in choosing and buying commodity house is negative.

Because of the fact that a large amount of money are usually investigated in construction project, the construction unit will consider both risk and capital recovery. There are rich experiences in the traditional construction where fixed models and ample examples are available. So the corresponding risk is smaller; Renewable energy buildings in the earlier stage of development on the other hand, need large amount of investments on purchasing equipment and the effectiveness on the short term is not obvious. In doing so, in the construction unit producer products the un- renewable energy building can only be passively accepted by consumer, and thereby reducing the renewable energy building development.

3.4 Opportunities Analysis

(1) The twelfth Five-Year Plan target of energy and emission reduction goals. The twelfth Five-Year Plan separately raised the energy conservation and emission reduction, and with the twelfth Five-Year Plan synchronously, which is the nation's strategic policy, reflects the nation's laying great emphasis on the work of energy conservation and emission reduction. Wen Jiabao comes up with the proposal of increasing the energy saving renovation investment in existence and promoting the development of new energy conservation of the building. Xian city also established energy saving and emission reduction plan during "The twelfth Five-Year Plan ": it will achieve a renewable energy construction application area of 6 million square meters or more, which will in turn enable energy saving of 130,000 tons on coal consumption , and reducing carbon emissions to 330,000 tons. At the end of "The twelfth Five-Year Plan", the renewable energy construction application area will account for 10% of that year's new civil building.

(2) The rapid development of China's economy. The rapid development of the China's economy provides the economic basis for renewable energy building development, which can in turn improve people's living standard on average. Generally the prices of renewable energy building commodity are higher than traditional building. When people are wealthy to choose and buy commodity house, they can not only consider their present interests, but also long-term interests and will then choose renewable energy construction after weighing real-time prices of commodity house and operation cost.

(3) The transformation of people's notion on life. The core contents of 'low carbon' are low carbon economy and low carbon life. As a way of life, "low carbon life" , which first rise from abroad ,is to try to reduce the amount of energy consumed during work and entertainment in people's lives so as to reduce emissions of carbon dioxide. Now this way of life has been quietly coming into China. Low carbon life is a kind of life attitude for ordinary people, and has become the trend of the new way to push people. In the construction industry, low carbon is to reduce the energy consumption in the process of building and running constructions, that is, to use renewable energy construction.

(4) The influence of the power market. On March 15, 2011, nuclear security issue brought about by Japan's strong earthquakes attracts people's focus. Taking into account the negative effect of nuclear power, China has suspended the development of nuclear power. With gradually abundance of China's power supply and country's attention of saving energy and reducing consumption, China began to intensify efforts to adjust the structure of the power industry. Coal-fired power industry's overall

efficiency has experienced certain down trend during the "eleventh five-year plan" period. The heavy loss of thermal power enterprise explains our state's decreasing position in coal-fired power industry. As for hydraulic power industry, from the current water resources development and utilization, our country basically has entered the later water and electricity market. Considering all the negative effects of different generating way: Nuclear power on air and environmental pollution, thermal power pollution of the atmosphere, water and electricity influences to aquatic organisms and Basin nature development, the state encourages all walks to the use of renewable energy, and reduces the use of conventional energy sources.

3.5 Threats Analysis

(1) The influence of the foreign market. Foreign construction energy conservation industry started early, and the corresponding policies are more perfect than those in our country, which has certain guidance and facilitate role to the development of building energy efficiency in our country. In the early stages of the building energy conservation industry development in our country, any development trend of the developed countries will cause huge impact on our country 's. These successful demonstration projects to construction provide experience for each subject: investors think in energy-saving building, considering the central financial subsidies and these price is higher than traditional commodity house, their profits is also higher than traditional building; The owner in the use of energy-saving building, can use the solar or ground source heat pump all year, and to reduce the cost of operation, but also improve the indoor environment within; The government take the demonstration project to propaganda ,encouraging people actively use renewable energy construction.

(2) The development of renewable energy building is slow. Although renewable energy building has obvious advantages, the renewable energy building develops slowly in general. This is the result of many factors. To promote renewable energy building's fast development, we must first make the related industry realize the importance of energy saving. Then we need to strengthen the publicity of building energy efficiency, making the various social mobilization works together to promote the development of renewable energy construction.

4 Conclusion

Through the analysis on development of renewable energy construction status of internal and external conditions in Xi'an , there are obvious advantages for development of renewable energy buildings. Development opportunity is greater than the challenges that we are facing. We should make full use of prominent interior advantages and numerous opportunities, increase investment to speed up development, complete relevant policy system, and promote the formation of supervision mechanism to speed up the development of renewable energy construction.

References

1. Xi'an urban and rural construction and the committee compiled. Xi'an renewable energy construction application of "The twelfth Five-Year Plan" development planning
2. Xi'an urban and rural construction and the committee compiled. Xi'an renewable energy building concentrated promotion plan
3. Kang, Y., Li, Y.: China's construction energy conservation barriers and countermeasures. Hvac HV&A 2, 33–36 (2006)
4. Zeng, F., Zhang, M.: Xi'an groundwater exploitation and protection countermeasures. Water Resources Planning and Design 6, 9–11 (2009)
5. Wang, H.: Xi'an groundwater resource development and utilization analysis. Ground Water 10, 375–377 (2005)
6. Xu, Y., Kuan, W.: North America region ground source heat pump technology and application. Chinse Chapter, International Ground Source Heat Pump Association

The Passenger Car Chassis Upgraded Based on Independent Suspension

Yang Wu, Li Ming, and Liang Zhi Hong

School of Mechanical Engineering & Automation, Xihua University, Chengdu 610039, China
yangWu123456020@sina.cn, {120162016,630824654}@qq.com

Abstract. Suspension is one of the important assemblies about modern vehicle, which puts frame and axle join up elastic flexibly. Its main task is to transfer function of all the force and moment between the wheel and frame, and alleviate the impact load which transfers to frame from uneven pavement, all of which is to ensure the normal operation of vehicle. And adopting independent suspension can reduce the spring load part weight greatly, reduced vibration load, and so improved vehicle running smoothness. Before, the use of passenger car suspension generally adopts a dependent suspension. As people pay attention to ride comfort, control stability, running smoothness, the original suspension system already cannot satisfy the requirement of people. Passenger car (Chengdu) company devotes to improve passenger car quality and increase passenger car products market share. Now this paper is upgrading chassis of CDL6606 model, changes the original plate spring suspension into that the front suspension adopts double wishbone independent suspension and the rear suspension uses few leaf springs.

Keywords: Double wishbone independent suspension, few leaf springs, upgrade, passenger car.

1 The Development Situation of Automobile Suspension

In recent years, with the development of high-speed road network at home and abroad and the Speed's improvement, more and more high demand is put forward for car suspension structure design. The development of foreign car suspension mainly revolves around cars. This is because of the high speed of cars and the high demand for comfort, steering stability[1]. But the suspension of passenger car and van are also in development. At present, although the suspension links of the car suspension system have various types, front suspension of middle or senior cars mostly uses the combination of wishbone suspension, Helical spring and shock absorber independent suspension ,the after suspension mostly still uses rigid axle suspension, passenger car suspension still uses plate spring primarily, but many domestic famous passenger car production enterprise will change suspension, as preferred suspension of luxury passenger car. The air suspension is gradually becoming favored for customers[2].

Y. Zhang (Ed.): Future Communication, Computing, Control and Management, LNEE 141, pp. 207–212.
springerlink.com

2 The Comparison of Independent Suspension and Rigid Axle Suspension

Rigid axle suspension, especially use parallel leaf spring as elastic components (and suspension link), because of its simple structure and reliable, easy to manufacture, maintenance convenient, when the wheel beats up and down, positioning parameter of steering wheel changed inconspicuously, tire abrades inconspicuously, and other advantages, which are widely used in the truck and some passenger car at present [3]. Some car's rear suspension also uses that structural style .But , because rigid axle suspension exists some serious flaws, for example it can not ensure sufficiently ride comfort and drive stability of vehicle when in high-speed, many vehicles especially cars adopt independent suspension, because that it has the following advantages: (1)The unspring mass of independent suspension lightened, ride comfort improved; (2)It can be installed with very soft spring so as to improve the ride comfort; (3)With regard to car's rear-wheel for FR type, it can fixed differential mechanism at the side of car body, thus reduce the between body under frame and the floor ,and the distance between rear seat and the floor ,as well as reduce the center of gravity of car .Relative above advantages, this suspension form has the below shortcomings: (1) Independent suspension's structure is complex, manufacturing cost is higher; (2) Auto's maintenance and its repair is difficult; (3) Front wheel alignment and wheel track often change when the passenger car is moving, which lead to big tire degradation[4].

3 The Structure Type and Analysis of Independent Suspension

Independent suspension accords to the wheel movement form is divided into: The suspension that Wheel swings in automobile transverse plane; the suspension that Wheel swings in automobile longitudinal plane; the suspension that Wheel swings in automobile oblique plane; the suspension that the wheel Slides along the dowel .Independent suspension accords to the structure characteristics of suspension link is divided into: double wishbone, singe trailing arm and so on[5].

3.1 The Development Situation of Automobile Suspension

Elastic element is one of the main part of the suspension .Its role is to support the above weight of the car suspension ,mitigate the impact loads that come from road surface and pass on to frame .At present the elastic component of suspension has various kinds ,for example, leaf spring, spiral spring, torsion bar spring, gas spring (air spring, oil and gas spring), rubber spring etc .It stores the energy mainly through material deformation, and then releases the energy slowly to mitigate the impact. So when analysis the performance of elastic component, the unit store energy is usually used to compare, store energy size relates to admissible stress, modulus of elasticity and the density of the material.

4 The CDL6606 Suspension Upgrade Scheme

The original passenger car chassis is rigid axle suspension Figure (1), elastic component is leaf spring, this suspension has never satisfied the requirement of smooth-going and control stability when in high-speed driving. According to overall dimensions and contoured design of car body (Figure 2), at present, front suspension adopts differ long double wishbone suspension (Figure 3).The rear suspension uses tapered-leaf spring (rigid axle suspension).

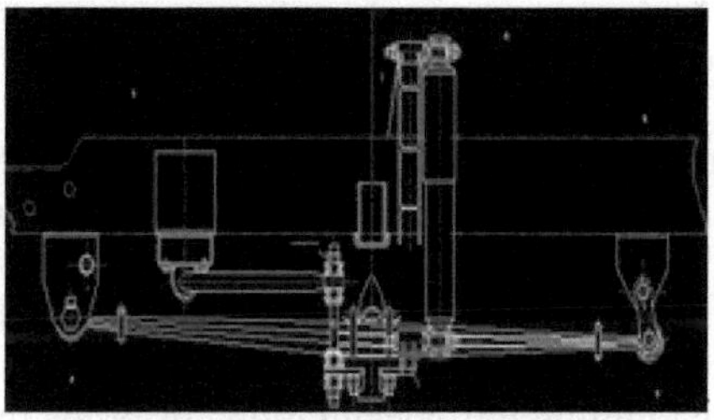

Fig. 1. Leaf spring suspension

Fig. 2. Body appearance

4.1 The Structure Type and Analysis of Independent Suspension

Because of leaf spring's simple structure, mature technology, it is widely used in van suspension and passenger car suspension. Its structural strength mainly impacts on Safety and Comfort [6].Compare to other elastic components , not only leaf spring plays the role of elastic component, but also can play the role of orientation component that decides the characteristics of motion of axle, wheel, and can transfer all kinds of force and torque. The friction between the pieces as well as between hinge departments can damp vibration, so it is still widely used as a elastic component at present. Its design mainly consider that leaf spring stress, the selection about Leaf spring main parameter(Spring load, static deflection and dynamic deflection, the arc high when the passenger car is fully loaded , spring length, the size of blade section and the number of slices etc) [7]. Less leaf spring (Figure 4) has some advantages of few slices, light quality, as well as good smooth, which have very important significance to realize the vehicle lightweight and improve vehicle Smooth and comfortable. Therefore exploiting less leaf which can bear high stress level and high intensity is the development trend at home and abroad. Meanwhile, design stress of less leaf spring must have more advance than multiple-leaf spring, in addition to that, less leaf spring, its own design stress level also rises ceaselessly, and the design stress of automotive spring has been raised from the original design stress 900MPa level to 1200MPa level.

In order to reduce vehicle quality and improve it's smooth, at home and abroad, less leaf spring is more and more used on the passenger car, which is composed of one or several pieces of longitudinal variable cross-section. Firstly, it is widely used in commercial vehicle and the passenger car, it uses variation thickness of section to keep equal strength characteristics, compared with the multiple-leaf spring, its quality decreased by 30 - 40 percent, and some even reach up to 50 percent.

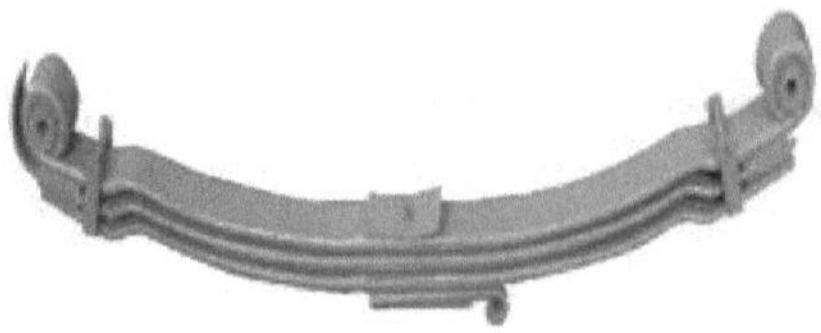

Fig. 3. Less leaf spring

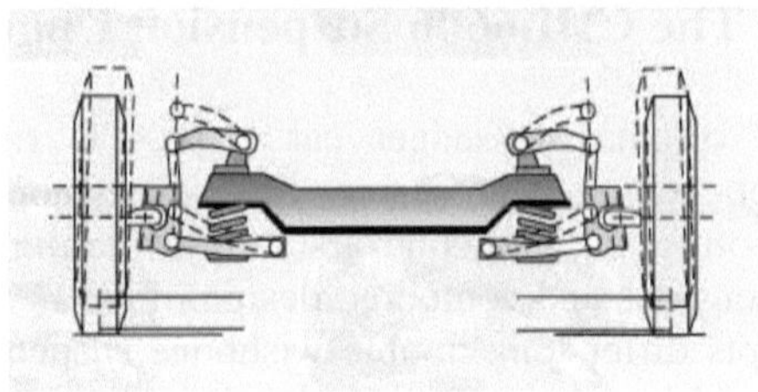

Fig. 4. Differ long double wishbone independent suspension

4.2 Independent Suspension Should Satisfy the Requirements

The requirement of front wheel independent suspension link : (1) When the load changes in suspension, suspension link should ensure that wheel tread doesn't make a conspicuous change, If wheel tread change conspicuously, which will cause the wheel wear seriously; (2) When the load changes in suspension, it should make front wheel positioning angle have reasonable variation characteristic;(3) When the load changes in suspension, wheel does not engender great longitudinal acceleration, In order to reduce the longitudinal impact, avoid the inertia torque is brought to bear on the steering knuckle, make the torque on the steering wheel change sharply; (4)when the car brakes, car body restrain "nod" action; (5)when the car accelerate, car body restrain "upward" action[8].

4.3 Differ Long Double Wishbone Independent Suspension

The length between the two swinging of double wishbone typed independent suspension can equal or differ. Equal length double wishbone independent suspension, when the wheel beats up and down, the wheel plane does not incline, but wheel tread changes conspicuously, which will increase the possibility of lateral sliding of wheel. Differ long double wishbone independent suspension, If two arms length select properly, which can make the length between master pin and wheel change inconspicuously, and wheel tread also changes inconspicuously, the small change of wheel tread can be supplied by the deformation of tire.

4.4 The Dipping Angle and Direction of Dip about Upper and Under Arm in Horizontal Plane

When guide mechanism is designed, wishbone should be also arranged correctly besides the choice of the size of upper and under arm and the ratio of its. when design position, the layout scheme old upper and under arm can be parallel to the horizontal plane, can be designed into a dip angle in the horizontal, and sloping direction can also have different choice(Figure 5), now a pair of steering mechanism is choused as a example ,we can analysis it and get its roll center, if the car body does not move ,the ground swings around the car body, the linear velocity direction of C and D point, which is the contact point between the ground and tire, Namely, the instantaneous center that the ground relative to the car body. If we presume that the wheel is rigid ,the wheel doesn't sideslip and also won't jump off the ground when the car body over, according to the above assumptions, the C and D point linear velocity on the ground is the same as the C and D point linear velocity on the tire which contacts to the same position on the ground ,therefore the vital question is how to get the velocity of C and D point which relates to car body .In order to analysis conveniently, the tire(including knuckle which connects the tire rigidly), upper arm, under arm and car body are labeled respectively with 1, 2, 3, 4 ,these four rods constitute four link mechanism. The 4 rod represents car body is motionless, it is easy to get instantaneous of "a" relates to 1 rod through the knowledge of principle of machinery, we can extend 2 rod and 3 rod which represents upper and under arm, the intersection of extension is instantaneous b, so the velocity direction of "c" on the tire can be ensured, which perpendiculars to the connection line of b and c point. Similarly, the speed direction of "d" (v_d) on the tire can also be found, it is perpendicular to connecting line of "a" and "b". If the speed direction of "c" and "d" which the tire contacts the ground are known, then the instantaneous swinging center that the ground relates to car body is on the point o of intersection of $\overline{ad}$ and $\overline{bc}$, This point "o" is roll center of car body, because the left and the right wheel is symmetrical each other, so point "o" must fall in automobile symmetrical center line (Figure 6).

By the above analysis, we can conclude that the center of the rob center with the arrangement of upper and under wishbone in a horizontal plane has a close relationship, as long as the change of arrangement angle of the upper and under in a horizontal plane, we can get different rob center position, the rob center position is usually evaluated by the height between the rob center and the ground, which affects the automobile driving performance, especially running smoothness and control stability, in short, the higher the distance of rob center, the shorter the distance from it to center of gravity, which can make the arm of force caused by lateral force short ,the result is to make car body side angle small. But, if rob center is too high to keep the wheel tread changes unobvious and reduce the attrition of tire, therefore the rob center should be chose appropriately and will not have big variation when the wheel changes. Through comprehensive consideration about car body factors, in order to avoid to produce larger interference, the arranging scheme that the upper and under wishbone parallels to horizontal line in horizontal plane is adopted.

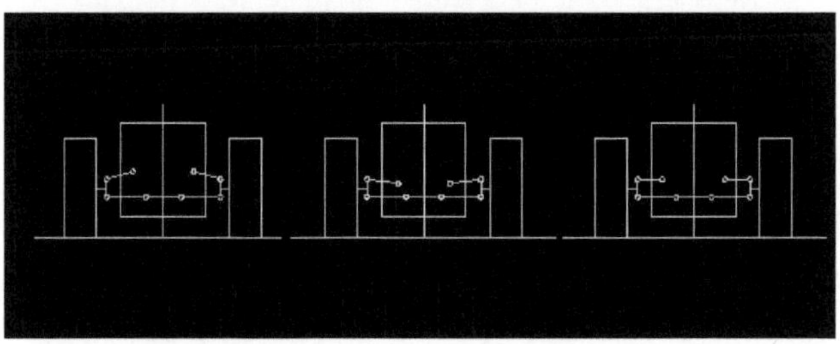

Fig. 5. The arrangement of upper and under arm in horizontal plane

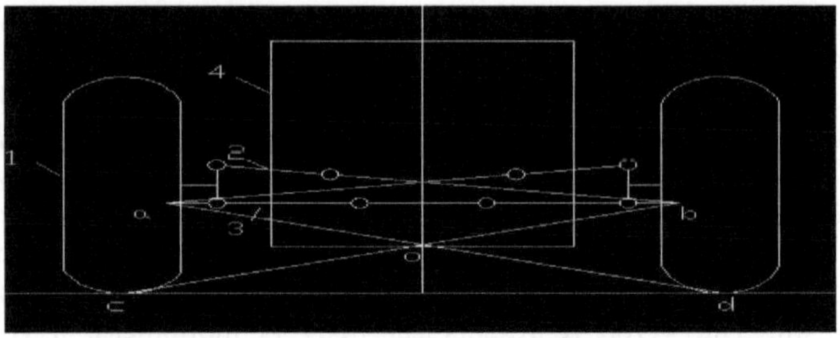

Fig. 6. The rob center of guide mechanism

5 Conclusions

This paper puts forward the original passenger car chassis upgrade scheme, through the scheme is proposed, the chassis suspension design thought is understood, especially ,the double wishbone suspension design has a detailed overview, the upgraded passenger car chassis is checked by experiments, its vehicle running smoothness and control stability have improved a lot, as people living standard is increasing day by day, in order to satisfy the need of the market ,the chassis will have further upgrade, for example, installing air spring suspension, adding slow speed device and so on.

References

1. Yu, Z.: Auto theory. Mechanic industry Press (2004)
2. Cai, X.: Automobile structure and principle. Mechanic industry Press (2004)
3. Ji, L., university: Automobile teaching and research section. Automobile design. Mechanic industry Press (1986)
4. Xiao, T., Zhi, H.S.: Automobile design. Mechanic industry Press (1990)
5. Zhang, office, Luo, Z.: Less leaf spring optimization design. Automobile Technology (2010)
6. Jiang, P.: Automobile suspension system simulation analysis and parameter optimization design, pp. 433–438 (2006)
7. Deng, Y.D., et al.: Car body design based on UG, pp. 41–52 (2005)
8. Ding, H., et al.: Less piece leaf spring mechanical properties of finite element analysis, pp. 101–107 (2009)

Application of Artificial Neural Network to Predict Real Estate Investment in Qingdao

Ping Zhang, Wenjing Ma, and Tiejun Zhang

College of Engineering, Ocean University of China, Qingdao, China
{zhangping084911,mawenjin163,zhangtiejun-699}@163.com

Abstract. As a research object of the paper, the amount of real estate investment in Qingdao is thought to be a nonlinear function of six parameters: city resident population, average wages of staff and workers, per capita consumer spending, commodity house sales, GDP and per capita disposable income. In this paper, we use MATLAB software to fit and predict the six parameters, and a time-series predicting model is set up by means of MATLAB toolboxes. The predicted values obtained can help government departments and real estate enterprises to formulate relevant policies and decisions.

Keywords: The Amount of Real Estate Investment, Neural Network Toolbox, Time Series, Predicting.

1 Introduction

The housing problem of urban residents is one of the critical livelihood issues, which is the government focus of concern. In order to solve the housing security problem, the government not only increases the administrative power to curb the rises of housing prices, but also enhances the construction of economically affordable housing, low-income housing, and low-rent housing.

Just under this background, it is necessary to study the amount of real estate investment: only if the government controls the capital input effectively, can the heat of real estate be adjusted, and the real estate bubble be reduced. However, there are many factors which influence the amount of real estate investment. In the past, people adopted linear mathematic model to predict the investment, but the accuracy cannot meet the requirement, and the effect is not ideal. Neural Network itself possesses the adaptive capacity and learning performance of Bionic Algorithm. In theory, it can approximate arbitrary nonlinear mapping, have excellent fault tolerance and stability. The predicted data are practical and reliable, which can be used by government departments and real estate enterprises to formulate decisions.

2 Neural Network Model Principle

An artificial neural network (ANN) is a model that simulates human nervous system. The model can simulate some intelligent behaviors of human brain, such as

Y. Zhang (Ed.): Future Communication, Computing, Control and Management, LNEE 141, pp. 213–219.
springerlink.com

consciousness, inspiration and imagery thinking. A neural network consists of an interconnected group of simple processing elements which mimic neurons in human brain. The typical neural network model is as follows:

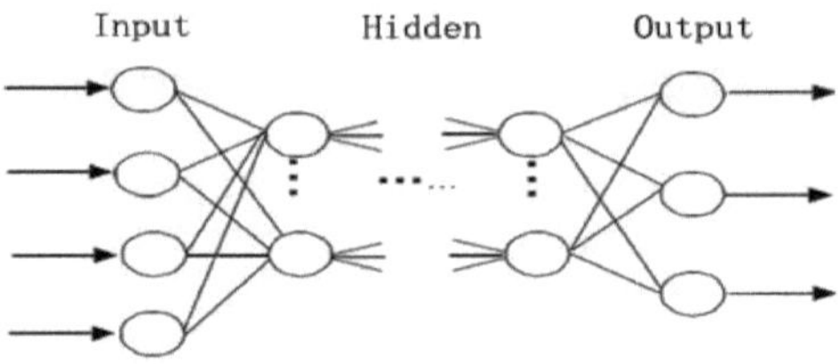

Fig. 1. A neural network model

3 Parameters Prediction

MATLAB Curve Fitting toolbox is used to fit six parameters, including city resident population, average wages of staff and workers, per capita consumer spending, commodity house sales, GDP and per capita disposable income.

Table 1. Historical data of six Parameters

Year	City resident population [10 Thousand]	Average wages [Thousand]	Per capita consumer spending [Thousand]	Commodity house sales [Million m^2]	GDP [Billion]	Per capita disposable income [Thousand]
2000	706.65	10.072	6.677	3.007	115.007	8.016
2001	710.49	11.426	6.849	3.870	131.608	8.731
2002	715.65	12.839	7.3440	4.291	151.817	8.721
2003	720.68	14.162	8.0562	4.532	178.042	10.0753
2004	731.12	15.950	9.0023	5.163	216.380	11.0888
2005	740.91	17.940	9.8835	7.399	269.550	12.9198
2006	749.38	20.244	11.9448	7.365	320.658	15.3281
2007	757.99	21.420	13.3758	8.333	378.652	17.8558
2008	761.56	23.296	14.9986	7.692	443.618	20.4644
2009	762.92	25.396	16.0799	12.619	489.033	22.3679
2010	763.60	28.549	17.5311	13.607	566.620	24.9981

Linear polynomial is used to fit the curves of the six parameters, the equations is as follows: f(x) = p1*x + p2.

Table 2. Parameters and Predicted Value

Parameters	p1	p2	R-square	2011Predicted Value	2012 Predicted Value
City resident population	6.605	-1.251e+004	0.9657	772.6550	779.2600
Average wages	1.804	-3599	0.9906	28.8440	30.6480
Per capita consumer spending	1.161	-2317	0.9625	17.7710	18.9320
Commodity house sales	0.9818	-1962	0.8819	12.3998	13.3816
GDP	46.08	-9.209e+004	0.9681	576.88	622.96
Per capita disposable income	7.768	−3530	0.95	25.4480	27.2160

4 Real Estate Investment Prediction

4.1 Network Structure Determination

The determination of artificial neural network refers to the selection of network layers numbers and the neuron nodes quantity of each layer. A typical BP network is a three-layer network, including input layer, output layer and hidden layer. From the data above, we can get the number of the neuron nodes in the input and output layer: six nodes in Input and one in Output. For most practical problems, one hidden layer is enough, so the hidden layer we use is also one; the neuron nodes number of the hidden layer can be gained from empirical equations, the equations used are as follows:

$$m = \sqrt{n+l} + a \tag{1}$$

$$m = \log_2 n \tag{2}$$

$$m = \sqrt{nl} \tag{3}$$

Where: m is the number of hidden nodes; n is the number of nodes in input layer; l is the number of nodes in output layer; a is a constant between 1 and 10.

4.2 Basic Procedures to Realize Neural Network Using MATLAB

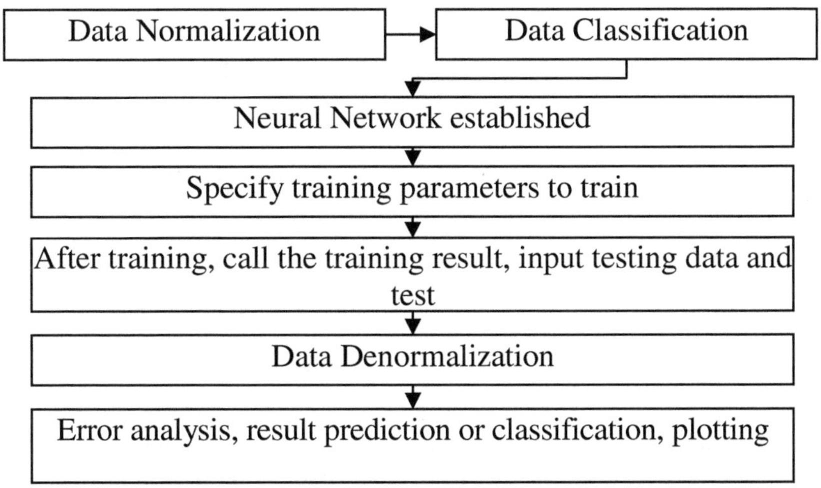

Fig. 2. Procedures to realize neural network using MATLAB

4.3 MATLAB Neural Network Toolbox

Neural network toolbox is based on artificial neural network theory, and use MATLAB language to construct activation function of typical neural network. In this way, the calculation of selected network output can change into invoking of activation function. In addition, based on various kinds of typical learning rules and network training process, subroutine program of network weights training can be programmed. The designers of neural network can call the programs relevant to neural network design and training in toolbox according to their own demands. In this way, they can free themselves from tedious programming work and concentrate on deliberating and solving the problems, thus the work efficiency and problem-solving quality can be improved.

At present, the latest version of MATLAB summarizes nearly all basic results of neural network, the network model involved including: Perceptions; Liner Filter; Backpropagation; Control Systems; Radial Basis Networks; Self-organizing Networks; Feedback Networks; Adaptive Filtering and Adaptive Training.

Time Series Prediction. Time Series, also called time progression, history plural or dynamic series, is a progression obtaining through arranging some statistical indicator values according to chronological sequence. Time Series Forecasting Method is to make and analyze time series, analogize or extend on the basis of evolution process, direction and trend which reflected by time series, so as to predict the potential development level in the next period of time. Time Series Prediction can be divided into three categories:

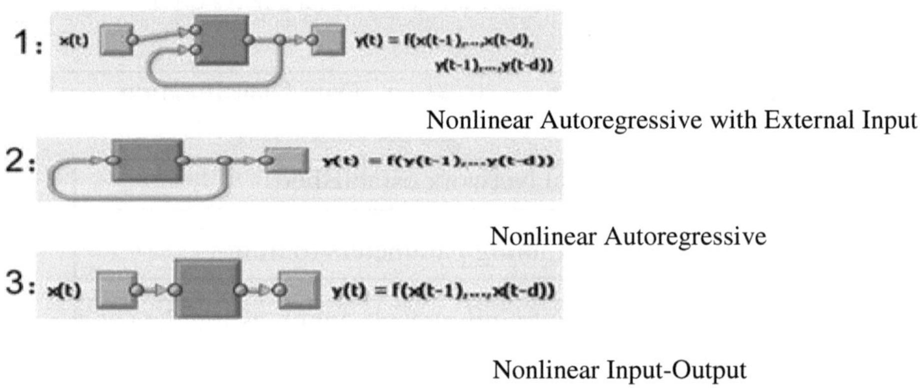

Fig. 3. Categories of Time Series Prediction

The prediction made in this paper is a time series prediction based on neural network. Neural network can be subdivided into static and dynamic network. Furthermore, dynamic network may be classified into feedback and non-feedback dynamic network. Here, we select the dynamic -feedback type. In addition, we need to split data into training data, validation data and testing data, the ratio among the three types of data can be set up on our own.

The specified procedures to predict using MATLAB toolboxes is as follows:

Start MATLAB software, open Editor Window, and input data:

p=[706.65 10.072 6.6770 3.007 115.007 8.0160; 710.49 11.426 6.8490 3.870 131.608 8.7310; 715.65 12.839 7.3440 4.291 151.817 8.7210; 720.68 12.597 8.0562 4.532 178.042 10.0753; 731.12 13.932 9.0023 5.163 216.38 11.0888; 740.91 16.015 9.8835 7.399 269.55 12.9198; 749.38 18.574 11.9448 7.365 320.658 15.3281; 757.99 21.419 13.3758 8.333 378.652 17.8558; 761.56 23.296 14.9986 7.692 443.618 20.4644; 762.92 25.396 16.0799 12.619 489.033 22.3679;763.6 28.549 17.5311 13.607 566.62 24.9981];

t=[67.5;92.52;103.7;127.8;162.7;223.8;268.4;322.4;373.1;459.5;602.4]; Press F5 to run.

Input "nnstart" in the Command Window, open the Neural Network Start Window; select "Time Series Tool" and choose the first kind of time series prediction; select "p" for Inputs, "t" for Targets and "matrix row" for Time Step, then press "Next"; select the ratios of the three data according to our own need, then "Next"; the number of hidden neurons is set up to be "15", then "Next"; click "Train" training network, then we can get the graph as follows:

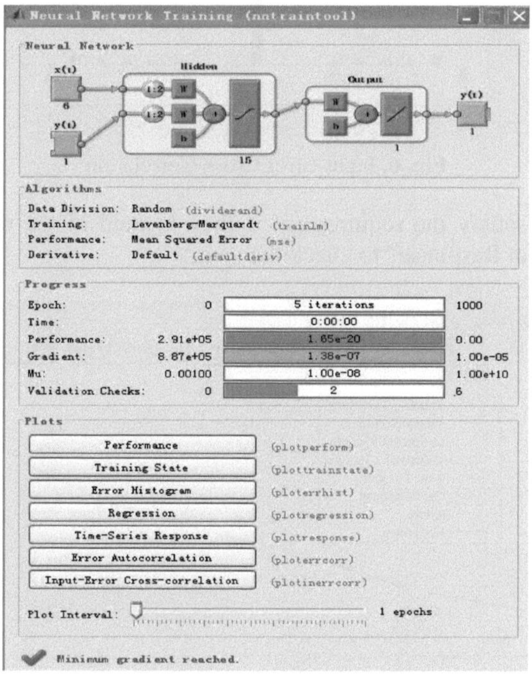

Fig. 4. Neural Network Training

Among all the indicators, the two indicators "Error Autocorrelation" and "Input-Error Correlation" are relatively important, which as shown in the following graphs:

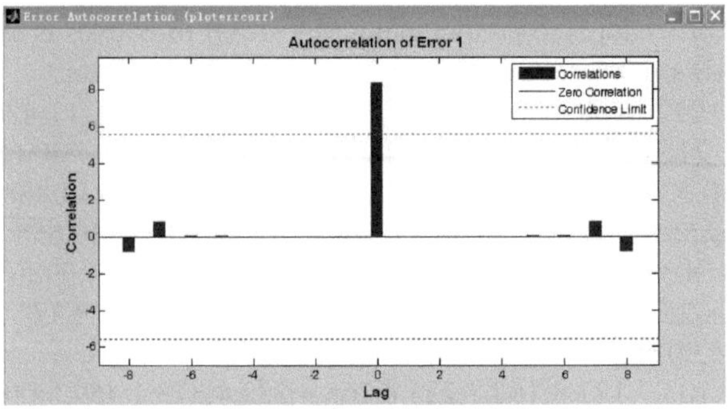

Fig. 5. Error Autocorrelation

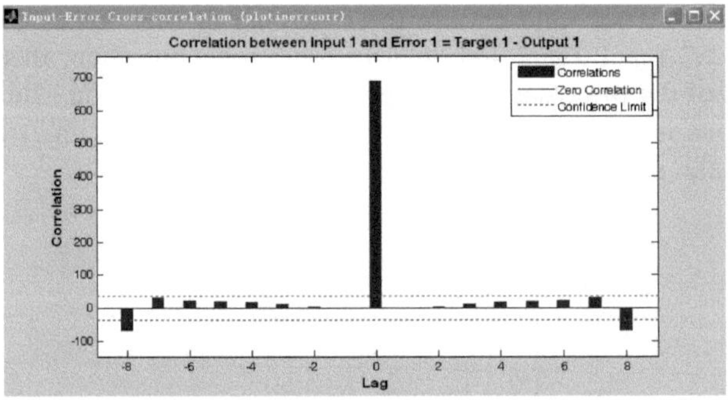

Fig. 6. Input-Error Cross-Correlation

The errors can satisfy the requirement if they located in the range with the two dash line, click "Plot Response" to check the error:

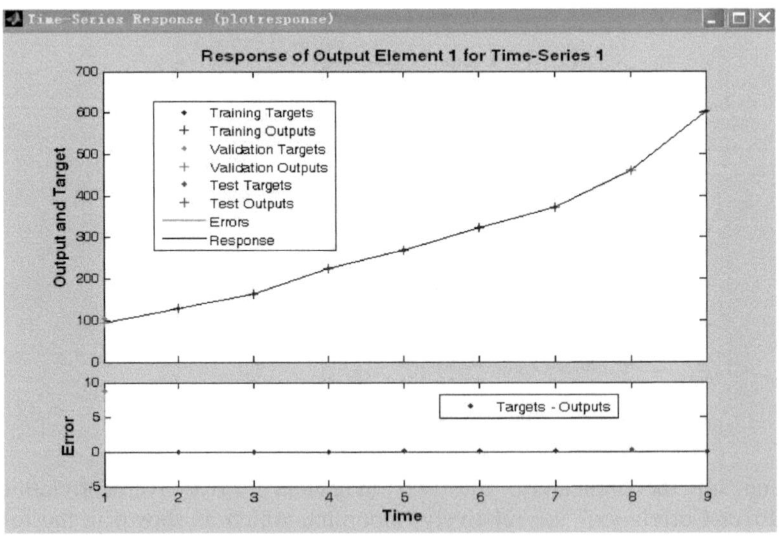

Fig. 7. Time-Series Response

If the result cannot meet the requirement, we need to retrain or adjust parameters until the result is conformance to requirements. Press “Next” when the result is satisfactory, click “Simple Script” to check the codes, then we can get the predicted value after running the program:

ys=103.6366 127.6955 163.2377 223.6702 268.4703 328.3422 373.1004 459.4968 602.39148 750.7179.

Take the historical and predicted values of relevant parameters into neural network to calculate again, and then we can get the predicted values for next year. The values are as follows:

Table 3. Predicted values of real estate investment

Year	2011	2012	2013
The predicted value of real estate investment [100 million RMB]	750.7179	816.2007	853.4147

Table 4. Errors between historical and predicted values

year	2002	2003	2004	2005	2006	2007	2008	2009	2010
errors	0.0634	0.1045	-0.5377	0.1298	-0.0703	-5.9422	-3.7575e-04	0.0032	0.0086

From the table above we can see the errors between the historical and predicted values are relatively small, so the predicted value we get through this network model is credible.

5 Conclusions

It is obvious that the neural network can solve nonlinear problem, the valued predicted is relatively accurate and the error is small, therefore, the method is practicable. In addition, the prediction process is on the basis of MATLAB toolboxes, codes can be generated by the software, which can decrease code-input time, increase work efficiency. What is more important, the forecasting process is simplified using this method: those MATLAB beginners can also use the useful toolboxes, which expand the target users.

References

1. Sha, L., Han, L.C.: Mathematics in Practice and Theory 33(11), 40–44 (2003) (in Chinese)
2. He, Q.H., Xie, L.L., Le, Y.: Journal of Huazhong University of Science and Technology (Urban Science Edition) 26(1), 101–104 (2009) (in Chinese)
3. Gong, P.: Research on the Real Estate Price Forecast Based on Artificial Neural Network, pp. 31–47. Jiangnan University, MS (2009) (in Chinese)

Test Requirements Simplification Based on Nonlinear Data Structure

Jianyuan Su and Xiaoming Wang

Research Center for Renewable Energy Generation Engineering,
Ministry of Education, Hohai University, Nanjing 210098, Jiangsu, P.R. China
hhsuwy@yahoo.com.cn

Abstract. Reducing the scale of test requirement suite is helpful for simplifying the relations between test requirement and test case, it can generate minimal test case suite rapidly, as is proposed in this paper. It bases on nonlinear data structure. Primarily, confirmed the test requirements and generated the initial test case suite according to program flow diagram, then drew the directed graph of relationships of test requirements, calculated the out-degree of various nodes in the directed graph, retained the nodes which out-degrees are bigger than one, structured test requirement tree according to these nodes and its related nodes, took out leaf nodes from the test requirement tree and put them into simplified test requirement suite. It is indicated by the example that this method could eliminate the relationships of equivalence and inclusion in test requirements automatically. It simplifies the test requirement suite effectively.

Keywords: Node, tree, graph, data structure, test requirement, out-degree.

1 Introduction

Test cases generation and test case suite simplification are the major work in the software test automatically. The method of test case determination is that the basic test cases are designed with black box method and the additional test cases are designed with white box method. The computer researchers must seek for the minimal test case suite through each kind of method after determining the test cases, and they often ignore of test requirement suite simplification, but simplifying the test requirement suite is useful for optimizing about test case suite. In reference 1, it set up the simplified model of test requirements and proposed some simplified algorithm, but it mainly got the relations of requirements via manual analysis. Concept Analysis algorithm could analyze the inclusion among test requirements and test cases simultaneously to simplify them, but it has complex construction [2]. The algorithm based on expansive technology could calculate the inclusion among test requirements with program analysis to simplify the requirement suite, which is mostly applied to unit test [3]. This paper proposed the method of test requirement suite simplification based on nonlinear data structure. The test requirement tree is constructed according to the directed graph of test requirements' relationships, and simplified test requirement suite is generated rapidly according to the test requirement tree.

Y. Zhang (Ed.): Future Communication, Computing, Control and Management, LNEE 141, pp. 221–226.
springerlink.com

2 Basic Definition

Data is the information which computer may save and process. Data structure is one subject which studies data element relates, it divides into linear data structure and nonlinear data structure. Tree and graph are nonlinear data structure.

2.1 Out-Degree and In-Degree

Graph G is composed by two sets V and R, whose definition of binary group of logical construction is as follows:

$$G = (V, R) \tag{1}$$

As shown in the graph, data element is usually named vertex, V is the vertex's finite nonempty set, R is the symplectic set between two vertexes in the graph. Define ki as degree of node vi, kiout as out-degree of node vi, kiin as in-degree of node vi, E(vi)out as arc suite about out-degree of node vi, E(vi)in as arc suite about in-degree of node vi, and then,

$$E(v_i)_{out} = \{\cup e_{ij} \mid \forall v_j \in V, e_{ij} \in E\}, k_{iout} =\mid E(v_i)_{out} \mid \tag{2}$$

$$E(v_i)_{in} = \{\cup e_{ji} \mid \forall v_j \in V, e_{ji} \in E\}, k_{iin} =\mid E(v_i)_{in} \mid \tag{3}$$

Define E(vi) as arc suite about degree of node vi, then,

$$E(v_i) = E(v_i)_{in} \cup E(v_i)_{out} \tag{4}$$

$$k_i = k_{iout} + k_{iin} \tag{5}$$

2.2 Leaf Node

Tree is the finite suite T(n≥1) contained n nodes, and satisfies: (1) There is one node called root, (2) Except root node, other nodes are divided into suites T1,T2……,Tm which does not intersect mutually, and each suite is also a tree in itself, these trees are called sub-tree. In the tree, leaf node is terminal node.

2.3 Equivalent and Inclusive Relations in Test Requirements

For test requirements ri,rj∈ R, if ri⊆ rj and rj ⊆ri, then ri and rj are the requirements of equivalence, marked ri≡rj. If ri∩rj=ri, then ri is named to be included in rj, marked as ri⊆rj.

2.4 Simplified Test Requirement Suite

Simplified test requirement suite is generated from primitive test requirement suite, and their test coverage ratio is same.

3 Thought of Test Requirement Suite Simplification

The thought of test requirement suite simplification based on nonlinear data structure is: according the structure of program flow diagram, confirm test requirements and generate initial test case suite, if test case Ti is satisfied with test requirement Rj, then in relation table of requirement/case S(T,R)m×n, element X(i,j)=1, otherwise X(i,j)=0. Then drew the directed graph of relationships of requirements, calculated the out-degree of various nodes in the directed graph, retained the nodes which out-degrees are bigger than one and put them into suite VR, analyzed the nodes relations in the directed graph, these nodes(ri,rj) are also in VR, found a path between ri and rj in directed graph, get all nodes in this path and the nodes behind ri and rj, added them into suite VR, structured test requirement tree according to these nodes and its related directed graph adjacent matrix, took out leaf nodes from the test requirement tree and put them into simplified test requirement suite RJ.

The generation step about simplified test requirement suite is as follows:

(1) save the adjacent matrix about the directed graph of relationships of test requirements.

(2) calculate out-degree of each node in the directed graph according to the adjacent matrix.

(3) put these nodes into suite VR, which out-degrees are bigger than one. If suite VR is empty, then return.

(4) structure test requirement tree.

(5) take out leaf nodes from test requirement tree, and put them into simplified test requirement suite RJ.

(6) delete the test requirements in RJ which satisfied all test cases.

The algorithm of test requirement suite simplification is as follows:

Input: the directed graph of relationships of test requirements;

Output: simplified test requirement suite RJ.

```
main ( )
{  int adjmx[m,n], char VR[k], char RJ[p];
   Input adjmx[m,n];
   Int i,od,r;
   For(i=1;i<=m;i++)
   {odi=outdegree(adjmx[i,n]);
If odi>1
{add ri to VR[k];}
}
   If VR[k]=NULL
   {return;}
   Creat testretree;
   Read tnode from testretree to RJ[p];
   For(i=1;i<=p;i++)
   {r=read(RJ[i]);
   If r satisfy all test cases
   {delete RJ[i];}
   }
   Output RJ[p];
}
```

4 Example Analysis

Through to one program flow diagram's analysis [5], there are 19 test requirements R={r1,r2,......rn}(n=19) and 12 test cases T={t1,t2,......tm}(m=12), the directed graph of relationships of test requirements is shown as figure 1.

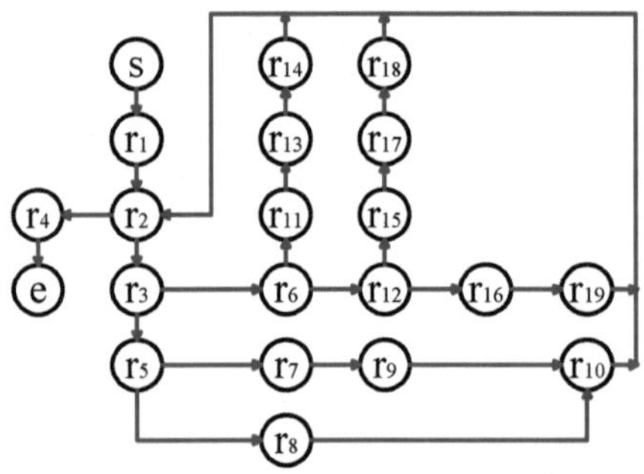

Fig. 1. Directed graph of relationships of test requirements

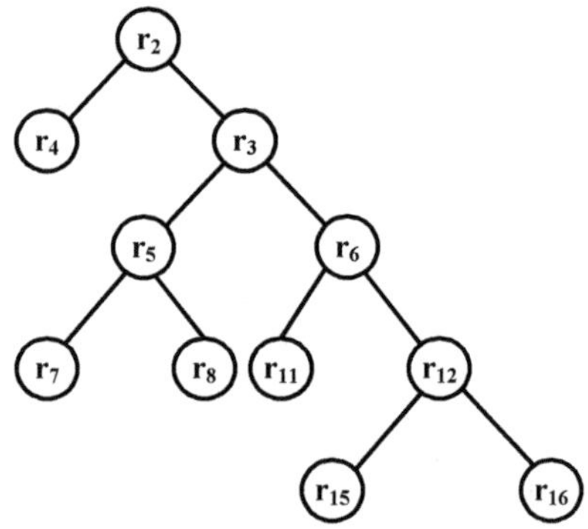

Fig. 2. Test requirement tree

In figure 1, the out-degree of various nodes is:

k1out=k4out=k7out=k8out=k9out= k10out=k11out=k13out=k14out=k15out= k16out=k17out=k18out=k19out=1;

k2out=k3out=k5out=k6out=k12out=2.

retained the nodes which out-degrees are bigger than one and put them into suite VR={ r2, r3, r5, r6 r12}, added the nodes into suite VR by the method of test requirement simplification in the paper, new suite is VR={ r2, r3, r4, r5, r6, r12, r7, r8, r11, r15, r16}, structured test requirement tree according to these nodes and its related directed graph adjacent matrix, as shown in figure 2, took out leaf nodes from the test requirement tree and put them into simplified test requirement suite RJ={ r4, r7, r8, r11, r15, r16}. Because r4 is satisfied with all test cases in S(T,R), must delete it, finally obtained the simplified test requirement suite RJ={ r7, r8, r11, r15, r16}.

By analysis of relationships between test requirements and test cases [1], the inclusions are r3⊆r1, r3⊆r2, r5⊆r3, r6⊆r3, r7⊆r5, r8⊆r5, r11⊆r6, r12⊆r6, r15⊆r12, r16⊆r12, the equivalent relations in r1~r19 are r1≡r2≡r4, r8≡r10, r7≡r9, r11≡r13≡r14, r15≡r17≡r18, r16≡r19, The result of calculation corresponds to the one of manual analysis, also to the one of reference 1. It eliminated the equivalent and inclusive relationships and reserved the mutually exclusive relationships and so on among test requirements.

The relationship between test requirements and test cases is S(T, R) in reference 5. In S(T, R),

t1(r1,r2,...r19)={1,1,1,1,1,0,1,0,1,1,0,0,0,0,0,0,0,0,0}.
t2(r1,r2,...r19)={1,1,1,1,1,1,0,1,0,1,1,0,1,1,0,0,0,0,0}.
t3(r1,r2,...r19)={1,1,1,1,0,1,0,0,0,0,1,1,1,1,1,0,1,1,0}.
t4(r1,r2,...r19)={1,1,1,1,0,1,0,0,0,0,0,1,0,0,1,1,1,1,1}.

t5(r1,r2,…r19)={1,1,1,1,1,0,0,1,0,1,0,0,0,0,0,0,0,0,0}.
t6(r1,r2,…r19)={1,1,1,1,0,1,0,0,0,0,0,1,0,0,0,1,0,0,1}.
t7(r1,r2,…r19)={1,1,1,1,1,0,1,0,1,1,0,0,0,0,0,0,0,0,0}.
t8(r1,r2,…r19)={1,1,1,1,0,1,0,0,0,0,1,1,1,1,0,1,0,0,1}.
t9(r1,r2,…r19)={1,1,1,1,1,1,0,1,0,1,0,1,0,0,1,0,1,1,0}.
t10(r1,r2,…r19)={1,1,1,1,1,1,1,0,1,1,1,0,1,1,0,0,0,0,0}.
t11(r1,r2,…r19)={1,1,1,1,1,0,0,1,0,1,0,0,0,0,0,0,0,0,0}.
t12(r1,r2,…r19)={1,1,1,1,1,1,1,0,1,1,0,1,0,0,1,0,1,1,0}.

After simplifying test requirement suite, S(T,R) is changed to S(T,RJ).

t1(r7,r8,r11,r15,r16)={1,0,0,0,0}.
t2(r7,r8,r11,r15,r16)={0,1,1,0,0}.
t3(r7,r8,r11,r15,r16)={0,0,1,1,0}.
t4(r7,r8,r11,r15,r16)={0,0,0,1,1}.
t5(r7,r8,r11,r15,r16)={0,1,0,0,0}.
t6(r7,r8,r11,r15,r16)={0,0,0,0,1}.
t7(r7,r8,r11,r15,r16)={1,0,0,0,0}.
t8(r7,r8,r11,r15,r16)={0,0,1,0,1}.
t9(r7,r8,r11,r15,r16)={0,1,0,1,0}.
t10(r7,r8,r11,r15,r16)={1,0,1,0,0}.
t11(r7,r8,r11,r15,r16)={0,1,0,0,0}.
t12(r7,r8,r11,r15,r16)={1,0,0,1,0}.

In S(T,RJ), r7 is related to (t1, t7, t10, t12), r8 is related to (t2, t5, t9, t11), r11 is related to (t2, t3, t8, t10), r15 is related to (t3, t4, t9, t12), r16 is related to (t4, t6, t8), minimal test case suite {t1,t2,t4} obtained based on S(T, RJ) are satisfied with all test requirements in S(T, R). S(T, RJ) kept the relation completeness between test requirements and test cases in S(T, R), and reduced the degree of complexity. Compared with S(T, R), utilizing S(T, RJ) to work out the minimal test case suite, could not only reduce the calculation workload, but also get more optimized test case suite rapidly.

This method has been performed on 8 examples, their program flow diagram is ordinal or circle or additional branch structure, and all can get the simplified test requirement suite [6], which demonstrates this method having better novel value.

5 Summary

The method based on nonlinear data structure to simplify test requirements was proposed in this paper which only must calculate the out-degree of various nodes in the directed graph of relationships of test requirements and structure test requirement tree and obtain simplified test requirement suite. This method is stable, easy to understand and realize and manual analysis and confirm, this paper gave the thought and calculation step, and with the simplification of test requirements, the calculation workload and complexity degree among test requirements will be reduced further, which is beneficial for getting the most optimized test case suite. However, its weakness is that it can't get the intersection among test requirements effectively, so further research should be gone on.

References

1. Zhang, X.F., Xu, B.W., Niu, C.H., Shi, L.: An approach for optimizing test suite based on testing requirement reduction. Journal of Software 18(4), 821–831 (2007)
2. Tallam, S., Gupta, N.: A concept analysis inspired Greedy algorithm for test suite minimization. In: Proceedings of the ACM SIGPLAN-SIGSOFT Workshop on Program Analysis for Software Tools and Engineering, pp. 35–42 (2005)
3. Marre, M., Bertolino, A.: Using spanning sets for coverage testing. IEEE Trans. on Software Engineering 29(11), 974–984 (2003)
4. Huang, W.L., Liu, Y., Zhong, Z.Q., Shen, Z.M.: Complex network based SMS filtering algorithm. ACTA Automatica Sinica 35(7), 990–996 (2009)
5. Chen, T.Y., Liu, M.F.: On the divide-and-conquer approach towards test suite reduction. Information Sciences 152(1), 89–119 (2003)
6. Guo, J.J., Gao, J.H.: Generation method of minimal test suite based on rataining redundant test cases. Computer Engineering 36(1), 45–48 (2010)

An Agent-Based Approach for Cost Evaluation Using ABC Method: Application to INOTIS Enterprise

Noria Taghezout[1], Sonia Ait Ameur[1], Amel Bekkal[1], and Bouabdellah Ascar[2]

[1] Computer Science department, Oran University, Algeria
IRIT-Toulouse University, France
taghezoutnour@yahoo.fr, noria.taghezou@irit.fr
[2] INOTIS Entreprise
bouabdellah.ascar@inotis.fr

Abstract. A more comprehensive definition of flexibility might be the ability of the enterprise to respond to variations more quickly, with lower costs, and less effects on system effectiveness. This paper develops a Production Control Decision Support System (PCDS) to assist in production control decisions. The proposed approach can help economic evaluation when product mix flexibility and the factors affecting flexibility are to be improved. It can also assist in making enterprise flexibility promotion decisions. The proposed model consists mainly of five components: Resource Agent, Cost Agent, COordinator Agent, Data Manager Agent and Maintenance Agent and a coordination protocol which ensures the exchange of messages among them. Agents interact to solve problems of resource allocation, conflict management, and better reaction during a resource's failure. We use Activity Based Costing (ABC) methodology to calculate the production costs. ABC is a process-oriented method which gathers information from the processes; it can be used to give managers the ability to match the resource needs with the available capacity as closely as possible and hence improving productivity. The Cost agent is responsible of the cost calculation. INOTIS enterprise is selected as an application domain in this study.

Keywords: Activity Based Costing (ABC) method, Agent-based approach, Cost Agent, Simulation, Tissue production, INOTIS.

1 Introduction

Basically, product cost systems measure a product's cost by integrating the price of a product's inputs and the technological coefficients of the production processes used in its manufacturing.

Traditional cost systems estimate the cost of resources used to create a product by tracing direct material and labor to products and allocating overhead based on one or more volume-based cost drivers. However, traditional cost systems have been criticized for their emphasis on meeting the requirements of financial reporting to the detriment of their usefulness for resource allocation decisions [6].

Y. Zhang (Ed.): Future Communication, Computing, Control and Management, LNEE 141, pp. 227–234.
springerlink.com

In the literature, many different cost estimating technique classifications can be found. Each researcher has developed his own classifications, making different assumptions and considerations. Many authors agree that the cost estimating methods can be broadly classified as follows: Qualitative techniques and quantitative techniques (for more details see [11]).

In particular, analytical techniques are based on a detailed analysis of the work required into the elementary tasks that constitute the manufacturing process. These are also termed as bottom up techniques where the cost data are collected from the smallest component levels and aggregated to the total product level. Activity Based Costing (ABC) is an example.

In fact, the ABC method continues to be the standard in terms of full cost evaluation methods. It is a costing model that identifies the cost pools, or activity centers, in an organization and assigns costs to products and services (cost drivers) based on the number of events or transactions involved in the process of providing a product or service. As a result, ABC can support managers to see how to maximize shareholder value and improve corporate performance [12].

As mentioned in [1], with the ABC model, precision depends on the multiplication of the number of activities this means that there is a greater need for collecting, processing and storing data. The multiplication of the number of activities has the added result of increasing allocation errors.

Production systems have complex interactions among tasks that require material flow and other coordination. In addition, the unexpected event such as demand variation and processing delay due to machine down time cause the system to become unstable. Solutions among entities become infeasible and need more coordination.

As a result, it can be difficult for a centralized control/scheduling approach to achieve its objective in highly dynamic and chaotic systems. To meet the need of communication and coordination between entities in the environment, a distributed scheduling algorithm has been developed in which the coordination mechanisms are adopted to resolve the conflicts or simple interactions [2, 4].

In practical manufacturing activities, the planning horizon is separated into equal periods and production decisions are made in each period. This is reasonable because the practical planning horizon of the manufacturing system is long, and in a long planning horizon correct and definite decisions cannot be made easily. We define flexibility at the internal production level as the ability of the manufacturing system to cope with changes such as product, process, load, and machine breakdown as considered in [8].

Our review and analysis of past research indicates that effective decision-making often relies on timely information monitoring, which aims to capture the newest status of the critical information items that are updated in different ways. The review also indicates that properly designed and implemented, agent-based control systems result in a performance that is flexible, robust, adaptive and fully tolerant, which are key factors for manufacturing success in the increasingly global market place [5, 7].

Regarding the issue of agent integration in flow-shop systems, we cite for example the work described in [10]. The authors considered here, the feasibility model of multi-agent scheduling on a single machine where each agent objective function is to minimize the total weighted number of tardy jobs. Lee et al. (2009) discussed a multi-agent scheduling problem on a single machine in which each agent is responsible for its jobs and wishes to minimize the total weighted completion time of its own set of jobs [11].

Leitao (2009) surveyed the literature in manufacturing control system using distributed AI techniques. He also discussed the reasons for the weak adoption of these approaches by industry and pointed out the challenges and research opportunities for the future [7].

The remainder of this paper is organized as follows: Section 2 describes the application area (INOTIS enterprise). Next, we present our approach: We firstly give a detailed description of the agents, then we describe the decision-making mechanism. The principles of ABC method are also given in Section 4. Finally, the simulation environment used to develop and test the proposed model is discussed in Section 5 and the paper finishes with concluding remarks in Section 6.

2 INOTIS Enterprise

INOTIS wants to be an international front-runner in the development, marketing and production of spunlace nonwovens for critical environments where contamination control is of vital importance.

Examples of typical markets

converters of surgical apparel and wound care products

converters for cleanrooms and clean production environments (e.g. pharmaceutical,food,electrotechnical)

converters for industrial applications where lint-free is of vital importance (aircraft, automotive surface preparation wipes, etc.)

With first class production capability of the latest generation, a clear market driven strategy and people who are skilled and determined to serve, the enterprise is well positioned to help their customers progress towards their own objectives.

INOTIS [3] is a privately owned company established in 2003. Production is projected to start early 2005. European customers will benefit from the support of a sales subsidiary in Paris, France.

Although the company and production plant are very new, the origins of the nonwoven activity date back much earlier. In the early 80's FANOTIS had been established close to Algiers. With the latest needlepunch technology available FANOTIS has been dedicated to different markets (decoration and geotextiles) and is therefore run entirely separate from INOTIS.

With its specific hydraulic entanglement of cellulose pulp and polyester fibers, INOTIS has created a productrange which can be designed in function of the specific market requirements. Thanks to the process and the materials used. Licial ® is a product superior to other nonwovens. Spunlacing allows an entanglement of fibers with no use at all of binders or adhesives. The resulting high purity of the fabric minimizes the possibility of product contamination and allergic reactions. We can summarize the production process in the following: First, the raw material will form a compact ball containing polyester fiber; it will be introduced in the fiber opener (see fig1.a). In this machine, polyester fiber will be weighed and wetted then it will be transported to the carding machine as it is shown in fig1.b. At this level, the carding operation will be executed on the fiber. After that, the lines of fiber will lead to the

Aqua jet where the sewing will be done at high pressure, thus the fiber lines will be superposed (see fig1. c). The obtained product will pass in a first dryer for drying and a brief inspection of manufacturing defects as it is shown in fig1.d. The scarf starts the next action in the process to define the type of tissue (hydrophilic or hydrophobic). Then the fabric passes through a second and final drying, it will be wound through the rolling machine around a cardboard tube (see fig1.e). The rolls of fabric will be weighed and taken to the stock.

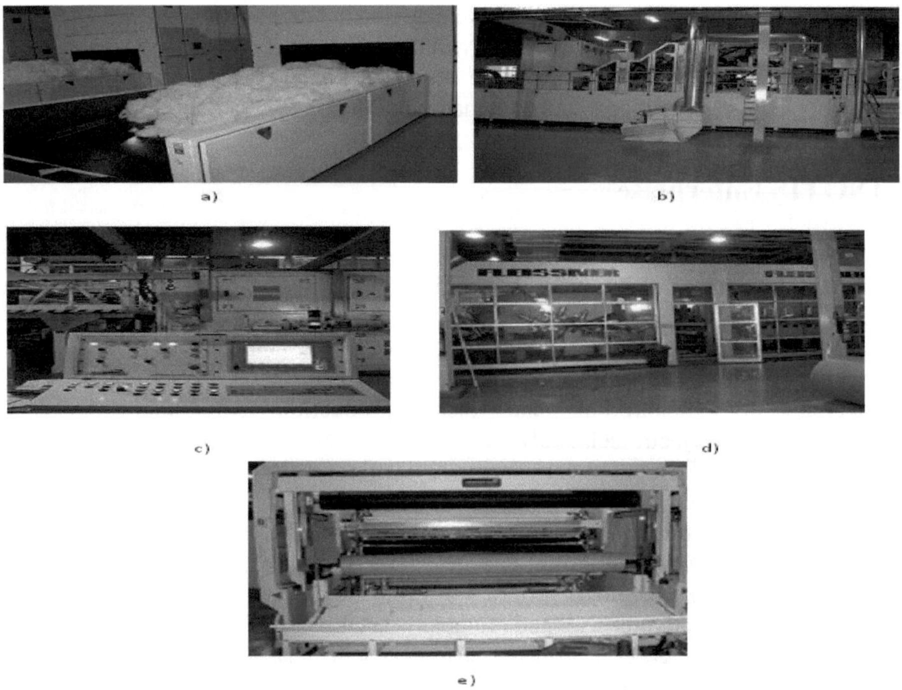

Fig. 1. General description of production process in INOTIS enterprise

Proposed approach

In this section, we present a general architecture of our multi-agent system. The proposed system mainly includes five components: Resource Agent (RA), Cost Agent (CA), Coordinator Agent (COA), Data Manager Agent (DMA), and Maintenance Agent (MA). Fig. 2 shows an overview of the architecture of the multi-agent system. Each of the five types of agent fulfills a particular role and is capable of performing a number of activities. An outline is now given of the main responsibilities of each agent:

Resource Agent (RA) represents enterprise's internal resource such as software, machine and workers. It comprises five functional modules: interface, proposal generator, Data Base, Knowledge base and control. RA will prepare a proposal according to its knowledge (including its capability, schedule, status, and cost) and send it. It interacts directly on the resource, each agent is responsible for a resource.

Cost Agent defines the performance indicators which are of two types: indicators for monitoring processes information and indicators for simulation. The CA establishes the scheduling performance for each decision objective criterion under different rules or a same algorithm.

COordinator Agent performs most of the autonomous problem solving: (1) receives user delegated task specifications from an Interface System, (2) interprets the specifications and extracts problem solving goals, (3) forms plans to satisfy these goals, (4) identifies information seeking sub-goals that are present in its plans, (5) decomposes the plans and coordinates with appropriate sub-systems for plan execution, monitoring, and results composition.

Database Management Agent (DMA) captures all the data and knowledge from databases and knowledge bases, comprises, manages and updates databases.

Maintenance Agent is responsible for solving problems that may occur on the resources under the control of the coordinator agent.

3 ABC Method Principles

ABC represents a paradigm shift from traditional cost systems. First, it is developed for managerial decision-making, independent of financial reporting requirements. Second, ABC models the physical transformation of resources into services by overhead activities and the consumption of these services by products. In effect, ABC reflects the physical flow of resources in order to trace how the costs of overhead resources are used to manufacture a product. Consequently, ABC results in more accurate product cost, as well as information about the technological coefficients measuring the ratio of a resource's inputs to the output it is used to produce [6]. With the costing based on activities, the cost of serving a customer can be ascertained individually. Furthermore, ABC implementation can help make employees to understand the various costs involved, which will in turn enable them to analyze the cost, identify the Value Added and Non Value Added Activities, implement the improvements and realize the benefits. The following table 1 summarizes the most important time in INOTIS production process, they significantly affect the final calculation of the total cost of production.

$$CP = \sum (nbUO \times cUO) + \sum (QMP \times cMP) \tag{1}$$

In eq.1, CP represents the product cost to calculate; In INOTIS enterprise, CP represents a roll of fabric.

nbUO is the number of work units, in our case, it is the time consumed by each inductor "transportation, employment, energy, " (indirect costs of production).

cUO represents the unit cost of each inductor.

QMP indicates the quantity of each raw material such as polyester, viscose, polypropylene, treated paper (already colored), Chemical (Fluorocarbon).

cMP represents the unit cost for each raw material

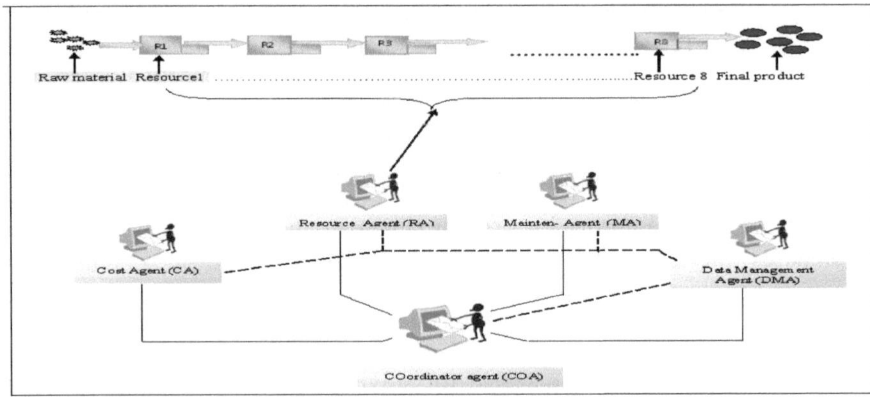

Fig. 2. Our agent-based approach

4 Simulation and Discussion

The MAS as well as the simulation environment was implemented in JADE (Java Agent DEvelopment framework) [8]. This latter also provides the agent with facilities to understand an Agent Communication Language (ACL) that expresses task requests and knowledge representation in a common format. Thanks to its distributed nature, JADE is particularly suited to create applications that run on distributed low cost architectures, such as blade, and therefore it perfectly meets the scalability requirement [8].

So, we tested the simulation program on INOTIS case study. Three kinds of packages have been developed: the plant network, the ABC code and the statistical package. The plant package consists of five agent types: RA, CA, COA, DMA, and MA. By using our approach, the coordination strategy may achieve more timely identification of important information updates using a controlled program implying more agents.

Table 1. The most significant idle time during production in INOTIS enterprise

Idle times	Duration (min)
The time to prepare the machines in the morning	60 min
The time of filtering water	50 min
The launch time machine	15min
After every 3 hours of work cleaning the perforated tape	30 min
Changing the paper roll	60 min
In case of occurrence of a fault in a piece of fabric: the estimated time to remove the damaged part	20 min
Handover period (teams 1 & teams2)	15 min
Downtime of machines	20 min
In case of power failure : Time to switch to another solution (group).	40 min

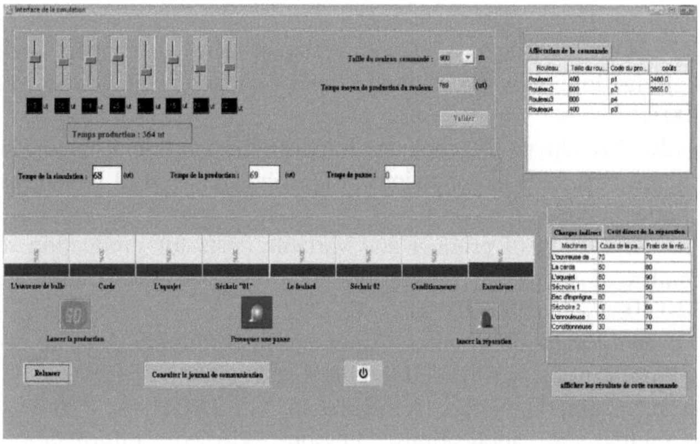

Fig. 3. Simulation interface

The statistical package collects the data at the end of every simulation run and it generates reports, histograms, graphs and statistical analyses. Two series of simulation experiments have been made. In the first series of experiments, the system uses the set of agents that are present and available at that time, to give the system a high degree of adaptability to the dynamic nature of the manufacturing aspect. In Fig. 3, one can see the simulation interface where all cost values are calculated by using ABC method. In the second series of experiments, we deal with the case having some disturbances; thus, we need to reschedule the tasks by involving only the agents directly affected, without disturbance to the rest of the agents that can continue with their work.

5 Conclusion

From our previous research, we understand that the structure oriented approach of traditional costing systems gives no decision support in allocating capacity to match resource needs. Over time this leads to cost inefficient organizations and poor profitability. ABC is viewed as more than a method for cost accounting - it invites to a whole new way of management. This method can direct attention towards the causes of costs (critical success factors) related to both cost objects and processes and not to mention the cost of quality. It details the causes of costs and gives a stronger indication of potential profitability. The system is under implementation, we intend to complete it and test it in the INOTIS enterprise.

References

1. de La Villarmois, O., Levant, Y.: Simplified Customer Profitability Evaluation: The equivalence based method. Cost Management, 28–34 (2009)
2. Mahesh, M., Ong, S.K., Ne, A.Y.C.: Communication and Coordination in Multi-Agent Manufacturing System. International Journal of Manufacturing Research 1(1), 59–82 (2006)

3. http://www.inotis.com
4. Lo Nigro, G., Noto La Diega, S.: Coordination policies to support decision making in distributed production planning. Robotics and Computer Integrated Manufacturing 19, 521–531 (2003)
5. Jenningsu, N.R.: Coordination techniques for distributed artificial intelligence. In: O'Harc, G.M.P., Jennings, N.R. (eds.) Foundations of Distributed Artificial Intelligence, pp. 187–210. Wiley, Chichester (1996)
6. Kee, R.: The sufficiency of product and variable costs for production related decisions when economies of scope are present. International Journal of Production Economics (2008), doi:10.1016/j.ijpe.2008.03.003,1-32
7. Leitao, P.: Agent-based distributed manufacturing control: A state-of-the-art survey. Engineering Applications of Artificial Intelligence 22, 979–991 (2008)
8. Bellifemine, F., Caire, G., Pogo, A., Rimassa, G.: JADE: A software framework for developing multi-agent applications: Lessons learned. Information and Software Technology 50, 10–21 (2008)
9. Lee, K., Choi, B.-C., Leung, J.Y.-T., Pinedo, M.L.: Approximation algorithms for multi-agent scheduling to minimize total weighted completion time. Information Processing Letters 109, 913–917 (2009)
10. Cheng, T.C.E., Ng, C.T., Yuan, J.J.: Multi-agent scheduling on a single machine to minimize total weighted number of tardy jobs. Theoretical Computer Science 362, 273–281 (2006)
11. Datta, P.P., Roy, R.: Cost modeling techniques for availability type service support contracts: A literature review and empirical study. CIRP Journal of Manufacturing Science and Technology 3, 142–157 (2010)
12. http://www.valuebasedmanagement.net/ABC

Power Engineering Investment Forecasting Based on Covering Rough Set

Lili Zhu, Qingyou Yan, and Cunbin Li

School of economic and management, North china electric power university, Beijing, 102206, China
lilizhu1223@yeah.net, qingyouyan@126.com, lcb999@263.net

Abstract. Investment forecasting is an important issue for power engineering's plan, investment and operation. In this paper, a new investment forecasting model is proposed based on covering rough set combined with neighborhood classifier theory. The classifier is used to classify the related historical data resulting in testing knowledge bases with certain testing precision. The knowledge base with the highest testing precision is chose to forecast the investment of new power engineering. The experiment results point out that the proposed model can forecast the power engineering investment effectually.

Keywords: Investment forecasting, covering rough set, power engineering.

1 Introduction

Investment forecasting plays an important role in solving the widespread problems in power engineering, such as out of control over investment, final accounts over budget as well as budget over estimate. Higher forecasting result will cause an effective investment and operation, while lower forecasting result will cause a fund shortage situation of the engineering. Due to influence by many factors, the power engineering investment decision always shows a complex rule. It is difficult to forecast investment accurately.

In recent decades, many of investment forecasting researchers employ various forecasting models in power engineering investment forecasting problem, such as regression forecasting model[1], unit cost model[2], artificial neural networks(ANN) [3, 4] and so on. The above-mentioned forecasting models all have their advantages and disadvantages. The regression forecasting model is simple to realize and it have a good linear trend fitting capability. However, it is criticized for its poor nonlinear fitting cumulative data. The unit cost model has many forms, and the most typical model is the BCIS, which is operated based on the data from the most similar and already done power engineering. Like the regression forecasting model, the BCIS is simple, but also have three defects as comparison characteristics, linear and subjective. Recently, as one of the most popular intelligence forecasting models, ANN has been successfully used in power engineering investment forecasting. According to related experiment, ANN outperforms the above two method. ANN has good nonlinear fitting capability, only a simple three layers neural network can get any an accuracy of a

Y. Zhang (Ed.): Future Communication, Computing, Control and Management, LNEE 141, pp. 235–240.
springerlink.com

nonlinear mapping. However, ANN need more expert knowledge and longer time of training for an appropriate structure than the above two models.

There are a lot of limitations of the above-mentioned methods no matter in the range or the accuracy of estimates to solve the investment forecasting of power engineering. The reasons are (1) the disadvantage of linear constraints makes it of poor nonlinear fitting capability, such as the regression forecasting model and the unit cost model; (2) the subjectivity of the decision parameter in the models leads to an arbitrary forecasting result, lack of objectivity. On aware of the defects of the existing models and the complexity of the power engineering investment forecasting, this paper proposed a new investment forecasting model, which is based on the covering rough set combined with the neighborhood classifier theory, with the aim to try to overcome the defects of existing models and improve the investment forecasting precision.

The remaining of this paper is organized as follows: Section 2 describes the theory of covering rough set and neighborhood classifier, followed with the working steps; section 3 gives a numerical example and corresponding result analysis. Finally, the conclusion is presented.

2 Description of Covering Rough Set Theory and Neighborhood Classifier

Rough Sets proposed by Z. Pawlak in about 1980s are a mathematical tool to deal with uncertainty. It can effectively analyze inaccurate, inconsistent and incomplete information, and has been successfully applied in many fields such as machine learning, knowledge acquisition, decision analysis, knowledge discovery, pattern recognition, expert systems and decision support systems. The marked difference of rough set theory from existing other uncertainty analysis theory like the fuzzy set theory and the probability statistics theory is that it is not required to provide any other prior information beyond data set needed. Therefore rough set theory is more objective than fuzzy set theory and probability statistics theory.

On the basis of classification mechanisms, rough set interprets classification as an equivalence relation of specific space. Since equivalence relations and the division of domain requirements are too strict, it limits the applications of the rough set theory in many practical cases. Therefore, it is considered to extend the equivalence relation and division of the domain. Zakowski extends division of the domain to covers, which extends Pawlak's rough set theory to cover generalized rough set theory. With the development of the generalized covering rough sets theory, a number of important results have been obtained [5]. Neighborhood classifier is an important one, which itself has high classification accuracy, with a wide range of applications[6]. The detail working steps are as follows:

Step 1. Given a specific norm m, calculate the distance between the sample vector S for the test

sample s and X_i

$$Des(s) = Des[X_i, S] = \|S - X_i\|_m, \quad i \le n \quad . \tag{1}$$

Step 2. Search for S 's δ neighborhood $\delta(S)$

$$\delta(s)=\{X_i | Des[X_i,S] \le \delta\}.$$

Step 3. Given threshold Δ, calculate the matching set F satisfying $dis(s)\ge\Delta$ in $\delta(S)$

$$F=\{x_i \mid Des(x_i)\ge\Delta, X_i\in\delta(S)\}. \tag{2}$$

Step 4. Calculate the estimated probability,

$$P(Des(s)\ge\Delta)=|F|\big/|\delta(S)|, \tag{3}$$

where $|F|$, $|\delta(S)|$ represent the length of the collection, respectively.

And δ is given by

$$\delta=\min(\Delta(X_i,S))+\omega\cdot range(\Delta(X_i,S)), \tag{4}$$

where $w\le 1$, $X_i(i=1,2,\cdots,n)$ are the training samples, $\min(Des[X_i,S])$ is the minimum distance between X_i and test sample vector S, $range(\Delta(X_i,S))$ is the range of values of $Des[X_i,S]$, δ depends on the local and global information of s.

3 An Experimental Analysis

3.1 Data Description

The historical data from an electric power company is partly listed in Table 1 below.

Table 1. The history data

	Terrain [/%]	Wire	Span [/m]	Iron tower	Cement Pole	Earthwork [/t·km^{-1}]	Hung [/t·km^{-1}]	Pull [/t·km^{-1}]	Insulator [/slice·km^{-1}]	Unit Investment [/103$]
1	100	3	341.9	5	39	236.68	0.56	0.27	89.33	33.06
2	65	3	308.0	9	7	480.09	0.13	0.20	91.97	28.34
3	100	1	284.4	12	20	295.51	0.15	0.14	0.69	27.61
4	30	3	398.6	79	104	199.00	0.45	0.00	88.16	29.72
...	...	...	...	...	...	...	...	...	...	...
250	50	3	273.9	12	36	178.22	0.22	0.00	149.57	27.57

3.2 Data Preprocessing

Map the historical data to interval $[0,1]$ through the maximum mapping method,

$$c_{new} = \frac{|c - c_{\max}|}{c_{\max} - c_{\min}} , \tag{5}$$

in which, c_{new} expresses the normalized value, and c is the original value, $c_{\max}$, $c_{\min}$ is the maximum and minimum value, respectively. The data preprocessed is listed in Table 2.

Table 2. The data preprocessed

	Terrain	Wire	Span	Iron tower	Cement Pole	Earthwork	Hung	Pull	Insulator
1	0	0	0.401	0.937	0.657	0.806	0	0	0.612
2	0.5	0	0.640	0.886	0.980	0	1	0.259	0.714
3	0	1	0.807	0.848	0.848	0.6115	0.954	0.482	1
4	1	0	0	0	0	0.9312	0.256	1	0.618
...	...	...	...	...	...	...	...	...	...
250	0.714	0	0.882	0.848	0.687	1	0.791	1	0.349

3.3 Neighborhood Calculation

The test sample is presented in following Table 3.

Table 3. The test sample

Terrain [/%]	Wire	Span [/m]	Iron tower	Cement Pole	Earthwork [/t·km^{-1}]	Hung [/t·km^{-1}]	Pull [/t·km^{-1}]	Insulator [/slice·km^{-1}]
90	3	304.69	4	47	242.14	0.62	0.48	105.62

Let $\delta = 1.030$, according to $\delta(S) = \{X_i \mid \|X_i - S\|_2 < \delta\}$, $i = 1,...,250$, and the neighborhood of the test sample is given as $\delta(S) = \{x_1, x_5, ...x_{248}, x_{250}\}$.

3.4 Investment Knowledge Base and Precision Analysis

The investment knowledge base is constituted by the preprocessed data and the historical investment value. The data is divided into two sets in each test, a train set and a match set, respectively. Considering the precision of the investment knowledge base, the mean from cycle test is used to lower the error resulting from the randomness of historical data. Set the cycle time as ten. The test precision is given by

$$\eta_i = \frac{|\bigcup match(Z_j)|}{|train[i]|}, Z_j \in train[i] , \tag{6}$$

where

$$match(Z_j) = \begin{cases} Z_j & \left|\bigcup m_i\right| / \left|\delta(Z_j)\right| \geq 1/2, \ M_i \in \delta(Z_j), \ Des(M_i) \leq Des(Z_j) \\ \Phi & other \end{cases} \quad . \tag{7}$$

Calculate the test precision by Matlab, which is presented as

$\eta = [0.7788, 0.7788, 0.7788, 0.7788, 0.7788, 0.8230, 0.7345, 0.7345, 0.7345, 0.9074]$

And the test set with the highest precision is set as the final investment knowledge base.

3.5 Investment Forecasting

The probability distribution of investment can be obtained by Eq.2 and Eq. 3, based on which the investment of the power engineering can be forecasted. With the selected investment knowledge base from the previous analysis, the probability distribution of the investment for the test sample is shown in Fig.1.

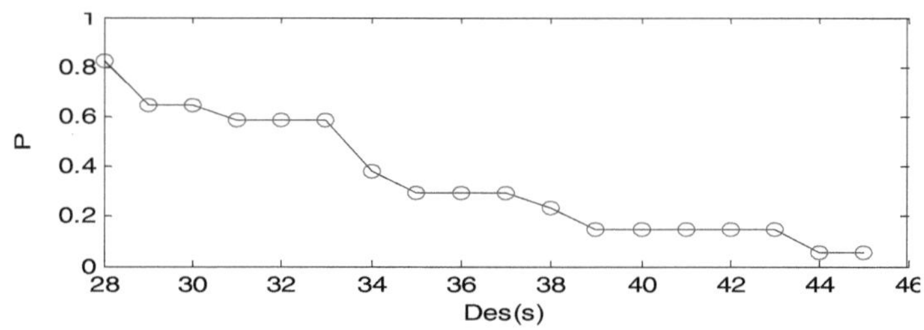

Fig. 1. The probability distribution of forecasted investment

As can be seen from the overall trend of curve in Fig. 1, the probability of the forecasted investment decreases with the threshold, which is consistent with the reality. From the curve point of view, for example, when $\Delta = 28$, it is shown that $P(Des(s) \geq 28) > 0.8$. The mean of the forecasted investment can be calculated as 27.79 . While the actual unit investment is 28.89 , indicating a forecasting model with relatively high prediction accuracy.

3.6 Comparing

For comparing, the results gotten by the regression forecasting model, unit cost model as well as ANN is 20.56, 21.35 and 23.65, respectively. It can be seen the forecasting value with the proposed method is better than any other model in this experiment. It is proven that the proposed method's performance may be superior.

4 Conclusion

A new approach to investment forecasting for power engineering is proposed based on covering rough set and neighborhood classifier theory in this paper. At first, historical data from existing down power engineering are normalized. Then, the neighborhood classifier with high precision characteristic is used to classify the historical data in order to establish the investment knowledge bases to estimate their average testing precision. Finally, the highest one is chosen to forecast the probability distribution of the investment of power engineering. At the ending of the paper, the feasibility and efficiency of the proposed model is proven by comparing with the existing other models. It is shown that the proposed model can reduced the disadvantages of other models in the aspects of mining rules hiding among the data objectively.

It should be noted that the validity of the model relies specially on the scale of the historical data available. In addition, the method is based on the experimental test, the result of which changes with values of parameters such as the measure of the neighborhood. The optimization of neighborhood is another direction worthy further study.

Acknowledgement. This research was conducted with the program for new century excellent talents in university in China (NCET-08-0772).

References

1. Park, J.H., Park, Y.M., Lee, K.Y.: Composite modeling for adaptive short-term load forecasting. IEEE Trans. Power Syst. 6, 450–457 (1991)
2. Intriligator, M.D., Bodkin, R.G., Hsiao, C.: Econometric models, techniques, and applications. Prentice-Hall, American (1996)
3. Chen, A.S., Leung, M.T., Daouk, H.: Application of neural networks to an emerging financial market: forecasting and trading the Taiwan stock index. Computers and Operations Research 30, 901–923 (2003)
4. Enke, D., Thawornwong, S.: The use of data mining and neural networks for forecasting stock market returns. Expert Systems with Applications 29, 927–940 (2005)
5. Bonikowski, Z., Bryniarski, E., Wybraniex, U.: Extensions and intentions in the rough set theory. Information Sciences 107, 149–167 (1998)
6. Hu, Q.H., Yu, D.R., Xie, Z.X.: Neighborhood classifiers. Expert Systems with Applications 34, 866–876 (2008)

Improving Wavelet Transform for the Impact-Echo Method of Non Destructive Testing

T. Bouden[1], M. Nibouche[2], F. Djerfi[1], and S. Dib[1]

[1] Laboratory of Non Destructive Testing, END-LAB, University of Jijel
BP°98, Ouled Aissa, 18000, Jijel, Algeria
[2] Bristol Robotic Laboratory, Department of Electrical and Computer Engineering, University of the West of England Frenchay Campus, Coldharbour Lane Bristol BS161QY UK
{bouden_toufik,djerfi.fares}@yahoo.com,
mokhtar.nibouche@uwe.ac.uk, samiradib@yahoo.fr

Abstract. The impact echo method, a technique based on the use of transient stress waves, has been used successfully for detecting flaws in many different types of concrete structures, including plate-like structures such as bridge decks, slabs, walls, beams, columns, layered structures and hollow cylindrical structures. Usually the received signals consist of components and additive structural noise. This noise often masks the signal of the defects and creates an embarrassment in its detection. Also, in thin samples the reflected signals are overlapping thus making detection of defects in the sample and accurate measurements impossible. It is thus necessary to enhance the visibility and resolution of the defects echoes by signal processing techniques for characterizing, detecting and locating the imperfections present in the used materials.

Keywords: Non destructive testing (NDT), impact echo, cement, mortar, material characterization, propagation velocity, defects detection and localization, thickness, time of flight (TOF), signal processing, and wavelet transform (WT).

In this paper experimental results have been presented to evaluate the effect of wavelet transform (WT) on ultrasonic signal analysis. In particular we proved that applying the WT, it has been possible to compress, filter the ultrasonic signal buried in noise, detect and locate the signal echoes, without any loss of accuracy in the time measurement. Using wavelet transform, we can then measure the time of flight (TOF) of the different echoes and the sample thickness. From measuring thickness and the TOF, it is easy to calculate the speed and longitudinal cross respectively, and then determine the elastic constants (Young modulus and Poisson's ratio) that allow us to describe the development of mechanical properties of materials and resolve the problems of material characterization and defects detection. The results have indicated the potential feasibility of the ultrasonic method and wavelets for ascertaining structural differences in materials.

Y. Zhang (Ed.): Future Communication, Computing, Control and Management, LNEE 141, pp. 241–247.
springerlink.com

1 Introduction

The Non Destructives Testing (NDT), play a very important role in the economy. They have become the essential tool for evaluating the mechanical properties of materials and detecting their defects. The new methods of NDT by ultrasounds of materials have much developed for a few years. In this work, we considered the pulse-echo method where obtained ultrasonic signal which must be processed by signal processing tools to compute the relative thickness (depth), or detect and localize defects and compute the material velocity. All these can determine the nature of the material if the latter is unknown and the position of different interfaces, so the size of the sample [1]-[12]. The objective here is to measure the TOF and the thickness between the front of the piece and its internal discontinuity (or back side) to locate the exact location of the echoes components signal. In the case of the paste cement sample, we estimate the velocity propagation of ultrasonic wave; this first piece has no flaws in its internal structure. And in the case of mortar sample, we detect and locate its defects. The tests were conducted in the laboratory of non destructive testing of Jijel University (NDT-Lab) in Algeria. Finally, the results are analyzed and the main conclusions of this paper are presented.

2 Related Work

Historically, Fourier spectral analysis has provided a general method for examining the global energy frequency distributions. As a result, the term `spectrum' has become almost synonymous with the Fourier transform of the signal. Partially because of its prowess and partially because of its simplicity, Fourier analysis has dominated the signal analysis efforts since soon after its introduction, and has been applied to all kinds of signal. Although the Fourier transform is valid under extremely general conditions, there are some crucial restrictions of the Fourier spectral analysis: the system must be linear; and the signal must be strictly periodic or stationary; otherwise, the resulting spectrum will make little physical sense. In the last decade many methods proposed for analyzing ultrasonic signals mainly apply analyses of frequency or time domain, but the non-stationary nature of ultrasonic signals, as with many real-world signals, makes those analyses unsatisfactory since their frequency components change with time [11]. Recently, many time-frequency analysis methods, e.g. short-time Fourier transform (STFT), Gabor transform or Wigner-Ville distribution as well as the wavelet-based approaches have been developed to analyze ultrasonic NDT signals [6], [12]-[16]. The wavelet transform allows you to filter the signal to detect and characterize the types of defects (planar or volumetric) of materials and improve the signal to noise ratio [1]-[4], [7]-[20].

3 Wavelet Transform

Above all the tool in time-frequency analysis, the famous one is the short-time frequency transform. The STFT's problem is that the fixed width of the window function lead to the resolution fixed. On the other hand, from the Uncertainty

Principle, we know that the product of the time domain resolution and the frequency resolution is constant, so we cannot have high resolution at both the time domain and frequency domain at the same time. Wavelet transform is one of the solutions to the above problem: by changing the location and scaling of the mother wavelet, which is the window function in Wavelet transform, we can implement the multi-resolution concept. Wavelet transform can capture both the short duration, high frequency and the long duration, low frequency information simultaneously. It is more flexible than STFT and particularly useful for the analysis of transients, aperiodicity and other non-stationary signal feature.

The Continue Wavelet Transform (CWT) can be defined by the coefficient of similarity between the signal and the wavelet used. The continuous wavelet transform is defined as follow:

$$\mathrm{W}T_{\mathrm{x}}(a,b) = \frac{1}{\sqrt{|a|}} \int_{-\infty}^{+\infty} x(t)\psi^{*}\left(\frac{t\text{-}b}{a}\right) dt = \langle x(t), \psi_{a,b}(t) \rangle \quad 1)$$

$$\psi_{a,b}(t) = \frac{1}{\sqrt{|a|}} \psi\left(\frac{t\text{-}b}{a}\right) \quad a,b \in R \; et \, a \neq 0 \quad (2)$$

Where:

$\Psi^*(t)$: the mother wavelet (basic wavelet function). The variable (1/a) gives the frequency scale and b the temporal location of an event. WTx(a,b): the signal energy of scale a at t=b. And ψa,b(t): the wavelet girl.

In the computation of the wavelet transform, a and b should be discretized. This is called the discrete wavelet transform (DWT). Obtaining bases discrete wavelet based on the theory of analysis multiresolution (AMR) proposed by S. Mallat [21], [22]. The scalogramme is expressed as the square of the module of the wavelet transform (Eq.1).

4 Proposed Method Using Wavelet

The proposed method for estimating the time of flight and detecting ultrasound signals is based on a choice of a wavelet analysis and a scale factor suitable in (Eq. 1). This estimate is done by calculating the coefficients of the most powerful relative to other factors (the coefficients they have maximum values). So, we can deduce the temporal and frequency positions corresponding to each echo that makes up the ultrasonic wave. Knowledge of time and frequency coordinates of each echo makes the estimation of TOF very easy [4], [8], [12]-[16]. Here Ci are the discrete wavelet coefficients.

$$E_i = Max(C_i) \quad (3)$$

5 Description of Experience and Results Discussion

In this section, we test wavelet transform method, on real ultrasonic signals, collected from prismatic samples of paste cement without defects and the mortar material with a synthetic defect.

This experience is based on the pulse echo technique (fig. 3), using two environments liquid/solid interface separated by size (6x6) cm. The device includes [3], a transmitter/receiver ultrasonic (Panametrics 5077PR, 606V), a transducer 1Mhz immersion, a digital oscilloscope (Tektronix TDS 1002) and a computer with data acquisition software (WaveStar).

During the experiment, the incident wave ultrasonic immersion transducer is normally intended to infringe on both sides of the used prism sample. Indeed, the transducer emits ultrasonic wave in a way that it falls vertically on the face of the sample to eliminate the effect of conversion of modes, so the echoes are received longitudinal waves. 4cm distance between the transducer and the sample.

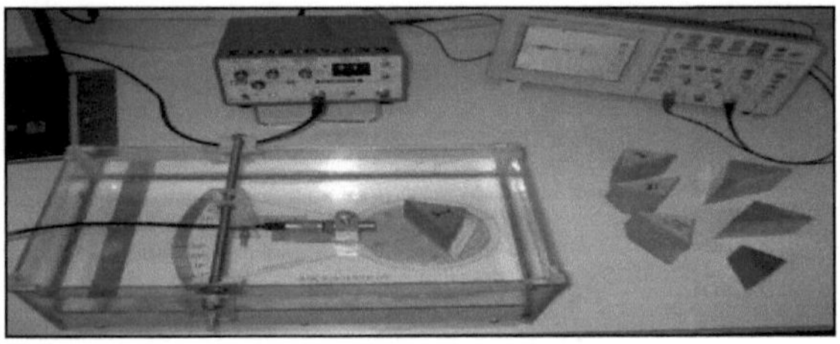

Fig. 1. Overview of the measuring device

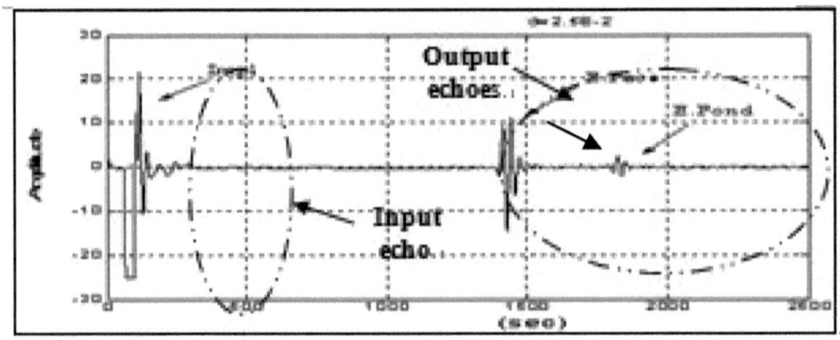

Fig. 2. The ultrasonic echoes signal reflected by paste cement sample (signal 01)

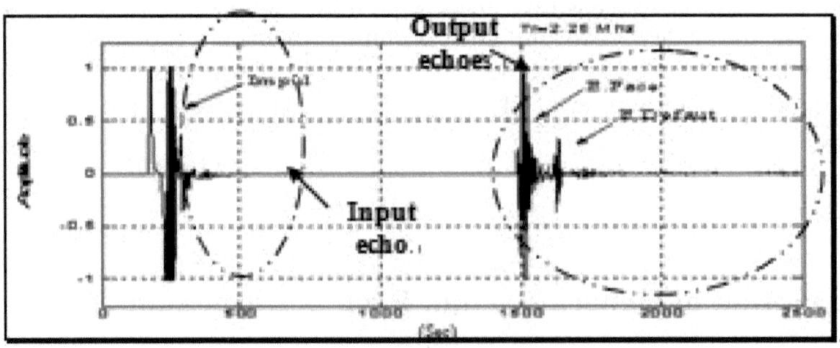

Fig. 3. The ultrasonic echoes signal reflected by mortar sample (signal 02)

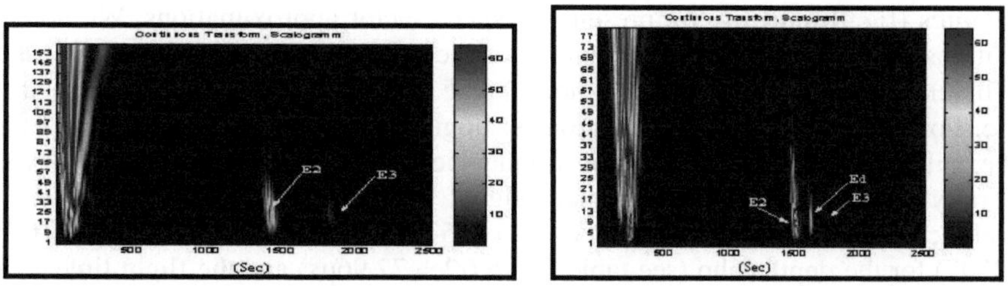

Fig. 4. Continuous wavelet transform of signal 01 and signal 02 (time-scale)

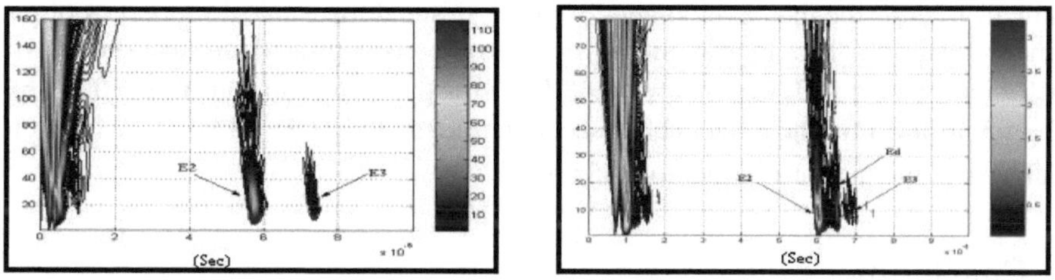

Fig. 5. Continuous wavelet scalogram (Time-scale edges) of signal 01 and signal 02

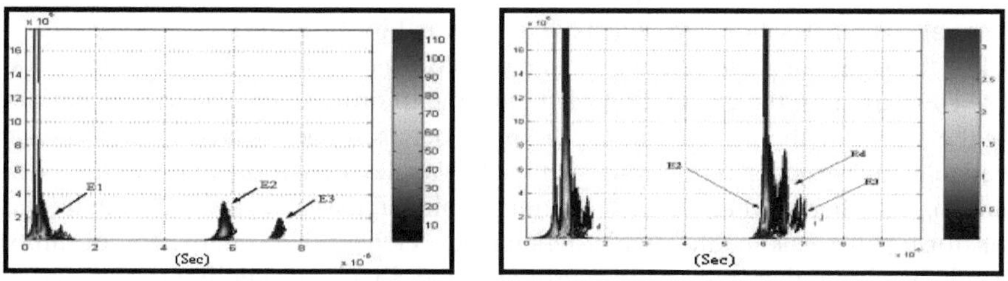

Fig. 6. Continuous wavelet transform (Time-frequency) of the signal 01 and the signal 02

Table 1. Calculating the paste cement velocity

Echos	Wavelet coefficient	Time μs	Frequency Mhz	Flight time	velocity
E.Face	C21 = 49.0241	57.68 μs	0.63776 Mhz	**15.28 μs**	**3272.3 m/s**
E.Depth	C26 = 10.0707	72.96 μs	0.68681 Mhz		

Table 2. Calculation of thickness defects in the mortar sample

Echos	Wavelet coefficient	Time μs	Frequency Mhz	Flight time	Thickness
E.Face	C9 =3.1372	60.48 μs	1.9841 Mhz	**4.92 μs**	**1.55 cm**
E.Defect	C10 =0.8346	65.40 μs	1.7857 Mhz		

Figures (fig. 4 until fig. 6), present the speed factor approximations. We note that variations of the signal after the contents of the original signal, where there is a low coefficient of variation in the non-noisy, and high coefficients of variation in the noise. For the time-scale representation (scalogram) shown in the figure (fig. 5), can make difference between the low frequencies and the high frequencies which generally represent. The paste cement signal echoes are concentrated around the wavelet coefficients (C21 = 49.0241) so (t1 = 57.68 μs) for the face echo and (C26 = 10.0707) for the depth echo (see table 1), so (t2 = 72.96μs) and the flight time TOF = 15.28 μs. From these found values, the paste cement velocity can be computed and finally it found equal to 3272.3 m/s and so we can determine the elastic constants that allow us to describe the development of properties and mechanical properties of paste cement material. The mortar signal echoes are concentrated around the wavelet coefficients (C9 =3.1372) so (t1 = 60.48 μs) for the face echo and (C10 =0.8346) so (t2 = 65.40 μs) for the defect echo (see table 2), and the TOF = 4.92 μs . The defect thickness in the mortar is equal to 1.55 cm.

6 Conclusion

In this paper, we made a study of wavelet transform and it's improving in the NDT material characterisation and defects detection and localisation.

For the signal of paste cement sample we have measured the longitudinal velocities of waves based on the calculation of flight time in order to characterize materials and for the signal of the mortar sample we have detected and located its defect. Among available time-frequency analysis methods, wavelet transform may be the best one and so, has been widely used for ultrasonic signal analysis. Although wavelet transform still has some inevitable deficiencies. Generally the frequency of ultrasonic signals is often rather high, and according to the Shannon sampling theory, large size samples are needed for detecting the defect and/or for characterizing the material in question.

References

1. Drai, R., et al.: Ultrasonics 38, 503–507 (2000)
2. Bettayeb, F., Rachedi, T., Benbartaoui, H.: Ultrasonics 42(1-9), 853–858 (2004)
3. Bouhadjera, A., Bouzrira, C.: NDT & Einternational 38, 135–142 (2005)
4. Bouden, T., Dib, S., Aissous, K., Grimes, M.: In: IEEE Ultrasonics Symposium, Roma Italy (2009)
5. Lin, L.M., Sansalone, M.: ACI Material Journal 3(93), 223–232 (1996)
6. Sansalone, M., Carino, N.J.: ACI Materials Journal 89(2), 175–184 (1989)
7. Malik, M.A., Saniie, J.: In: Ultrasonics Symposium, vol. 1, pp. 701–704 (1996)
8. Dib, S., Bouden, T., Aissous, K., Grimes, M.: MENDT Bahraine (2009)
9. Okamoto, J., Adamowski, J.C., Tsuzuki, M.S.G., Buiochia, F., Camerini, C.S.: Autonomous system for oil pipelines inspection. Mechatronics 9(7), 731–774 (1999)
10. Reher, K., Beller, M., Willems, H., Barbian, O.A.: In: IEEE Ultrasonics Symposium, vol. 1, pp. 665–671 (2002)
11. Yamani, A., Bettayeb, M., Ghouti, L.: Ultrasonics 35(7), 525–531 (1997)

12. Abbate, A., Frankel, J., Das, P.: In: IEEE Ultrasonics Symposium, vol. 1, pp. 751–755 (1995)
13. Abbate, A., Koay, J., Frankel, J., Schoeder, S.C., Das, P.: IEEE Trans. Ultrason., Ferroelect., and Freq. Contr. 44(1), 14–25 (1997)
14. Legendre, S., Goyette, J., Massicotte, D.: NDT and E International 34(1), 31–37 (2001)
15. Qidwai, U., Costa, A.H., Chen, C.H.: Insight 41, 700–703 (1999)
16. Michalodimitrakis, N., Laopoulos, T.: In: Proc. 18th IEEE Instr. and Meas. Tech. Conf., vol. 1, pp. 589–594 (2001)
17. Huang, N.E., Shen, Z., Long, S.R., Wu, M.C., Shih, H.H., Zheng, Q., Yen, N., Tung, C.C., Liu, H.H.: Proc. of the Roy. Soc. of London A 454, 903–995 (1998)
18. Wu, Z., Huang, N.E.: Proc. Roy. Soc., London A (December 2002)
19. Cexus, J.C.: Doctorat thesis, Rennes1 University French (2005)
20. Yimei, M., Peiwen, Q., Qi, Z.: 17th Wor. Conf. on NDT, Shanghai, China, October 25-28 (2008)
21. Mallat, S.: IEEE Trans.On Patt. Ana. and mach. Int. 11, 674–693 (1989)
22. Mallat, S.: Academic Press (1989)

A Hybrid Intelligent Algorithm for Mixed-Variable Optimization Problems

Zhang Yunqiang, Tan Ying, Sun Chaoli, and Zeng Jianchao

Complex System and Computational Intelligence Laboratory,
Taiyuan University of Science and Technology, Taiyuan, Shanxi, 030024, China
{zyq19840120,clsun1225}@163.com, tanying1965@gmail.com,
zengjianchao@263.net

Abstract. A hybrid intelligent algorithm (HIA), in which MPSO with feasibility-based rules proposed by Sun et al. [1] is used as a global search algorithm and simulator annealing (SA) is used to do local searching, is proposed in this paper to solve mixed-variable optimization problems. MPSO can obtain the values of the non-continuous variables very well for mixed-variable optimization problems. However, the imprecise values of continuous variables brought the inconsistent results of each run. A simulator annealing algorithm is proposed to find optimal values of continuous variables after the MPSO algorithm finishes each independent run, in order to obtain the consistent optimal results for mixed-variable optimization problems. The performance of HIA is evaluated against two real-world mixed-variable optimization problems, and it is found to be highly competitive compared with other existing algorithms.

Keywords: Hybrid intelligent algorithm, particle swarm optimization, simulator annealing.

1 Introduction

A mixed-variable problem is a mathematical programming problem with multiple variables and nonlinearities in the objective function and/or the constraints. Mixed-variable optimization problems are difficult to solve for the values of the different kinds of variables. Generally, a mixed-variable optimization problem can be expressed in mathematical form as follows:

Where $f(\vec{x})$ is the objective function, $\vec{x} = (\vec{x}^c, \vec{x}^z, \vec{x}^{i'}, \vec{x}^d)^T$ is the vector of solutions, $\vec{x}^c = (x_1^c, \ldots, x_{n_c}^c), \vec{x}^z = (x_1^z, \ldots, x_{n_z}^z), \vec{x}^{i'} = (x_1^{i'}, \ldots, x_{n_i}^{i'})$ and $\vec{x}^d = = (x_1^d, \ldots, x_{n_d}^d)$ are the vector of n_c continuous, n_z zero-one, n_i integer and n_d discrete variables, respectively. The total number of variables $n = n_c + n_z + n_i + n_d$. The problem also accounts for m inequalities $g(\vec{x})$ and l equalities $h(\vec{x})$. $g(\vec{x}) \leq 0$ and $h(\vec{x}) = 0$ are called performance constraints, and $x_{k1}^{cl} \leq x_{k1}^c \leq x_{k1}^{cu}, k1 = 1, 2, \ldots, n_c$ is called boundary constraints (or upper and lower bounds constraints) of each continuous variable. $\Re^z, \Re^{i'}$ and $\Re^d$ are respectively represented

Y. Zhang (Ed.): Future Communication, Computing, Control and Management, LNEE 141, pp. 249–256.
springerlink.com

the set of zero-one, integer and discrete. The set composed by all solutions which satisfy all boundary and performance constraints is called the feasible region. Generally, equality constraints are hard to satisfy in the mixed-variable optimization problems. Therefore, an equality constraint is often transformed into inequality constraint using tolerance allowed $|h_j(\vec{x})| \leq \delta$ in order to find the right solution of the problem as much as possible.

$$\begin{aligned}
\min\quad & f(\vec{x}) \\
s.t.\quad & g_i(\vec{x}) \leq 0 \quad i = 1, 2, \ldots, m \\
& h_j(\vec{x}) = 0 \quad j = 1, 2, \ldots, l \\
& \vec{x} = (\vec{x}^c, \vec{x}^z, \vec{x}^i, \vec{x}^d)^T \\
& \quad = (x_1^c, \ldots, x_{n_c}^c, x_1^z, \ldots, x_{n_z}^z, \\
& \qquad x_1^i, \ldots, x_{n_i}^i, x_1^d, \ldots, x_{n_d}^d) \\
& x_j^{cl} \leq x_j^c \leq x_j^{cu}, j = 1, 2, \ldots, n_c \\
& x_j^{zl} \leq x_j^z \leq x_j^{zu}, j = 1, 2, \ldots, n_z \\
& x_j^{il} \leq x_j^i \leq x_j^{iu}, j = 1, 2, \ldots, n_i \\
& x_j^{dl} \leq x_j^d \leq x_j^{du}, j = 1, 2, \ldots, n_d
\end{aligned} \tag{1}$$

Some research work have been reported to solve mixed-variable optimization problems. Taken altogether, there are two classes of optimization methods for solving mixed-variable problems: deterministic search methods [2-4] and stochastic search methods [5-9]. The main drawback of the deterministic search methods is that they make strong assumptions on the continuity and differentiability of objective functions, while stochastic search algorithms have none of these requirements. Therefore, more and more complexity optimization problems are solved by stochastic search methods.

Sun et al. [1] has proposed a modified particle swarm optimization (MPSO) for solving mixed-variable optimization problems. In MPSO, all non-continuous variables are valued according to the flying direction and magnitude of the velocity for each particle. The experimental results showed that the MPSO algorithm can obtain the values of non-continuous variables very well, however, the difference between each independence run is mainly resulted from the values of continuous variables. In order to improve the imprecise of the MPSO, simulated annealing (SA) is proposed to do local search once after MPSO finishes an independence run.

The rest of this paper is organized as follows. Section 2 reviews the modified particle swarm optimization. In Section 3, a hybrid intelligent algorithm for mixed-variable optimization problems is presented. Section 4 provides experimental results and gives some discussions. Finally, Section 5 outlines conclusions.

2 Modified Particle Swarm Optimization

The method to deal with the constraints should be chosen firstly in solving mixed-variable optimization problems. Generally, there are three kinds of methods to solve constraints: the penalty method, the constraint-preserving methods and the feasibility-based rules method. In the MPSO algorithm, the feasibility-based rules was proposed to solve the constraint violation.

2.1 The Feasibility-Based Rules

Referring to [10], feasibility-based rules employed in MPSO are described as follows:

(1) Any feasible solution is preferred to any infeasible one.

(2) Between two feasible solutions, the one having better objective function value is preferred.

(3) Between two infeasible solutions, the one having smaller constraint violation is preferred.

where the constraint violation in MPSO is calculated as follows:

$$viol(\vec{x}) = \sum_{i=1}^{m} \max(0, g_i(\vec{x})) + \sum_{j=1}^{l} \max(0, abs(h_j(\vec{x}))) \tag{2}$$

Based on the above criteria, in the first and third cases the search tends to the feasible region rather than infeasible one, and in the second case the search tends to the feasible region with good solutions. In brief, such a simple rule aims to obtain good feasible solutions.

According to the feasibility-based rules, feasible solutions are always considered to be better than infeasible ones, which may lead to pressure on selection of feasible solutions and result in premature convergence. In order to improve the diversity of the swarm so as to avoid premature convergence, an improved evolutionary mode are proposed in MPSO.

2.2 Evolutionary Mode of MPSO

Generally, the average velocity of the swarm can reflect the fly direction and speed of the swarm, especially with the increasing iterations. So in MPSO, the reversed average velocity of the swarm is considered in the updating of each particle's velocity, which is expected to expand the search ranch so as to improve the diversity of the swarm and to avoid the premature converge. The new updating equations for velocity and position are shown as follows:

$$\begin{aligned} v_{id}(t+1) =& \omega(v_{id}(t) - \omega' \bar{v}_d(t)) + c_1 r_1 (p_{id}(t) - x_{id}(t)) + \\ & c_2 r_2 (p_{gd}(t) - x_{id}(t)) \end{aligned} \tag{3}$$

$$x_{id}(t+1) = x_{id}(t) + v_{id}(t+1) \tag{4}$$

Where $\bar{v}_d(t)$ represents the average velocity of the swarm on dimension d which is calculated as follows:

$$\bar{v}_d(t) = \sum_{i=1}^{n} v_{id}(t) \tag{5}$$

3 A Hybrid Intelligent Algorithm for Mixed-Variable Optimization Problems

Particle swarm optimization was proposed to solve continuous optimization problems, which is unable to solve mixed-variables problems directly. So how to value the non-continuous variables is another important aspect in solving the mixed-variable problems. In MPSO, Sun et al. proposed to value the non-continuous variables according to the flying direction and magnitude of velocity. The experimental results showed that the proposed method can obtain the values of non-continuous variables are good. For detailed information please refer to Ref. [1]. However, seen from the experimental results, the precision of the results on some problems with continuous variables are not well. By analysis, we found that the main difference between each independence run is resulted from the values of continuous variables. So in this paper, a hybrid intelligent algorithm is proposed to solve mixed-variable optimization problems, in which MPSO is mainly used to do global search, and simulation annealing is used to be as a local search algorithm in order to improve the precise of the algorithm.

The description of HIA is shown as follows:

Begin
  Parameter setting, including the swarm size n, inertia weight ω, cognitive parameter c_1, social parameter c_2 and turbulence parameter ω', $t = 0$;
  Initialize positions: For each dimension of particle i, random generate a value in the upper and lower bounds. Set $int(x_{id}(t))$ as the value of integer variable and the closest discrete value as the value of discrete one;
  Initialize velocity for each particle;
  Calculate the violation and fitness of each particle;
  Generate the best history position of each particle: $\vec{p_i}(t) = \vec{x_i}(t), i = 1, 2, \ldots, n$;
  Generate the best history position of the swarm: find $\min f(\vec{p_i}(t))$ where $viol(\vec{p_i}(t)) = 0$, set $\vec{p_g}(t) = \vec{p_i}(t)$ if $f(\vec{p_i}(t))$ is minimal;
  While the stopping criterion is not met
    For each particle i in the swarm
      Updating the velocity and position using (3) and (4);
      Call the handling procedure for different kinds of variables;
      Calculate the violation value using (2);
      Calculate the fitness value;
      Update the best history position of particle i according to feasibility-based rules;
    End for
    Update the best history position of the swarm: find $\min f(\vec{p_i}(t))$ where $viol(\vec{p_i}(t)) = 0$, set $\vec{p_g}(t) = \vec{p_i}(t)$ if $f(\vec{p_i}(t))$ is minimal;
  End While
  call SA algorithm;
End

The pseudocode of the SA algorithm is shown as follows:

Begin
Parameter setting, including initial temperature T_0, the initial position $\vec{x}_0 = \vec{p}_g$, the iteration
number L on temperature T; $T = T_0$; $k = 1$;
Set the current best solution $f(x_{best}) = f(\vec{x}_0)$
While ($T > T_{min}$)
For i:=1 to L
Produce a new candidate solution $\vec{x}_{new}$, which is in the feasible region, in the
neighberhood of the current best solution $\vec{x}_{best}$;
Calculate the fitness value of the new produced solution $f(\vec{x}_{new})$;
If ($f(\vec{x}_{best}) > f(\vec{x}_{new})$)
$\vec{x}_{best} = \vec{x}_{new}$;
End If
End For
$k = k + 1$;
$T = \frac{T_0}{k}$
End While

Pay attention that in the fitness calculation of the local search algorithm, all values of non-continuous variable is treated as constants, which are just same to the optimal results obtained by the MSPO algorithm. It might also be noted that in our simulator annealing algorithm, only when the fitness of the new produced solution is absolutely better than that of the current best solution, the current best solution can be replaced.

4 Experimental Results and Some Discussions

Two real-world mixed-variable optimization problems, which included continuous and non-continuous variables and frequently employed in the literature, are used to evaluate the performance of HIA. The parameters in MPSO are set as follows. The inertia weight ω is decreased linearly from 0.9 down to 0.4; the perturbation coefficient is changed sinusoidally, with its amplitude of 0.5. The cognitive parameter is set to decrease linearly from 3.6 down to 1.8 and the social parameter increase linearly from 0.2 up to 1.8 so as to fight against prematurity with the reversed average velocity of the swarm. Total number of individuals is 40, the maximum number of iterations is 10,000 per run and 30 independent runs are executed for each example. In the simulator annealing algorithm, the parameters are set as follows: the initial and the minimal temperature T_0 and T_{min} are respectively set to 0.1 and 0.001; the iteration number L is set to 1000.

Example 1: Pressure Vessel Design

Pressure vessel design is a common benchmarking problem for mixed-variables optimization problems presented by Sandgren [11]. The objective of this problem is to minimize the manufacturing cost of the pressure vessel, which is a combination of material cost, welding cost and forming cost. The design variables are specified as

$$\vec{x} = (x_1, x_2, x_3, x_4)^T = (T_s, T_h, R, L)^T$$

where x_1 is the spherical head thickness, x_2 is the shell thickness, x_3 and x_4 are the radius and length of the shell, respectively. In the design variables, x_3 and x_4 are continuous, while x_1 and x_2 are discrete values and integer multiplies of 0.0625 inch respectively.

The best optimal results of this problem obtained by different stochastic search algorithms are shown in Table 1. It can be seen that all PSO-based algorithms obtained better results than other stochastic algorithms. Table 2 shows the results comparison on all PSO-based algorithms, though the best results of MPSO and HIA are not good as that of HPB, the mean results and the standard deviations are all better than HPB, especially our proposed HIA obtained consistent best optimal solutions for each run.

Table 1.

EP[12]	EA[13]	GA[14]	PSO[15]	HPB[9]	MPSO[1]	HIA
1.000	0.9345	0.8125	0.8125	0.8125	0.8125	0.8125
0.625	0.5000	0.4375	0.4375	0.4375	0.4375	0.4315
51.1958	48.3290	40.097398	42.0984456	42.09893	42.098446	42.098446
90.7821	112.6790	176.654047	176.636595	176.6305	176.636792	176.636792
-0.0119	-0.00475	-0.00002	0.000000	0.00000	0	0
-0.1366	-0.038941	-0.035891	-0.0358808	-0.03587	-0.03588083	-0.0388083
-13584.5631	-3652.87683	-27.886075	0.000000	0.00000	-1.166400×10^{6}	-1.166400×10^{6}
-149.2179	-127.321	-63.345953	-63.363404	-63.69484	-63.36321	-63.36321
7108.6160	6410.3811	6059.94634	6059.7143	6059.65457	6059.718932	6059.718932

Table 2. Comparison of HIA with other PSO-based approach

	PSO[15]	HPB[9]	MPSO[1]	HIA
Best	6059.7143	6059.65457	6059.718932	6059.718932
Mean	6289.92881	6060.08	6059.727988	6059.718932
Worst	-	-	6059.764215	6059.718932
Standard deviation	305.87	0.02194	0.003559	2.335115×10^{-8}

Example 2: Spring Design

The second example is to minimize the volume of a compression spring under constant loading. It is a common real-world optimization problem which involves discrete, integer and continuous design variables. The design variables are specified as

$$\vec{x} = (x_1, x_2, x_3) = (d, D, N)$$

The spring wire diameter, x_1, may have only discrete values according to available stand spring steel wire diameters shown as follow:

$$x_1 \in \{0.009, 0.0095, 0.0104, 0.0118, 0.0128, 0.0132,$$
$$0.014, 0.015, 0.0162, 0.0173, 0.018, 0.020, 0.023,$$
$$0.025, 0.028, 0.032, 0.035, 0.041, 0.047, 0.054,$$
$$0.063, 0.072, 0.080, 0.092, 0.105, 0.120, 0.135,$$
$$0.148, 0.162, 0.177, 0.192, 0.207, 0.225, 0.244, 0.263,$$
$$0.283, 0.307, 0.331, 0.362, 0.394, 0.4375, 0.500\}$$

The outside diameter, x_2, is a continuous variable and the number of spring coils, x_3, is an integer variable. For its mathematical model please referred to Ref. [9].

The optimal results of this problem obtained by different algorithms are shown in Table 3, and Table 4 shows the comparison results between different PSO-based algorithm. Seen from these two Tables, it can obviously be seen that all stochastic algorithms can find the best optimal solutions except the genetic algorithm [13]. The comparison results between different PSO-based algorithm showed that our proposed HIA can obtain consistent optimal results for each run, the mean result and the standard deviation for this problem is much better than other PSO-based algorithms.

Table 3. Optimal solutions of the coil compression spring design

	GeneAS[13]	DE[6]	PSO[15]	HPB[9]	MPSO	HIA
x_1	0.283	0.283	0.283	0.283	0.283	0.283
x_2	1.226	1.22304101	1.22304101	1.22301421	1.223041	1.223041
x_3	9	9	9	9	9	9
g_1	-713.510	-1008.8114	-1008.8114	-1011.6168	-1008.629	-1008.653
g_2	-8.933	-8.9456	-8.9456	-8.9457	-10.55277	-10.55278
g_3	-0.083	-0.083	-0.083	-0.0830	-0.083	-0.083
g_4	-1.491	-1.777	-1.777	-1.7769	-1.776938	-1.776959
g_5	-1.337	-1.3217	-1.3217	-1.3216	-1.321774	-1.3217
g_6	-5.461	-5.4643	-5.4643	-5.4643	-5.946424	-5.946429
g_7	0.0000	0.0000	0.0000	0.0000	0.000000	0.000000
g_8	-0.009	0.0000	0.0000	0.0000	-0.000115	0.000000
$f(\vec{x})$	2.665	2.65856	2.65856	2.65850	2.658555	2.658555

Table 4. Comparison of DPSO with other PSO-based approach

	PSO	HPB	MPSO	HIA
Best	2.65856	2.65850	2.658555	2.658555
Mean	2.738024	2.6985	2.658601	2.658555
Worst	-	-	2.659459	2.658555
Standard deviation	0.107061	0.0239	0.000030	3.151773×10^{-11}

5 Conclusions

A hybrid intelligent algorithm, in which particle swarm optimization and simulator annealing algorithms are cooperated to find the best consistent optimal solutions for mixed-variable optimization problems. Feasibility-based rules are used as the handling mechanism for constraint conflict. The proposed HIA make good use of the method of MPSO to value the non-continuous variables, and the good property of SA in local searching to find the continuous optimization problems.

Experimental results showed that our proposed HIA can obtain the best consistent optimal results for mixed-variable optimization problems.

Acknowledgements. This work is supported in part by Youth Foundation of Taiyuan University of Science and Technology under Grant No. 20103012.

References

1. Sun, C., Zeng, J., Pan, J.: A modified particle swarm optimization with feasibility-based rules for mixed-variable optimization problems. International Journal of Innovative Computing, Information and Control 7(6), 3081–3096 (2011)
2. Land, A.M., Doig, A.G.: An automatic method of solving discrete programming problems. Econometrica 26, 497–520 (1960)
3. Bremicher, M., Papalambros, P.Y., Loh, H.T.: Solution of mixed-discrete structural optimization problems with a new sequential linearization algorithm. Computers & Structures 37(4), 451–461 (1990)
4. Praharah, S., Azarm, S.: Two-Level Nonlinear Mixed Discrete-Continuous Optimization-Based Design: An Application to Printed Circuit Board Assemblies. Journal of Electronic Packaging 114, 425–435 (1992)
5. Rao, S.S., Xiong, Y.: A Hybrid Genetic Algorithm for Mixed-Discrete Design Optimization. Transactions of the ASME 127, 1100–1112 (2005)
6. Lampinen, J., Zelinka, I.: Mixed Integer-Discrete-Continuous Optimization By Differential Evolution - Part 2: a practical example. In: 5th International Mendel Conference on Soft Computing, Proceedings of MENDEL 1999, pp. 77–81 (1999)
7. Lampinen, J., Zelinka, I.: Mixed integer-discrete-continuous optimization by differential evolution. Part 1: the optimization method. In: 5th International Mendel Conference on Soft Computing, Proceedings of MENDEL 1999, pp. 71–76 (1999)
8. Kitayama, S., Arakawa, M., Yamazaki, K.: Penalty function approach for the mixed discrete nonlinear problems by particle swarm optimization. Struct. Multidisc. Optim. 32, 191–202 (2006)
9. Nema, S., et al.: A Hybrid Particle Swarm Branch-and-Bound (HPB) Optimizer for Mixed Discrete Nonliear Programming. IEEE Transactions on Systems, Man and Cybernetics–Part A: Systems and Humans 38(6), 1411–1424 (2008)
10. Pulido, G.T., Coello, C.A.C.: A constraint-handling mechanism for particle swarm optimization. Congress on Evolutionary Computation (2004)
11. Sandgren, E.: Nonlinear integer and discrete programming in mechanical design optimization. Journal of Mechanical Design 112(2), 223–229 (1990)
12. Cao, Y., Wu, Q.: A mixed variable evolutionary programming for optimisation of mechanical design. Int. J. Eng. Intell. Syst. Electic. Eng. Commun. 7(2), 77–82 (1999)
13. Deb, K.: Gene AS: A robust optimal design technique for mechanical component design. Springer, Heidelberg (1997)
14. Coello Coello, C.A., Mezura Montes, E.: Use of Dominance-Based Tournament Selection to Handle Constraints in Genetic Algorithms. In: Intelligent Engineering Systems through Artificial Neural Networks (ANNIE 2001), pp. 177–182 (2001)
15. He, S., Prempain, E., Wu, Q.H.: An improved particle swarm optimizer for mechanical design optimization problems. Engineering Optimization 36(5), 585–605 (2004)

Annotating Taiwan Cancer Registry to caDSR for International Interoperability

Shin-Bo Chen[1], Chien-Yeh Hsu[2], and Chi-Hung Huang[3]

[1] Graduate Institute of Medical Informatics Taipei Medical University, Taipei, Taiwan
meok.chen@gmail.com
[2] Graduate Institute of Medical Informatics and Center of Excellence for Cancer Research (CECR) Taipei, Taiwan
cyhsu@tmu.edu.tw
[3] En Chu Kong Hospital, New Taipei City, Taiwan
huangchihung@yahoo.com.tw

Abstract. It is very difficult to exchange and integrate the data among different cancer institutions, in order to discover useful biomedical information for the cancer community. We developed a cancer Biomedical Informatics Grid™ (caBIG™) Silver level compliant cancer registry database system, called the gridTCR (grid of the Taiwan Cancer Registry), which integrates the demographic data, clinical history, pathology data, and clinical outcome data including treatment, recurrence and vital status from cancer registry databases in Taiwan. In this manuscript, we will be developing the common data elements using vocabulary standards, ontology and semantic modeling methodology. The Taiwan Cancer Common Data Element Project (TCCDEP) developed 40 data elements to annotate the cancer registry data collected. The aim of this project is to creat a core set of data elements for annotating the cancer registry data and achieve the interoperability over the caBIG community. We describe the process required to develop the model, the caDSR CDEs, and the results of the modeling effort. We address difficulties we encountered and modifications for solution. Currently, the Taiwan cancer registry CDEs are released and available in CDE browser for reusing. Furthermore, we will extend our CDEs to daily clinical practice and trials, along with how the methods were used to fully implemented in hospitals and cancer research centers in Taiwan.

Keywords: Cancer Registry, caDSR, Common Data Elements, caBIG™, interoperability.

1 Introduction

The huge amounts of biomedical data are stored in various formats and accessed through numerous interfaces. It is a crucial task for Data integration and exchange in cancer research. Data elements play an important role in data integration. The NCI supports a broad initiative to standardize the common data elements (CDEs) used in cancer research data capture and reporting. The Taiwan Cancer Registry (TCR), established in 1979, is organized and funded by the Health Department of the central

Y. Zhang (Ed.): Future Communication, Computing, Control and Management, LNEE 141, pp. 257–263.
springerlink.com

government. The primary goal of TCR is to survey the incidence of cancer in Taiwan. The aim of the Taiwan Cancer Common Data Element Project (TCCDEP) is to facilitate convergence towards a common metadata standard in Taiwan cancer registry data. The project is implemented using a set of open source software and tools developed by the NCI, such as the caCORE SDK and caGrid. The experience of building, learning and using the open toolkit, Cancer Data Standards Repository (caDSR), developed by the National Cancer Institute Center for Bioinformatics, NCICB, in the USA is reported. The caDSR is a metadata repository including CDEs used by NCI-sponsored organizations. The object of this work is to develop a database of metadata for medical data elements, referred to as the TCCDEP and to establish a common classification of data elements used in cancer registry.

2 Taiwan Cancer Registry

Most cancer patients in Taiwan are diagnosed and treated in hospitals with 50 or more beds. As far as cancer treatment is concerned, all major medical centres have oncology departments and the specialty of oncology requires board certification. There are ten major medical centres and one comprehensive cancer hospital in Taiwan. Hundreds of regional and district general hospitals are also actively involved in cancer diagnosis and treatment. Among 31,532 physicians and surgeons in Taiwan, there are 170 medical oncologists.

The Taiwan Cancer Registry, a population-based cancer registry, was founded in 1979. Hospitals with greater than 50-bed capacity which provide outpatient and hospitalized cancer care are recruited to participate in reporting all newly diagnosed malignant neoplasms to the registry. The registry is organized and funded by the Department (Ministry) of Health of the executive branch of the central government. The National Public Health Association has been contracted to operate the registry and organized an advisory board to standardize definitions of terminology, coding, and procedures of the registry's reporting system. The registry's primary goal is to survey the incidence of cancer in Taiwan. It also participates in planning and evaluation of cancer control and prevention programmes. Cancer incidence data appear each year in a special bulletin and in an annual registry report published by the Ministry of Health. Analyses and observed trends are published in annual reports or in medical journals. The registry also provides a database regarding cancer for various research efforts [1].

3 The Cancer Biomedical Informatics Grid

The caBIG™ is a network of information and tools. It allows institutions and individuals to easily share, integrate, and analyze data. A “fully interoperable” IT network has been developing by caBIG™ that will supercharge the entire field of cancer research. CaBIG™ is an informatics infrastructure that will connect teams of cancer and biomedical researchers to enable them to better develop and share tools and data in an open environment with common standards. CaBIG™ will create a voluntary virtual network that links individuals and institutions both nationally and internationally, effectively forming a World Wide Web of cancer research. The caBIG™ pilot program was launched in 2003 with more than 50 NCI-designated cancer centers working in partnership with NCI to develop the vision, approach, and structure of caBIG™.

4 Method

4.1 TCCDEP Development and UML Information Model

In this manuscript, we will be developing the common data elements using vocabulary standards, ontology and semantic modeling methodology. The CDEs included demographic data, clinical history, pathology data, and clinical outcome data including treatment, recurrence and vital status. These CDEs will be further enhanced to data sets across the participating cancer institutes to facilitate and supplement translational research. The Taiwan Cancer Registry (TCR) model and standard will be use as the basis for an electronic data standard repository to metadata or data descriptors. We created four main data categories for annotation of the CDE development process. The four main categories were: 1) Patient Identifier, which records basic patient demography; 2) Cancer Identifier, which records which records basic overall information including the clinical history data; 3) First Course treatment; 4) Outcome. Sub-categories and additional data elements for each of the four main groups are described in Fig 1.

Information models provide an abstract formal representation of the conceptual or physical entities in a domain. caBIG™ Information models are constructed as UML class diagrams. The UML is a non-proprietary language for constructing, visualizing, and documenting the artifacts of software engineering. [2].

We used Enterprise Architect (EA) [3] as the UML modeling tool for this project due to its low cost and high performance. During modeling, each UML class, attribute and enumeration value was annotated with a human readable definition of the semantic meaning of that class or attribute. These definitions were created as tagged values in UML, which provides a method for adding additional information to a UML stereotype. Models annotated with these definitions were then saved in the XML Metadata Interchange (XMI) [4] format for further processing. We designed and implemented the gridTCR system based on the caBIGTM paradigm. We utilize the open source software such as EA (the UML modeling tool) , caCORE SDK 4.1.1, caGrid 1.3, and Introduce (the caGrid service creation toolkit) [5], etc. The process workflow for the system design is shown in Fig2.Generatingca DSRCDEsfromUML.

I. Patient Identifier
1. Medical Record Number
2. Name
3. ID Number
4. Sex
5. Date of Birth
6. Residence Code

II. Cancer Identifier
1. Age at Diagnosis
2. Sequence Number
3. Class of Case
4. Date of First Contact
5. Date of Initial Diagnosis
6. Primary Site
7. Laterality
8. Histology
9. Behavior Code
10. Grade/Differentiation
11. Diagnostic Confirmation
12. Date of First Microscopic Confirmation
13. Vital Status
14. Cancer Status
15. Cause of Death

The process of creating semantic metadata as annotations in the caDSR is facilitated through the caCORE Software Development Kit (SDK) – a set of software tools. The key components are Semantic Connector, Semantic Integration Workbench (SIW) and UML Loader. These tools are used to transform UML information models into caDSR-structured metadata. Each tool performs a specific role in the transformation process.

III. First Course treatment
1. Date of First Surgical Procedure
2. Surgical Procedure of Primary Site at this Facility
3. Date of RT Started
4. Institute of RT
5. Chemotherapy at This Facility
6. Date of Chemotherapy Started at This Facility
7. Hormone Therapy at This Facility
8. Date of Hormone Therapy Started at This Facility
9. Immunotherapy at This Facility
10. Date of Immunotherapy Started at This Facility
11. Hematologic Transplant and Endocrine Procedure at This Facility
12. Date of Hematologic Transplant and Endocrine Procedure Started at This Facility
13. Palliative Care at This Facility

IV. Outcome
1. Date of First Recurrence
2. Type of First Recurrence
3. Date of Last Contact or Death

Fig. 1. c gridTCR CDE categories. Four main categories of CDEs and the sub-data types collected.

Detail description of each CDE under these four main categories and its sub-groups can be found in the CDE Browser.

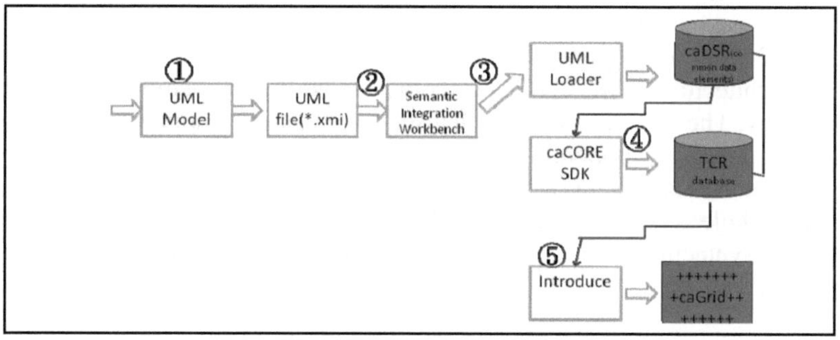

Fig 2. Process flow for the gridTCR system design

The UML information model exported to XMI is first processed by the Semantic Connector application. All UML element names for classes and attributes are matched to NCI Thesaurus concepts. Using the SIW tool, developers can manually revise the automated matches to produce a semantic annotation for each UML entity. Semantic Connector is applied again to produce an annotated XMI file, which contains all of the metadata derived from the original UML model in addition to the semantic information derived from the semantic annotation process. The final annotated XMI file is then used to review all associations and data types.

5 Result

The TCCDEP developed 40 data elements to annotate the cancer registry data collected. In this project, we describe the process required to develop the model, the caDSR CDEs, and the results of the modeling effort. We address difficulties we encountered and modifications for solution. The caBIG™ grid project, gird model of Taiwan Cancer Registry (girdTCR), using the caCORE tools to define data elements for cancer registry has been shown to caBIG™™ UML model project. Currently, the Taiwan cancer registry CDEs are released and available in CDE browser for reusing. Furthermore, we will extend our CDEs to daily clinical practice and trials, along with how the methods were used to fully implemented in hospitals and cancer research centers in Taiwan.

We can review these CDEs using the CDE Browser at https://cdebrowser.nci.nih.gov/CDEBrowser/ or the UML Model Browser at http:// umlmodelbrowser.nci.nih.gov/umlmodelbrowser/ The Browser is a curator tool for CDEs in reusing.

6 Conclusion

A prototype of caCORE-based TCR data standard has been established. The freely available TCR CDEs developed by TCCDEP are robust, based on caDSR, and are ISO 11179 compliant. TCR CDEs can be used as a basis of electronic data standards for cancer research data management and exchange. The gridTCR system is developed based on the caBIGTM paradigm. It demonstrates the effectiveness of rapid application development for building 'caCORE-like' data system using model driven architecture (MDA) and n-tier distributed computing architecture methodology. The system is built with a set of open source applications that are easily accessible to any application development team. We deploy our caCORE SDK-generated data system on the caGrid infrastructure using the Introduce toolkit to make the whole system caBIGTM silver level compatible. The gridTCR system realizes the syntactical and semantic interoperability among multisite which host the cancer data.

References

1. Taiwan Cancer Registry, `http://tcr.cph.ntu.edu.tw/main.php?Page=N1`
2. Unified Modeling Language, `http://www.uml.org`
3. Enterprise Architect, `http://www.sparxsystems.com.au/ea.htm`
4. XML Metadata Interchange (XMI), `http://www.omg.org/technology/documents/formal/xmi.htm`
5. Introduce, `http://cagrid.org/wiki/Introduce`

6. Tobias, J., Chilukuri, R., Komatsoulis, G.A., Mohanty, S., Sioutos, N., Warzel, D.B., et al.: The CAP cancer protocols–a case study of caCORE based data standards implementation to integrate with the Cancer Biomedical Informatics Grid. BMC Med. Inform. Decis. Mak. 6, 25 (2006)
7. Covitz, P.A., Hartel, F., Schaefer, C., De Coronado, S., Fragoso, G., Sahni, H., Gustafson, S., Buetow, K.H.: caCORE: a common infrastructure for cancer informatics. Bioinformatics 19, 2404–2412

A Comprehensive Behavioural Process Model for Design Synthesis

Yi-Min Deng[1,a], Yongsheng Ma[2,b], Meiyan Wang[3,c], and Liang Shi[1,d]

[1] Department of Mechanical Engineering, Faculty of Mechanical Engineering and Mechanics, Ningbo University, Ningbo City, Zhejiang Province 315211, P.R. China
dengyimin@nbu.edu.cn, sliang212@163.com

[2] Department of Mechanical Engineering, Faculty of Engineering, 4-9, Mechanical Engineering Building, University of Alberta, Edmonton, Alberta T6G 2G8, Canada
yongsheng.ma@ualberta.ca

[3] Department of Precision Machinery and Precision Instrumentation, University of Science and Technology of China, Hefei city, Anhui Province 230027, P.R. China
yiqun@mail.ustc.edu.cn

Abstract. Design synthesis is a critical activity for the development of design solutions. Although there are already quite some design synthesis models for the conceptual phase of product design, there still lacks a comprehensive model that takes into account extensive aspects of behavioural process characteristics, which are essential for the development of design solutions. In this paper, the characteristics of behavioural process are investigated and a number of inherent relations and characterization within a behavioural process are identified, including temporal relation, loop relation and state characterization, in addition to the existing causal relation. A comprehensive behavioural process model is subsequently developed, where the identified relations and characterization, together with the behavioural process development procedure are incorporated. The model can be used in explicitly exploring the behavioural process characteristics and thus assisting the designers in synthesis for design solutions. A design case study is also presented to illustrate as well as to verify the proposed methodologies.

Keywords: Behavioural process, Design synthesis, Relation and characterization.

1 Introduction

Design synthesis includes the identification of one or more design solutions consistent with the design requirements and any additional requirements identified during synthesis [1]. To assist design synthesis, it is generally necessary to exploit the relevant behaviour information. This is because behaviour is a bridge to link the gap between function and physical structure [2]. For example, in the Function-Behaviour-Structure design model proposed by Qian and Gero [3], behaviour is categorized as the expected behaviour that is desired by the design in fulfilling a required function, and the actual behaviour that is performed by the product structure. Expected behaviour is derived from a formulation or specification of the design problem. Actual behaviour is derived

Y. Zhang (Ed.): Future Communication, Computing, Control and Management, LNEE 141, pp. 265–270.
springerlink.com

from analysis of a database of existing structures. In this way, a synthesis method is proposed to generate the actual behaviour from the expected behaviour. Other behaviour related synthesis models include the Function-Behaviour-State design model [4,5], the causal behavioural process design model [6-8], and so on.

It is noted that all these synthesis models have exploited one or two aspects of behavioural characteristics but none has incorporated all. Besides, there could be more than those identified aspects of characteristics, which are also useful for design synthesis. For example, in the Function-Behaviour-State model [4,5], behaviour is regarded as a physical state transition of the behaviour actor. In the Causal Behavioural Process model [6-8], it is regarded as the physical interactions between the behaviour actor and its environment, including the input actions to the behaviour actor and the output actions from it; hence the synthesis model exploits only the causality information of behaviours.

To address this problem, this paper shall make an investigation of all aspects of behavioural characteristics and propose a comprehensive behavioural process model to assist the development of conceptual design solutions.

2 Behavioural Process Characteristics

By studying a number of mechanical and electromechanical devices, we noticed that there exist some inherent relations and characterization within behavioural processes.

2.1 Causal Relation

A behavioural process consists of a number of individual behaviours. Each behaviour exhibits a number of input actions and output actions. Between two adjacent behaviours, an output action from the proceeding behaviour actually works as one of the input actions to the succeeding behaviour, hence the proceeding behaviour can be regarded as the "cause" for the occurrence of its succeeding behaviour. Together, the individual behaviours form a network, which is referred to as a Causal Behavioural Process (CBP) [6-8]. Among all the input actions and output actions, some input actions may come from the environment, while some output actions may be delivered to the environment. These are the input and output actions of the entire behavioural process. The causal relations between adjacent behaviours of a behavioural process characterize the causality of the behavioural process. This is one of the most essential characteristics of a behavioural process, given its critical role in establishing the dependency between individual behaviours thus the formulation of the behavioural process.

2.2 Temporal Relation

Each individual behaviour of a behavioural process shall occur for a certain amount of time; it could be long or just a short instance. Multiple behaviours may occur simultaneously or one after another. Hence, there exists temporal relation among individual behaviours, which indicates the concurrency or sequential characteristics of the behaviours. This information is useful for design synthesis. For example, when two or more individual behaviours need to utilize a same energy or material resource, the temporal relation information may help the designers in determining how the individual behaviours should use the resource, including the sequence, the duration, as well as the concurrency, and so forth. Temporal relation may be utilized not only in design synthesis, but also in design evaluation and verification [9].

2.3 Loop Relation

For some utility devices such as the automatic mounting devices, automatic press fitting devices, electric or pneumatic nailing devices, etc., there is generally such a phenomenon within the behavioural process of the devices in that, all the movable components of the devices will return to the initial state after finishing their corresponding behavioural process. This is to ensure that the device can be ready for another round of behavioural process. In doing this, the behavioural process shall have a trigger behaviour and a revert behaviour, where the former is used to trigger the start of a single behavioural process, while the latter is used to revert the moving components of the device to its initial physical status, including position, orientation, velocity, etc. As such, the relevant behaviours have for a loop relation. In fact, there could be multiple loops within such a behavioural process, hence they are also referred to as phenyl loops. There are also some rules and characteristics regulating the phenyl loop relations [10].

2.4 State Characterization

The above relations all characterize mutual relations between individual behaviours of a behavioural process. However, an individual behaviour itself shall also exhibit some useful characteristics which are useful for design synthesis. In fact, the behaviour actor may exhibit multiple physical states during the occurrence of the behaviour. The state transition may be triggered by different working conditions; or they may be triggered by the other individual behaviours of the behavioural process. The state characterization of behaviours shall be discussed further later.

3 Behavioural Process Model for Design Synthesis

With the above relations and characterization of behavioural process, we propose a comprehensive behavioural process model for design synthesis, as is illustrated in Fig.1. This model consists of three layers of design information, namely, function layer, behaviour layer and structure layer. In the function layer, design synthesis is materialized as a function decomposition, from which a function structure proposed by Pahl and Beitz [11] would be generated. For example, Fig. 1 shows that the overall function F0 is decomposed into three sub-functions, F1 – F3. The sub-functions are then mapped to structure either directly, such as F3 is mapped to S3; or via a single behaviour, such as F2 is mapped to S2 via behaviour B8; or it may need to develop a behavioural process first before an implementing physical structure can be identified, such as the behaviour process (behaviours B1-B7) for mapping F1 to S1.

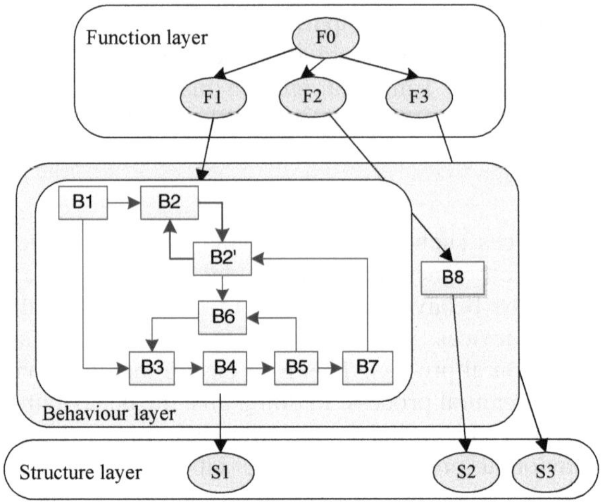

Fig. 1. Illustration of the proposed design synthesis model

In the design synthesis process, a direct mapping is always carried out first for all the decomposed sub-functions. If this fails, then mapping via a single behaviour is tried. Either direct mapping or mapping via a single behaviour may be assisted by past design knowledge stored a priori. This is often referred to as the design catalogue approach. If, however, it is found that it is not feasible for a single behaviour to fulfil a required function, then the development of a behavioural process will be necessary. During this process, the aforementioned causal, temporal and loop relations, as well as the state characterization can be applied to assist the development of a feasible behavioural process, where the behaviour actors of all the constituent individual behaviours can subsequently be constructed to offer a physical solution to the corresponding function.

For example, in developing the behavioural process shown in Fig.1, a backward reasoning process proposed by Deng et al. [6-8] may be applied to identify individual behaviours B7, B5, B4 and B3, where the causal relations between the adjacent individual behaviours are utilized. In developing B6, the loop relation is exploited to ensure that the corresponding part of the device can revert to its initial physical status each time the behavioural process is fulfilled. From B7 to B1, there are two individual behaviours B2 and B2', both of which are actually exhibited by a same physical structure, but are associated with different physical states. Hence, they are also referred to as state behaviours. With the driving input from B1, B2 changes the physical state from one to another. B7 is the trigger behaviour of this behavioural process. B2' resets the structure state and generates a functional output to B6.

4 Case Study

In this section, we shall study a simple electric nailing device to illustrate how the proposed behavioural process model can be applied for design synthesis. The overall function of the device is to strike nails into an object quickly and repeatedly, indicated as F0. By applying the proposed model, functional decomposition should be carried out first to develop a function structure of the design problem. After that, each sub-function shall be mapped to structure either directly or via a single behaviour or a behavioural process. To simplify our discussion, we shall study the development of a behavioural process, as is shown in Fig.2, for one of the sub-functions "to strike the nail with a working head and revert the working head", indicated as F1. By exploiting past design experience, an individual behaviour B1 is developed, which is to generate a magnetic force with a magnetic field. The driving input to B1 is that of the electric power supply, which is readily available as has been specified in the given design problem. With the output actions from B1, i.e. the striking force, B2 is developed, which is to strike the nail with a working head. Hence, part of the functional requirement of F1 has been fulfilled.

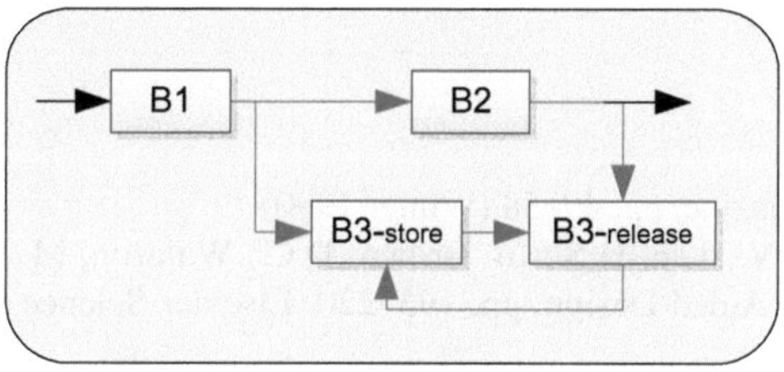

Fig. 2. Behavioural process development of a design case

To fulfil the remaining functional requirement specified in F1, we need to provide a yet another force to push the working head back after each striking action. In doing this, the state behaviours proposed in the previous section may be employed. We may, for example, utilize energy storage mechanism [12] by using a compression spring to deliver the two state behaviours, i.e. B3-store and B3-release, as are shown in Fig. 2. B3-store is the behaviour of the spring in storing energy (the spring is being compressed) when the working head is moving forward; whereas B3-release is the behaviour of releasing energy when the working head is moving back. This solution effectively addressed the design requirement in that, there is no additional outside energy required to provide the revert action; besides, the mechanism is relatively simple.

Further more, it is necessary to exploit the temporal relations within the developed behavioural process to verify whether the developed structural solution can work appropriately. Still further, when combining all of the developed behavioural processes (which are for the decomposed sub-functions respectively) to form the overall behavioural process of the design, it is necessary to apply the loop relation to ensure that the whole device can revert to its initial status after each striking action (not just the working head – the next nail should also be in position for the next striking action). Due to the space constraint, these are not further discussed.

5 Conclusions

In the previous sections, we have studied the characteristics of the behavioural process and proposed some additional inherent relations and characterizations within a behavioural process other than the existing causal relation, including temporal relation, loop relation and state characterization. These characteristics are then utilized to develop a comprehensive design synthesis model, where the development of behavioural process to assist function to structure mapping is elaborated. A simple case study demonstrates that the model is useful in assisting designers to develop effective design solutions. Further work is needed in applying the model to more design case studies and if necessary, to modify and refine it, so that it can be applied to more complex design problems.

Acknowledgements. This work was supported by the National Natural Science Foundation of China (Grant No. 50875133). It was also supported by a project titled Design Methodology Research for Assembly Systems from the Foreign Experts Affairs Bureau of Ningbo, China.

References

1. Maher, M.L.: AI Magazine, pp. 49–58 (Winter 1990)
2. Gero, J.S., Tham, K.V., Lee, H.S.: In: Brown, D.C., Waldron, M., Yoshikawa, H. (eds.) Intelligent Computer Aided Design, pp. 193–220. Elsevier Science Publishers B.V, North Holland (1998)
3. Qian, L., Gero, J.S.: Artificial Intelligence for Engineering Design. Analysis and Manufacturing 10, 289–312 (1996)
4. Tomiyama, T., Umeda, Y., Yoshikawa, H.: Annals of the CIRP 42, 143–146 (1993)
5. Umeda, Y., Ishii, M., Yoshioka, M., Shimomura, Y., Tomiyama, T.: Artificial Intelligence for Engineering Design, Analysis and Manufacturing 10, 275–288 (1996)
6. Deng, Y.-M., Tor, S.B., Britton, G.A.: Engineering with Computers 16, 36–52 (2000)
7. Deng, Y.-M., Tor, S.B., Britton, G.A.: Journal of Engineering Design 11, 347–375 (2000)
8. Deng, Y.-M.: Artificial Intelligence for Engineering Design, Analysis and Manufacturing 16, 343–362 (2002)
9. Deng, Y.-M.: In: Proceedings of the 2nd International Conference on Mechanical and Electronics Engineering, Kyoto, Japan, August 1-3, vol. 1, pp. 351–355 (2010)
10. Shi, L., Deng, Y.-M.: In: Proceedings of the 2nd International Conference on Mechanical and Electronics Engineering, Kyoto, Japan, August 1-3, vol. 1, pp. 346–350 (2010)
11. Pahl, G., Beitz, W.: Engineering Design: A Systematic Approach. Springer, London (1988)
12. Wang, M.-Y., Shen, L.-G., Deng, Y.-M.: Advanced Science Letters 4, 1–5 (2011)

Research and Application of Contactless Electrical Energy Transmission System

Huang Ke, Zhang Liping, Kong Xianwang, Yang Yongkui, and Wang Junwei

Beijing Special Engineering and Design Institute, P.O.Box 4702#6,
Chaoyang, Beijing, China
huangkenudt@163.com

Abstract. Contactless electrical energy transmission system realizes power transmission to electrical equipment with no electrical or physical connection by the perfect combination of inductive coupling technique and electronics .This paper introduces the basic principle of contactless electrical energy transmission, establishes the mathematic model based on the loosely coupled circuit, and analyzes the influence factor of the transmission efficiency. Finally combined with the engineering application, this paper foresees the perspective application of this system in harsh environments.

Keywords: Contactless electrical energy transmission, loosely coupled, mathematical model, reflected impedance.

1 Introduction

The traditional supply methods need electrical connection between power supply system and the load, but in some harsh environments (such as chemical industry, underground mining, underwater operations, etc.), especially supplying power for moving load, it is unsafe due to wear and bare conductors and other constraints, so the security and reliability of the system is low, and the application is greatly restricted. Contactless electrical energy transmission system is a loosely coupled inductive system using air as the coupling media, which generates energy to electrical equipment through the high frequency radiation way without any electrical connection. The system has the advantages of mobile and flexible, without environmental impact, maintenance, etc., can provide clean, stability and security electric power to the moving load in harsh environments.

This paper introduces the structure and principle of contactless electrical energy transmission system, establishes the mutual inductance model, analyzes the influence factors of the transmission efficiency, and describes the coupling effect using the induced potential between the primary and secondary subsystems.

2 System Structure and Principle

Contactless electrical energy transmission system is based on the theory of electromagnetic induction principle, which core part is the radiation and reception of

Y. Zhang (Ed.): Future Communication, Computing, Control and Management, LNEE 141, pp. 271–277.
springerlink.com

high frequency electromagnetic energy, achieving by the loosely coupled transformer. The primary and secondary windings of the traditional transformer coil on the common closed iron core, as shown in Fig. 1 (a). Contactless electrical energy transmission system has a separate inductive coupled circuit, the primary and secondary windings are wound on different magnetic structures to realize the energy coupling without any physical connection between the power and the load. According to the relative motion between the primary and secondary subsystems, there are three different kinds of inductive coupled power transfer systems, as shown in Fig. 1 (b), (c), (d) .

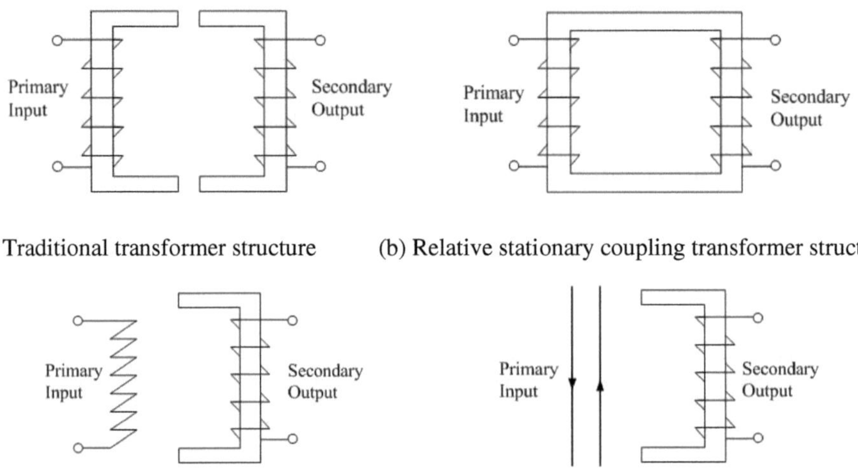

(a) Traditional transformer structure (b) Relative stationary coupling transformer structure

(c) Rotating coupling transformer structure (d) Linear motion coupling transformer structure

Fig. 1. Transformer structure

Under the action of the applied voltage, primary side of the loosely coupled transformer produces alternating magnetic flux, because the magnetic potential mainly descends in the loosely coupled transformer's air gap, so the magnetic flux density can be calculated by the following equation:

$$\int H ds = \frac{2sB}{\mu_0} = I_s \tag{1}$$

Where, s is the length of the air gap, Is is the excitation current, B is the magnetic flux density, μ0 is the magnetic permeability.

According to the law of electromagnetic induction, the induced potential of secondary coil is:

$$U_{is} = 4.44fN_s A_{Fe} B = \frac{4.44\mu_0 fN_s A_{Fe} I_s}{2s} \tag{2}$$

Where, Ns is the number of winding of the secondary coil in the loosely coupled transformer, AFe is the ferrite core cross-sectional area, f is the power frequency.

From Eq.1, Eq.2 we can know, the transmission media between the primary and secondary coils of the loosely coupled transformer in contactless electrical energy transmission system is air. The air permeability of the gas is very small and there is a big gap between primary and secondary windings, then the transformer is in loosely coupled state and the magnetic flux leakage is large, so the coupling coefficient is much lower than traditional transformers. It must reduce the circuit power transmission losses. In order to generate enough potential to satisfy the needs of the load, the feasible method is to increase the frequency of the primary alternating current in order to increase the frequency of alternating magnetic field to obtain larger secondary induced potential. The high frequency alternating current is usually in the range of 10kHz ~ 100kHz. Changes of the magnetic field will cause strong electromagnetic induction between the primary and secondary subsystems, so we can obtain the induced electromotive force in the secondary subsystem according to the electromagnetic induction law and Lenz's law.

The structure of the contactless electrical energy transmission system is shown in Fig. 2, there is no physical connection between the primary and secondary subsystems. When the system works, the two or three phase alternating current power is rectified and filtered into direct current (DC) power, the DC power provides input for the high frequency inverter to get the high frequency alternating current which is inputted to the primary side, then can obtain the induced current with the corresponding frequency in the secondary side, achieving to supply the load by regulating the voltage, current and phase.

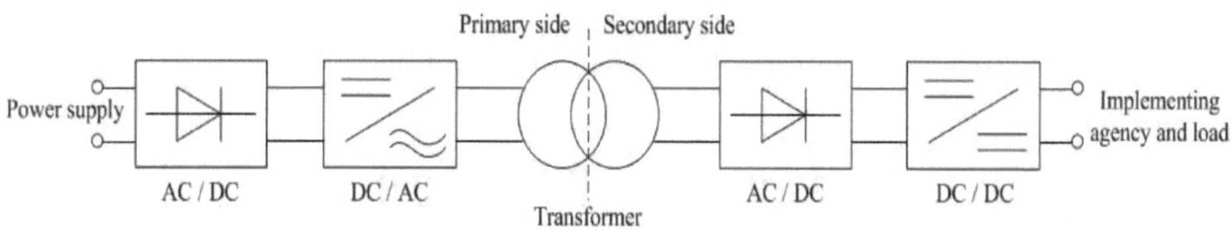

Fig. 2. Components of contactless electrical energy transmission system

The primary winding and secondary winding of loosely coupled transformer are separate, and can maintain in relative stationary or movable state, especially in the areas of logistics mobile delivery, the primary coil is usually extended for a long circuit, the secondary coil is wound on an open magnet, so the two coils allow a long distance relative movement. And the primary winding can be radiated electromagnetic energy to multiple secondary winding, to power multiple loads simultaneously in order to meet different application requirements.

3 System Model

Because the primary and secondary subsystems in contactless electrical energy transmission system are in the loosely coupled state, the traditional transformer model based on magnetic components is no longer suitable, so we use the coupling mutual inductance model to analyze the system's electromagnetic relations, while introduce the reflected impedance of secondary subsystem to describe the power transmission capacity. This model does not need to separate the coupling inductance and leakage inductance, fully considering the leakage inductance and magnetizing inductance influence on the system.

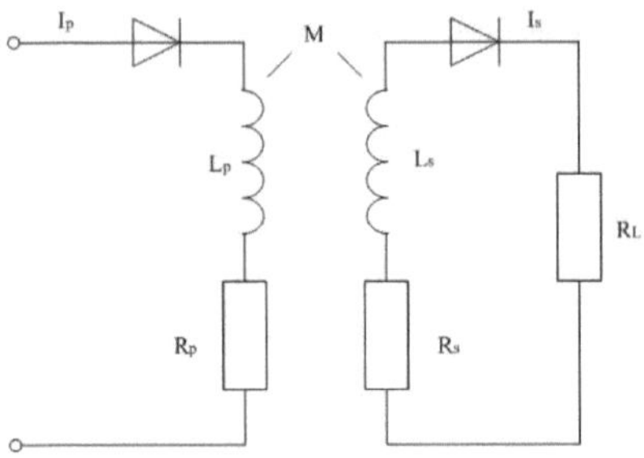

Fig. 3. Mutual inductance model of coupled structure

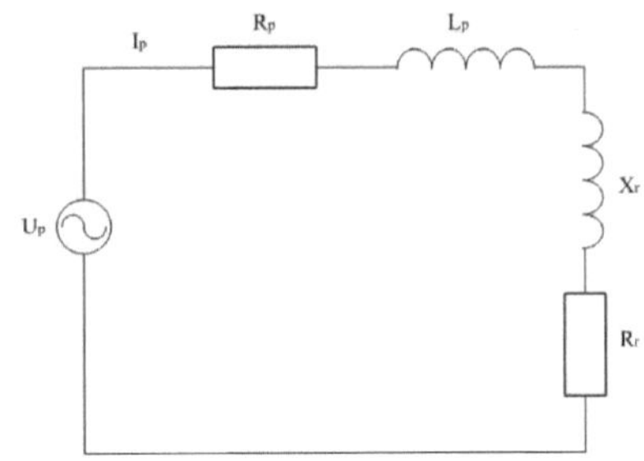

Fig. 4. Equivalent circuit

The structure of mutual electromagnetic coupling model is shown in Fig. 3, Rp and Lp is the primary winding parameters, Rs and Ls is the secondary winding parameters, RL is the load resistance. The current of primary winding is Ip, and the voltage is Up. M is the mutual inductance between the primary and secondary magnetic coupling winding, ω is the angular frequency carrying in the primary winding.

As shown in Fig. 3, we can get the following circuit equations:

$$\dot{I}_p\, j\omega L_p + \dot{I}_p R_p - \dot{I}_s\, j\omega M = \dot{U}_p \quad (3)$$

$$\dot{I}_s\, j\omega L_s + \dot{I}_s R_s + \dot{I}_s R_L = \dot{I}_p\, j\omega M \quad (4)$$

Further we can get the equivalent circuit (as shown in Fig. 4) of electromagnetic structure coupled inductor, reflected impedance embodies the secondary subsystem's influence on the primary subsystem. From Eq. 3 and Eq.4, we can get:

$$Z_r = R_r + jX_r = -\frac{\dot{I}_s\, j\omega M}{\dot{I}_p}. \quad (5)$$

$$R_r = \frac{\omega^2 M^2}{(R_s + R_L)^2 + \omega^2 L_s^{\,2}}(R_s + R_L) \quad (6)$$

$$X_r = -\frac{\omega^3 M^2 L_s}{(R_s + R_L)^2 + \omega^2 L_s^{\,2}} \quad (7)$$

Where, ω and the supply frequency f have the following relationship:

$$\omega = 2\pi f \quad (8)$$

The reflected impedance directly reflects the performance of power transmission, the plural power absorbed by the reflected impedance is equal to the power absorbed by the secondary subsystem. From Eq.5~8 we get, the reflected impedance is negative. Rr(ω) is an increasing function, as the frequency increases, the reflected impedance is gradually increasing; and when the frequency is a constant with a specific load, the reflected impedance will reach its maximum; and while the load increases, the reflected impedance and reactance response are tend to zero.

The reflected resistance directly reflects the active power of the transmission system. Output power of the load depends on the load resistance, when RL= ωLs - Rs, the output power reaches its maximum, as shown in Eq.9:

$$P_{o\_max} = \frac{\omega M^2 I_p^2}{2L_s} \tag{9}$$

Contactless electrical energy transmission system is a loosely coupled structure, there is a big air gap between the primary and secondary circuits, on the one hand to achieve the power supply between the subsystems, make up the inherent defects of traditional power transmission. On the other hand, the existence of the large air gap causes a similar effect compared the system leakage to the initiation, and even higher than the initiation, has limited the power transmission capacity and efficiency. And its output capacity depends on the load impedance, output power is greatly limited. Therefore in the applications, we usually take effective compensation measures on the primary and secondary subsystems to improve the stability and power factor, at the same time to reduce the voltage and current stress of the power converter.

4 Engineering Application

Contactless electrical energy transmission system has taken the academic and engineering community's attention with its continuous development and improvement, and has been widely applied in transmission equipment, robot, stage system and other areas.

In Dali Theater (Yun Nan Province, China), there is a big car wagon with horizontal movement, which path is arc, and its traveling schedule is 56 meters and working in harsh environments (open-air, moist). The designers decided to use the Movitrans, one contactless electrical energy transmission system, to drive and control the car wagon which structure is shown as Fig. 5.

The Movitrans transforms the standard 3-phase 380 VAC, 50HZ alternating current to 25kHZ intermediate frequency power inputting to the voltage source converter TPS and current source converter TAS, and through the inductive pickup THM transmits the energy to the mobile converter TPM, then outputs 500 VDC and 24 VDC for the use of the inverter and motor. The practice proved that this system can adapt the harsh environments with the advantages of easy maintenance and no noise, can satisfy the equipment's requirement of random movement and achieve the good application effect.

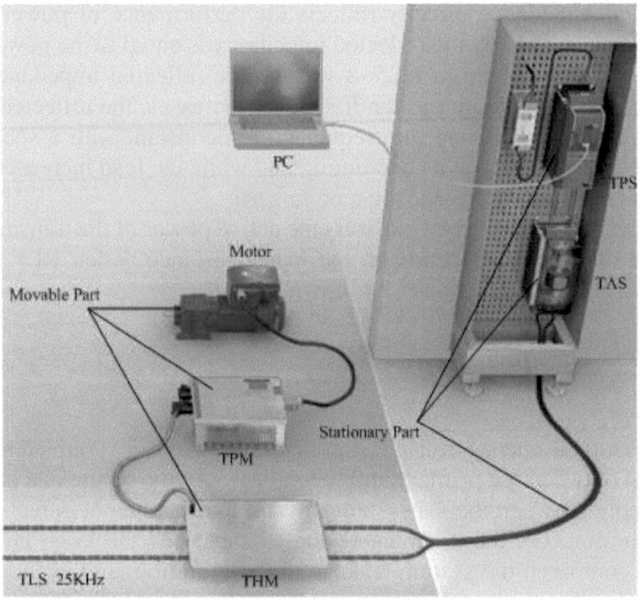

Fig. 5. Application example structure

5 Conclusion

Contactless electrical energy transmission system realizes power transmission to moving electrical equipment, which is safe, fast, no noise, easy maintenance, etc., and is suitable for high-speed movement or harsh environments applications. This article describes the principle of the contactless electrical energy transmission system, and builds its mutual inductance coupling model to analyze its transmission characteristics. With the development of technology, the structure and performance can be further optimized, this system will be able to get more widely application.

References

1. Wu, Y., Yan, L., Huang, C., Xu, S.: Performance analysis of new contactless electrical energy transmission system. Advanced Technology of Electrical Engineering and Energy 22(4) (October 2003)
2. Klontz, K.W., Divan, D.M., et al.: Contactless power delivery system for mining applications. IEEE Trans. On Industry Applications (1995)
3. Zhang, L., Yao, R.: Development and Prospective Application of Contactless Power System. Electrical Technology 1 (2004)
4. Wang, P., Tang, H.: The Technique Analysis based on Inductive Electric Power Transmisio. Microprocessors 3 (June 2006)
5. Sheng, S., Du, G., Zhang, B.: The Design of Induction Coupling No-Touch Power Transfer System. Telecom Power Technologies 24(5) (September 2007)

6. Mao, S., Wang, H.: Characteristic Ayalyses for Separate Transformer in Contactless Inductive Power Transfer System. The World of Power Supply (May 2006)
7. Sun, Y., Lou, P., Wu, L.: Research on Application Platform of Contactless Supply System. Industrial Control Computer 22(3) (2009)
8. Ou, Y., Yang, M., Zhu, S.: Research on Contactless Mobile Power Technology. Journal of Hunan University (Natural Sciences) 33(1) (February 2006)

Rapid Manufacturing: An Industrial Revolution or a Passing Fad?

Sarat Singamneni

School of Engineering, AUT University, Auckland, New Zealand
sarat.singamneni@aut.ac.nz

Abstract. Some of the Rapid Prototyping (RP) processes have seen enormous improvements in the recent past and are racing towards Rapid Manufacturing (RM), the ability to produce end use parts directly from CAD files. While this sounds quite exciting, considering the time-to-market constraints constantly faced by manufacturing engineers, the key question to be answered is whether the processes are truly capable of achieving the required quality standards. This paper addresses this question considering the mechanisms of material consolidation in a couple of RP processes that are currently considered sure candidates for RM. The conclusion is that RM is a possibility, but depends on the outcomes of research and development activities yet to be undertaken in many directions.

Keywords: Rapid prototyping, manufacturing, fused deposition modeling and laser sintering.

1 Introduction

Selective laser Sintering (SLS), Selective Laser Melting (SLM) and Fused Deposition Modeling (FDM) are the common RP processes that have crossed the threshold to become qualified as future RM processes [1]. There are other, indirect routes to rapid manufacturing, such as rapid tooling and rapid casting applications made possible by one or the other RP processes, resulting in considerable amount of savings in time, but this paper is concerned with rapid manufacturing referring to the direct production of end use parts from digital data, without any complex tooling or intermediate processing [2]. The actual material manipulation proceeds with the point-by-point, line-by-line and layer-by-layer consolidation of the raw material to build 3D shapes. Depending on the technique used, the power source could be an ultra violet light, a laser beam or an electron beam with raw materials varying from powder ceramics or metals to fused polymers.

In spite of numerous benefits, there are both material and process limitations, if functional parts are required to be produced. The point-by-point processing results in porosity and defective internal structures that may be detrimental to the mechanical characteristics of the part. While these questions on the final part characteristics are still at large, there has been a lot of research interest on different aspects of rapid manufacturing. Driven mainly by the rapid manufacturing research team at

Y. Zhang (Ed.): Future Communication, Computing, Control and Management, LNEE 141, pp. 279–283.
springerlink.com

Loughborough University, various issues such as the possible benefits of RM , unlimited design freedom and the consequential manufacture for design as against the traditional design for manufacture and cost analyses comparing RM and its traditional counter have been reported apart from a few other publications [3-7].

These initial research reports find RM as a potential means of solving many problems faced by current manufacturing industries. The ability to produce rapidly means the much needed oxygen to the struggling small to medium scale industries in terms of meeting market changes and ever increasing demand to shorten manufacturing lead times. Further, there will be consequential influences on several aspects of factory organization and management strategies and approaches such as group technology might become superfluous while agile and reconfigurable manufacturing systems might need a complete reconsideration. However, the inherent weaknesses of all potential processes are still at large, and the feasibility of RM can only be decided based on an evaluation of these. This requires a closer look at the material processing techniques and the microstructures achieved as a result.

2 Fused Deposition Modeling

Fused Deposition Modeling allows thermoplastic polymers in the form of a filament to be processed layer by layer to form 3D shapes from CAD files. It was reported by Mansour and Hague [4], that this process has the capability to compete with the injection moulding process at low to medium production levels. In FDM, the ABS filament is heated to a semisolid state and is deposited line by line in each layer of the part. This results in a characteristic road structure within each layer. The thermal energy of the semi-molten material drives the bond formation between successive roads [8]. The resulting material is a composite of bonded ABS filaments and voids and the extent of bonding significantly influences the mechanical characteristics. When the liquefier and envelop temperatures are sufficiently maintained, inter-road bond formation happens more efficiently. The filaments gradually cool to room temperature in a transient manner and the mechanics of inter-road and interlayer bond formation is a complex process and quite dynamic in nature.

Bellehumeur et al. [8] proposed a sintering mechanism leading to the formation of a neck as the means of achieving the bond between adjacent roads It was shown that most sintering occurs within seconds after extrusion and deposition. The rate and level of wetting will be influenced in a complex way by several process parameters. Notable factors being extrusion temperature, envelop temperature, and convection conditions. Rodriguez et al. [9, 10] analysed the mesostructures resulting from different deposition patterns in FDM. It was clearly stated that the void geometry and the extent and geometry of the bonding between individual roads have a large influence on the anisotropy and magnitude of the strength developed by the material. The results also showed significant mesostructural influence on the stress-strain response. Loss of load-carrying capacity due to the presence of voids plus a loss of molecular orientation during the FDM extrusion process contribute to the decrease in stiffness and strength. All these observations point to the fact that FDM parts are not comparable to those of traditional injection moulded parts in terms of mechanical and physical characteristics.

3 Selective Laser Sintering and Melting

SLS and SLM are the other potential candidate RP processes for RM.. Based on binding mechanisms, Kruth et al. [11] classified SLS and associated processes such as Direct metal Laser Sintering (DMLS) can be classified into four classes: solid state sintering, chemically induced binding, liquid phase sintering-partial melting, and full melting. Phenix systems are based on solid state sintering that occurs at temperatures between half and the melting temperature of the work material powder. Adjacent particles are fused into each other when the laser strikes, leading to a neck formation. The inter granular gaps result in micro and macro porosities at the end and the processes cannot withstand the test for being suitable for the rapid manufacturing of critical parts. Similar drawbacks are associated with Chemical bonding methods used for sintering. Several commercially available technologies use liquid phase binding, in which a structural and a binding material constitute the resulting material structure developed either from two different materials or in some cases, the same material. The molten binder when finally solidified, forms the network of binding material which supports particles of the structural material, but the process is lengthy in terms of the need for elaborate post sintering secondary processes and so cannot justify rapid manufacturing

Polymer coated metal powder particles (Directsteel and Laserform) can also be sintered to form green parts which will then be processed by de-binding and infiltration. SLS of polymer coated sand particles as used on SandForm (DTM, EOS GmbH) systems also falls under this category, but is mainly used in the direct production of expendable sand moulds. EOS process also allows part building by partial melting of constituent powder particles, and will allow powders of different materials to be combined together. There have been improvements in all these processes, and mechanical properties comparable to traditional methods are reported by Kruth et al. [11], but the parts are still inferior in terms of internal microstructural characteristics and inter particle binding mechanisms and cannot be compared to those from traditional manufacturing processes such as machining, forming and casting.

Full melting of powder particles aimed at complete fusion and subsequent solidification to form a continuous metal structure is attempted in Selective Laser Melting (SLM). A high powered laser is used to fuse fine particles of metal powders together. Typical layer thicknesses range from 20 to 100 μm. Gas atomized metallic powders of titanium, stainless steel, cobalt chrome and tool steels are processed. Fully dense parts are possible to be produced and the internal structure of the resulting material is continuous and free of porosity and comparable to the structures possible in casting. This is a true rapid manufacturing technique in that functional parts of required mechanical properties can be produced directly from CAD files.

The downside of the process however is the very nature of processing in SLM. A high energy heat source melts a tiny part of the powder bed at a time and rapidly moves from one point to the other. This rapid melting and solidification and the dynamic movement of the heat source result in internal cavities and micro cracks. These issues have not yet been investigated, as in the case of other traditional manufacturing processes and the true applicability of this can only be judged based on result of scientific investigations, if any in this direction. The prerequisite in this case is that the thermodynamics of the complex thermal fields of the SLM process needs to be analysed and understood.

4 Discussion

It is evident; FDM in its current form cannot be claimed as a process suitable for rapid manufacturing. The final part characteristics are much inferior compared to the injection moulded counterparts. However, the current literature offers plenty of evidence suggesting the suitability of FDM for direct production of end use parts. Elaborate experimental investigations leading to a complete understanding of the mechanism and mechanics of material consolidation in FDM is necessary to be undertaken. Further, there is a significant influence of the road structure on the mesostructure and subsequent mechanical characteristics of final printed parts. This requires experimental and numerical investigations leading to establishment of the influence of the direction of lay on the final part performance. The scheme of deposition is yet another major area of investigation. Whether it is the flat layer or the curved layer or a combination of both needs to be worked and in most practical cases, a judicial mixture might work the best, but a scientific understanding of all aspects requires to be developed.

SLS, though brings parts significantly closer to expected levels of quality in terms of dimensional tolerances and mechanical properties, the internal structure is still inferior compared to the parts produced by traditional means. And most of these disqualify as true RM processes. Selective Laser Melting on the other hand has the capability to construct fully dense parts with micro level material continuity. However, the thermal phenomena are not understood completely, and there is always the possibility of micro cracks and residual stresses and strains. While heat treatment processes might eliminate some of these problems, the thermodynamics of the process needs to be understood first, followed by experimental evaluation of the influences of process parameters on the microstructure of final parts. Experimental investigations leading to an understanding of the influence of constituent factors [12, 13, 14] and finite element modeling studies [15, 16] have already been attempted. Further studies in this direction and a critical evaluation and compilation of results are needed before moving towards true RM using SLM.

5 Conclusion

Rapid Manufacturing offers promising solutions to many problems currently faced by the manufacturing industry. Not all RP processes can actually be used for RM. FDM for polymers and SLM for metal and ceramic processing are potential candidates for RM. There are definite shortcomings associated even with these processes. Evaluation of the mechanism and mechanics of material processing, experimental and numerical estimation of influences of constituent factors, development of appropriate deposition strategies need research and development to qualify RM as a promising approach to real manufacturing applications. Otherwise, it will turn out to be another passing fad, instead of being an industrial revolution.

References

1. Hopkinson, N., Hague, R., Dickens, P.: Rapid Manufacturing: An industrial revolution for the digital age. John Wiley & Sons (2006)
2. Singamneni, S., Mc Kenna, N., Diegel, O., Singh, D., Neitzert, T., St George, J., Roy Choudhury, A., Yarlagadda, P.: Rapid casting: A critical analysis of mould and casting characteristics. Australian Journal of Mechanical Engineering 7(1), 33–41 (2009)
3. Hopkinson, N., Dickens, P.: Rapid prototyping for direct manufacture. Rapid Prototyping Journal 7(4), 197–202 (2001)
4. Mansour, S., Hague, R.: Impact of rapid manufacturing on design for manufacture of injection moulding. Proc. Of Instn, of Mech. Engrs, Part B: J. Engineering Manufacture 217, 453–461 (2002)
5. Hopkinson, N., Dickens, P.: Analysis of rapid manufacturing-using layer manufacturing processes for production. Proc. Of Instn, of Mech. Engrs, Part C: Journal of Mechanical Engineering Science 217, 31–39 (2003)
6. Hague, R., Mansour, S., Saleh, N.: Material and design considerations for Rapid Manufacturing. Int. J. Prod. Res. 42(22), 4691–4708 (2004)
7. Hague, R., Mansour, S., Saleh, N., Harris, R.: Materials analysis of stereolithography resins for use in Rapid Manufacturing. Journal of Materials Science 39, 2457–2464 (2004)
8. Bellehumeur, C., Li, L., Sun, Q., Gu, P.: Modelling of bond formation between polymer filaments in the fused deposition modelling process. Journal of Manufacturing Processes 6(2), 170–178 (2004)
9. Rodriguez, J.F., Thomas, J.P., Renaud, J.E.: Characterisation of the mesostructure of fused deposition acrylonitrile butadiene-styrene materials. Rapid Prototyping Journal 6(3), 176–185 (2000)
10. Rodriguez, J.F., Thomas, J.P., Renaud, E.J.: Mechanical behaviour of acrylonitrile butadiene styrene (ABS) fused deposition materials, experimental investigation. Rapid Prototyping Journal 7(3), 148–158 (2001)
11. Kruth, J.P., Mercelis, P., Van Varenbergh, J., Froyen, L., Rombouts, M.: Binding mechanisms in selective laser sintering and selective laser melting. Rapid Prototyping Journal 11/1, 26–36 (2005)
12. Montero, M., Roundy, S., Odell, D., Ahn, S.-H., Wright, P.: Material Characterization of Fused Deposition Modeling (FDM) ABS by Designed Experiments. In: Annual Conference on Society of Manufacturing Engineers, Cincinnati, OH, pp. 1–21 (2001)
13. Yadroitsev, I., Bertrand, P., Smurov, I.: Parametric analysis of selective laser melting process. Applied Surface Science 253, 8064–8069 (2007)
14. Badrossamay, M., Childs, T.H.C.: Further studies in selective laser melting of stainless and tool steel powders. International Journal of Machine Tools and Manufacture 47, 779–784 (2007)
15. Crespo, A., Vilar, R.: Finite element analysis of the rapid manufacturing of Ti-6Al-4V parts by laser powder deposition. Scripta Materialia 63, 140–143 (2010)
16. Dong, L., Makradi, A., Ahzi, S., Remond, Y.: Three-dimensional transient finite element analysis of the selective laser sintering process. Journal of Materials Processing Technology 209, 700–706 (2009)

X-FEM Studies on an Inclined Crack in a 2-D Finite Piezoelectric Media

R.R. Bhargava and Kuldeep Sharma

Department of Mathematics, Indian Institute of Technology Roorkee, Roorkee-247667, India
rajrbfma@iitr.ernet.in, kuldeeppurc@gmail.com

Abstract. An inclined crack problem is investigated for a poled finite two-dimensional piezoelectric plate using X-FEM. Intensity factors and energy release rate make the subject matter of investigations. The parametric studies for these are presented with respect to angle of inclination and length of the crack, and electromechanical loadings. It is noted that mechanical intensity factors are independent of electric loading while electric displacement intensity factor depends on electromechanical loading. The study shows that local energy release rate may be considered as a fracture criteria for 2-D piezoelectric media.

Keywords: Crack, Electric-displacement intensity factor, Extended finite element method, Interaction integral, Piezoelectric ceramics, Stroh formalism.

1 Introduction

Piezoelectric ceramics by now have unambiguously proved their utility as a key component of sensors/actuators/transducers in high tech apparatus. Due to the electromechanical coupling these materials behave differently as compared to ordinary materials. Consequently it is imperative to investigate the fracture behavior of such materials.

Allik and Hughes [1] were the first to apply FEM to investigate viberations in piezoelectric ceramics. Kumar and Singh [2] investigated crack propagation and energy release rate under combined mechanical and electrical loadings. Kuna [3] gave a detailed survey of different techniques to solve piezoelectric fracture mechanics problems and present status of work done till date in piezoelectric.

The extended finite element method (X-FEM) proposed by Moës et al. [4] has proven to be a very efficient tool for the numerical modeling of cracks in LEFM. A detailed review of X-FEM applied to various problems in engineering given by Belytschko et al.[5].More recently Be′chet et al. [6] has applied X-FEM to investigate the cracked 2-D piezoelectric material having arbitrary polarization direction.

The paper reports a study on effects of an inclined impermeable crack on intensity factors and energy release rates for a finite two-dimensional piezoelectric plate. Six basis functions together with interaction integral are employed to obtain the solution. A case study is presented for PZT-5H piezoceramic.

Y. Zhang (Ed.): Future Communication, Computing, Control and Management, LNEE 141, pp. 285–290.
springerlink.com

2 Extended Finite Element Methodology

The X-FEM exploits the partition of unity property of FEM, which allows local enrichment functions to be easily incorporated into a finite element approximation.

Considering the impermeable boundary conditions at the crack surface, the extended finite element approximation for the displacement and electric potential can be written as follows

$$u^h(x,y)=\sum_{I\in N}N_I(x,y)u_I+\sum_{I\in N^{cr}}N_I(x,y)\big(H(f^h(x,y))-H(f_I)\big)a_I+\sum_{I\in N^{TIP}}N_I(x,y)\sum_{k=1}^{6}\left(F^k(r,\theta,a_k^{re},a_k^{im})-F^k(x_I,y_I,a_k^{re},a_k^{im})\right)b_I^k \quad (1)$$

$$\phi^h(x,y)=\sum_{I\in N}N_I(x,y)\phi_I+\sum_{I\in N^{cr}}N_I(x,y)\big(H(f^h(x,y))-H(f_I)\big)c_I+\sum_{I\in N^{TIP}}N_I(x)\sum_{k=1}^{6}\left(F^k(r,\theta,a_k^{re},a_k^{im})-F^k(x_I,y_I,a_k^{re},a_k^{im})\right)d_I^k \quad (2)$$

where $H(f(x,y))$ is a modified Heaviside step function

$$H(z_1)=\begin{Bmatrix}-1 & if\ z_1<0\\ +1 & if\ z_1>0\end{Bmatrix} \quad (3)$$

And the shape functions, $N_I(x,y)$, are isoparametric linear quadrilateral element shape functions. The column matrices u_I and ϕ_I are the nodal displacements and electric potential respectively, and a_I, b_I^k and c_I, d_I^k are the additional parameters. $F^k(r,\theta,a_k^{re},a_k^{im})$ are the standard six basis functions for the crack tip.

The total energy release rate (GT), mechanical energy release rate (GM) and local energy release rate (GL) has been calculated using KI, KII as mechanical IFs; KIV as electrical displacement IF

$$G_T=\frac{1}{2}\left(\frac{K_I^2}{C_T}+\frac{K_{II}^2}{C_L}+2\frac{K_IK_{IV}}{e}-\frac{K_{IV}^2}{k}\right),\qquad G_M=\frac{1}{2}\left(\frac{K_I^2}{C_T}+\frac{K_{II}^2}{C_L}+\frac{K_IK_{IV}}{e}\right),$$

$$G_L=\frac{1}{2}\left(\frac{K_I^2}{C_T}+\frac{K_{II}^2}{C_L}+2\frac{K_IK_{IV}}{e}+\frac{C_T}{e^2}K_{IV}^2\right)$$

where $C_T, C_L, e\ and\ k$ are evaluated from the Irwin matrix Y defined in [4].

3 The Problem

A poled piezoelectric plate of width 2m and height 2h is weakened by a straight crack of length 2a. The crack is inclined at an angle α with x-axis and is free of stresses and electric charge. The plate is poled along y-direction. The boundary of the plate $y=\pm h$ is subjected to a uniform constant stress $\sigma_{yy}=\sigma_\infty$ and electric displacement $D_y = D_\infty$. The boundaries $x = \pm w$ are free of stresses and charges. The entire configuration is schematically depicted in Fig.1.

4 X-FEM Solution

The plain strain case is considered with tensile loading $\sigma_{yy} = \sigma_\infty = 10\,[\mathrm{MPa}]$ and electric-displacement $D_y = D_\infty = 0.001[\mathrm{c/m^2}]$. The material considered for the analysis is PZT-5H, and material constants are taken from Liu [7]. To analyze the problem numerically, a structured mesh with four noded

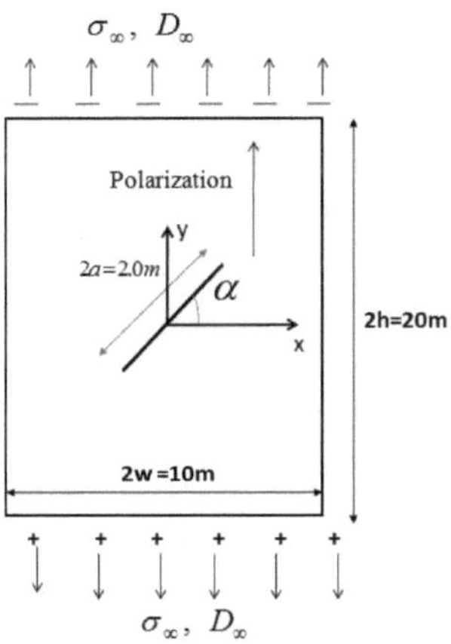

Fig. 1. Model geometry of the problem

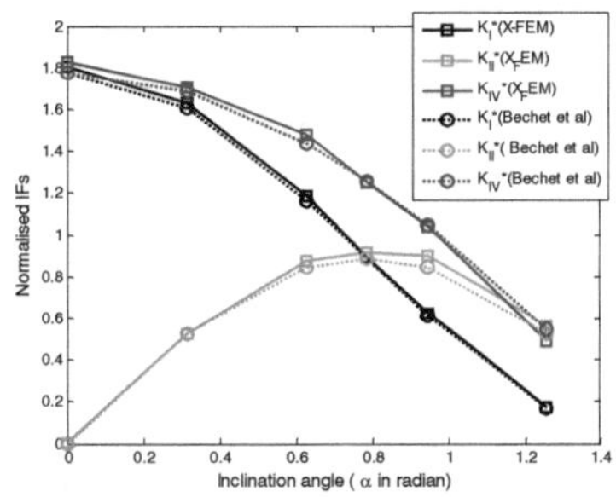

Fig. 2. Variations of normalized IFs versus α

isoparametric elements and a non-alignment case has been considered. The IFs are calculated on a domain radius equal to 0.8a. The convergence study for the case of an inclined crack in the given domain requires us to take a mesh with 59x119 elements.

4.1 Validation with Bechet et. al[6]

In this subsection IFs are calculated for different inclination angles of the crack using X-FEM. The results of normalized IFs are compared with those of Bechet et. al [6]. The normalized results of IFs obtained using X-FEM closely follow the results of [6] as shown in Fig. 2.

4.2 Effect of Electromechanical Loadings

Results for IFs and energy release rates are obtained for an inclined crack using X-FEM having different inclination angles as depicted in Fig.3. The uniform tensile loading σ_{yy} = 10[Mpa] and five cases of electrical displacements

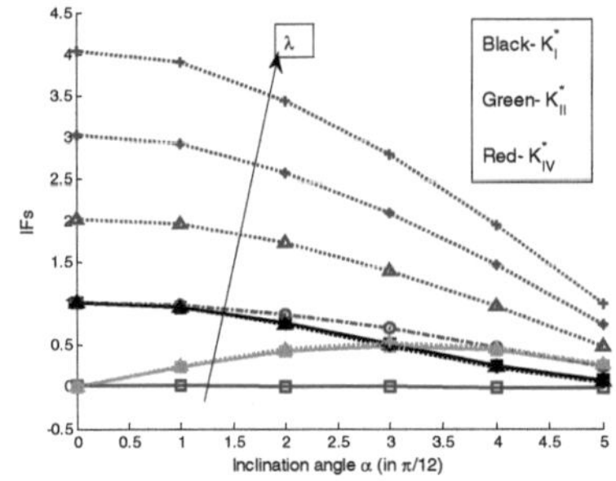

Fig. 3. Variations of normalized

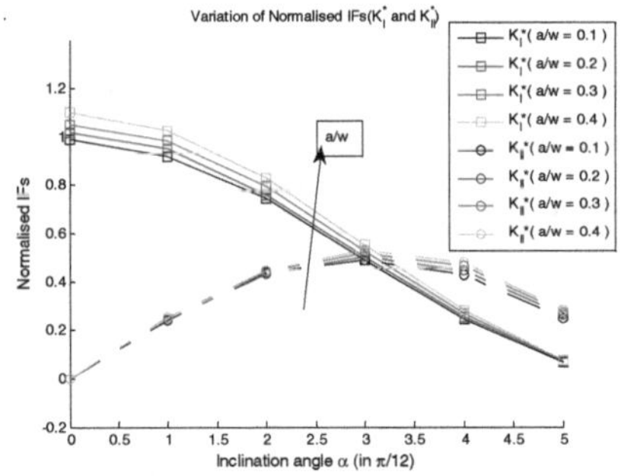

Fig. 4. Normalized IFs (K_I* andK_{II}*) with respect to α respect to α for different crack lengths

$D_y = \lambda((e_{33}/c_{33})\sigma_{yy})$ when (i) $\lambda = 0$ (ii) $\lambda = 1$ (iii) $\lambda = 2$ (iv) $\lambda = 3$ (v) $\lambda = 4$ are investigated. The normalized IFs are calculated using.

$$K_I^* = K_I / (\sigma_{yy}\sqrt{\pi a}), K_{II}^* = K_{II} / (\sigma_{yy}\sqrt{\pi a}), K_{IV}^* = K_{IV} / ((e_{33}/c_{33})\sigma_{yy}\sqrt{\pi a}).$$

Figure 3 depicts the variation of normalized IFs (KI*, KII* and KIV*) with respect to inclination angle α, for the five cases of electromechanical loadings. The results show that normalized IFs KI* and KII* are almost independent of the electrical loadings. Whereas the normalized electric displacement intensity factor KIV* depends upon the electrical loading and it increases with an increase in electric displacement loadings. It may also be noted that IFs converge for larger value of inclination angle.

4.3 Coupled Effect of Inclination Angle and Crack Length

In this subsection results of normalized IFs are presented with respect to inclination angle and for considering different crack lengths. Here, uniform tensile loading σ_{yy} = 10[Mpa] and electrical loading $D_y = ((e_{33}/c_{33})\sigma_{yy})$ is considered. Figures 4 and 5 show that normalized IFs KI*, KII* and KIV* increase with respect to increase in crack length. It is seen that the difference in normalized IFs with respect to crack length also depend upon the inclination angle. In case of normalized KI*, this difference decreases with respect to the increase in inclination angle. But the difference increases with increase in inclination angle in case of normalized KII*. Figure 5 shows that variation of normalized KIV* with respect to crack length is same as in case of normalized KI*.

4.4 Effect of Electrical Loadings on Different Energy Release Rates

Figure 6 shows the effect of electrical loadings on normalized total energy release rate, normalized mechanical energy release rate and normalized local energy release rate. The study is carried out for three different inclined crack orientations $\alpha = 0, \pi/4$ and $\pi/3$. It is shown in Fig. 6 that total energy release rate decreases with respect to increase in electrical displacement which is in contradiction to experimental results.

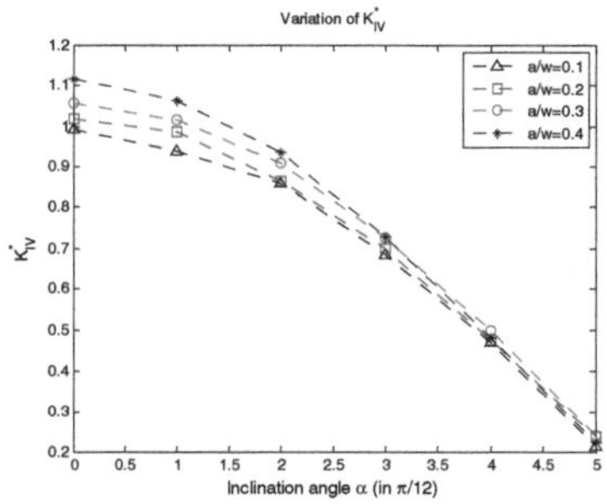

Fig. 5. Variations of K_{IV}* with respect to α

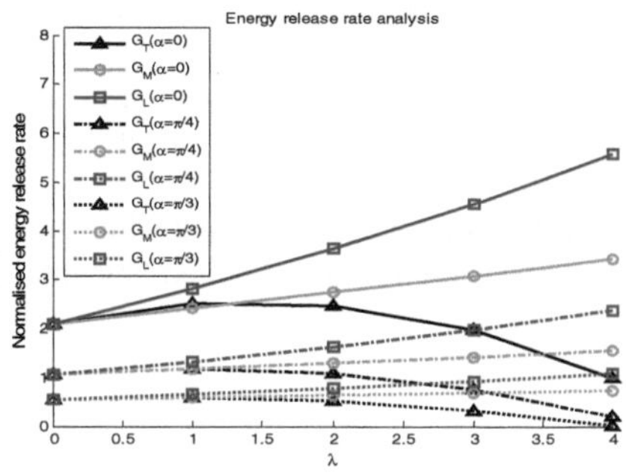

Fig. 6. Energy release rates versus electrical loadings

Both mechanical energy release rate and local energy release rate increases with increased electrical loadings. The results are in agreement with the results of Liu [7].

5 Conclusions

The conclusions are

- It is concluded that as the crack inclination angle is increased the variations in the IFs with respect to the parametric studies decreases.
- The normalized IFs KI* and KII*are independent of the electrical loadings whereas KIV* strongly depends on it and it increases with the increase in electrical loadings. It may be clearly observed from GM, GL expressing that GM is zero when KI is zero while GL has a positive value and increases with increase in electrical loading. Consequently, it could be taken as fracture criteria.

Acknowledgement. The authors are grateful to Prof. R.D. Bhargava (Senior Professor and Head, retd., Indian Institute of Technology Bombay, Mumbai, India) for the encouragement throughout the course of this work.

References

1. Allik, H., Hughes, T.J.R.: Finite element method for piezoelectric vibration. International Journal for Numerical Methods in Engineering 2, 151–157 (1970)
2. Kumar, S., Singh, R.N.: Crack propagation in piezoelectric materials under combined mechanical and electrical load. Acta Materialia 44, 173–200 (1996)

3. Kuna, M.: Fracture mechanics of piezoelectric materials-Where are we right now? Engineering Fracture Mechanics 77, 309–326 (2010)
4. Moës, N., Dolbow, J., Belytschko, T.: A finite element method for crack growth without remeshing. International Journal for Numerical Methods in Engineering 46, 131–150 (1999)
5. Belytschko, T., Gracie, R., Ventura, G.: A review of Extended/Generalized finite element methods for material modeling. Modelling and Simulation in Materials Science and Engineering 17, 043001 (2009)
6. Béchet, E., Scherzer, M., Kuna, M.: Application of the X-FEM to the fracture of piezoelectric materials. International Journal for Numerical Methods in Engineering 77, 1535–1565 (2009)
7. Liu, T.J.C.: Anomalies associated with energy release parameters for cracks in piezoelectric materials. Theoretical and Applied Fracture Mechanics 51, 102–110 (2009)

Development of Flow Stress Model Based on Internal Variables

Jarosław Nowak, Dmytro S. Svyetlichny, and Łukasz Łach

AGH University of Science and Technology, Faculty of Metals Engineering and Industrial Computer Science, al. Mickiewicza 30, 30-059 Kraków, Poland
nowak@metal.agh.edu.pl, svetlich@metal.agh.edu.pl, lukasz-lach@wp.pl

Abstract. In the paper a flow stress model based on the dislocation theory in consideration of the recrystallization is briefly presented. The model consists of two parts: the classic model of the dislocation evolution and recrystallization model. The latter deal with various types of recrystallization as the same process rooted in nucleation and grain growth. The results of the model parameters identification and the simulation are presented in this paper. Then disadvantages of the model are considered and new proposal for improvement the model are presented. The results of preliminary simulation are presented as well.

Keywords: Flow stress, dislocation density, recrystallization.

1 Introduction

Problem of choosing the stress-strain function has been widely discussed in the scientific literature [1]. For analytical description of yield stress, depending on temperature, strain rate and stress is commonly used. There are also adaptive dynamic models and artificial neural networks [2]. Among several models that describe the material behavior under the loading, the flow stress model based on the Taylor dislocation theory has been an important function in the material science. There are known models of dislocation evolution developed by Kocks [3], Roberts [4] and Yoshie [5]. This model has an advantage when a non-stationary process takes place. In the processes when the temperature or strain rate varies in a wide range, the models based on the dislocation theory give the correct results. It can be explained by the fact that these models use internal variables to describe the state of the material. Models based on the internal variables allow one to describe the recovery and the recrystallization not only during the deformation, but also between the consequent deformations. It allows the correctly evaluation the flow stress properly in the initial stage of the following loadings. The models are based not only on one internal parameter, but also there have been proposed models with two [6], three [7] internal variables or the dislocation density distribution function [8]. The introduction of the processes of a static, metadynamic and dynamic recrystallization in the dislocation model as additional term of the main equation

Y. Zhang (Ed.): Future Communication, Computing, Control and Management, LNEE 141, pp. 291–296.
springerlink.com

complicates the model significantly and the problem still remains unresolved. Static models that take into account the dynamic recrystallization, most are based on the concept of Sellars [9]. One of them is a model developed by Davenport [10].

The objectives of the paper are the description of the new model based on internal variables in consideration of the recrystallization and the presentation of its ability to predict the flow stress during deformation.

2 The Flow Stress Model Based on the Theory of Dislocations and Recrystallization

Model [11, 12] was developed on the basis of the Taylor dislocation theory [13] and the KJMA (Kolmogorov-Johnson-Mehl-Avrami) theory of recrystallization [14], which combines a model of yield stress with the microstructure evolution. The paper presents the results of studies on stress only, do not focus on modeling the microstructure evolution.

In the model the following equation is used:

$$\sigma = \sigma_0 + \alpha G b \sqrt{\rho} \tag{1}$$

where σ0 - stress necessary to move dislocation in the absence of other dislocations, α - constant, G - shear modulus and b - Burgers vector, ρ – dislocation density.

It is generally assumed that during the deformation the evolution of the dislocation density is dependent on two components:

$$\frac{d\rho}{d\varepsilon} = U(\varepsilon) - \Omega(\rho) \tag{2}$$

Here, U(ε) is regarded as the generation and the storage of the dislocation (hardening), while the term Ω(ρ) contributes an annihilation of the dislocation (recovery), ε - strain. In [11] it is proposed to take into account the strain rate $\dot{\varepsilon}$, temperature T and dislocation density ρ in the following form:

$$\dot{\rho} = w_1 \dot{\varepsilon}^{w_2} \exp\left(\frac{Q_1}{RT}\right) - w_3 \dot{\varepsilon}^{w_4} \rho^{w_5} \exp\left(\frac{Q_2}{RT}\right), \tag{3}$$

where R - gas constant, Q1, Q2 - activation energy, w1-w5 - coefficients.

It should be noted that (3) contains no term corresponding to the recrystallization. It is assumed [11] that the introduction to the model (2) an additional term corresponding to the recrystallization is inconsistent and creates an internal disagreement within the model. The additional term can include dynamic recrystallization, and it works well during deformation, but it is impossible to use after the deformation. It complicates the description of static and metadynamic recrystallization. Introduction to model different logical conditions also appears to be unfounded. Solving the problem is possible using, for example, cellular automata [15]. The model proposed in [11] considers the dynamic, metadynamic and static recrystallization as the same process that involves the nucleation of new grains growth. In the model average dislocation density ρav is used. For the current value, it is the relation (2) for the dislocation density ρ non-recrystallized grains and the recrystallized fraction χ is used as follows:

$$\rho_{av} = \rho(1-\chi) \tag{4}$$

In the model, this fraction χ is common to all kinds of recrystallization. After a full static or metadynamic recrystallization, it reaches a value of 1, while the full dynamic recrystallization, it remains less than 1. Details about this model can be found in [12].

3 Examples of Simulations

There is an analysis of this model [11, 12] and comparison with the model based on the concepts Sellars [9, 10] presented in the papers [16, 17]. Some results of this comparison are shown in Figs.1-2. Symbols on the graphs represent data from plastometric tests, dashed lines are the results of simulations using the Sellars' model, while the solid lines are the results of simulations using the model described above. More detailed results and conclusions are presented elsewhere [16, 17].

4 Motivation for Development of the Model and Proposals

The comparison of model calculations with the physical modeling in simulator Gleeble revealed two major discrepancies of the model with actual results, when the deformation with a wide range of veried conditions is carried out.

The first difference can sometimes be seen in the case when strain rate decreases. Then, an increase of flow stress (dislocation density) can follow an initial fall. That can occur when decrease of strain rate is not significant and stress is far enough from its peak or saturation. To clarify the situation is helpful graphs presented in Fig. 3. Symbols (squares) represent measurements. This effect is impossible to obtain within the model according to equations (2) and (3) (solid line in Fig.3), because ordinary differential equation of the first order with constant coefficients gives solution when solved variable can decrease or increase only, but not an increase follows a decrease or vice versa.

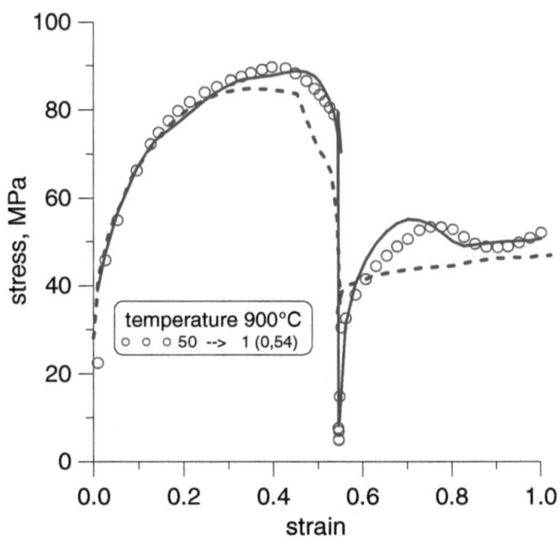

Fig. 1. Results of plastometric tests and models simulation at the temperature of 900°C with step decreasing of strain rate [17]

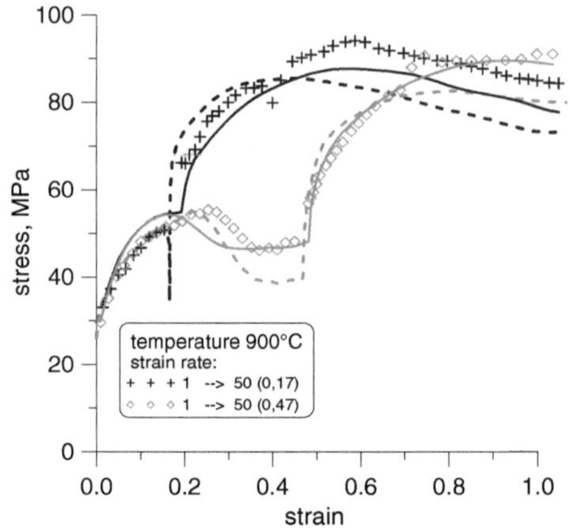

Fig. 2. Results of plastometric tests and and models simulation with step increasing of strain rate [17]

The second difference is associated with the change of dislocation density in the pauses between the deformations, which affects the initial value of stress in the subsequent deformation. It can be explained by the several processes: faster one leading not to full restoration, and slower other ensuring a full recovery.

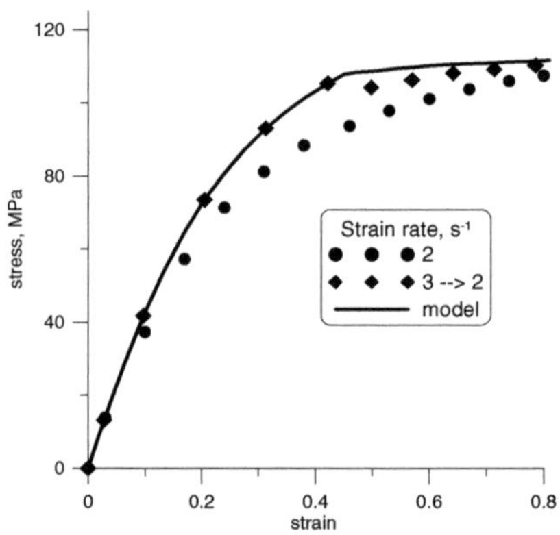

Fig. 3. Results of testing of the model with step decreasing of strain rate

The model (2) - (3) containing an internal variable is proposed to extend by a second internal variable, responsible for strengthening associated with the evolution of the dislocation structures.

In the paper [18] all dislocations were represented by the forest dislocations and mobile ones. The stress depends on dislocation density forming the forest of dislocations [19]. In the paper [6] stress is related to the density of dislocations in the cell walls and in their interior. Besides, the volume fraction of cell walls f is introduced, it depends on strain. Mobile dislocations are additionally taken into account in [7], whereas the mobile dislocations directionally dependent are considered in [20]. A use two kind of dislocation instead one (forest dislocations and mobile dislocations) can extend the capabilities of the model to obtain the different kinetics of evolution of the dislocation density, which has an effect on stress, but this effect is not substantial. Mobile dislocations play a supporting role in the model [7]. The evolution of dislocation runs very similarly in the walls and cells interior (mainly differing in scale), therefore the constant fraction of the wall volume does not affect much on the stress. In the papers [6, 20] the main attention paid on evolution of dislocation structure or texture. In the present study somewhat simplified approach are applied. For evolution of dislocation structure the same equation, as in [6], is used:

$$f = f_\infty + (f_0 - f_\infty)\exp(-\varepsilon / \varepsilon_t). \tag{5}$$

where f0 - the initial value of f, f∞ - saturation value at large strain, εt determines the rate of evolution of cellular structure.

When material deforms, the volume fraction of the walls reduces, but the dislocation density in the wall grows and stress increases, while the dislocation density in cells interior remains almost the same. So, equation (2) is rewritten into the following form:

$$\frac{d\rho}{d\varepsilon} = U(\varepsilon) - \Omega(\rho) + Sf \cdot \tag{6}$$

where S - the excess dislocation density associated with the evolution of dislocation structure, while the meaning of f0 and f∞ are different from those proposed in [6]. For the annealed microstructure f0 = 0, and f∞ should depend on temperature. Then equation (3) is of following form:

$$\dot\rho = w_1 \dot\varepsilon^{w_2} \exp\left(\frac{Q_1}{RT}\right) - w_3 \dot\varepsilon^{w_4} \rho^{w_5} \exp\left(\frac{Q_2}{RT}\right) + w_6 \dot\varepsilon \exp\left(\frac{Q_3}{RT}\right) - w_7 \exp\left(\frac{Q_4}{RT}\right) \tag{7}$$

The last term of the equation (7) introduces a static recovery. If dynamic recovery (the second term) affects during the deformation only, the static recovery acts not only during deformation, but also after that, and together with recrystallization lead to softening of the material in pauses between loadings.

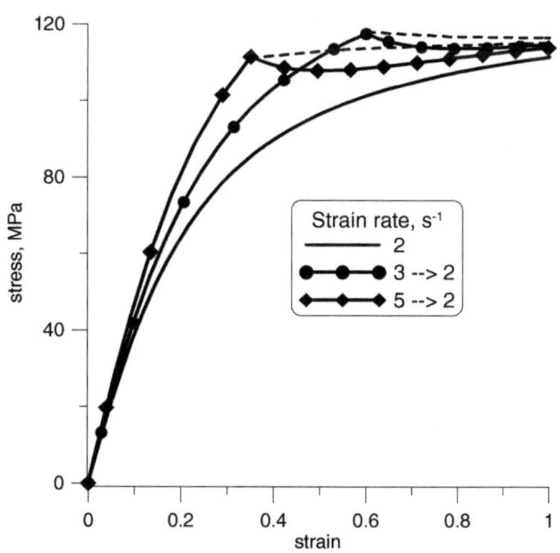

Fig. 4. Results of testing of the model with step decreasing of strain rate. Continues line – new model, dotted line – previous model.

Some examples of simulations of the stress with a use the proposed model are shown in Fig.4. Deformation with the varied strain rate is modeled. There can be seen the stress increase follows a decrease (solid lines with symbols). For comparison the effect of the previous model (dashed line) is also shown.

5 Summary

The paper deals with an analysis of the flow stress model used so far, its advantages and disadvantages. A modification of the model is proposed on the basis of analysis of models with two and three internal variables. Results from the first part of this paper confirm the high efficacy of the model to describe the flow stress in wide range of the deformation conditions. But some discrepancies in particular conditions were the basis to include additional part into the model, which take into account other phenomena in the material. New approach is presented and the preliminary simulations show good capabilities of the new solution. It is planned to carry out further experimental tests and a development of an identification method of the model parameters.

Acknowledgements. Support of the Polish Ministry of Education and Science is greatly appreciated (Grant No. N N508 620140).

References

1. Grosman, F.: J. Mater. Proc. Technol. 64, 169 (1997)
2. Svyetlichnyy, D.: Arch. Metall. 45, 435 (2000)
3. Kocks, U.F.: J. Eng. Mater. Technol. 98, 76 (1976)
4. Krauss, G.: Roberts in Deformation, Processing and Structure. ASM, Metals Park (1984)
5. Yoshie, A., Morikawa, H., Onoe, Y.: Trans. ISIJ 27, 425 (1987)
6. Estrin, Y., Toth, L.S., Molinari, A., Breachet, Y.: Acta Mater. 46, 5509 (1998)
7. Roters, F., Raabe, D., Gottstein, G.: Acta Mater. 48, 4181 (2000)
8. Sandström, R., Langeborg, R.: Acta Metal 23, 387 (1975)
9. Sellars, C.M., Tegart, W.J.M.: Sci. Rev. Met. 63, 731 (1966) (in French)
10. Davenport, S.B., Silk, N.J., Spark, C.N., Sellars, C.M.: Mat. Sci. Techn. 16, 539 (2000)
11. Svyetlichnyy, D.S.: ISIJ Int. 45(8), 1187 (2005)
12. Svyetlichnyy, D.S.: Mater Sci. Techn. 25, 981 (2009)
13. Taylor, G.I.: Proc. Roy. Soc. A145, 362 (1934)
14. Kolmogorov, A.N.: Izv. Izv. AN SSSR 3, 355 (1937) (in Russian)
15. Svyetlichnyy, D.S.: Comp. Mater. Sci. 50, 92 (2010)
16. Svyetlichnyy, D., Pidvysots'kyy, V.: Rudy Metale R 50, 560 (2005) (in Polish)
17. Svyetlichnyy, D.S., Pidvysotskyy, V., Kuziak, R.: In: Szeliga., D., Pietrzyk, M., Kusiak, J., Szczawnica, Akapit (eds.) Informatyka w Technologii Metali KomPlasTech (2006) (in Polish)
18. Kubin, L., Estrin, Y.: Acta Metallur. 38, 79 (1990)
19. Majta, J.: Deformation and properties. Micro-alloyed steels: Selected problems. In: AGH Uczelniane Wydawnictwa Naukowo-Dydaktyczne, Poland, Kraków (2008) (in Polish)
20. Peeters, B., Van Houtte, P., Hoferlin, E., Aernoudt, E.: Int. J. of Plasticity 17, 819 (2001)

Invasion Detection Algorithm Research and Application

Zhao Yanwei

Department of Information Engineering Jilin Business and Technology College, Changchun 130062, China
zhyw_68@163.com

Abstract. This paper adopts multi-agent technology and designs an adaptive distributed IDS model. It has good adaptive ability and better response and treatment on the network environment and changeful attack methods, also adopt various mining algorithm combining detection mechanism to void the new invasion behavior. The updating mechanism of rule library adopts LFU to eliminate the old rules set, effectively improve the misuse detection of matching speed. Meanwhile, the source of alarm, secondary detection can effectively avoid the several times alarm phenomenon caused by the same attack, greatly reduce the rate of false alarms.

Keywords: Invasion Detection, Distributed, Adaptive.

1 Introduction

As is known to all, with the development of information technology and network technology, while people enjoy the various conveniences, more and more problems of network security have become ubiquitous. Currently, there are many IDS (invasion detection system), but these systems do not have adaptability. When the environment of network have changed, the system is difficult to change according to it. So the result is large omission and material misstatement of the invasions [1]. To improve the security of the network, recently the adaptive distributed IDS have become a hot issue.

Traditional distributed IDS using distributed acquisition, centralized processing or hierarchical processing send processing results from low-level node to top-level node. This kind of design exists obvious deficiency, mainly appears in the following aspects: concentrated analysis of components with higher loading, may become the chock point of IDS and lead to the failure of the single point. Hierarchical analysis decreases the real-time of IDS. Traditional system rule library adopts simple database stores, each match starts from scratch scanning databases, for the update of rule library do not have new methods, the efficiency of the matching is low. The problem with the same attack leading high false rate of alarming is not solved efficiently. In the aspect of mining algorithm, traditional system [2] mostly used single algorithm, which will not adapt to different invasion environments and lead to low detection rate, bad adaptability etc. While the system under attack collapse, no good dynamic election algorithm to make it run normally in time.

Y. Zhang (Ed.): Future Communication, Computing, Control and Management, LNEE 141, pp. 297–305.
springerlink.com

According to the above analysis, in order to solve the existing problems and improve the performance and the security of IDS, this paper presents an adaptive distributed IDS model based on multi-agent.

2 Overall Structure of Adids

ADIDS (Adaptive Distributed Invasion Detection System Based on Multi-agent) is a multi-agent model under the distributed environment. Each agent is independent, divisional and mutual cooperation. Due to the multi-agent collaboration structure without controlled center, avoid the problem of single failure and bring about the operation of the data source distribution and data analysis [3]. The whole platform structure (see Fig.1 platform structure) including seven components: central console (CC), management agent (MA),analysis agent (AA), communication agent (CA), reaction agent (RA), rule library (RL) and data warehouse (DW).

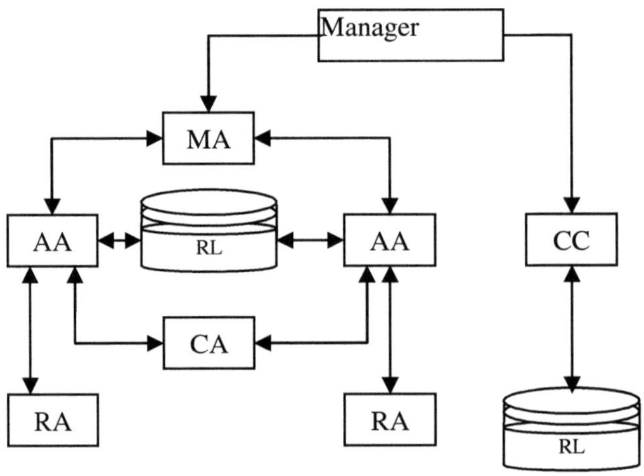

Fig. 1. Platform structure

ADIDS is different from the traditional DIDS (distributed invasion detection system) which is an immutable structure of integration. It is a flexible DIDS and adaptively changes internal state of components of IDS. Analysis agent and reaction agent can dynamically adjust their states according to the need of the detection for the purpose to achieve real-time online adaptive detection task [4]. At the same time, analysis agent can intelligently invoke the data mining algorithm according to the specific circumstances of invasion detection to improve the adaptability of the whole system.

Summing up the above, ADIDS can be distributed in any type of network hosts, reaction agent receive data packets flowing through the host, after that transmit the packets to analysis agent or dock received preliminary examination and identify obvious malicious packets [5]. Finally, transmit the detection results and unanalyzed packets to analysis agent, and then analysis agent will analyze and mine the suspected data. The judging result attackable or not is sent to management agent and submit to the manager through the console for the last decision.

3 Characteristics of Agents

3.1 Central Console

The central console is responsible for providing users with a monitoring invasion interface [6]. Users can understand the control of all parts of operation and the security situation of the whole network, and then to supply management agent with relevant control instructions. Receive alarm transmitted over from management agent and show them to the user. The central console is the top control unit of system structure shown in Fig.2.

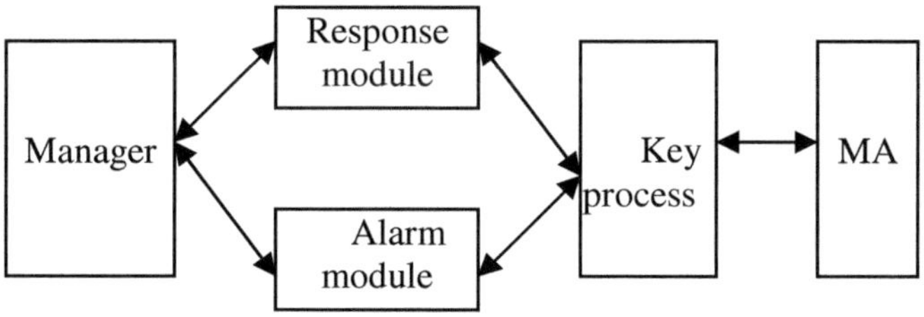

Fig. 2. Internal structure of CC

3.2 Reaction Agent

In principle, reaction agent widely collects all kinds of data from the monitoring environment [7], it could be from the packets of the network or system audit trail of the host. Reaction agent has two working state: one is a senior state receiving packets and preliminary inspection packets, in this state, reaction agent response to keep flowing through packet preprocessing, then through simple detection and filtration, identify the obvious malicious packets and all detected reports and undetected packets will be submitted to the analysis agent. The other state is responsible for only receiving packets and its transmission to analyze the agent of low-level state [8]. Before the start of reaction agent, its appearance at the beginning of the preset for senior state, executive receiving data and the primary testing are two functions. When reaction agent itself load heavy reach the setting value, it automatically reduced to only be responsible for receiving data of low-level state. When low-state data load is higher than the setting value, reaction agent through communication interface [9], transmit the receiving part of packets to adjoining reaction agent to lighten the burden of this analysis agent. Fig.3 shows that the internal structure of reaction agent.

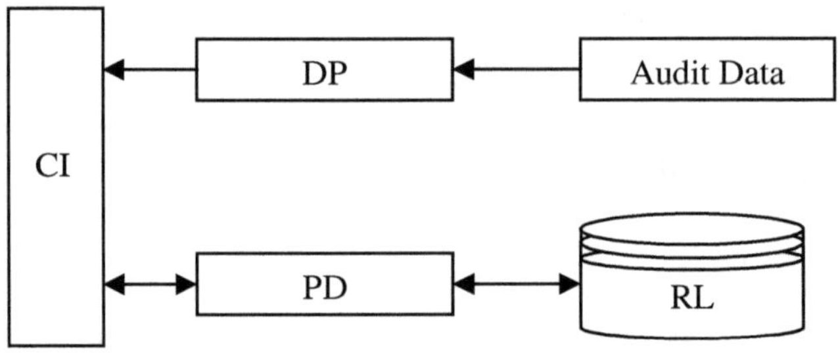

Fig. 3. Internal structure of RA

Communication interface (CI): it is the interface of internal mechanism and outside information communication, its main function is to send the processing results of audit data to analysis agent and update itself rule library [10] through receiving analysis agent. Data preprocessing: mainly to mark the audit data standardization. Preliminary detection: preliminarily analyze the standardized data and detect obvious aggression or abnormal behavior, and then will transmit test results through the communication interface to analysis agent. Rule library: mainly store the necessary knowledge of preliminary detection and analyzing the feedback of the analysis agent with the central console initialization.

3.3 Analysis Agent

Analysis agent receives network packets and the detection report from the reaction agent, through the data misuse detection and abnormal detection, judge the security state of monitoring network [11] and alarm users. Fig.4 shows that analysis agent can cooperate mutually in the process of detection.

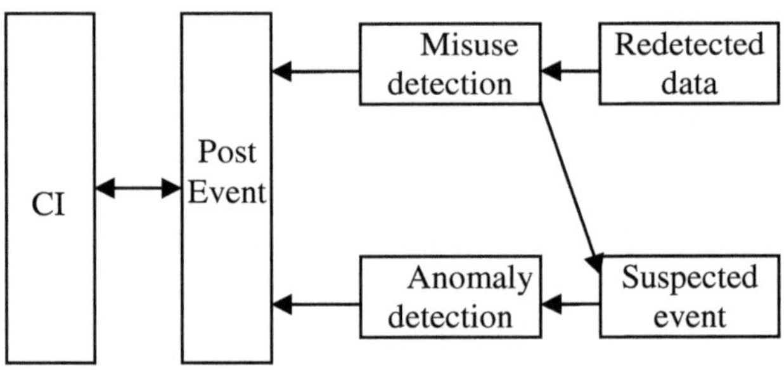

Fig. 4. Process of detection

Each internal analysis agent is divided into misuse detection and anomaly detection according to the detection technology. Misuse detection [12] is mainly used to accurately measure the known invasion active, emphasis its detection speed. If misuse detection can be judged as aggression, exception detection does not have to be invoked. Anomaly detection mainly used to forecast unknown invasion activities, it has stronger and more comprehensive inspection and analysis ability, emphasis is on the detection accuracy. When misuse detection fails to detect whether a packet is malicious or not, list it as suspicious anomaly detection [13].

The updating mechanism of rule library of misuse detection adopts LFU (least frequently used algorithm). It pursues detection speed, so this set of module rules adopts special storage ways and each rule library set the counter [14]. The most times rule sets of recent match succeeds store in the top of rule library when it has no storage space, eliminate the old rule sets with LFU.

Anomaly detection module [15] carried out detection using data mining algorithm. The main data mining algorithms applied to anomaly detection are correlation analysis, sequence analysis, cluster analysis, classification analysis, rough set analysis and outlier analysis algorithm etc. Firstly, obtain the connection pattern using association rules algorithm [16] recorded feature attribute, and then record correlation time through further analyzed connected sequence analysis, thus obtains the sequential patterns and digs out the normal behavior patterns. To filter packet with normal behavior patterns, thus obtains higher purity invasion data. Combine test with the new invasion data selecting one or more algorithm of clustering analysis, classification analysis, rough set analysis and outlier analysis, exhume abnormal results to the user and deposit exhumed new rule collection to rule library. Fig.5 shows that mining process of anomaly detection.

The choice mechanism of mining algorithm [17] is as follows, when anomaly detection module mining algorithm reaches a heavy load, adjust the algorithm first (load directly influences the performance and speed detection of algorithm).If the adjusted algorithm can reduce the load, you do not need to shunt data packets to other mining algorithm, otherwise proceed to shunt. When the analysis accuracy of mining algorithm is insufficient, the converted algorithm is higher than the previous algorithm. The selecting evaluation formula is defined as

$$f(x) = \sum_{n=1}^{4}\left(\frac{LFn}{LBn} + \frac{ABn}{AFn}\right) / n$$

LFn is the host load of mining algorithm at first; LBn is the host load for x time after changed to some analysis algorithm;ABn is the analysis accuracy for x time changed to some analysis algorithm; AFn is the default analysis accuracy. If $f(x) > 1$, mining algorithm can choose to convert the algorithm; If $f(x) <1$, mining algorithm does not undertake the analysis algorithm conversion; When AFn of a mining algorithm reach the setting value P, it does not joint various mining algorithm, only when $AFn < P$, to make the two or more algorithms combination.

Many alarms for the same invasion cause misstatement of high rate of false-alarm phenomenon, analysis Agent [18] will change mining object, mine second alarm information source, so as to filter the false alarm, thereby eliminating mistakenly police, and improve the system detection accuracy [19].

3.4 Communication Agent

Communication Agent is specially used for analysis of telecommunication of analysis agent, its function basically has the following aspects: data transfer of analysis agent, suspicious events, collaborative communication, rules and sharing etc. Each analysis agent [20] is independent testing analysis for the unit, when collected information cannot be judged as abnormal behavior, communication agent and other analysis agent is necessary in order to improve the analysis speed and accuracy.

In the initialization, management agent allocates a mailing address and communications key (public-key encryption algorithm) for each analysis agent, and stored in address episode of communication agent, as communication identification of analysis agent. Fig.5 shows the flow chart of communication agent.

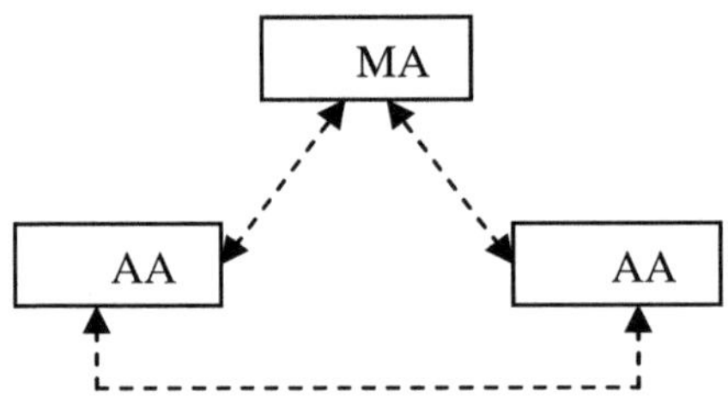

Fig. 5. Flow chart of CA

Communicate between agent A and agent B

Application between agent A and communication agent and communication between agent A and analysis agent B.

Communication agent first check key submitted by analysis agent A to identify its identity, then check the identity of analysis agent B.

To give the addresses of analysis agent B to analysis agent A and inform the analysis agent A and B communications.

Start communication between Agent A and Agent B.

3.5 Management Agent

Management agent is responsible for coordinate management of all behavior of analysis agent and comprehensive analysis to the submitted detection report [21],and the results submit to the central console to complete security tasks of security system.

When network segment within the management agent is attacked, to ensure the whole system back to normal operation, other management agents will dynamically elect a lightest load management agent to replace the original management agent. Will classic Bully algorithm was improved, regard this system dynamic election algorithm, the algorithm flow chart is shown in Fig. 6 and the specific steps as follows

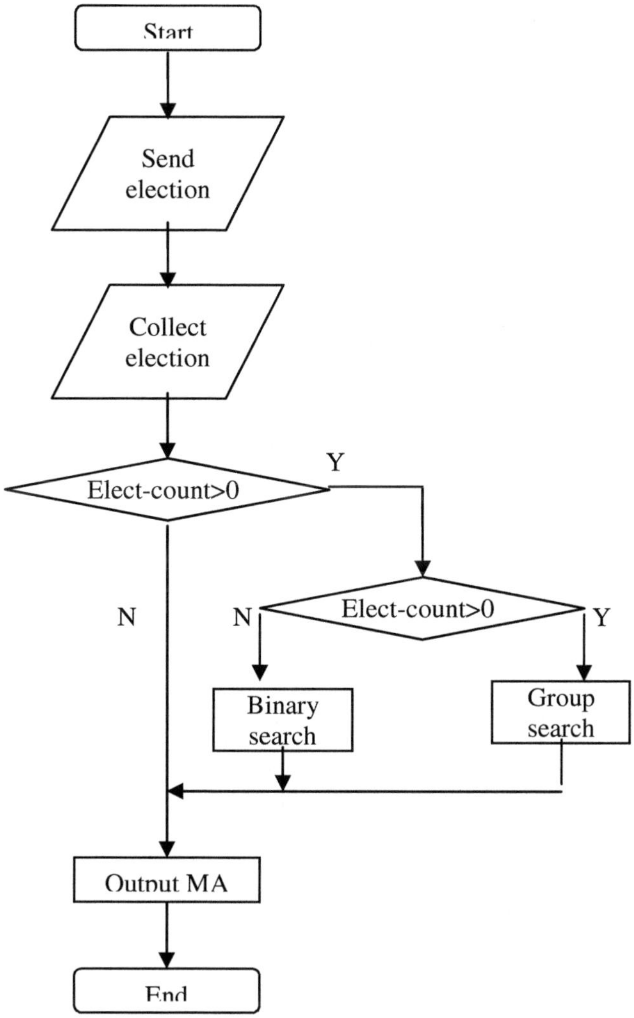

Fig. 6. Improved flow chart of Bully

Step 1: Sponsors send an election (Election (X, Y)) to all other management agent, X is load, Y is address.

Step 2: If all other management agent within network segment discover their own load less than X, launch the election and sends his election to management agent, otherwise exit elections.

Step 3: If the initiated management election received no election, then it will reproduce itself to replace original management agent, otherwise exit and turn to Step 4.

Step 4: Initially management agent will receive the election using binary search method or grouping search method lookup load the smallest management agent, when the election number < P then to use binary search method, or use several grouping search method.

4 Conclusion

When the system is attacked or collapsed, dynamic election algorithm of management agent with the improved algorithm Bully timely recovery system, make the normal operation of the system has good anti-destroy ability and self-restore ability.

Acknowledgement. The authors gratefully acknowledge the funding Jilin province Education Science "Eleventh Five Years" Scheme Item(2010222,2010224).

References

1. Pinzón, C., Navarro, M., Bajo, J.: AIDeM: Agent-Based Intrusion Detection Mechanism. In: Demazeau, Y., Dignum, F., Corchado, J.M., Bajo, J., Corchuelo, R., Corchado, E., Fernández-Riverola, F., Julián, V.J., Pawlewski, P., Campbell, A. (eds.) Trends in PAAMS. Advances in Intelligent and Soft Computing, vol. 71, pp. 347–354. Springer, Heidelberg (2010)
2. Miguel, A., Sanz-Bobi, Castro, M., Santos, J.: IDSAI: A Distributed System for Invasion Detection Based on Intelligent Agents. IEEE, doi:10.1109/ICIMP.2010.8
3. Ghit, B., Voicu, O., Pop, F., Cristea, V.: Distributed Agent Platform with Invasion Detection Capabilities. IEEE, doi:10.1109/INCOS.2009.72
4. Mo, Y., Ma, Y., Xu, L.: Design and Implementation of Invasion Detection Based on Mobile Agents, vol. 2. IEEE (August 2008)
5. Vokorokos, L., Baláž, A., Chovanec, M.: Distributed Detection System of Security Intrusions Based on Partially Ordered Events and Patterns. In: Rudas, I.J., Fodor, J., Kacprzyk, J. (eds.) Towards Intelligent Engineering and Information Technology. SCI, vol. 243, pp. 389–403. Springer, Heidelberg (2009)
6. Shyu, M.-L., Sainani, V.: A Multiagent-based Invasion Detection System with the Support of Multi-Class Supervised Classification. In: Data Mining and Multi-agent Integration. Springer (2009)
7. Gu, G., Fogla, P., Dagon, D., Lee, W., Skoric, B.: Towards an Information-Theoretic Framework for Analyzing Intrusion Detection Systems. In: Gollmann, D., Meier, J., Sabelfeld, A. (eds.) ESORICS 2006. LNCS, vol. 4189, pp. 527–546. Springer, Heidelberg (2006)
8. Lin, H.-S., Pao, H.-K., Mao, C.-H., Lee, H.-M., Chen, T., Lee, Y.-J.: Adaptive Alarm Filtering by Causal Correlation Consideration in Intrusion Detection. In: Nakamatsu, K., Phillips-Wren, G., Jain, L.C., Howlett, R.J. (eds.) New Advances in Intelligent Decision Technologies. SCI, vol. 199, pp. 437–447. Springer, Heidelberg (2009)
9. Patil, N., Das, C., Patankar, S., Pol, K.: Analysis of Distributed Invasion Detection Systems using Mobile Agents, vol. 254, pp. 1255–1260. IEEE (2008), doi:10.1109/ICETET
10. Wang, Y.: A sort of multi-agent Cooperation Distributed Based Invasion Detection System. Modern Computer (2008)
11. Martin, J., Ragsdale, D., Surdu, J.: A hydrid approach to the profile creation and invasion detection. In: Proc.of DARPA Information Survivability Conference & Exposition (2001)
12. Chen, A.A.: Common Invasion Detection Framework, January 17 (2002), `http://seclabs.Cs.ucdavis.edu/cidf`
13. Du, X., Smolka, S.A., Cleaveland, R.: Local Model Cheeking and Protocol Analysis. Software Tools for Technology Transfer
14. Teng, H.S., Chen, K., Lu, S.C.-Y.: Adaptive real-time anomaly detection using inductively generated sequential paterns. In: Proceedings of 1990 IEEE Symposium on Security and Privacy, Oakland, pp. 278–284 (May 1990)

15. Zheng, C., Jiang, T.: Research on Inetanet network security technology based on intelligent firewall. Computer Engineering and Applications, 156–158 (2005)
16. Yang, X., Su, J.: Research on Invasion Detection technology based on Protocol Analysis. Computer Application Research, 108–110 (2004)
17. Wang, C.: Research and desing on invasion detection system in the computer network security. Engineering Master Dissertation (2007)
18. Snort, J.K.: Solution of invasion detector, pp. 35–53. Publishing House of Machinery Industry, Beijing (2005)
19. Xie, X.: Computer Network, pp. 110–111. Publishing House of Electronics Industry (2005)
20. U.S. National Security Agency Releases. In: The Technical Framework of Information Assurance, pp. 46–57. Beijing China Electronical Software publishing house (2004)
21. Allen, J., Christie, A., et al.: State of the Practice of Invasion Detection Technologies. Technical Report, Networked Systems Survivability Program, pp. 47–85 (2000)

Application Layer Protocol Identification Algorithm Design

Liao Xiaoju, Yin Mingyong, and Lu Hai

64 MianShan Road, MianYang, SiChuan Province, China 621900
iaoxj@caep.ac.cn, yinmy@caep.ac.cn, luhai@caep.ac.cn

Abstract. It is great significance for network administrators, researchers, service providers and users to accurately identify network bandwidth occupied by what application and who. It is the foundation for network planning, troubleshooting, intrusion detection, traffic monitoring, accounting management, and user behavior analysis. Traffic analysis based on full data has high requirements of storage space and CPU processing power. Traffic analysis based on NetFlow requires network equipment to support NetFlow. The availability of former and the accuracy of the latter are challenged when data traffic up to 1Gbp or higher.In this paper, we consider user preferences, server address database, application layer protocol features and NetFlow technology when designing the application layer protocol identification algorithms, which can improve the availability and accuracy of identification, experimental results it works well..

Keywords: NetFlow, user preferences, traffic identification, server address database.

1 Introduction

It is significant for network administrators, researchers , service providers and users to accurately identify network bandwidth is occupied by what application and who. It is the foundation for network planning, network troubleshooting, intrusion detection, traffic monitoring, accounting management, and user behavior analysis. However, with the continuous improvement of network bandwidth and new applications are emerging, the following two issues become more prominent: First, The feasibility of identification: current corporate network backbone bandwidth has been increased to 1Gb or 10Gb, this high-speed high-bandwidth network produces millions of traffic flows in one minute, how to design algorithms to handle large and growing number of network traffic is the most important issue; Second, the recognition validity. In the past network environment, the vast majority of network traffic is Web, FTP, SMTP, Telnet and other protocols, but in recent years, the application layer protocol are more complex than in the past, the traditional protocol of the traffic in total traffic proportion is less and less. On the contrary, P2P, streaming media, online games and other new application protocol, have occupied more than 60% of network traffic. More importantly, these new protocols and norms are often not open. Therefore, protocol recognition algorithm must address how to correctly identify these complex protocols.

Y. Zhang (Ed.): Future Communication, Computing, Control and Management, LNEE 141, pp. 307–311.
springerlink.com

Traffic analysis based on full data has high requirements of storage space and CPU processing power. Traffic analysis based on NetFlow requires network equipment to support NetFlow. The availability of former and the accuracy of the latter are challenged when data traffic to 1Gbp or higher.

2 Application Layer Protocol Identification

In this paper, the application layer protocol identification algorithms, considering user preferences, the server address database, application layer protocol features and NetFlow technology, can improve the availability and accuracy of identification, experimental results also prove this point.

2.1 Analysis of User Preferences

When users use the network will show a certain regularity, their behavior are fixed on a certain period, such as the use Thunder or BT download, play games, watch movies and so on. The recognition accuracy based on preference can increase more than 5 percentage points,We need grab an early time period (usually 24 hours, best if stored to allow one week) as a training set of preference data, and extract each object (corresponding to each IP address) of the preferred characteristics for the convenience of protocol identification in implementation.

2.2 Server Address Database

Common applications are C / S, B / S, P2P model, which P2P applications will query the server about the news of the seed address in the initial application stage, so there will be C / S connectivity in the beginning. We can increase the recognition accuracy by storing game servers, Web servers, streaming media server's IP addres, port information into a dictionar

2.3 Analysis of Application Layer Protocol Features

There are many private protocols in addition to common Web, Ftp, Pop3, SMTP, P2P and other protocols,, application layer protocol analysis is mainly based on the characteristics of the source and destination IP address, port, transport protocol type. It is a typical scanning attack such as opening multiple ports to a number of different destinations IP on port 135.Games application show more complex features , some show similar characteristics with the Web, in which case you need to use other means for further analysis, such as packet size based and arrival time of the traffic.

3 Application Layer Protocol Identification Algorithm Design

This design algorithm is based on user preferences, server address database, application layer protocol feature and NetFlow technology. The deployment environment is as follows:

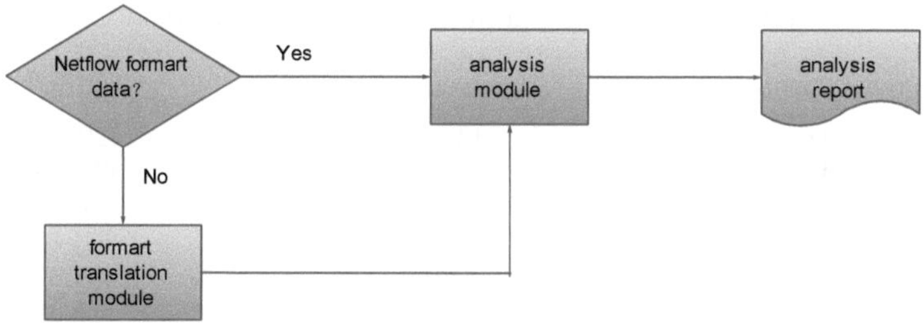

Fig. 1. The deployment environment

According to the difficult of identify, Application layer protocol can be divided into Obvious characteristics, easily identifiable protocols, such as ports less than 1024 common application: DNS, SMTP, POP3, Active FTP, SSH, HTTPS, etc ; Features visible, but not high degree of certainty, such as QQ, MSN, SQL, etc., the port showed some regularity; Features not obvious, but high degree of certainty, such as P2P, gaming and streaming media. Needs special attention to HTTP applications, some of the P2P software will uses HTTP to bypass the firewall. Therefore, we need combination with application preferences, server dictionary and packet size to identify the application protocol. Algorithm flow is as follows:

1) Is it an new data stream ? If yes then the next step, if no update traffic statistics counters and time stamps.
2) Is it a well-known traffic such as DNS, Telnet, SMTP, POP, ICMP? if yes, update traffic statistics counters and time stamps; otherwise the next step.
3) Determine whether the port is greater or less than 1024, further analysis is needed for port 80 traffic, because many P2P applications use port 80 as the tunnel. For other ports below 1024 can be directly determined if it is already in the IANA assigned port list, otherwise go to the next step.
4) For port 80 traffics, query the server dictionary, it can be determined if it is associated with the known web site traffic, or need to go to the next step.
5) Ports greater than 1024, if it is the common applications such as QQ, MSN, analyze the average packet size and packet interval, combined with the server dictionary, user preferences to determine, or need to go to the next step.
6) If it still is not sure, you need heuristic analysis through combining user preferences, the server dictionary, packets size and arrival time.

4 Experimental Results

In the experimental environment, in order to comparative analysis, we used Snort signature with full-load method and our algorithm to identify the same packet, the experimental data captured from the really network, the size is 10Gb in 10 minutes. Table 1 and Table 2 are the results of two methods of identification:

Table 1. The results signature and load based algorithm

Protocol name	Percentage	Traffic(byte)
HTTP	42.1%	4208.012M
P2P	30.1%	3013.303M
UnKnown	13.1%	1212.454M
RTSP	3.5%	351.121M
HTTP Proxy	3.4%	341.301M
HTTPS	3.3%	328.353M
RTP	0.7%	71.223M
QQ	0.7%	70.237M
Game	0.4%	41.683M
MSN	0.3%	23.894M
Other	2.4%	24.124M

Table 2. The results through our recognize algorithm

Protocol name	Percentage	Traffic(byte)
HTTP	39.1%	3905.012M
P2P	33.6%	3362.303M
UnKnown	10.1%	1089.454M
RTSP	4.8%	483.213M
HTTP Proxy	3.5%	352.121M
HTTPS	3.4%	341.301M
Straming Media	1.3%	134.353M
QQ	0.7%	71.223M
MSN	0.5%	52.232M
FTP	0.4%	38.683M
Other	2.6%	246.894M

Data can be seen from the table, the paper design of the application layer protocol recognition algorithm compared to the payload-based recognition algorithm can identify more P2P applications, only 10% of unknown traffic, recognition rate of 90% complete; recognition accuracy can reach 80%.

5 Conclusion

Our algorithm can reach more than 80% recognition rate, especially the identification of P2P from the experimental results. But there are still some flow cannot be identified, this is because of too little data captured and a lot of useful information lost.

References

1. Jiang, J., Papavassiliou, S.: A Network Fault Diagnostic Approach Based on a Statistical Traffic Normality Prediction Algorithm. In: Proceedings of IEEE GLOBECOM, 20031291822922
2. Marina, T., Ji, C.: Adaptive Thresholding for Proactive Network Problem Detection. In: IEEE International Workshop on Systems Management, Rhode Island, Newport, 10821161 (1998)

3. Goldman, A.: Anomaly Detection Based on an Iterative Local Statisticals Approach. IEEE Signal Processing 84(7), 44024431 (2004)
4. Mahoney, M.V.: Network Traffic Anomaly Detection Based on Packet Bytes. In: SAC 2003. ACM, Melbourne (2003)
5. Hellerstein, J.L., Zhang, F., Shahabuddin, P.: Characterizing Normal Operation of a Web Server: Application to Workload Forecasting and Problem Detection
6. Yang, W., Gu, L.: Time Series Analysis and Dynamic Data Modeling. Press of Beijing University of Science and Engineering, Beijing (1981)

A Novel Intrusion Detection Method Base on Attribute Reduction Algorithm

Zhao Yanwei

Department of Information Engineering Jilin Business and Technology College. Changchun 130062, China
zhyw_68@163.com

Abstract. Intrusion detection is for the detection of intrusions and attacks by collecting and analyzing network behaviors, security logs, audit dataset and other Internet information and some key points in a computer information system, to inspect the network or system for the presence of the security violating strategy behavior and the signs of an attack. This article through the use of data sets KDDCup99 data set on the intrusion detection, and simulation of its data pretreatment, attribute reduction, and intrusion rules to generate, lastly, test data sets testing experiment, through test of rate and by mistake examining rate of the experimental results were analyzed and compared, in the whole intrusion detection process, attribute reduction algorithm for intrusion rules generation has played a key role, which also determines the intrusion detection performance, therefore, choose a good effect of attribute reduction algorithm is particularly important. The paper analyzes current attribute reduction algorithm, and finally puts forward the improved one kind of attribute reduction algorithm for intrusion detection.

Keywords: Rough sets, Intrusion detection, Attribute reduction, Attribute importance rate.

1 Introduction

Rough sets theory is a new mathematical tool for handling the vague and uncertainty proposed by the Poland scholar Z.Pawlak in 1982[10][11]. It is regarded as extremely vital significance for the artificial intelligence and the cognitive science, which provides a theoretical framework for the machine learning, data mining, knowledge acquisition, pattern recognition and approximate reasoning and other areas of information processing. Based on this, it makes a certain theoretical feasibility for the rough set applied in intrusion detection field..

2 The Overview of Intrusion Detection

Intrusion detection is for the detection of intrusions and attacks by collecting and analyzing network behaviors, security logs, audit dataset and other Internet information and some key points in a computer information system, to inspect the

Y. Zhang (Ed.): Future Communication, Computing, Control and Management, LNEE 141, pp. 313–320.
springerlink.com

network or system for the presence of the security violating strategy behavior and the signs of an attack [1]. Intrusion detection system as a kind of positive safe protection technology which provides real-time protection for internal and external attack and attack of the incorrect operation of, response before the invasion in the network system harm intercept [6]. Therefore it is considered for the second security gate following with the firewall. It greatly improves the network system security without affecting the performance of network situations by monitoring network [3][4].

3 The Summary of Rough Sets

In order that Rough sets can be applied to intrusion detection, we first give some related definitions of Rough set by introducing reference [2] as follows:

Definition 1. Given set A, A subset of all of the power set is called A collection, recorded as P (A) or 2A , which is $P(A) = 2^A = \{X \mid \forall X \subseteq A\}$.

Definition 2. Given set X, Its power sets is 2X, and is called collection of random element.

Definition 3. Suppose U is objects composed of non-empty finite sets, called a domain. Domain and a subset of any U, called the domain and a concept or U category.

For formulation sake, $\varnothing$ is considered as void conception. In the domain of any U are called clusters about U subset of abstract knowledge, abbreviation knowledge, actually, it expressed in individual classification. One division of par(U) is A domain and the U and U the loophole set cluster, if meet the requirement :

(1) $A_i \neq \varnothing$ (2) $A = \bigcup_{i=1}^{m} A_i$ (3) $A_i \cap A_j = \varnothing$, $i \neq j, 1 \leq i, j \leq m$, then says subset clusters is a collection of a kind of division.

Definition 4. If the relationship R based on set A meets these conditions as follows:(1) Reflexive property, namely $\forall x \in A, xRx$ (2) Symmetrical property, namely $\forall x, y \in A,$ if xRy, there must exist yRx; (3) transitive property, namely $\forall x, y, z \in A,$ if $xRy \wedge yRz$, there must exist xRz .Then we call that relationship R is a Equivalence relation based on set A.

Definition 5. Given a domain and the U and U on cluster of equivalence relation S, says binary group, SK = (U) was about domain U a knowledge base.

Based on the above relevant definition, domain and the equivalence relation is actually represents division and knowledge, and knowledge base said the domain by equivalence relation is derived on the various knowledge, i.e. division or classification model, also represents the classification of domain and ability.

Definition 6. Given knowledge base, SK = (U), including U for domain, S for domain U on equivalence relation is cluster, $\forall X \subseteq U$ and domain U a equivalence relation about knowledge R X definition subset of lower approximation $\underline{R}(X)$ and the upper approximation $\overline{R}(X)$ respectively is:

$$\underline{R}(X)=\{x|(\forall x\in U)\wedge([x]_R\subseteq X)\}=\cup\{Y|(\forall Y\in U/R)\wedge(Y\subseteq X)\}$$

$$\overline{R}(X)=\{x|(\forall x\in U)\wedge([x]_R\cap X\neq\varnothing)\}=\cup\{Y|(\forall Y\in U/R)\wedge(Y\cap X\neq\varnothing)\}$$

According to the definition 6, we come to following conclusions:

Set $bn(X)=\overline{R}(X)-\underline{R}(X)$ is called X r. edge boundaries $pos_R(X)=R(X)$ called X R. positive field $neg_R(X)=U-\overline{R}(X)$ called X R. negative domain.

4 Current Attribute Reduction Algorithm and the Improved One

4.1 The Current Attribute Reduction Algorithm

Current attribute reduction algorithm from the surface of the reduction of definition, from the attribute sets starting point, to random conditional attribute decision table $DT=(U,C\cup D,V,f)$ for deleting, making $pos_{IND(C-\{a\})}(D)=pos_{IND(C)}(D)$.

It founded which justify attribute a can be deleted, or it can't be . So repeatedly to execute this process, until can't delete the current all condition attribute concentration of any attribute.

Common attribute reduction algorithm is described below:

input: decision table $DT=(U,C\cup D,V,f)$

output: the red, a relative reduction of DT

step1:make red=C;

step2: $\forall a\in C$, if $\forall a\in red$ then $mark(a)=0$, or ($mark(a)=1$;

step3: choose a, which meets the condition $\forall a\in red$ and $mark(a)=0$, random, then make $mark(a)=1$, if $pos_{IND(red-\{a\})}(D)=pos_{IND(red)}(D)$ then delete attribute a from red, $red\Leftarrow red-\{a\}$, jump to step2,or jump to step4;

step4: if $\exists a\in red$ and $mark(a)=0$, then jump to step3,otherwise red is for requests and return the value of red, procedure ends.

Through the algorithm analysis, we can conclude that when the number of decision table condition attribute under the circumstances of less, the method can get a relatively good reduction results, but when decision-making system is bigger, and may not necessarily can get a satisfactory result, so the method of generality is restricted.

4.2 The Improved Attribute Reduction Algorithm

The reduction algorithm of section 4.1 is the relatively common attribute reduction algorithm, and exists certain disadvantages, on this basis, this paper introduced a attribute importance function, put forward an improved attribute reduction algorithm.

first, introduce some definitions on relative attribute importance:

Definition 7. Given decision table $DT=(U,C\cup D,V,f)$, $B\subseteq C$, the Relative approximation classified quality of attribute subset B to condition attribute set C is as follows:

$$\gamma_B(D)=\frac{|pos_B(D)|}{|pos_C(D)|} \tag{1}$$

Through the above definition, now put forward relevant one kind of improved attribute reduction.

Definition 8. Given decision table $DT=(U,C\cup D,V,f)$, $B\subseteq C$, $b\in C-B$ the Importance degree of b to set B is:

$$sig_{BD}(b)=\gamma_{B\cup\{b\}}(D)-\gamma_B(D) \tag{2}$$

from Formula (1)and Formula (2),it can conclude:

$$sig_{BD}(b)=\frac{|U|-|pos_B(D)|}{|pos_{B\cup\{b\}}(D)|} \tag{3}$$

Because attribute b is only one of the attributes in the decision table, if only considering b in their attributes concentration of important degree, this is one-sided, through formula (3) not only considered the attribute b relative to the importance degree of concentration in attribute, also considered the b in all the attributes concentration matter degrees.

$$sig=\sqrt{sig_{BD}(b)^2+sig_{CD}(b)^2}\left(0\le sig\le\sqrt{2}\right) \tag{4}$$

Through the above definition, now put forward relevant one kind of improved attribute reduction algorithm, the ideas for empty sets starting point, circulation from condition attribute concentrated elected sig's biggest attribute, joining into reduction set. Algorithm described as follows: input: decision table (, ,,) $DT=(U,C\cup D,V,f)$;

output: the red, a relative reduction of DT

step1:make $red=\varnothing$;

step2: calculate $r=\gamma_C(D)$ to random $a_i\in C$, calculate $sig_{CD}(a_i)$;

step3: make sig1=0, i=0, index=0;

step4:calculate $sig(a_i)=\sqrt{sig_{BD}(a_i)^2+sig_{CD}(a_i)^2}$, if $sig1<sig(a_i)$, then make $sig1=sig(a_i)$, index=i, i=i+1 Cycle until all a_i calculated;

step5: if sig1==0 it shows that all of the a_i are redundant attributes, the procedure ends;

step6: $red=red\cup\{a_{index}\}$, then calculate $\gamma_{red}(D)$,if $\gamma_{red}(D)==r$, then it shows that the attribute sets in red are as important as set C ,namely that red is for request ,the procedure ends;

step7:make $C=C-\{a_{index}\}$, jump to step4;

step8:the procedure ends , ouput the red.

This algorithm is to find the biggest sig per cycle a trip, join into reduction set red and then continue to search for the maximum attribute degrees besides last sig in the condition attributes set,until $\gamma_{red}(D)=\gamma_C(D)$ then procedure ends. the time complexity of the algorithm is $O(N^2)$, Although this algorithm on the time complexity is relatively large, but it has an obvious advantage that it can find out which are the most important conditional attribute, regarded as the most important in the attribute reduction field.

5 Experiments Result and Analysis

In order to take analysis and testing to the experimental results, this paper introduce DR (detection rate) and FDR (fake detection rate) to the intrusion detection system of performance assessment, the definitions about DR and FDR are as follows:

$$DR=\frac{samples\ of\ \det ced\ attacks}{samples\ of\ total\ attacks} \tag{5}$$

$$FDR=\frac{normal\ connections\ mistaken\ for\ attacks}{normal\ connections} \tag{6}$$

Table 1. The DR of seven times' test (%)

Attack Type	test 1	test 2	test 3	test 4	test 5	test 6	test 7
Normal	99.3	97.6	98.5	98.3	97.9	99.1	98.3
DoS	93.6	91.4	91.7	91.9	92.3	91.4	91.6
Probe	89.7	87.6	88.5	95.4	89.3	98.6	95.6
R2L	88.1	79.2	77.2	65.2	67.5	78.6	79.1
U2R	79 .5	72 .1	74.5	69.4	79.6	68.3	76.4

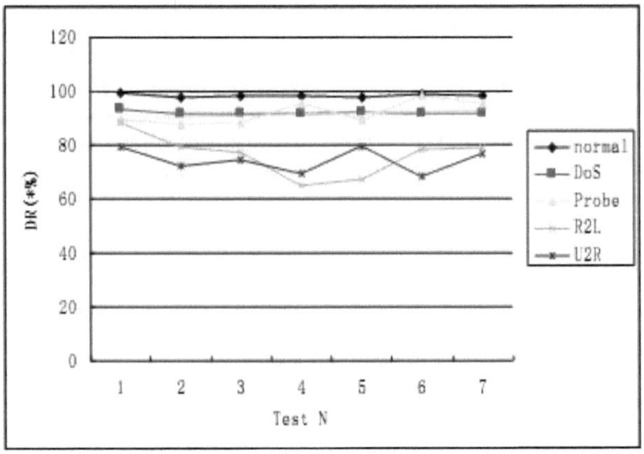

Fig. 1. The DR of seven times' test

Table 2. The FDR of seven times' test (%)

Attack Type	test 1	test 2	test 3	test 4	test 5	test 6	test 7
Normal	2.7	4.4	6.5	4.7	3.1	2.9	3.7
DoS	5.5	5.6	8.3	8.1	7.7	8.6	8.4
Probe	20.4	15.4	19.5	18.6	20.7	21.4	14.4
R2L	26.1	26.8	22.8	34.8	32.5	21.4	28.9
U2R	27.3	31.9	25.5	30.6	33.4	31.7	28.6

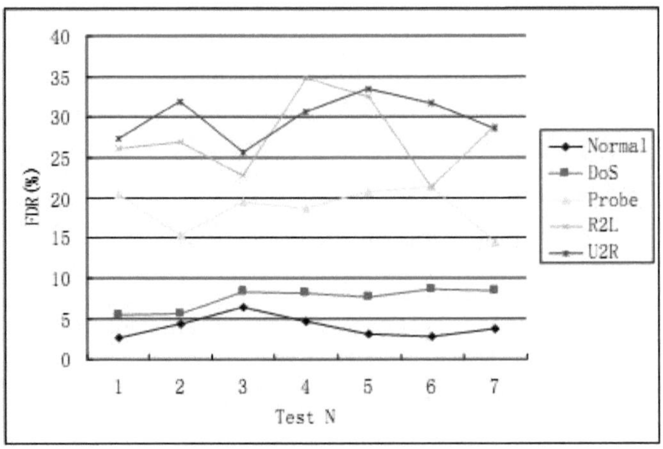

Fig. 2. The FDR of seven times' test

From the Figure 1 and Figure 2,the intrusion detection algorithm based on Rough sets of this paper has very good inspection rate for normal connection, DoS attack and Probe attack, and the seven times 'test results is also relatively smooth. But the detection rates to R2L attack and U2R are relatively low, and for seven times the test result of the fluctuating range is larger. Appearing such result, we can analyze it from two respects: first, due to the selection of datasets contain normal connection records, DoS attack records and Probe number are relative more, so the invasion rules to these three kinds of connection are relatively completely, and the inspection rate is natural relatively higher; Second, viewing from these attacks' type characteristics of themselves, R2L attacks and U2R attacks usually disguised as some legitimate user's identity in order to attack, making its data record features similar with the characteristics of normal connection data, so the detection algorithm lost effect in a certain extent.

In order to validate that the suggested intrusion detection algorithm based on Rough sets is effective, we choose the seven times' average detection rate of the test data to compare with the detection rates from the reference [9] intrusion detection algorithm, the results are as table 5.3 shown.:

Table 3. Comparison to the detection rates of algorithms(%)

Attack Type	Reference [9]	The average detection rate of this paper
Normal	not given	98.42
DoS	99.10	91.99
Probe	87.64	92.1
R2L	72.55	76.41
U2R	73.12	74.25

Through the table 3 we can conclude that the overall detection rates of this paper's intrusion detection algorithm based on Rough sets is superior to reference [9], especially for detection rates of R2L, U2R,which is difficult to detect, this paper has obtained the good effect, so this algorithm is effective.

Acknowledgement. The authors gratefully acknowledge the funding Jilin province Education Science "Eleventh Five Years" Scheme Item(2010222,2010224).

References

1. Tang, Z., Li, J.: Technology of Intrusion Detection, pp. 6–9. Tsinghua University Press, Beijing (2004)
2. Miao, D., Li, G.: Rough Sets Theory Algorithms and Applications. Tsinghua University Press, Beijing (2008)
3. Liu, Y., Li, X., Liao, X., et al.: Intrusion Detection Based on Data Mining. Chongqing University Journal: Natural Science Edition 25(10), 128–131 (2002)
4. Mao, J., Jiang, X.: Intrusion Detection Technology Based on Data Mining. The Modern Electronic Technology (06), 25–27 (2004)

5. Guo, S., Li, X.: Model Formatted by Intrusion Detection Online Rules. Computer Journal 29(09), 1523–1531 (2006)
6. Cao, L., Zhang, Z.: Research on Intrusion Detection Technology. Network Security Technology and Application (03), 18–20 (2009)
7. Pei, X.: A Rough Set Relative Reduction Calculation Method. Computer Science 33(2), 185–187 (2006)
8. Xu, F.: An Improved Core Attribute Calculation Method. Computer Engineering and Application 13, 57–59 (2006)
9. Jiang, S.: Research on Intrusion Detection Algorithm Based on Clustering, pp. 187–204. Science Press, Beijing (2008)
10. Pawlak, Z.: Rough sets. International Journal of Computer and Information Science 11(5), 341–356 (1982)
11. Pawlak, Z.: Rough classification. International Journal of Human-Computer Studies 51, 369–383 (1999)

Design of Coreless Winding Machine Electrical Control System

Wu Yaohui, Shao Pengfei, and Wei Rencan

Zhejiang Wanli University, Ningbo Zhejiang 315100, China
yaohuinb@yahoo.com.cn, cbb_bb@126.com, 56510005@qq.com

Abstract. The winding of oblique coil type coreless motor is the most difficult for the winding machine, it need strict requirement for the mechanical structure and electrical control systems, but it has the greatest market value. Through the analysis of the working principle about the oblique coil type coreless winding machine, this paper designed the main circuits of the Electrical Control System and joint debugged them, solved the key issues about the timing synchronization between various circuit modules, dividing control, software flexibility and so on. The system used in the production line works well, and winded coreless motor can meet the coil tidy, uniform density and other requirements.

Keywords: Coreless, Motor, Winding machine, Electrical control system.

1 Introduction

Currently, a large literature on Coreless winding machine has been actively studied. For example, a new control mode of transformer winder that have a separate drive between main shaft and the mechanism for arranging the wires was given to realize the flexible adjustment of the winding parameter[1]. By electronic inductor and microchip IC adjusting damper torque, combining with the simple structure design, the winding machine realizes the determination of automatic adjustment[2]. AC Servo System was used to wind and position the armature, Automatic tension control system was introduced to enhance the capacity of winding tension control[3].

But Most of these studies remain in the theoretical design phase. Some designed winding machines seem to wind a variety of coreless motors, but haven't efficiency and cost advantage for any kind of coreless motor. Market research found that, the winding of oblique coil type coreless motor is the most difficult for the winding machine, it need strict requirement for the mechanical structure and electrical control systems, but it has the greatest market value. This paper designed the electrical control system of the coreless winding machine, the main contents include: (1) study of coreless winding machine's working principle. (2) The design of control circuit of the main shaft mechanism. (3) The design of control circuit of the arrange-line mechanism. (4) The design of control circuit of the dividing motor. (5) The design of other control circuits.

Y. Zhang (Ed.): Future Communication, Computing, Control and Management, LNEE 141, pp. 321–327.
springerlink.com

2 Study of Coreless Winding Machine's Working Principle

To wind oblique coil type coreless motor, the designed mechanism is shown in Figure 1, the height of the winded coreless motor is equal to the horizontal distance between the crochet hooks ab.

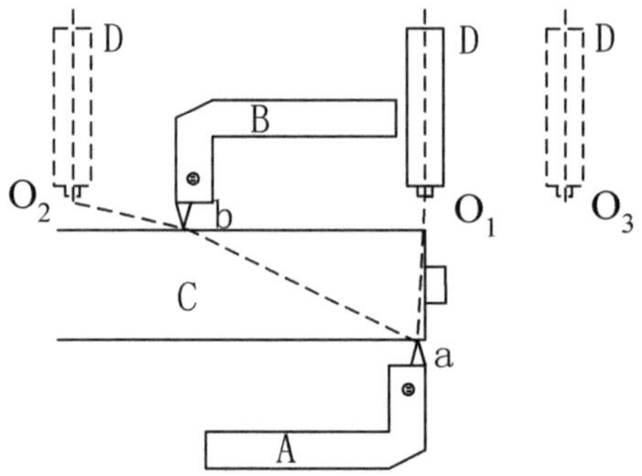

Fig. 1. Schematic diagram of coreless winding machine

Beginning, the main shaft rotation mechanism AB is in the horizontal position, arrange-line pole D is over crochet hook a. Go through pole D and bypass crochet hook a, the wire is fixed. Then, driven by the main shaft motor, mechanism AB counterclockwise rotates (seen from right to left). At the same time, driven by the arrange-line motor, pole D moves left at constant speed. When the rotation angle of mechanism AB is equal to 90, pole D reaches just over crochet hook b. At this time, one side of the wire is hooked at crochet hook a, the other side is in touch with crochet hook b. When the rotation angle is between 90 and 135, pole D moves left at lower speed. When the rotation angle is equal to 135, pole D reaches the most left. At this time, the wire is hooked between a, b and O2, in triangular form. When the rotation angle is between 135 and 180, pole D moves right at lower speed. When the rotation angle is equal to 180, pole D reaches just over crochet hook b, half oblique coil is winded at this moment. When the rotation angle is between 180 and 270, pole D moves right at constant speed. When the rotation angle is equal to 270, pole D reaches just over crochet hook a. When the rotation angle is between 270 and 315, pole D moves right at lower speed. When the rotation angle is equal to 315, pole D reaches the most right. At this time, the wire is hooked between b, a and O3, in triangular form. When the rotation angle is between 315 and 360, pole D moves left at lower speed. When the rotation angle is equal to 360, pole D reaches just over crochet hook a, one oblique coil is winded at this moment. At this time, mechanism AB and pole D are stopped at the origin, dividing shaft rotates one divide driven by dividing motor, then next cycle starts.

3 The Design of Control Circuit of the Winding Machine

3.1 The Design of Control Circuit of the Main Shaft Mechanism

The control circuit is shown in Figure 2.

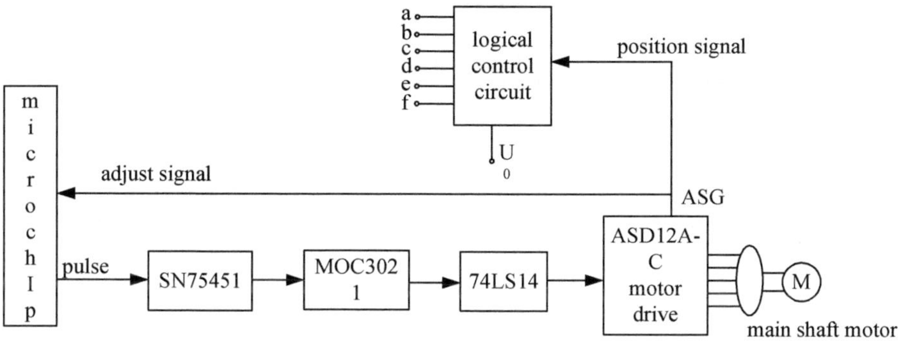

Fig. 2. Control circuit of the main shaft mechanism

To control the main shaft motor speed and direction, first an adjustable pulse sequence is output by microchip, then amplified by line amplifier SN75451, then electrically isolated by optocoupler MOC3021, then shaped by chip 74LS14, finally input to motor drive. When the control circuit of the winding machine was designed, all timing was based on the rotation of the main shaft rotation mechanism AB, while the rotation angle was associated with the current pulse count. In theory, the current timing can be inferred by monitoring the number of pulse output from microchip. For example, if mechanism AB rotates one cycle by 5000 pulses, then it rotates 90 degree by 1250 pulses. The fact is not the case. Taking into account mechanical resistance and other incidental factors, when the microchip outputs one pulse, the main shaft will not turn one angle possibly. The pulse is called "invalid pulse". At that case, only by monitoring the number of pulse output from microchip, all subsequent actions will lead to confusion. Here, only by monitoring pulse output pin ASG of motor drive, looking on effective pulse feedback signal from pin ASG as valid pulse, then all timing synchronization can be ensured.

3.2 The Design of Control Circuit of the Arrange-Line Mechanism

Arrange-line mechanism must keep pace with main shaft mechanism, see Figure 1. For example, when the rotation angle of main shaft rotation mechanism AB is equal to 90, pole D reaches just over crochet hook b, while the angle is indirectly controlled by pulse. As shown if Figure 2, the count of valid pulse is based on effective pulse feedback signal of pin ASG. The feedback signal is processed through the logic control circuit, then converted into logical signal a~f and U0, which control the rotation of arrange-line motor. The conversion relationship table is shown in table 1, U0 is set 1 while mechanism AB rotate 360 degree.

Table 1. Conversion relationship table

AB rotation	ASG Pulse number	a b c d e f
0～90	1～1250	1 0 0 0 0 0
90～135	1251～1875	0 1 0 0 0 0
135～180	1876～2500	0 0 1 0 0 0
180～270	2501～3750	0 0 0 1 0 0
270～315	3751～4375	0 0 0 0 1 0
315～360	4376～5000	0 0 0 0 0 1

The rotation speed and direction of arrange-line motor is also controlled by pulse, which is controlled by the above seven logic signal control. The count of valid pulse is based on effective pulse feedback signal of pin ASG of arrange-line motor drive, as shown in Figure 3.

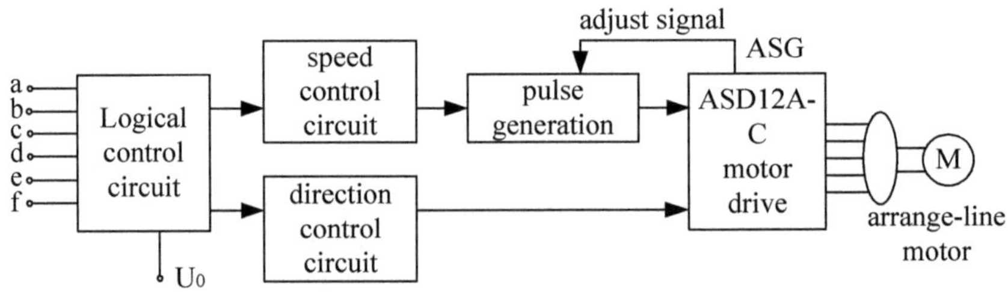

Fig. 3. Control circuit of the arrange-line mechanism

3.3 The Design of Control Circuit of the Dividing Motor

When main shaft motor and arrange-line motor works, the dividing motor doesn't work. Only when main shaft rotation mechanism AB rotates one cycle and pole D moves one back and forth, that means the winding machine has winded one oblique

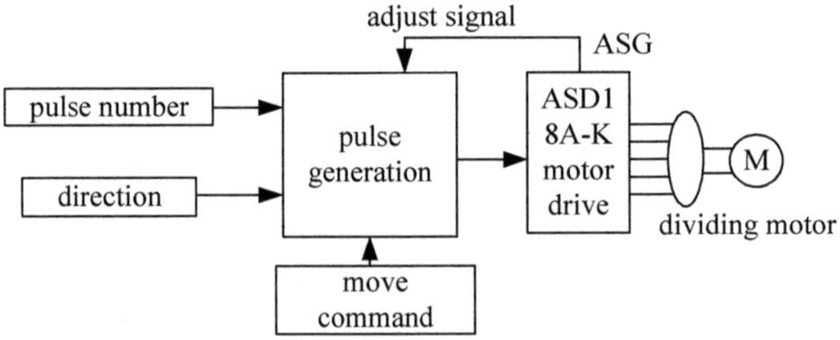

Fig. 4. Control circuit of the dividing motor

coil, round mold C need to turn a small angle driven by the dividing motor. Thus, the second oblique coil can be next to the first oblique coil. While the small angle depends on the ratio of wire diameter and the diameter of round mold C. The rotation of dividing motor is also controlled by pulse, the count of valid pulse is also based on effective pulse feedback signal of pin ASG of dividing motor drive, as shown in Figure 4.

3.4 The Design of Other Control Circuits

Number of coils form one group, each group has a tap. Generally, coreless motor has several taps. The tap is winded by crochet hook. The movement of crochet hook, the cut of tap, the drying and shaping of coreless cup before stripping line, all these need specific control circuit coordinated by microchip, shown in Figure 5.

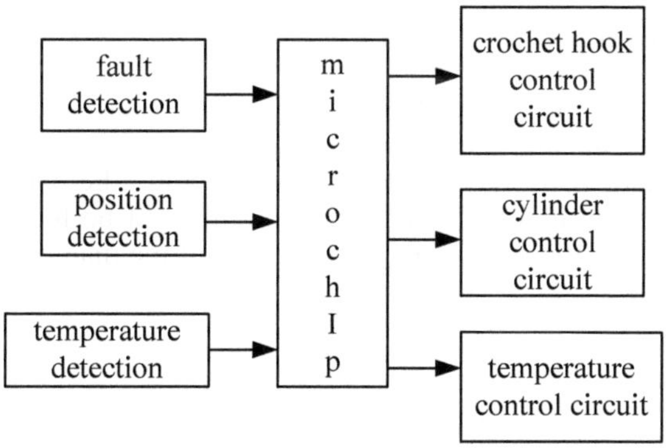

Fig. 5. Other control circuits

4 The key Issues of Electrical Control System Design

The electrical control system of the coreless winding machine is very complex, it has many circuit modules, strict timing synchronization between each module need to be ensured. In addition, many modules need theoretical calculations, sudden circuit or mechanical failure need to be considered during the actual operation of the machine. Above all, the key issues are timing synchronization between various circuit modules, dividing control, software flexibility and so on.

4.1 Timing Synchronization between Various Circuit Modules

Because the design of the winding electrical control system is practice-oriented product development, winding, arrange-line, dividing and other actions can not be disorders, strict timing synchronization between each module need to be ensured. Here, all timing was based on the rotation of the main shaft rotation mechanism AB, while the rotation angle was associated with the current pulse count. The count of valid pulse is based on effective pulse feedback signal of pin ASG of each motor drive, it is based on the actual engineering experience to be understood.

In the control circuit of main shaft mechanism, the output signal of pin ASG of motor drive is as the calibration signal to control the output pulse of microchip, as shown in Figure 2. If the microchip output pulse doesn't consist with the output signal of pin ASG of the motor, which means the occurrence of "invalid pulse", the microchip need to output the "invalid pulse" again.

In the control circuit of arrange-line mechanism, the output signal of pin ASG of arrange-line motor drive is as the calibration signal of pulse generation of arrange-line motor, as shown in Figure 3. If "invalid pulse" occurs in arrange-line motor, the control circuit of main shaft mechanism will pause the pulse output, waiting for synchronous with arrange-line motor. Similarly, the output signal of pin ASG of dividing motor drive is as the calibration signal of pulse generation of dividing motor, as shown in Figure 4. If "invalid pulse" occurs in dividing motor, the control circuit of main shaft mechanism and arrange-line mechanism will both pause the pulse output, waiting for synchronous with dividing motor.

4.2 Dividing Control

If the winding machine has winded one oblique coil, round mold C need to turn a small angle driven by the dividing motor. Thus, the second oblique coil can be next to the first oblique coil. But if the angle is too small, the winded coils will be overlapped. If the angle is too big, the winded coils will be few and scattered. The small angle depends on the ratio of wire diameter and the diameter of round mold C, as shown in Figure 6.

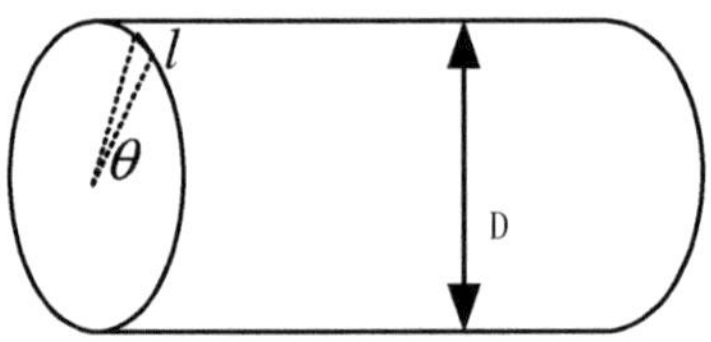

Fig. 6. The dividing of coreless motor

The small angel of round mold C is set θ, the corresponding arc length is l. As compared with the diameter D of round mold C, l can be negligible, then

$$l = \frac{\theta \pi D}{360} \tag{1}$$

The arc length l is just the diameter d of the wire covered, then

$$d = l = \frac{\theta \pi D}{360} \Rightarrow \theta = \frac{360 d}{\pi D} \tag{2}$$

Assuming the motor need N pulses to rotate one cycle, then the rotation angle degree corresponding one pulse is 360/N, then the needed pulse number corresponding one degree is N/360, pulse number corresponding degree is

$$n = \frac{N}{360}\theta = \frac{N}{360} \bullet \frac{360d}{\pi D} = \frac{Nd}{\pi D} \tag{3}$$

So, if diameter d of the wire and the diameter D of round mold C is set, pulse number n the dividing motor need corresponding degree can be calculated. If the number of pulses is preset to the control circuit of dividing motor, the desired results can be gotten. If pulse number N is much bigger, then dividing control will be much finer. Actually, N can be set according to situational effect, until the winded coreless motor is satisfied.

4.3 Software Flexibility

Throughout the electrical control system design, software is the control center, and the software's flexible design can promote the hardware. Software can real-time adjust the winding tension servo system during the winding process. As shown in Figure 1, when the rotation angle degree of mechanism AB is between 135 and 180 or between 315 and 360, pole d moves back, which will inevitably lead to looseness of the wire hooked at a、b、O. At this time, tension servo system is needed to control the tension of the wire. The rotation of the motors in the electrical control system design is controlled by pulse output, which can be adjusted flexibly by software, including the gap of pulse width. Software design also involves a lot of computing process, for example, the calculation of dividing control as formula 3. Besides, the running state of the winding machine need to be monitored by software, once the wire is break, or cylinder action is out of order, or temperature does not meet the requirements, then software system should shut down the machine itself.

5 Summary

Through the analysis of the working principle about the oblique coil type coreless winding machine, this paper designed the main circuits of the Electrical Control System and joint debugged them, solved the key issues about the timing synchronization between various circuit modules, dividing control, software flexibility and so on. The quality of winding and arrange of the wire is improved. The system used in the production line works well, and winded coreless motor can meet the coil tidy, uniform density and other requirements.

References

1. He, H., Hui, J.: Digital Design of Transformer Winding Machines. Electric Drive, 72–76 (May 2010)
2. Xiong, J.-S.: The design and realization of adaptive type winding machine. Applied Technology, 133–134 (March 2010)
3. Wang, Y., Zhao, C.: The Design of A Twin Flyer Winding Machine Control System Based on PLC. Control and Testing, 72–74 (July 2009)
4. Yuan, X.: Development of fine wire winding machine. Modern Manufacturing Technology and Equipment, 8–9 (May 2010)
5. Zhao, X.: Automatic winder machine equipment development. Equipment for Electronic Products Manufacturing, 41–45 (November 2010)

Kinematic Simulation of Rescue Robot

Weichun Zhang[1], Baohao Pei[2], Jie Chen[2], Xianbin Du[2], Bingbing Ma[2], and Xiao Wang[3]

[1] Shandong University of Technology, Zibo, Shandong Province 255049, China
Zhangwc@sdut.edu.cn
[2] Shandong University of Technology, Zibo, Shandong Province 255049, China
Peibaohao@163.com
[3] Zibo Vocational Institute, Zibo, Shandong Province, 255314, China
teacherwx556@126.com

Abstract. The Pro/E three dimensional modeling, the MECH/Pro model transformation and the ADAMS rigid body kinematics simulation were conducted based on the rescue robot developed by our laboratory. The trail of the robot hand and the parameters of limit operating location were obtained. The results provide important data support for the improvement of the rescue robot.

The robot in this research is refirred by one crawler hypraulic excavator, and it can fullfil these tasks of dealing with an emergency, transporting, forcible entry. The robot can also fullfil tasks of going forward and backward, estending and shortening its hand, capturing things and so on when it is controled in distant place by operator. Howerer, the working range of robot hand can not content the need of rescue. It needs to do some kinematics simulation on robot hand by platform of spftware, for the improvement of hydraulic system and kenematics simulation in the future.

Keywords: computer application, simulation, rescue robot, kinematics.

1 Kinematic Analysis of Rescue Robot

Working device of the robot is the main object of the study, such as pig 1(the pionts from A to H are link joints). The parts of 1,12 are fixed on the substrate, going up and down of the hand is operated by the combination of hydro-cylinder in the number 2 and 11; the revolving of paw is operated by assembly in number 10; opening and closing of the hand is operated by hydro-cylinder in number 7. Fig 2 is the kinematic model of working device.

Y. Zhang (Ed.): Future Communication, Computing, Control and Management, LNEE 141, pp. 329–335.
springerlink.com

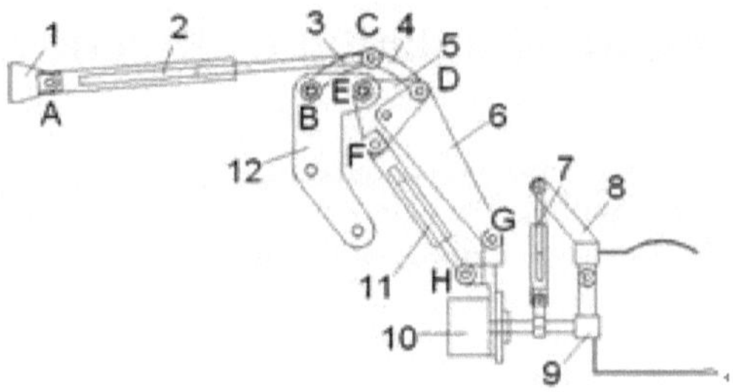

1-fixed link joint of substrate 2-combination of big hydro-cylinder 3-long connecting rod 4-short connecting rod 5-triangle connecting rod 6-forearm 7-hydro-cylinder of paw 8-top paw 9-down paw 10-motor 11-combination of small gydro-cylinder 12-big arm

Fig. 1. Hand structure of rescue robot

Working device is divided into 3 basic modules, there are rotaty actuator mechanism(ABC and FGH), plane motion mechanism of rigid body(DEFG and GHI) and hingebar(BCDE), as Fig 2.

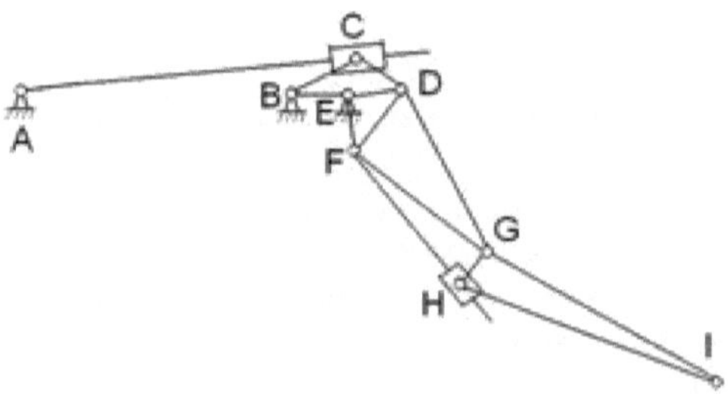

Fig. 2. Kinematics model of working device

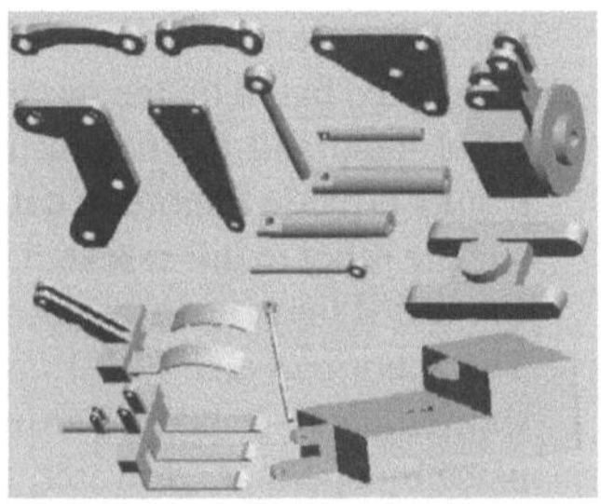

Fig. 3. Three-dimensional diagram of basic parts

Kinematics model of each universal basic modules is builded minutely in references[2], the idea of whole kinematic analysis system are as follows:1) the movement locus of point C can be obtained according to rotaty actuator mechanism(ABC);2) the movement locus of point D can be obtained by points B, C and E according to hingebar;3) the movement locus of point G can be obtained by points F and D, the movument locus of point F can be obtained by point E and D according to module of rigid body analysis;4) the movement locus of point H can be obtained according to module of rotaty actuator mechanism(FGH);5) the movement locus of point I can be obtained by points G and H according to module of figid body analysis.

In a word, the location and locus of point I can be obtained by assemblies in number 2 and 11.

2 Rigid Body Modeling of Rescue Robot

2.1 Drawing and Assembling of Parts

The working directory is setted, units are selected for MMNS and parts are drawed according to its basic character such as extending. 16 basic parts are drawed by software of Pro/e as Fig 3.

2.2 Skill in Assembling

There are more or less underirabilities after assembled one part then another part is assembled with it, there is probably that "allowed assume" is selected by software Pro/e itself. We can try to unselect "allowed assume" when the underirability appeared. Contrast "allowed assume" is selected with that is unselected as Fig 4.

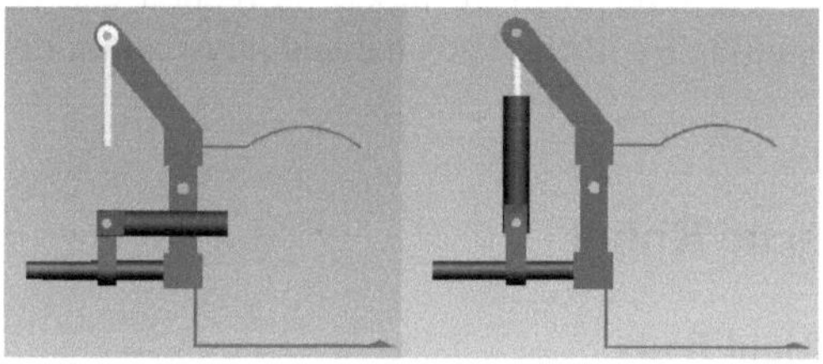

Fig. 4. Assembling of robot paw

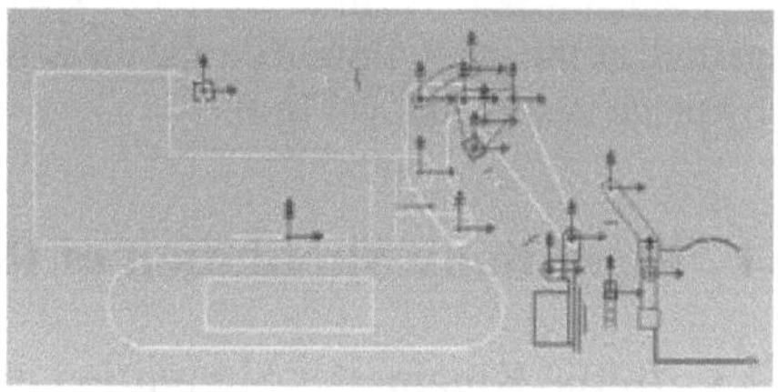

Fig. 5. The location of markers

3 Model of Rigid Body Is Exported

3.1 Basic Setting

The whole model *.asm of rescue robot that is builded in pro/e is opened, the unit, density and quality are setted as follow:1) setting units: clicking compile>set in menu bar after assembly is opened and then selecting units in the menu manager, that is to say the menu manager is opened, after that building a new system of unit as MMKS. Clicking application program>mechanica in menu bar and then clicking convert, by then, all units of the assembly are transformed to the defined system of unit and the whole data can be transferred correctly; 2) setting material and density: clicking compilation>setting>material>steel>add in menu bar, and clicking setting>quality information, then desity of parts is imported; 3) defining rigid body: clicking set up mechanism>rigid bodies>great>by selection according to MECH/Pro, and 20 rigid bodies are defined by 16 parts; 4) defining Markers: clicking set up mechanism>markers>great according to MECH/Pro and 15 markers are defined. The location of these markers are showed as Fig 5.

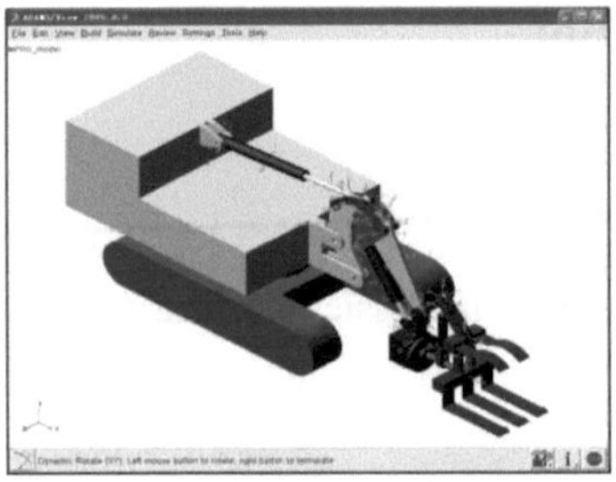

Fig. 6. The model in ADAMS

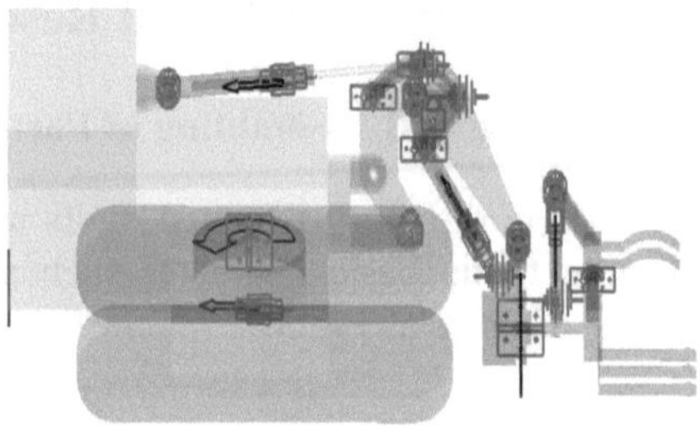

Fig. 7. Restrain and drive in the model

3.2 Exporting Model

Clicking interface>adams/view>output type>gemetry quality>done return by MECH/Pro, then *.adm file that mechanism are described and *.cmd, *.mpr file which is needed when Asams/View is exported and data of pro/e model is readed can be getted, at the same time Adams/View is opening by itself. The changed physical model by pro/e is showed as Fig 6.

4 Rigid Body Simulation of Rescue Robot

4.1 Initial Kinematics Simulation of Rigid Body

The analysis type of mechanical system is ascertained by the freedom of this mechanical system[3][4]. What is kinematics simulation for mechanical system, in a word, the characteristics of motion for system is considered only when there is no freedom of this system. The kinematics simulation of rigid body is carried out when this system has no freedom and each hdraulic cylinder working in turns, with the purpose that ultimate working paremeter can be obtained and practical applicability as well as usability of the model can be checked.

4.1.1 Adding Restrain and Drive

There are 19 rigid bodies with the exception of grand in all, the parts that have the same movement locus need to be merged and 16 rigid bodies can be getted. There are 96 freedom need to be restrained in this system because each body has 6 freedom. The model is restrained and drived showing as Fig 7.

4.1.2 Checking Model

The correctness of the model need to be checked after defining restrain and drive, then the button verify the model that lower right corner of ASAMS is clicked, after that the test report will popup, the test report shows 16kinetic bodies, 4 Cylindrical jionts, 6 hinge joints, 4 sphere joints, 4 translational jionts, 2 fixed joints, 6 drives and 0 freedom. That is to say, there is no redundant restrain and the model can be operated for kinematics simulation.

4.2 Rigid Body Kinematics Simulation of Rescue Robot

The kinematics simulation is operated under this circumstance of 0 freedom and no redundant restrain with the purpose that locomotibity of the mechanism can be measured and the max radius of clean-up, the min radius of gyration, the lowest depth for fetching and the max operational altitude can be obtained. The principle of kinematics analysis is: working in turn, avoiding intervene and so on. The specific step is showing as follows:

4.2.1 Adding Drive

Step function is weited on the base of measuring travel for each hydraulic cylinder according to the principle of working in turns. There are 6 motions need to be simulated for the max coverage area, it contains big hydraulic cylinder extening, small hydraulic cylinder shrinking, big hydraulic cylinder shrinking, small hydraulic cylinder extening, big hydraulic cylinder returning and small hydraulic cylinder returning.

1) the step function for journey of big hydraulic cylinder: STEP(time, 0, 0, 1, 20)+STEP(time, 3, 0, 10, -420)+STEP(time, 12, 0, 18, 400) ; 2) the step function for journey of small hydraulic cylinder: STEP(time , 1, 0, 3, 80)+STEP(time, 10, 0, 12, -180)+STEP(time, 18, 0, 20, 100).

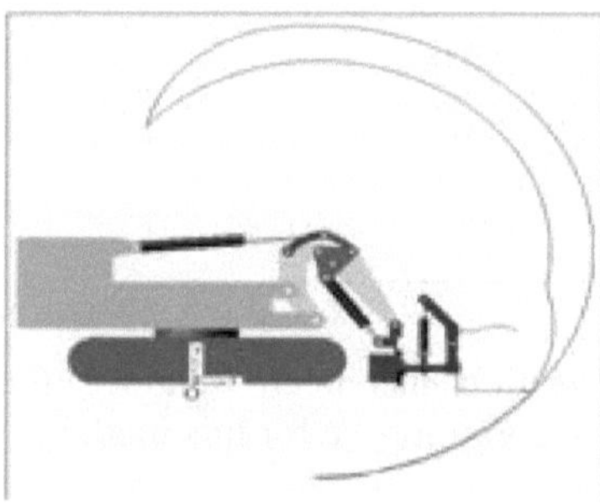

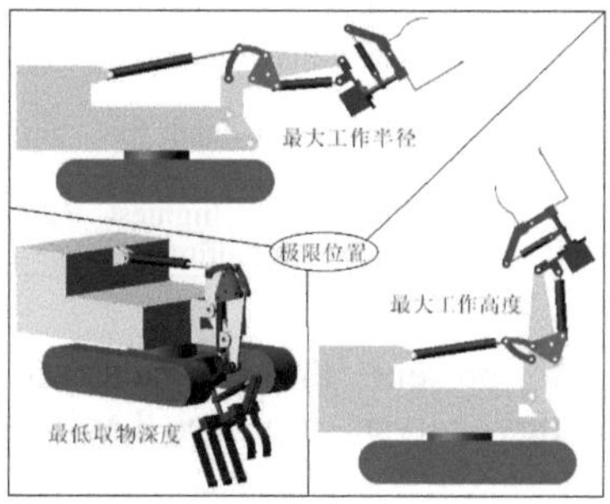

Fig. 8. Movement locus of defined gauge point **Fig. 9.** Operative mode in the limited location

4.2.2 Kinematics Simulation

Setting simulative time and step, defining gauge point by fingertip in lower paw. As a result the movement locus of the defined gauge point can be obtained and shown as Fig 8, the operative mode in the limited location is shown as Fig 9.

4.3 Result of Simulative Analysis

The base point is defined by the projective point that is projiected on ground by substrate and centre of gyration in track, a coordinate graphs is drawn as that X-axis shows horizontal range from gauge point to centre of gyration and Y-axis shows length of perpendiculars from gauge point to centre of gyration. The coordinate graphs is shown as fig 10(Y=400mm means the length of perpendiculars from centre of gyration to ground).

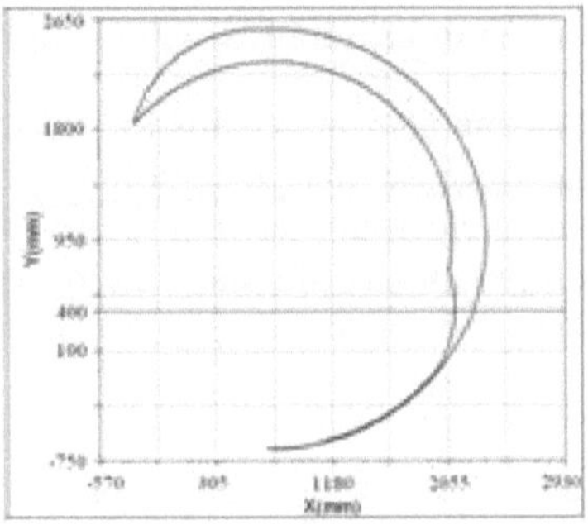

Fig. 10. Movement locus of fingertip

The limitive coordinate is read through a tool for chart statistic and 4 woeking parameters in the limit are obtained.

Table 1. Data of simulation and measure

Measured projecr	Data of simulation (mm)	Data of measure (mm)
The max working radius	2341.5	1514.3
The min gyroradius	1070	1500
The lowest depth for fetching	663.4	300
The hightest operational altitude	2569.6	2008

Up to present, the parameter in limited location has been obtained through kinematics analysis, it proves that it has much space to improve for this model.

5 Conclusion

One shovel crawler is transformed to rescue robot, which is fitted more for working. The rescue robot paw is simulated by kinematics analysis through the idea of modularization and the result of simulation can provide some good suggestion for involved professional, the movement locus and parameter in limited location of the model is obtained from the simulation, as a result the simulation forming the foundation for improving the whole system in the future.

References

1. Nan, D.: Simulation and optimization of working device of loader based on ADAMS. Construction Mechanization (7), 38–40 (2007)
2. Wang, Y., Song, L.: Kinematic ssimulation and CAD study of hydraulic excavator. Construction Mechanization (2), 24–27 (2006)

3. Li, J., Xu, M., Yang, J.: Design and Kinematics Analysis of Bomb-disposed Robot Manipulator. Machinery & Electronics (3), 66–69 (2009)
4. Zhao, Y., Fang, Y., Nie, J.: Study on Kinematics Analysis of 5-dof Robot Arm. Science Technology and Engineering 9(6), 1571–1573 (2009)
5. Cao, X., Si, A., Wang, G., et al.: Hydraulic-mechanical Coupling Simulation Analysis of Wheel Loader Operation Mechanism. China Mechanical Engineering 12, 2461–2464 (2006)

The Optimal Design and Modal Analysis for a Truck Crane Frame

Weichun Zhang[1], Baohao Pei[2], Bingbing Ma[3], Jie Chen[4], and Xianbin Du[5]

[1] Shandong University of Technology, Zibo, Shandong Province 255049, China
[2,3,4,5] Shandong University of Technology, Zibo, Shandong Province 255049, China
Zhangwc@sdut.edu.cn, Pei baohao@163.com

Abstract. The three-dimensional model of the frame is carried out by the ANSYS Workbench software, which the finite element model is built on. The static analysis and modal analysis are carried out to the truck crane's frame. The results of the analysis provide theoretical basis for the optimum design and weight lighting of the frame.

Keywords: crane frame, finite element, static analysis, optimum design, modal analysis.

1 Introduction

The finite element analysis on a truck crane frame[1] is one of the important issues in the field of truck design. The simulation by finite element becomes an important method in the design of crane frame, according to the developing of computers. Stress, strain and vibration frequency of truck crane frame can be mastered in all kinds of conditions. Which provides scientific grounds for devising and reaches the purpose of reducing costs and shortening cycles, lightweight, enhance the quality of design.

2 Frame Model

2.1 Geometrical Model

The geometrical model of truck crane frame is built by ANSYS. The whole frame is simplified suitably for the purpose of satisfying engineering research, and some small parts are omitted (such as chamfering, ostiole, convex plate and reinforcing rib). The built model likes fig 1.

Y. Zhang (Ed.): Future Communication, Computing, Control and Management, LNEE 141, pp. 337–342.
springerlink.com

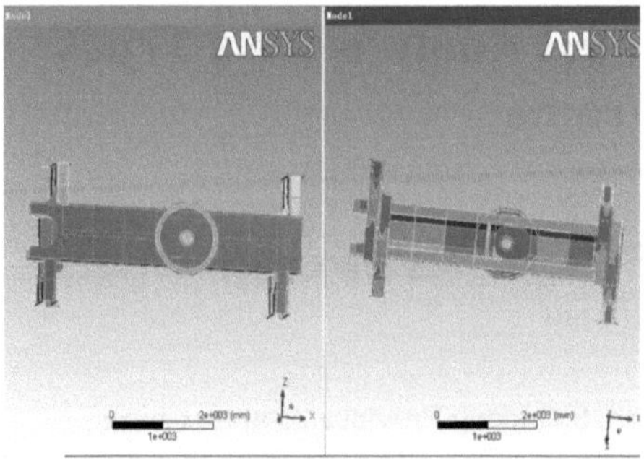

Fig. 1. Geometrical model of truck crane frame

Fig. 2. Finite element model of the frame

2.2 Finite Element Model

The geometrical model is divided into finite element model, which likes the fig 2, and the total nodes are 72924, the total elements are 71116.

2.3 Load and Restraint

Inertia force: vertical load (G1+GQ) is added to the location of bolt in the middle of the frame, and torque M is added to the location of centre of gyration and the web in the frame. Load and restrain is added as fig 3.

Condition 1: both ends of standing leg are restrained in order to restrain all of the DOF. Condition 2: the location of A, B, D is restrained.

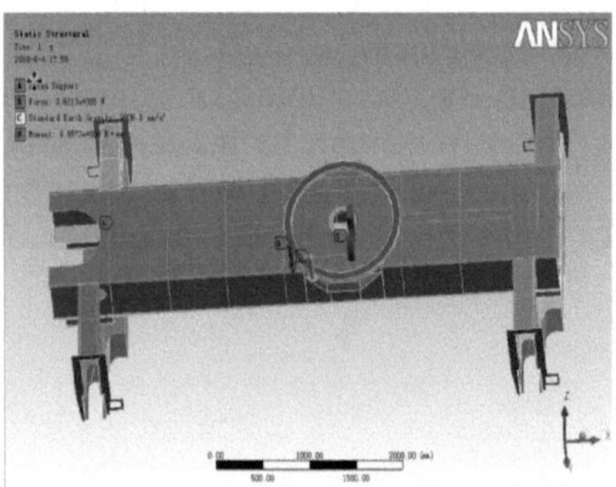

Fig. 3. Load and restraint of the model

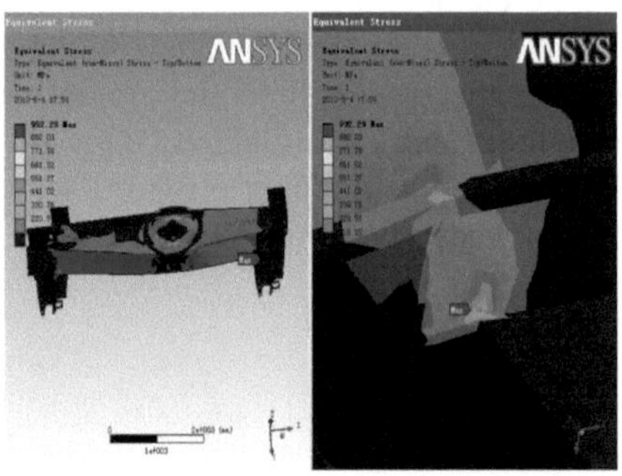

Fig. 4. Equivalent stress of the frame in condition 1

2.4 Result of Calculation

Load are added on the basis of condition 1, the result of analysis as fig 4, 5, 6.

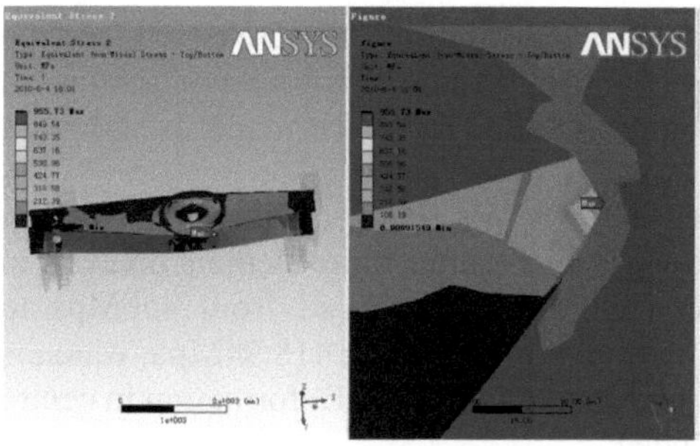

Fig. 5. Equivalent stress in the connection

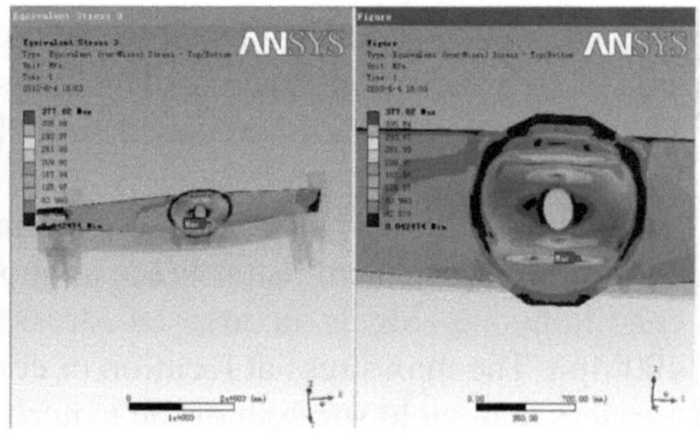

Fig. 6. Equivalent stress in the centre of gyration

The result shows stress distribution of model satisfies actual situation, which proves that finite element model is correct. The max stress 992.29Mpa appears at connection of fixed box in left behind and web of main frame(as Fig. 4), stress in connection of web and batter brace is reach to 955.73Mpa(as Fig.5), stress in the centre of gyration is not admissible stress(as Fig. 6). Stress in connection of fixed box in left behind and web of main frame and connect of web and batter brace is too big, because of stress concentration. So it is important to eliminate the stress concentration effect.

Deformation in centre of gyration is noteworthy problem, deflection of frame should not more 0.5% than lengthways span of active outrigger. The max value of deflection is 24mm because lengthways span of frame is 4800mm. It can fit the bill in condition one, the max value of total deformation in centre of gyration is 7.6612mm.

3 Optimization Scheme Is Raised, Checked and Evaluated for One Crane Frame

3.1 Optimization Scheme Is Raised

It is some problem finding according to finite element analysis in condition one and condition two: 1. Stress in connection of fixed box in left behind and web of main frame and connect of web and batter brace is too big; 2. Stress in active outrigger and clapboard of main frame is lesser; 3. Stress in U-sheet is lesser.

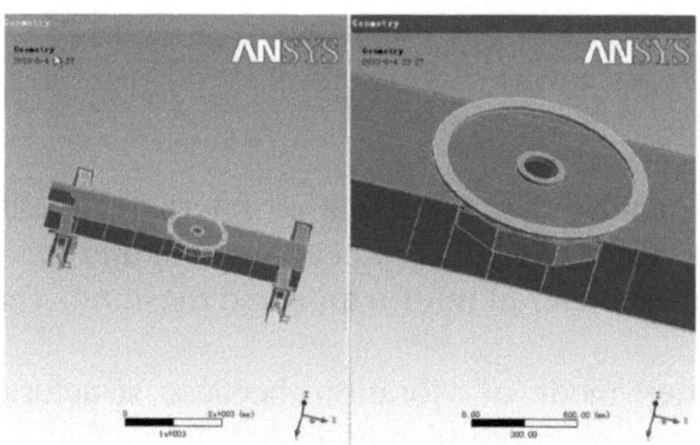

Fig. 7. Optimized model

Fig. 8. Load and restraint of optimized model

Some optimization is raised according to problem finding: 1. Increase connection area of web and batter brace; 2. Thickness of U-sheet is reduced from 8mm to 6mm; 3. Thickness of clapboard is reduced from 16mm to 8mm.Optimized model is shown as Fig 7.

Optimized model is loaded and restrained as Fig 8. We obtained equivalent stress in connection of web and batter brace as showing in Fig. 9 after finite element analysis in condition one. Stress in area of stress concentration is reduced from 950Mpa to 400Mpa. The max stress at location of centre of gyration is reach 418.94Mpa, which is also less than yield strength of the material and shown in Fig.10. Deformation in centre of gyration is 7.3962mm, which is also reduced.

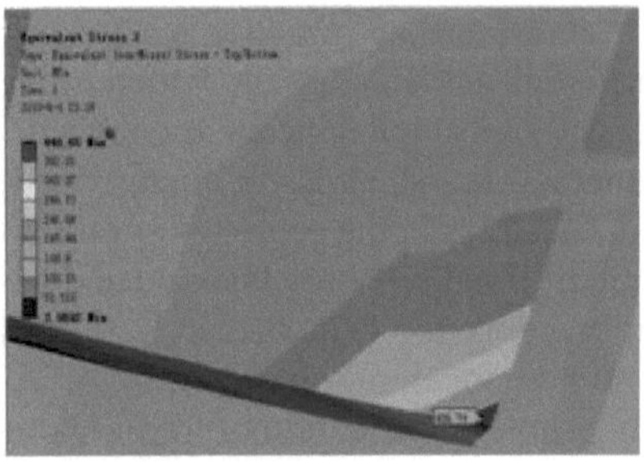

Fig. 9. Equivalent stress in the connection of web and batter brace

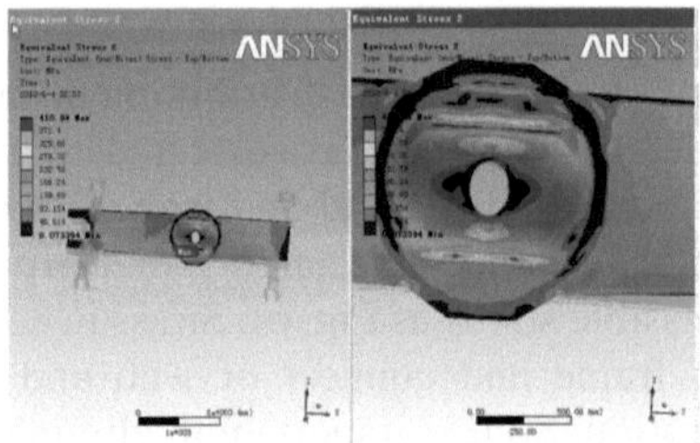

Fig. 10. Equivalent stress in the centre of gyration in optimized model

Stress concentration in connection of web and batter brace is disappear through improvement of frame structure. Stress is increased but the value of it still less than admissible stress after being lightweight. So the improvement of frame structure is correct and the weight of frame is reduced by 83.6kg.

4 Model Analysis

Dynamic property of physical construction is decided by vibratory model parameter, such as frequency of structure, damping and mode of vibration. The advantage and disadvantage for dynamic properties of structure is decided by result of model analysis, because natural frequency of system is not affected by any other factors.

4.1 Solving and Result of Model Analysis

The main problem is to calculate model parameter when it is out of external force in model analysis of frame, that is in freedom status. So it is not need restrain and load. Lead finite model to simulation module, define material information and meshing. The size of reseau is 25mm in this paper.

Dynamic property is decided by low step mode of vibration, because structural power is affected by low step mode of vibration is more than high step. It is calculate anterior 10 step mode of vibration in this paper and frequency of each step is shown as tab 1.

Table 1. Inherent property of frame

degree	Natural frequency/Hz
1-5	<1
6	1.2001
7	42.007
8	49.292
9	50.065
10	53.387

The value of frequency in anterior 5 step is less than 1Hz, so ignore it and analysis modal frequency from 6 step to 10 step. Mode of vibration in each step is shown as Fig. 11.

4.2 Results Analysis of Model Analysis

Idling speed of the engine NTC-290-QY25 is 575—650rpm and driving frequency range is 28.75—32.5Hz; Rated revolution is 2100r/min and driving frequency is about 105Hz; Normal working speed is 1300—2100r/min and frequency range is 65—105Hz.

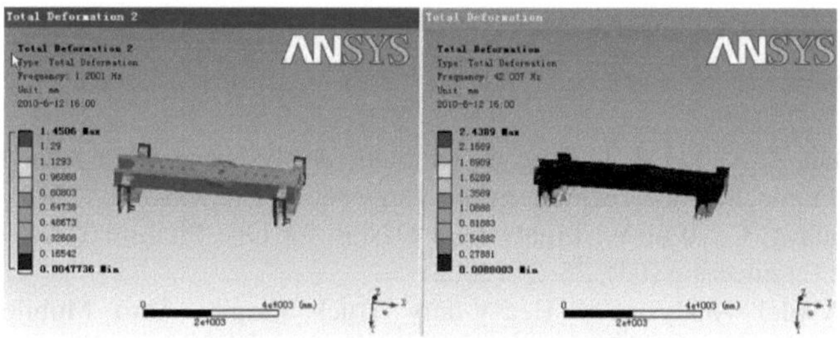

Mode of vibration in 6 and 7 step

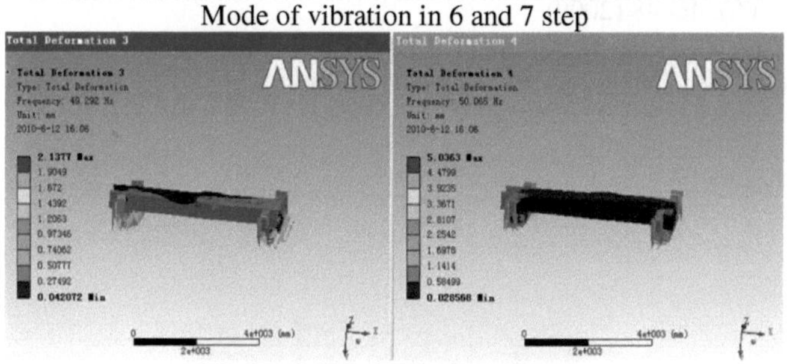

Mode of vibration in 8 and 9 step

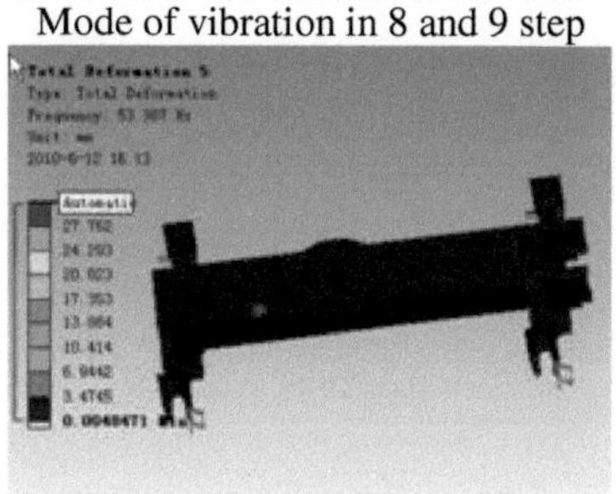

Mode of vibration in 10 step

Fig. 11. Mode of vibration in each step

The aspect of exciter checking is shown as follows:(1) checking exciter of idling operation and uneven road. The value of frequency in step 1 is about 42.007Hz after loaded in frame, and driving frequency from smooth road is about 20Hz. So the frame of truck crane resonated is impossible; (2) checking of normal working condition. Frequency in 7—10 step is about 50Hz according to model analysis, which is less than frequency in idling condition. That is sub-synchronous resonance can not be happened.

The model parameter of frame is good and structure of frame is reasonable according to the analysis in this paper.

5 Summary

Finite element model is built for one truck crane frame, as well as static state and dynamic analysis is operated in this paper. Section of frame is optimized and lightweight as well as strength and rigidity is improved, and sub-synchronous resonance can not be happened, which proves that the structural design is correct and offers rationale for structural optimized design and lightweight.

References

1. Chen, T., Han, Z.-X., Lei, J.-J., Feng, Y.-K.: Static Analysis and Mode Analysis of a Kind of 6470 SUV Frame. Science Technology and Engineering (1), 565–566 (2010)
2. Qian, L., Guang, C., Wu, Y.: Lightweight Design for One Pick-up Truck Frame. Beijing Automotive Engineering (03), 25–27 (2008)
3. Wei, L.: Model Analysis of Heavy-duty Truck Frame. Auto Mobile Science and Technology (09), 46–48 (2009)
4. Long, K., Zuo, Z., Feng, H.: Strength Analysis and Improvement of a Heavy-duty Truck Frame. Automobile Technology (04), 24–26 (2008)

Finite Element Analysis of Some Anchor Windlass Based on ABAQUS

Zhuang Hong[1], Li Qiong[2], Zhang Jian[1], and Tang Wen-xian[1]

[1] School of Mechanical Engineering, Jiangsu University of Science and Technology, Zhenjiang, Jiangsu, 212003, China
[2] Zhenjiang Just Marine Equip. Co., Ltd, Zhenjiang, Jiangsu, 212013, China
zhuanghong_work@163.com, zifeimu2006@163.com, zhjian127@163.com, tangwenxian@163.com

Abstract. The reliability of anchor windlass is strongly related to weigh anchor, cast anchor, moor and bollard of ships. In this paper, finite element model of some anchor windlass was presented by using HyperMesh software, which was solved by ABAQUS/Standard. And stress distributions were given under supporting load case. The results showed that the stresses of the anchor windlass were lower than the allowable stress, and finite element analysis of the anchor windlass could provide theoretical guidance for anchor windlass design.

Keywords: Anchor windlass, Supporting load, FEA, ABAQUS.

1 Introduction

As an important part of the ship deck machinery, anchor windlass is used for anchor windlass, anchor, and mooring. The reliability of the anchor windlass directly related to weigh anchor, cast anchor, moor and bollard of ships. During the anchor windlass designing process, many load cases, such as gravity, winding drum load, vice roll tension load, working load and supporting load, should be included. However, supporting load is the biggest among them, which plays a very important role during the windlass working process.

With the rapid development of the numerical theory and computer technology, finite element method has been widely used in machine design and analysis of the windlass. The pipe laying drum shaft anchor was studied by Zhou Zhihong, transition radius and the distance from the step keyway root and grooves on the level of intensity was analyzed. But the drum shaft was just imposed equivalent and the model was too simplified [1]. Fang Zhenxing analyzed windlass base and the strength of the local structure of the hull using ANSYS software. However, other parts of the windlass were not considered [2]. The stress and displacement of some windlass were studied by Wang Hong. And the mechanical performances of the reinforced and un-reinforced deck hull were compared finally. But the real interactions between the components were ignored because of using structural elements [3]. Hu Fucai et al. analyzed strength and stiffness of drum and base of the windlass, the transient temperature field of the brake system, and the stress and failure mechanism of the head screws. Besides, related experiments were present to validate the analyzing results. And they pointed that the

Y. Zhang (Ed.): Future Communication, Computing, Control and Management, LNEE 141, pp. 343–347.
springerlink.com

traditional design method was very conservative [4-6]. Recently, the chain wheel gear box was investigated by FEA, which was compared with stress experiment. And the thickness of the gear box cover, reinforcement plates and other support plate was reduced reasonably [7]. However, research about mechanical performance of the windlass under supporting load condition was rarely found.

Therefore, taking some windlass as example, three-dimensional model of the windlass was proposed firstly by commercial software Pro/E. Then the FE model was built by HyperMesh software and translated into input file, which was solved by ABAQUS/Standard. Finally, mechanical performance was analyzed under supporting load condition.

2 Structure and Principle Analysis for Windlass

The length, width, and height of the anchor windlass are 8.6m, 1.2m and 1.6m, respectively. As shown in the Fig.1, the windlass mainly includes hydraulic motor (1), main spindle (2), transmission gears(3), chain wheel (4), winding drums (5,6), vice drum (7), windlass base (8), bracket (9,12,15,18), steel box (10,11,13,14,16,17) and other components (such as the bracket cover, bolts, etc). The clutch and the main spindle are made of 45-steel shaft, brake material is made of 16-Mn, the transmission gears and gear shafts are made of 40Cr, the steel box, wind drums, and the brackets are made of Q235. The diameter of the chain wheel is 44mm, and the supporting load is 691 KN.

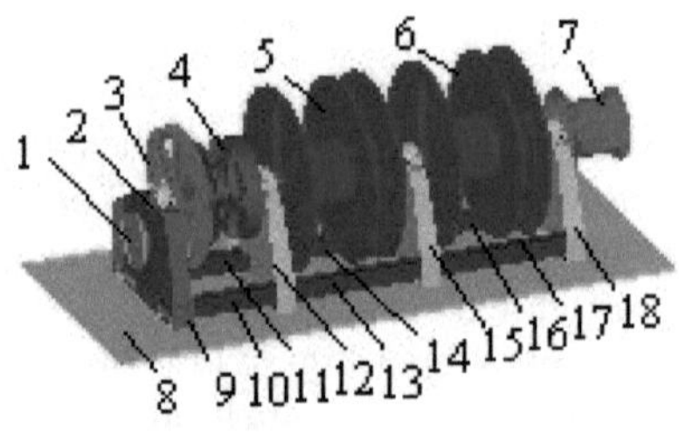

Fig. 1. Schematic diagram of the windlass

The working principle of anchor windlass is shown in Fig.1. The main spindle was driven by hydraulic motor. Then the motion was translated into other components by the corresponding clutch. In this way, weigh anchor, controlling cable could be carried out.

3 FE Model of the Windlass

There-dimensional CAD model was built based on two-dimensional drawings provided by cooperators (Fig.2), firstly. Small features, such as small chamfer, fillet and weld seam, were ignored in the model, so did the effects of the brake. An equivalent disk plate was used take place of hydraulic motors. They had the same weight and centroid. FE model of the windlass was proposed based on HyperMesh software. Because of the complexity of the chain wheel, the structure was meshed with tetrahedral elements. The other components were mesh by many brick elements with compatible mode (C3D8I) and a few of wedge elements (C3D6). In the thickness direction of the brackets, base plates and ribs, at least two grid layers was used, to prevent shear locking phenomenon. The mesh consists of 280959 elements and 375531 nodes.

As the relative displacement between the bolt and its contacting component was very small, the interactions between them were defined by using small sliding surface to surface contact mode. The lower surfaces of the brackets were fixed. And the gravity of the model was added at 9.8N/m2.

Supporting load (691KN) was also considered in the analysis. In the supporting load case, the clutch was pulled out. So the supporting load was only needed to define in the chain wheel with a coupling constrain (as shown in Fig.3). The finite element model of the windlass was translated into the input file. Finally, ABAQUS/Standard solver was used to solve the stiffness equation.

Fig. 2. Finite element model of the windlass

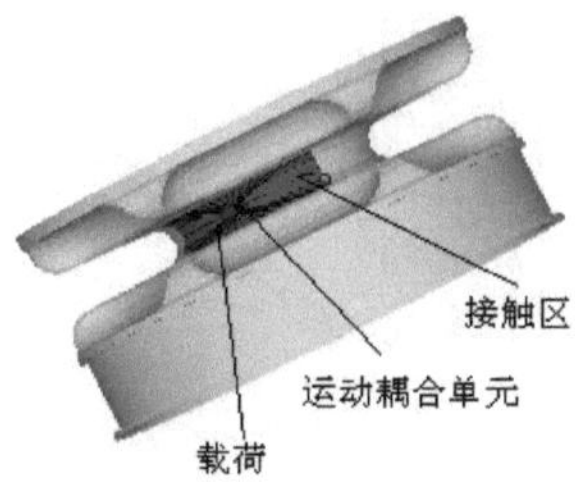

Fig. 3. The maximum load chain wheel

4 Results and Discussion

As shown in the Fig.4, the anchor unit in support of the load distribution of the overall equivalent stress, the stress of spindle 2#, 9# and 12# bracket and its cover, 10#, 11#, 13# and 14# strengthen plate is bigger, but other parts is less, so these key components are analyzed importantly.

Fig. 4. The overall stress cloud windlass

5 Axis Analysis

As shown in the Fig.5, the stress of the spindle is biggest, the size is 290MPa; because it is contacting with the chain wheel and close to 12# bracket Fig5 (b) is the maximum of the point through with spindle axis parallel to the direction of the stress curve, it can be seen that spindle peak stress is at the connection of various of bracket and chain wheel, the stress at connection with the chain wheel is bigger, and other locations are small(less than 100MPa). At the appearance of two peak stress in the vicinity of the chain wheel, Section stress axis cloud as shown in Fig.5 (a).As shown in the Figure, the stress contours is distributed into an oval, elliptical axis and chain wheel load was vertical, short axis is parallel with the chain wheel load, the stress on the outer surface of spindle is bigger, but internal stress is less. The stress on the axis total is lower than its allowable stress (355MPa), so the spindle meets design requirements under the load strength.

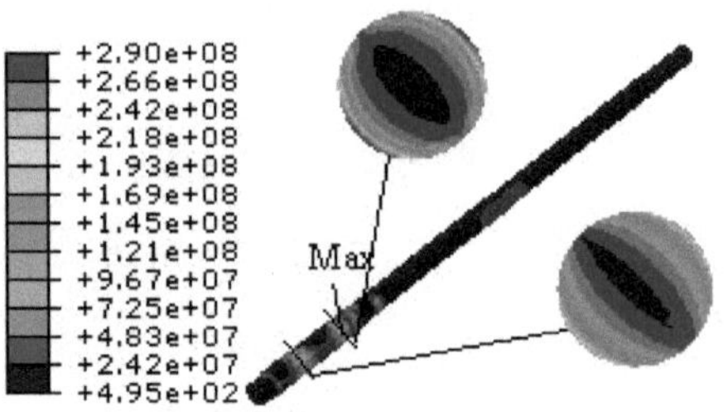

(a) Stress contour (Pa)

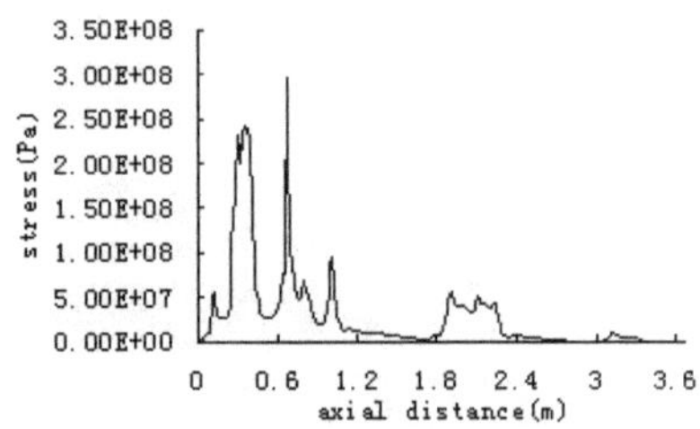

(b) Axial stress

Fig. 5. Spindle stress distribution

6 Bracket Analysis

As shown in the Fig.6, the stress distribution of 9# and 12# stand when windlass works in support of load, the maximum stress of 9# and 12# stand was 126MPa and 148.7MPa, symmetrical distributing on the side of Intermediate bracket close to the chain wheel. In addition, stress concentration is on the upper left of bracket 9#, maximum 118.41487MPa. Support the overall stress is lower than its allowable stress (157MPa) [8], it meets design requirements in support of the work load support strength.

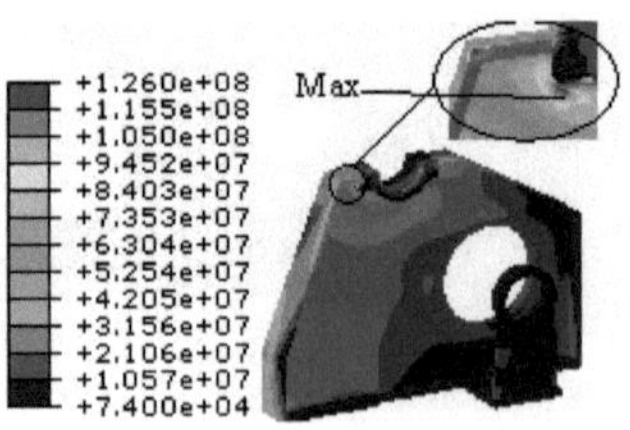

9# stand

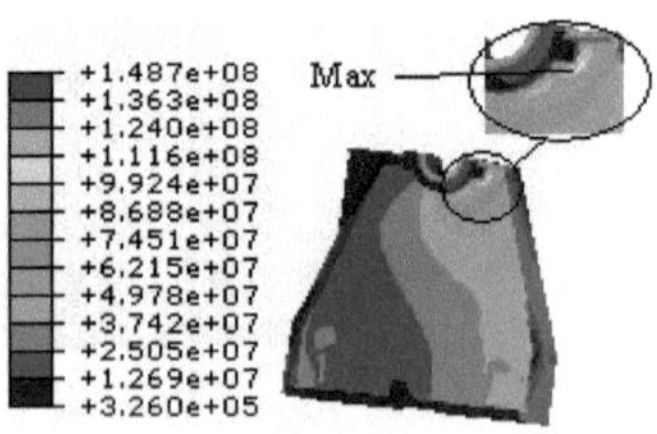

12# stand

Fig. 6. Stress contour support (Pa)

7 Analysis of Other Parts

As shown in Table1, Maximum stress of the equivalent stress of windlass bracket cover and strengthen the board, the maximum stress of 10#, 11#, 13# and 14# strength plate was in connect with the 12# stand.

It can be seen that these parts meet the design requirements under the support loads.

Table 1. The maximum stress of machine parts anchor (MPa)

Object	Maximum stress	Allowable stress
9# Bracket cover	59.9	157
12# Bracket cover	115.4	157
10# Stiffener	59.2	157
11# Stiffener	30.0	157
13# Stiffener	75.3	157
14# Stiffener	41.6	157

8 Conclusion

In this paper, the three-dimensional CAD model was built by Pro/E and finite element model was built by HyperMesh, and ABAQUS/Standard solver was used to solve the model calculation, the force was researched in the situation of windlass working under the support load. As shown in the results, the strength of the anchor windlass were in the range of allowable values under the support load, the finite element analysis windlass can provide theoretical guidance for the design.

References

1. Zhou, Z.H., Zheng, X.P., Fan, C.: Petroleum Machinery 30(7), 24–26 (2002) (in Chinese)
2. Fang, Z.X., Zhang, L.D.: Guang Dong Shipbuilding 3(1), 40–42 (2008) (in Chinese)
3. Wang, H., Liu, A.L.: Jiangsu University (Natural Science) 22(3), 1–4 (2008) (in Chinese)
4. Hu, F.C., Zhou, Y., Xiang, Y., Yang, J.G.: Ship & Ocean Engineering 36(2), 53–56 (2007) (in Chinese)
5. Hu, F.C., Zhou, Y., Xiang, Y., Yang, J.G.: Ship & Ocean Engineering 29(4), 9–12 (2007) (in Chinese)
6. Hu, F.C., Zhou, Y., Xiang, Y., Yang, J.G.: Hoisting and Conveying Machinery (25), 51–54 (2007) (in Chinese)
7. Guo, H., Xiang, Y., Hu, F.C., Zhou, Y.: Wuhan University of Technology (Transportation Science & Engineering). 34(4), 827–830 (2010) (in Chinese)
8. China Classification Society. steel ship classification rules. China Communications Press (2006)

FCM-Based Model for Network Security Assessment

Shihui Cheng[1,2]

[1] Wuhan University of Technology, Wuhan, China
[2] Henan Institute of Education, Zhengzhou, China
cshgjl@sina.com

Abstract. A FCM-based model for network security assessment was proposed. According to classified security criteria of information system, factors, criteria, and metrics for network security evaluation were designed. By building hierarchical assessment models and setting reasonable weights, networks' security situation was calculated qualitatively and quantitatively. Theoretical analysis and validation tests show that the model is feasible, and it has objective, comprehensive, and practical features. Thus, it provides a good solution to security evaluation of computer information systems.

Keywords: Computer network, security assessment, FCM.

1 Introduction

The assessment of computer network security is not a simple matter [1, 2, 3]. It requires a detailed review and analysis of the lowest-level implementation characteristics of the selected technology together with the ability to map significant behaviors and failures up to their architectural relevance. Generally, network security assessment tools offer efficient testing models that people can adopt, refine, and reuse to create proactive defensive strategies to protect their computer and information systems from the threats that are out there, as well as those still being developed. One of the most important tasks of network security assessment systems is to find the weaknesses and vulnerabilities of computer information systems from physical security, administrative security, data security, and technology security in accordance with the classified criteria [4, 5, 6].

The existing developed network security assessment methods and tools have this or that defects. Nessus is a great tool designed to automate the testing and discovery of known security problems, but it does not take the security level and safety management strategies of the target into account [7]. The Common Criteria for Information Technology Security Evaluation (CC) is a technical basis for an international agreement, and the CC is the driving force for the widest available mutual recognition of secure IT products. However, it does not have the possibility of direct action [8, 9].

This paper proposed a FCM-based model of network security assessment, called FNSAM. The remaining of the paper is organized as follows. Section 2 discusses the model principles. In Section 3, theoretical analysis and validation are provided. Section 4 contains the conclusion and future work.

Y. Zhang (Ed.): Future Communication, Computing, Control and Management, LNEE 141, pp. 349–355.
springerlink.com

2 Model Principles

Factor-Criteria-Metrics (FCM) model is a common level of software quality metrics, and it is usually constructed in a tree-like fashion. Within FCM, each factor is composed of lower-level criteria, the criteria are easier to understand and measure than the factors and the actual measures (metrics) are proposed for the criteria. The computer network security assessment and FCM model have a certain similarity, and the Correspondence between FCM and FNSAM is illustrated in Table 1.

Table 1. Correspondence between FCM model and FNSAM

FCM Model	FNSAM
Factors	Assessment factors (including physical security, administrative security, data security, and technology security)
Criteria	Assessment criteria. Such as the detection of weak passwords within the factor of technology security
Metrics	Metrics for network security assessment. Such as the number of user accounts whose password length is less than 6

2.1 Model Architecture

Similar with FCM, the tree-like architecture of the proposed model is illustrated in Fig. 1.

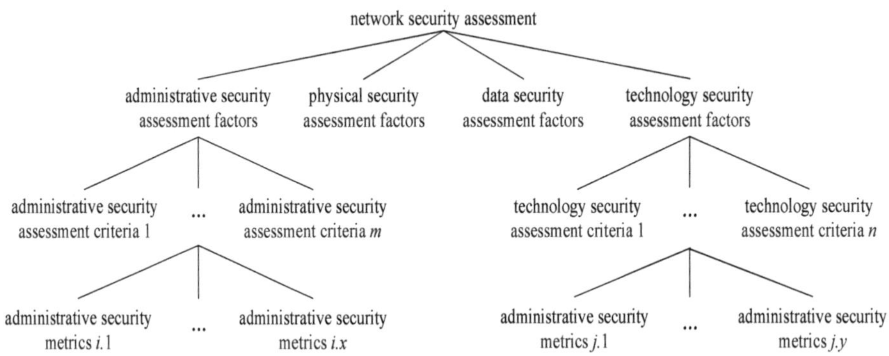

Fig. 1. Hierarchy of FNSAM

Fig. 1 describes the pertinent relationships between network security assessment factors and their dependent criteria. So, the assessment factors can be measured in terms of the dependent criteria. In practice, assessment factors selecting should take into account the factors detailed below:

Meeting the requirements of the target network security level
Considering management and physical security

Taking data authenticity, confidentiality, and integrity into account
Technical security, such as authentication, access control, audit, and etc.

In FNSAM, the term "assessment criteria" means the standards on which a judgment can be made regarding the suitability of a network, given the characteristics or features of the network that may have impact on information system safety. Being the different security level of e-governments, e-military affairs, e-commerce, and the campus networks, the appropriate, detailed, and comprehensive assessment criteria are particularly important to the assessment result. Additionally, the network metrics that an organization uses for evaluating network security have a big impact not only on the networks that get chosen but also on the networks that get proposed. Within this paper, the measurements are obtained through interviews and penetration testing.

In order to describe the implementation method of FNSAM conveniently, an instance of E-government outward network is illustrated in Table 2.

Table 2. Factor-Criteria-Metrics instance of E-government outward network

Security level	Assessment factors	Assessment criteria	Metrics
The second level (level of Protection guidance)	User authentication of technology security	uniqueuser ID	Whether user ID is unique or not
		User identification	Whether it can correctly handle the user login information or not
			Whether it can identify the legitimate user's identity or not
		Weak passwords	The account number of initial password
			The account number of password-length<6
			The account number of password does not meet complexity requirements
		ability of anti- counterfeiting	Whether the network can detect false user login information or not
			Whether the network can prevent error login actions or not
			Whether the network can detect copied user login information
			Whether the network can prevent the login action using copied login information or not
		error processing ability for user login failed	Whether the network limits the user login attempts or not
			Whether the times threshold of login attempt are reasonable or not
			Whether the network can lock the account while user attempt login times are more than the specified threshold

2.2 Model Theory

From the instance illustrated in Table 2, we know that there are two types of measurement values, that is the logical type and numeric type. Through setting thresholds, the numerical measurement results can be turned into logical type easily. Within FNSAM, the author assumes that the logical truth as 1, and logical false value as 0.

In the proposed model, in order to achieve quantitative evaluation, let N denote the set of natural numbers, F represent the set of all the assessment factors of FNSAM, and $|F| = n (n \in N)$. The formal definition of the i th $(1 \le i \le n)$ assessment factor (f_i) is described as follows.

$$f_i = \langle v_{f_i}, w_{f_i} \rangle \tag{1}$$

where v_{f_i} and w_{f_i} represent the measurement result and weight of f_i.

From Eq. 1 and the definition of F, the formal definition of F is illustrated in Eq. 2.

$$F = \{ f_i \mid f_i = \langle v_{f_i}, w_{f_i} \rangle, \sum_{i=1}^{n} w_{f_i} = 1, i = 1, 2, \cdots, n \} \tag{2}$$

We suppose that the assessment criteria number of f_i is $n'(n' \in N)$, C_{f_i} and C represent the assessment criteria set of f_i and all the assessment criteria of FNSAM. The formal definitions of C_{f_i} and C are given as below.

$$C_{f_i} = \{ c_{f_{ij}} \mid c_{f_{ij}} = \langle v_{f_{ij}}, w_{f_{ij}} \rangle, \sum_{j=1}^{n'} w_{f_{ij}} = 1, i \in \{1, \cdots, n\} \wedge j = 1, 2, \cdots, n' \} \tag{3}$$

$$C = \bigcup_{i=1}^{n} C_{f_i} \wedge C_{f_j} \bigcap C_{f_k} = \Phi, j \neq k \wedge j, k \in \{1, \cdots, n\} \tag{4}$$

where $v_{f_{ij}}$ and $w_{f_{ij}}$ denote the measurement result and weight according to the j th $(1 \le j \le n')$ assessment criterion of $f_i (1 \le i \le n)$, respectively.

Let $n''(n'' \in N)$ represent the metrics number of $c_{f_{ij}}$, $M_{c_{f_{ij}}}$ and M represent the assessment metrics set of $c_{f_{ij}}$ and all the metrics of FNSAM. The formal definitions of $M_{c_{f_{ij}}}$ and M are described in Eq. 5 and Eq. 6, respectively.

$$M_{f_{ij}} = \{ m_{f_{ijk}} \mid m_{f_{ijk}} = \langle v_{f_{ijk}}, w_{f_{ijk}} \rangle, \sum_{k=1}^{n''} w_{f_{ijk}} = 1, i \in \{1, \cdots, n\} \wedge j \in \{1, \cdots, n'\} \wedge k = 1, 2, \cdots, n'' \} \tag{5}$$

$$M = \bigcup_{j=1}^{n'} M_{f_{ij}} \wedge M_{f_{il}} \bigcap M_{f_{im}} = \Phi, i = 1, 2, \cdots, n \wedge l \neq m \wedge l, m \in \{1, 2, \cdots, n'\} \tag{6}$$

where $v_{f_{ijk}}$ and $w_{f_{ijk}}$ represent the measurement result and weight of $m_{f_{ijk}}$, respectively.

From the above theory we know that $v_{f_{ijk}}$ can be obtained through interviews and penetration testing. At the same time, the weight of w_{f_i}, $w_{f_{ij}}$ and $w_{f_{ijk}}$ can be gotten by expert scoring, level of analysis, factoring, and other methods. Moreover, the calculation methods of $v_{f_{ij}}$ and v_{f_i} are illustrated in Eq. 7 and Eq. 8.

$$v_{f_{ij}} = \sum_{k=1}^{n''} v_{f_{ijk}} \times w_{f_{ijk}} \tag{7}$$

$$v_{f_i} = \sum_{j=1}^{n'} v_{f_{ij}} \times w_{f_{ij}} \tag{8}$$

In conclusion, the overall assessment result of the target network is calculated by Eq. 9.

$$v = \sum_{i=1}^{n} v_{f_i} \times w_{f_i} \tag{9}$$

Within FNSAM, the network security situation (nss) is divided into five cases, and it is illustrated in Eq. 10.

$$nss = \begin{cases} very \quad safe & v \in [0.8, 1.0) \\ safer & v \in [0.6, 0.8) \\ safe & v \in [0.4, 0.6) \\ unsafe & v \in [0.2, 0.4) \\ very \quad unsafe & v \in [0.0, 0.2) \end{cases} \tag{10}$$

3 Theoretical Analysis and Verification Tests

3.1 Theoretical Analysis

In theory, FNSAM is feasible. The correctness and feasibility of FNSAM is specifically analyzed as follows.

Firstly, the theoretical basis for the presented model is correct. The assessment factors of FNSAM include physical security, administrative security, data security, and technology security. So, the model is in line with the requirements of CC, GB 17859, GB/T 18336, GB/T 20984, ant etc.

Secondly, the model is easy to design and implement. From Fig. 1 and Table 2, we know that the architecture of FNSAM is hierarchical, the measurement criteria are detailed, and all the metrics can be measured quantitatively.

Finally, the universality of the model is good. The model metrics and their weights can be appropriately adjusted according to business needs. Therefore, FNSAM can be used for e-governments, e-military affairs, e-commerce, and etc.

3.2 Verification Tests

To verify the validity of FNSAM, we tested it in the Network Security Assessment System of FoundLight. Within the testing experiments, the assessment factors and criteria was designed as in Table 3 (Because of space limitations, the metrics were not listed).

Table 3. Assessment factors and criterion of the testing experiments

Assessment factors	Elements of assessment criteria	
physical security	✧ Physical location of choice ✧ Physical Access Control ✧ Anti-theft ✧ Anti-Tampering ✧ Lightning	✧ Fireproof ✧ Waterproof ✧ Moisture proof ✧ Anti-static ✧ Electromagnetic protection
Administrative security	✧ Post design of networks administration ✧ Staffing of networks administration ✧ Authorization and approval of managers ✧ Communication and cooperation ✧ Audit and Inspection ✧ Safety management system and its development, release, review and revision	✧ Management of personnel recruitment, education, training, and leaving the post ✧ Operation and maintenance management of network environment, assets, media, equipments, monitoring, security, malicious code protection, passwords, changes, and emergency plans ✧ Security incident handling
data security	✧ Data Integrity ✧ Data confidentiality	✧ Data backup ✧ Data recovery
technology security	✧ Architecture, network segment, access control, security audit, border integrity check, intrusion and malicious code prevention, equipments protection of networks	✧ Identification, autonomy and mandatory access control, security audit, system protection, residual information protection, intrusion prevention, malicious code protection, and resources control of hosts ✧ Authentication, access control, security audit, residual information protection, integrity and confidentiality of communications, non-repudiation, software fault tolerance, resources control, and code security of application systems

Testing experiments were carried out in E-government extranets, the assessment factors, criterion, and metrics were chosen according to the classification criteria for E-government information security. In our experiments, the weights were gotten by expert scoring. Experiment results show that the proposed model is feasible, and it has objective, comprehensive, and practical features.

4 Conclusion and Future Work

The FCM-based model for network security assessment can evaluate the security of computer networks and information systems from physical security, administrative security, data security, and technology security. Moreover, the proposed model is in line with the requirements of CC, GB/T 20984-2007 [10], GB/T 18336-2008 [11], and GB/T 25058-2010 [12]. The model can implement qualitative and quantitative security assessment for computer networks and information systems. Theoretical analysis and validation tests show that the model is feasible, and it has objective, comprehensive, and practical features. Thus, it provides a good solution to security evaluation of computer information systems.

The value of the metric threshold, and the assessment factors' weight should be tested more in future. At the same time, a large number of penetration testing tools should be explored and investigated in the process of computer network security assessment.

References

1. Sun, F.X.: Journal of Networks 6, 255–272 (2011)
2. Sun, F.: Gene-Certificate Based Model for User Authentication and Access Control. In: Wang, F.L., Gong, Z., Luo, X., Lei, J. (eds.) Web Information Systems and Mining. LNCS, vol. 6318, pp. 228–235. Springer, Heidelberg (2010)
3. Klarreich, E.: Nature 415, 468–470 (2002)
4. Sun, F.: Advanced Materials Research 187, 148–154 (2011)
5. Sun, F., Wu, Z.: Advanced Materials Research 182, 66–71 (2011)
6. Sun, F.: Advanced Materials Research 179-180, 1333–1337 (2011)
7. Information on, `http://www.nessus.swri.org/`
8. Information on, `http://www.commoncriteriaportal.org/`
9. Li, T.: Science in China. Seri. F: Information Science 51, 1475–1486 (2008)
10. Information on, `http://www.zhunbiao.com/guojiabiaozhun/GBT_20984-2007.html`
11. Information on, `http://www.shtec.gov.cn/art/2011/3/23/art_234_1446.html`
12. Information on, `http://www.standardcn.com`

Energy Regeneration System for Hybrid Electric Vehicle Based on CAN Bus

Li Ran[1], Wu Junfeng[1], and Li Gechen[1,2]

[1] Dept. of Automation, Harbin University of Science & Technology, Harbin, China
zeemoolr@126.com, wu_jf@hrbust.edu.cn
[2] CENS Energy-Tech Co. Ltd. Hangzhou, China
ligechen@126.com

Abstract. It has a strong request to real-time communication and sharing the information between each subsystem of hybrid electric vehicle (HEV). The mainstream of constructing network of hybrid electric vehicle applies the CAN Bus system. It follows the standard SAE J1939 protocol, uses the characteristics of the motor and inverter. In order to achieving superior braking, the total braking moment is divided into energy regeneration braking, mechanical breaking and ABS breaking, according to the actual situation, which are divided into different proportion. Finally, in the road test on the actual data collection, verification of the hybrid electric vehicle CAN bus control network hardware and software modules designed with the academic value and practical application.

Keywords: HEV, SEA 1939 Agreement, CAN Bus, Energy Regeneration.

1 Introduction

HEV uses fuel and electricity driver way, however, how to improve energy efficiency and extend the renewal drove mileage, become a key problem that is solved urgently. Energy regeneration is the main measure to solve this problem, including braking energy regeneration and taxiing energy regeneration. This article focuses on mixed braking energy regeneration system based on CAN bus [1]-[3].

As the CAN bus communication applies in HEV, just using two roots of signal line can achieve sharing among speed, the engine speed, acceleration, the brake pedal, the accelerator pedal and gear signal as well as communication among the various control systems. This communication is designed to provide transport harness simple and reliable, saving copper, and easy maintenance.

It is necessary to achieve safe and efficient feedback deceleration, but also with the traditional friction brake work. Considering the bus running, brake pedal position, battery SOC, the battery charging power, electrical power generation capacity and other factors, HEV bus could save battery power and reduce the mechanical brake wear [4].

Y. Zhang (Ed.): Future Communication, Computing, Control and Management, LNEE 141, pp. 357–364.
springerlink.com

2 The Vehicle System Structure

With the requirements of environment protection and energy security continues to improve, and gradually increased vehicle facilities, original point-to-point control mode in electric control system will cause huge wiring systems, aggravated contradiction between the bulky constraints and car limited space. Meanwhile the anti-jamming of wires is poorer, especially in the bus starts, brake, high-frequency power supply charge to transfer data in high noise environments, there will be instability in signal transmission, or even error.

The system is composed of eight node, including: power train controller, motor controller, charge machine control system, AMT, ABS, EPS, battery management system, combination instrument system. Bus communication rate is 250Kb/s. The CAN bus network message structure is saw in Table 1, among them, the priority is three bit, can have eight priorities; R is general fixed as 0; DP is fixed to 0.

Table 1. Allocation Table of Identifier

IDENTIFIER 11BITS											SRR	IDE	IDENTIFIER EXTENSION 18BITS																	
PRIORITY			R	DP	*PDU FORMAT(PF)*						*SRR*	*IDE*	*PF*		*PDU SPECIFI(PS)*								*SOURCE ADDRESS(SA)*							
3	2	1	1	1	8	7	6	5	4	3			2	1	8	7	6	5	4	3	2	1	8	7	6	5	4	3	2	1
28	27	26	25	24	23	22	21	20	19	18			17	16	15	14	13	12	11	10	9	8	7	6	5	4	3	2	1	0

The system block diagram and node SA address are shown in Figure 1, among them, 208~231 is pre-setting address for highway transport [5]-[6].

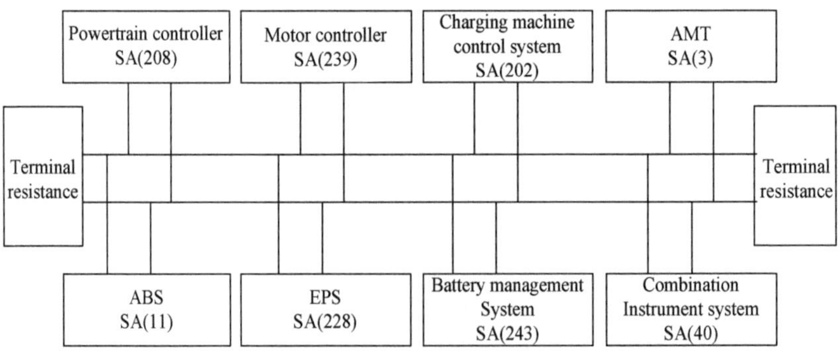

Fig. 1. System structure diagram

3 Braking Energy Regeneration System

This system will introduce the CAN bus into the energy regeneration system of HEV bus, and use the work characteristics of motor and inverter to achieve energy regeneration mode that combined ABS mechanical braking with motor regenerative braking [7].

Mechanical braking produces braking torque through the brake pressure system driving sluice piece liquid in the way of rubbing brake piece. ABS braking when the vehicle deceleration reached the ABS threshold node, when activated, ABS system is activated and independent control of the vehicle's braking process. The motor regenerative braking is divided into high speed braking and low speed braking. High-speed brake is called as the motor speed is exceeded to the ideal unloaded speed when bus is downhill, and then motor control node controls motor power; Low speed brake is that changing mechanical energy into electrical energy at the braking stage, and producing braking torque, with energy stored in the battery pack.

In work condition of braking energy regeneration, keeping feel of brake pedal and driving flexibility is the first consideration, therefore, according to the operation condition, it divides into motor energy regeneration and rear wheel mechanical braking, electrical energy regeneration and mechanical braking of front wheel and rear wheel braking together , ABS antilock braking.

System uses the LiFePO4 power battery, which has high rate charging characteristics of power, and can improve the proportion of regenerative braking. HEV bus is positioned as city buses, which has frequent braking action in bus routes, but only a small part needs very big braking torque. When the power train node detects the braking signal, it shares information through CAN network with motor control node and battery management system node, to determine the proportion of the energy feedback braking and rear mechanical braking torque; Power train node detects the speed, if it failed to meet the requirements, the front mechanical braking launched; At high speed, front wheel and rear wheel all braking may occur holding in death, vehicle sliding, bus is not easy to control, and then through the ABS node complete regenerative braking.

4 System Hardware Design

Motor speed, speed, acceleration, the brake pedal position in system are detected by motor control node and power train node, and make the running information shared through a CAN bus.

4.1 Related Detection Signal Design

The motor speed is measured by pulse coder. Motor every turning week produces 1,024 A, B rout square wave signals whose phase difference is 90 °, after 74HC14 Schmitt toggle reshaping, connecting the QEP unit of TMS320LF2407, confirm the speed and direction of rotation. The design uses MEMS accelerometer to measure acceleration, and through the I2C way send into power train node μPDF70F3239 controller.

The brake signal is detected by the valve opening sensor installed in the brake master cylinder. Valve opening sensor has a linear relationship with pedal position, and sends the voltage signal into the μPDF70F3239, 12 A/D converters, in the power train node controller.

4.2 CAN Network Design

This core of the system is the CAN network design. The design takes the motor controller node as an example, and in detail introduces the hardware design of network interfaces.

In the motor controller node, the design chooses TMS320LF2407 which TI launched with a chip CAN port. Its core uses the Harvard architecture, with dedicated hardware multiplier. It uses its programmable with dead zone PWM to realize motor vector control. The design chooses low power consumption SN65HVD230 as the drive chip between controller and the physical bus, to improve difference sending ability of bus and the difference receiving ability of CAN controller in severe environment, such as over-voltage, high temperature, and high total membrane.

Enhance the anti-interference ability of the motor controller, TMS320LF2407 connected with SN65HVD230 through high-speed light-coupler HCPL600, realized the electrical isolation between each CAN node on the bus, both sides of light-coupler achieves independent power supply, and a fully isolation[8].

Schematic diagram is shown in Figure 2. The SN65HVD230 CANH and CANL pins each with a 5Ω resistor connected with the CAN bus, play a role in the current limiting, and avoid suffer from over-current impact. CANL and CANH parallel the two 30pF small capacitances with GND, to filter out high-frequency interference on the bus and certain electromagnetic radiation ability. CANL and CANH respectively connects reverse a protect diode with the ground, when the can bus has a higher complex voltage, can have certain over-voltage protection function.

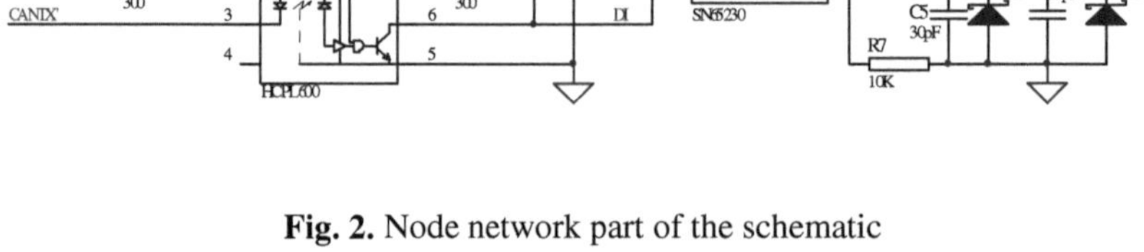

Fig. 2. Node network part of the schematic

4.3 Powertrain Node Design

The center handler of system takes the 32-bit processorμPDF70F3239 of car rank, with 4 CAN mouths, is the hub of control system -gateway. Through CAN communication and direct way receive from the different ECU signals and according to preset control strategy to send control signal, to accomplish the switch and switching process coordinated control in different working conditions of HEV bus.

The ABS controller of system used 32-bit controller TMS320F28016 with the CAN interface. Responsible for driving braking part pump, carry out mechanical braking and ABS braking and can complete itself fault diagnosis.

The maximum oscillator tolerance of oscillator of CAN controller is 1.58%. CAN bit rate is divided into four parts: the synchronization section, communication section, phase segment 1 and phase segment 2. Synchronization section uses for synchronizing different nodes on the bus, which was defined as a time range. Communication is used to compensate physical delay time within the network segment. The phase segment is used to compensate for the error of edge phase [9]-[10].

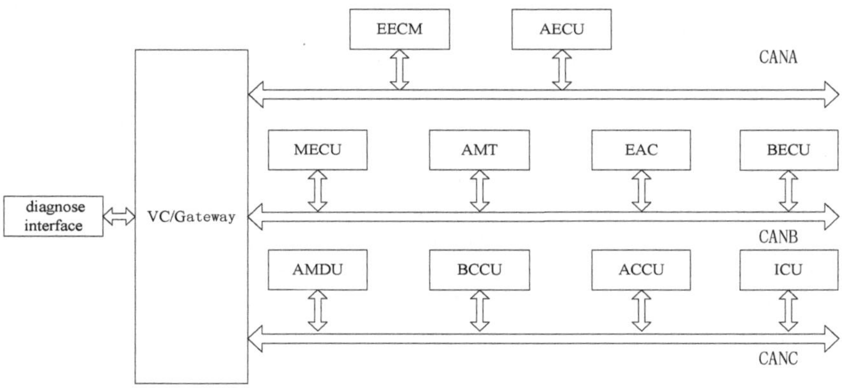

Fig. 3. CAN-Bus topography of electric vehicle

CAN bus node design work: the CAN bus system of electric transit, including the master node, the node of lights, air conditioning node, the node of door, instrument node and so on. CAN bus electric bus topology is shown in Figure 3. Master node is the hub of the entire vehicle control system, which is responsible for all data communication with other nodes and for message processing, and accept input from the driver's switch. The child node will detect a variety of content such as digital and analog, package deal with all the necessary information into the packets that a message sent to the master node. According to the information contained in the message, the master node corresponding to process, then packaged into a message sent to the child node implementation. CAN bus network nodes and part description of the function ID information distribution are shown in Table 2.

Table 2. The node function and ID information table

Num.	CAN node name	CAN ID address
1	Gear symbols	0x18FFCE00 Byte 1~6
2	Front and rear fog lamp switch symbol	0x18FECA00 Byte 1~4
3	Left and right turn signal sign	0x18FECA00 Byte 5~8
4	Oil warning light symbols	0x18FEEE00 Byte 1
5	Fuel gauge	0x18FEEE00 Byte 5、6
6	Water temperature gauge	0x18FEEE00 Byte 1、2
7	Oil pressure gauge	0x18FEEF00 Byte 3、4
8	Speedometer	0x18FF83D0 Byte 1、2

5 Program Architecture Design of Energy Braking Regeneration System

The system is a large system composed of multiple controllers in Figure 4, in order to strengthen the portability of program and multiple embedded engineers work together, take module structure. The whole framework contains the main program, the analog quantity sampling subroutine, the digital quantity sampling subroutine, data sending subroutine, data accept subroutine, node self-inspection program, interrupt subroutine module and so on.

The car speed signal obtains by motor speed signal. The motor controller timer T0 timing packs the speed data frames to the bus. Node self-inspection program is composed of two components that are automatic detection on electricity and automatic detection on the timer timing, to ensure all the unit work normally. For the real time and reliability of program, each node embedded real-time operating system μC/OS-II.

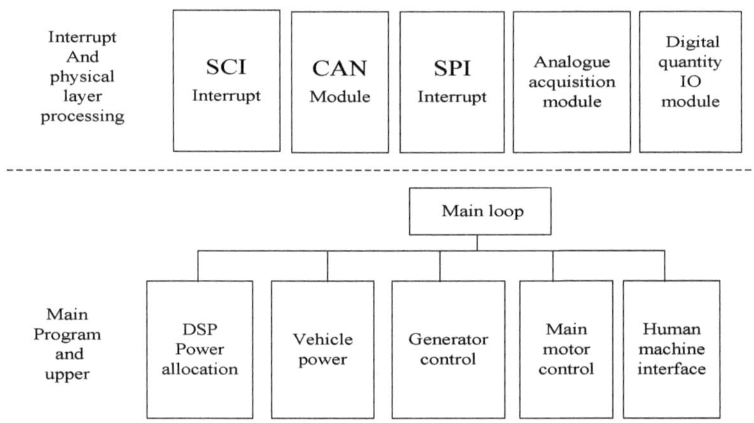

Fig. 4. System software design process

6 Design of Verification

Vehicle braking energy feedback system software and hardware design is completed, the stability of the system functions, reliability and consistency of data transmission verification. After the system is powered on, select the electric bus passenger car on a battery pack module that records every 1 second module, a battery voltage information, the cumulative record of the electric buses in operation in about 6000 sets of data, data from the CAN Bus receive buffer. And collecting the results recorded as follows in Table 3. At the same time using CAN bus analyzer to synchronize data on the bus recorded.

Table 3. System structure diagramReal-time data of the single package of the battery voltage (Unit V)

Voltage / Time	1-20	21-40	41-60	… …	5081-6000
1	26.403	26.449	26.388	…	25.197
2	26.417	26.410	26.437	…	25.141
3	26.413	26.413	26.417	…	25.082
4	26.429	26.421	26.388	…	25.029
5	26.437	26.410	26.370	…	24.974
6	26.429	26.381	26.350	…	25.022
7	26.429	26.350	26.343	…	25.053
8	26.478	26.347	26.339	…	25.118
9	26.512	26.335	26.343	…	25.154
10	26.502	26.323	26.339	…	25.193

Bus analyzer to collect the results and CAN bus receive data buffer that after comparing data synchronization, real-time from the vehicle CAN node sends a large number of packets of information accuracy, stability, and no abnormal communication, such as the number of lost, missing Frame, the number of such strings. The stability of system functions, reliability and consistency of data transmission are in line with requirements. The compare of the data between reality and receiving is shown in Figure 5.

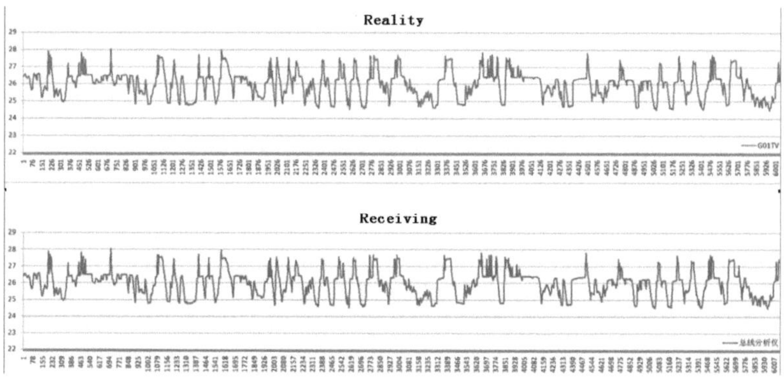

Fig. 5. The compare of the data between reality and receiving

7 Conclusion

This system develops on the basis of the braking system in the bus. In HEV bus operational phase, data transmission is reliable and stable. Statistics data of actual operation do show that the system can make the sequel mileage of the city HEV bus increase by 20%. The efficiency of the energy regeneration optimal is a dynamic optimization process, the algorithm is very complex, and can not conclude the one-plan-fits-all empirical formulas by off-line, so establishing a model refer to adaptive system in the power train node, to the important parameters for online identification.

References

1. Song, Q., Wu, L.: Design of CAN-Bus Intelligent Node. Automation Technology and Application, 9–11 (2008)
2. Chen, X.: The Application in the large-and-middle-scale passenger train of CAN bus. Electronic Technology, 52–63 (2007)
3. Yao, S., Wei, J.: The Application of SAE J1939 Protocol in the HEV Communication. Journal of Hubei Automotive Industries Institute, 19–22 (2003)
4. Liu, H., Song, L.: Data Collection and Display System Based on CAN-bus of Rear Engine of Coach. Vehicle & Power Technology, 37–41 (2002)
5. Philips Semiconductors, C.A.N.: Specification Version 2.0, Parts A and B (1992)
6. Guo, X., Zhang, J.: Modeling and Simulation of a Regenerative Braking System in HEV. Journal of Wuhan Automotive Polytechnic University, 116–120 (2005)
7. Xue, D.: Realization of CAN bus communication module based on SJA1000. Automation & Instrumentation, 54–56 (2008)
8. Cheng, S., Tu, T.: The Simulated Studies of the Hybrid Electric Vehicle's Energy Matching, 73–75 (2001)
9. Chou, Q., Wu, Y.: Base on the example to study power flow control strategy for HEV, 69–71 (2009)
10. Zhang, Y., Yang, L.: The Control Strategy of Energy Regeneration for Electric Vehicle. Automotive Engineering, 25–27 (2005)

The Probabilistic Approach to Modification of Seismic Linear Response Spectra

Cada Zdeneka, Kala Jirib, Salajka Vlastislavc, and Kanicky Viktord

Brno University of Technology, Faculty of Civil Engineering, Veveri 95, Brno, Czech Republic
{cada.z,kala.j,salajka.v,kanicky.v}@fce.vutbr.cz

Abstract. The paper deals with the seismic linear response spectrum modification based on a probabilistic approach. The modification improves the correctness of the deterministic seismic analysis of building structures using the response spectrum method. In that case, the structure response depends strongly on computed natural frequencies. The proposed response spectrum modification accounts for uncertainties in response due to uncertainties in frequencies caused by variances of computation model parameters. The response spectrum is modified so, that with a selected probability the given spectral values should not be exceeded. The variance of natural frequency values is obtained by modal analysis of a stochastic model of the structure.

Keywords: Seismic analysis, seismic response spectrum, deterministic analysis, response uncertainty, stochastic model, modified response spectrum.

1 Introduction

The dynamic behaviour of building structures subjected to seismic action is commonly analyzed using the deterministic approach. The computation model is designed using the finite element method. The time domain response to defined non-stationary ground motion can be computed using direct integration of structure equations of motion. For many reasons, procedures based on structure vibration mode superposition are preferred. The response spectrum method is very frequently applied [1, 2, 3]. The structure response depends strongly on computed natural frequencies. Due to variances of characteristic model parameters, particularly when soil-structure interaction is considered, high values of variance of computed frequency values could be expected. Hence, possible variation of the natural frequency value in relatively wide range must be respected in the response analysis. Consequently, responses of quite different levels could be obtained (see illustration in Fig. 1).

Uncertainties in computation of responses could be reduced by selecting appropriate configuration of parameters of the deterministic computation model. This way is lengthy and results are not guaranteed. Application of a stochastic model is limited

Y. Zhang (Ed.): Future Communication, Computing, Control and Management, LNEE 141, pp. 365–371.
springerlink.com

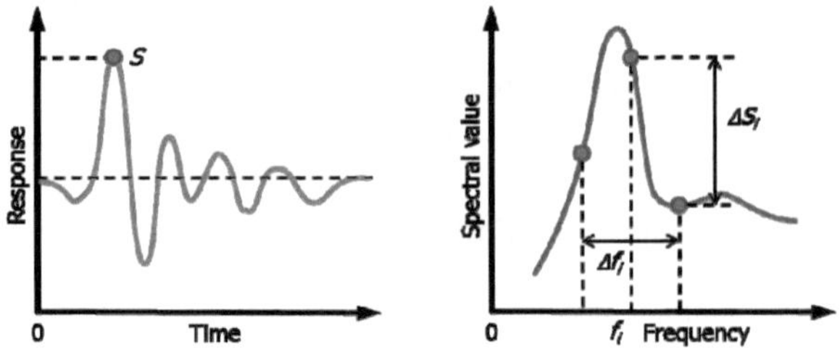

Fig. 1. Variation of response spectrum values due to change of model frequency

due to pretentious computations. Another approach to account for uncertainties in response consists in modification of the given response spectrum. The modification of the response spectrum reduces e.g. the effects of random coincidences of computed frequencies with extremes of the spectrum course. Controlled broadening and lowering of the spectrum is recommended in [4]. The proposed response spectrum modification reduces effectively the uncertainties in response due to uncertainties in frequencies caused by variances of computation model parameters. The modification is based on the probabilistic description of natural frequencies of the analyzed structure, obtained by modal analysis of a stochastic model of the structure.

2 Probabilistic Distribution of the Natural Frequency Value

The proposed response spectrum modification procedure will be explained using a case problem involving a thirteen-story panel building structure. The modal analysis has been performed applying an advanced finite element computation model (see Fig.2). Solid and shell finite elements from the ANSYS program package have been used in modeling the structure. Solid massless elements have been used to model the layered subsoil. The characteristic parameters decisive for the analysis of first bending modes of vibration of the model have been selected after performing a parametric study [7] of the dynamic behavior of the structure. Moduli of elasticity of soil layers have been investigated experimentally. Effective densities of structure parts are based on the analysis of variations of masses of panels and those of floor live loads. Probability distribution types for all considered parameters have been determined (beta-type for soil, normal for panel densities, log-normal for live load). Corresponding mean values, standard deviations and coefficients of variation have been computed. Mean values of modulus of elasticity of soil layers [8] with indicated standard deviations are shown in Fig. 3. Probability distributions of dominant horizontal bending frequencies fx and fy of the parametric model have been determined using the ANSYS program [5] (modal analysis) in combination with the optiSlang program [6] (LHS-method of generation of input parameters, analysis of output data). Results of computation are shown in Fig. 4.

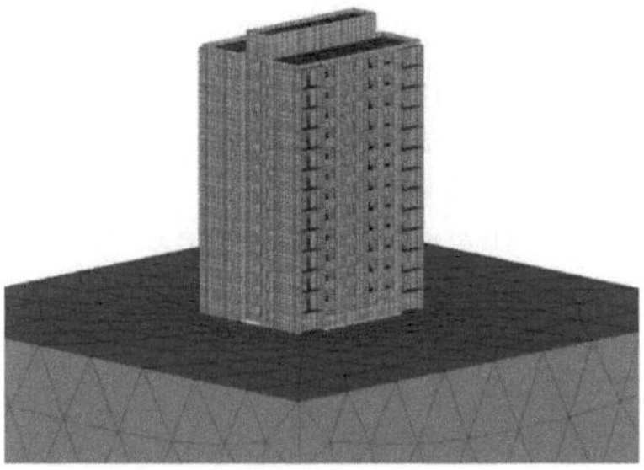

Fig. 2. Finite element model of the panel building

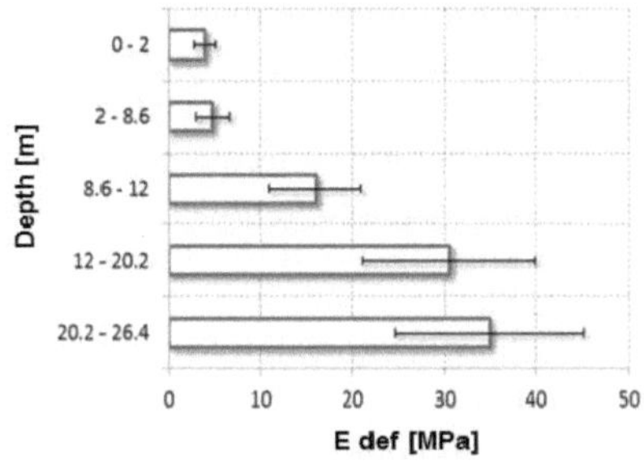

Fig. 3. Modulus of elasticity of soil layers

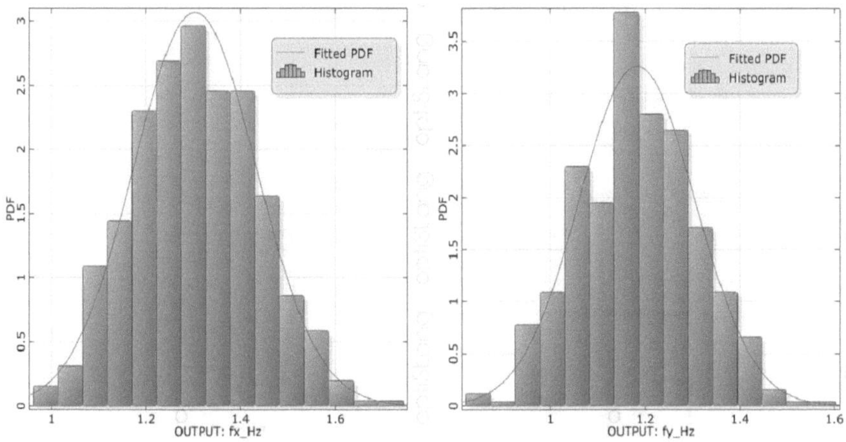

Fig. 4. Probability distribution of dominant natural bending frequencies fx and fy of the model

Stochastic analysis has been performed also at effective modal masses (meffx, meffy) and effective stiffnesses (keffx, keffy) computed from the relation (Eq. 1).

3 Modification of Response Spectra

3.1 The Proposed Procedure

The procedure assumes, that the following is known:

- linear site-specific seismic response spectrum,
- probabilistic distribution of natural frequencies,
- position of the selected frequency in the probability function,
- required probability of non-excedence of the spectral value.

The repeated steps of the procedure of modification are as follows:

1 - selection of the frequency fi,
2 - construction of the probabilistic distribution function of fi,
3 - repeated steps for spectral values Sj in interval 0 to Smax:
a - determination of frequency range where spectral values exceed Sj,
b - computing area pj under the probabilistic function of fi,
c - marking the point [Sj, pj] of the distribution function of exceedence of the spectral value Si,
4 - reading of the value Si for the given probability of exceeding p from the distribution function,
5 - fixing the point [Si, fi] for the modified response spectrum,
6 - return to step 1 to select next frequency.

The modification procedure is represented graphically in Fig. 5. For known probabilistic distribution of natural frequencies of the model, the modified response spectrum ensures no exceedence of spectral value at given probability p.

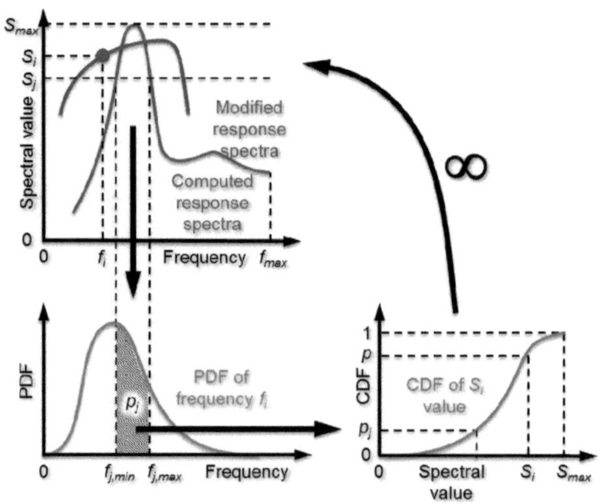

Fig. 5. Response spectrum modification procedure

3.2 Change of the Response due to the Change of the Frequency

The natural frequency fi is related to effective mass meff,i and effective stiffness keff,i:

$$fi = 1/(2\pi).(keff,i/meff,i)1/2 \qquad (1)$$

Assuming constant acceleration response spectrum, change in response follows any change in model parameters, i.e. meff,i and keff,i. The following relations between the change of natural frequency f and change of response u or structural force fk can be derived:

$$umod/udet = (fdet/fmod)2, fk,mod/f k,det = <(fdet/fmod)2; 1> = (fdet/fmod)e, \text{ for } e = <0; 2>. \qquad (2)$$

Suffix det denotes values obtained by deterministic analysis, suffix mod corresponds to probabilistic distributions. Relations are illustrated by Fig. 6. The choice of the exponent e depends on statistical characteristics of natural frequencies fi and corresponding meff,i and keff,i. If the variability of keff,i is several times less then that of meff,i, exponent e approaches to the value 2. In the reversed case, e approaches to 0. When the change of response is accounted for, the modification procedure step 3a has to be changed. Instead of the initial spectrum Sini(f), the spectrum

$$Smod,i(f) = (fi/f)e.\ Sini(f) \qquad (3)$$

is used.

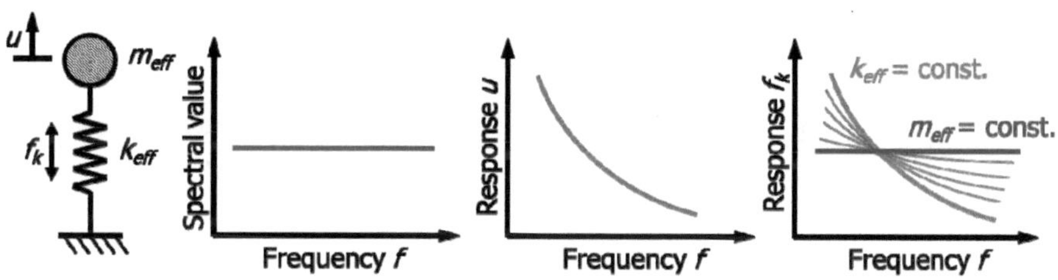

Fig. 6. Change of response due to change of the frequency or due to change of model parameters

4 Modified Response Spectra Features

Selected cases of modified response spectra are shown in Fig. 7. The parameter p is the probability of exceeding the distribution function and e is the exponent of the response change due to frequency change. Normal distribution of frequencies is assumed, with mean value corresponding to frequency fi and with coefficient of variation v.

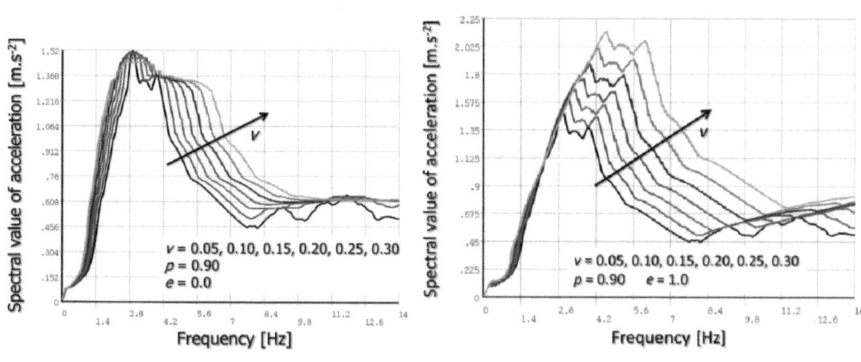

a) b)

Fig. 7a Change of response due to change the frequency or due to change of model parameters

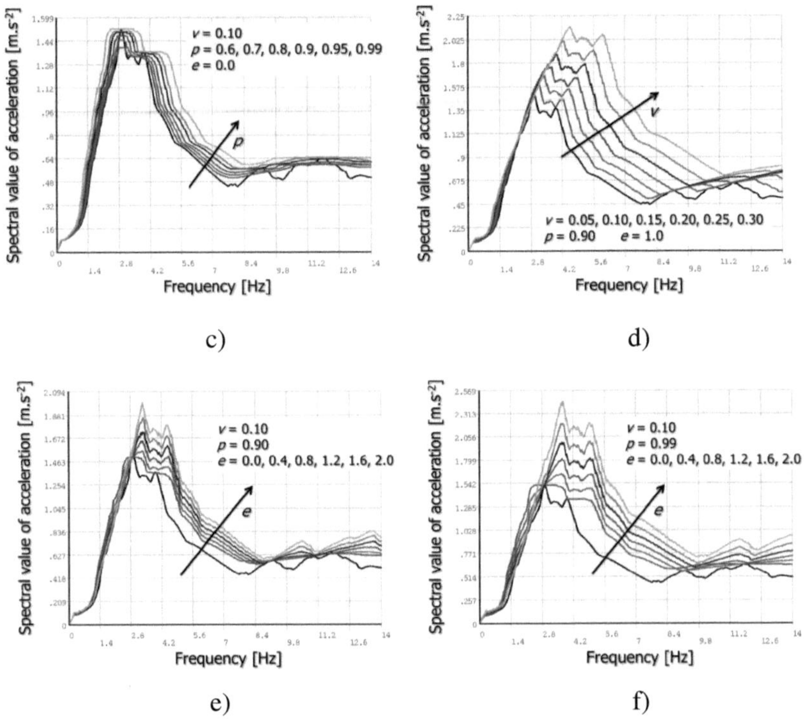

Fig. 7b Change of response due to change the frequency or due to change of model parameters

5 Modified Response Spectrum for the Analyzed Panel Building

The modified response spectrum corresponding to parameters of the analyzed panel building is shown in Fig. 8. The 95 % probability of exceeding the distribution function has been selected.

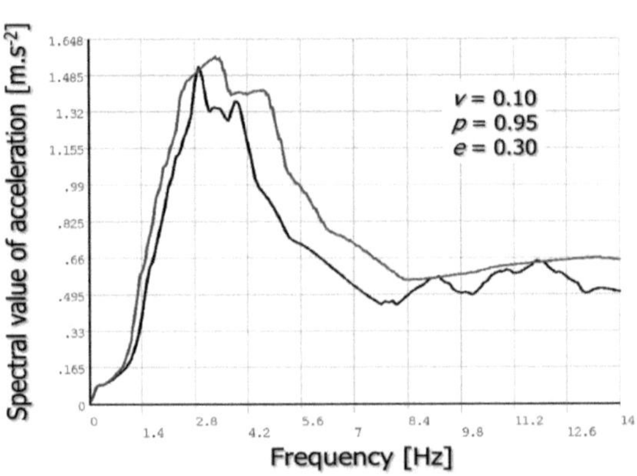

Fig. 8. Response spectrum of the model as generated (black) and modified (red)

6 Conclusion

The proposed response spectrum modification procedure ensures correctness of the deterministic seismic analysis of building structures using the response spectrum method. The modification reduces effectively the uncertainties in response due to uncertainties in frequencies causes by variances computation model parameters.

Acknowledgment. This outcome has been achieved with the financial support of the Ministry of Education, Youth and Sports of the Czech Republic, project CIDEAS No. 1M0579 and project of GACR 103/09/2007 The effect of technical and natural seismicity on structural reliability of buildings.

References

1. Chopra, A.: Dynamics of structures. Theory and Applications to Earthquake Engineering. Prentice-Hall, New Jersey (2001)
2. Gupta, A.K.: Response Spectrum Method. In: Seismic Analysis and Design of Structures, CRC Press, United States of America (1992)
3. Braja, M.D.: Fundamentals of Soil Dynamics. Elsevier, New York (1984), ISBN 0-444-00705-9
4. ASCE: ASCE 98-4, Seismic Analysis of Safety Related Nuclear Structures and Commentary. American Society of Civil Engineers (2000)
5. ANSYS, Inc. Release 12.1 Documentation for ANSYS. SAS IP, Inc. (2009)
6. Dynardo GMBH: OptiSlang 3.1.4 Documentation, Weimar, Germany (2010)
7. Čada, Z., Hradil, P., Mrózek, M., Salajka, V.: Parametric study on dynamic behaviour of a multi-storey building including soil-structure interaction. In: 17th International Conference, Engineering Mechanics, pp. 91–94 (2011)
8. Míča, L., Boštík, J., Bůček, J., Rusina, R.: A circular base plate - Part 1, Statistical analysis of geomechanical variables, Geotechnika (March 2008) (in Czech), ISSN 1211-913X

Study on Vibration and Noise of Hybrid Electric Vehicle

Xiong Jian-qiang[1,2], Huang Ju-hua[1], and Liao Qun[2]

[1] School of mechanical & electrical engineering, Nan Chang University, Nanchang 330031, China
[2] School of mechanical & electrical engineering, Jiang Xi Yu Zhou Scientific & Technological Institute, Xinyu 338029, China

Abstract. Structural characteristics of hybrid electric vehicle (HEV) are studied in this paper, due to the differences of its structure compared with single fuel vehicle, HEV has great changes in its sources and characteristics of vibration and noise. On the basis of analyzing the sources and characteristics of HEV,s vibration and noise, this paper put forward methods of controlling its vibration and noise. Finally difficulties and development trends of HEV,s vibration and noise are summarized.

Keywords: HEV, Vibration and Noise, Electromotor, Engine.

1 Introduction

HEV is a new development branch of energy-saving technology of automobile industry in recent years, and its development is highly emphasized by the world,s giant automobile enterprises. However little study on HEV,s vibration and noise has been carried on at present. In view of that, this paper focus on the changes of HEV,s structure and key components compared with single fuel vehicles, which result in the changes of sources and characteristics of its vibration and noise, and the methods of controlling its vibration and noise are also put forward. Finally the controlling difficulties and development trends of HEV,s vibration and noise have been summarized.

2 Characteristics of HEV,s Structure and Vibration and Noise Sources

2.1 Analysis of HEV,s Structural Characteristics

HEV is a new energy-saving and environmental friendly vehicle, whose driving depends on the energies provided by engine and electromotor. Due to this characteristic, HEV has great changes in mass、 structure and key components in order to adapt to the requirements of its multi energies driving.

Y. Zhang (Ed.): Future Communication, Computing, Control and Management, LNEE 141, pp. 373–382.
springerlink.com

The following figure 1 is HEV,s simple structure:

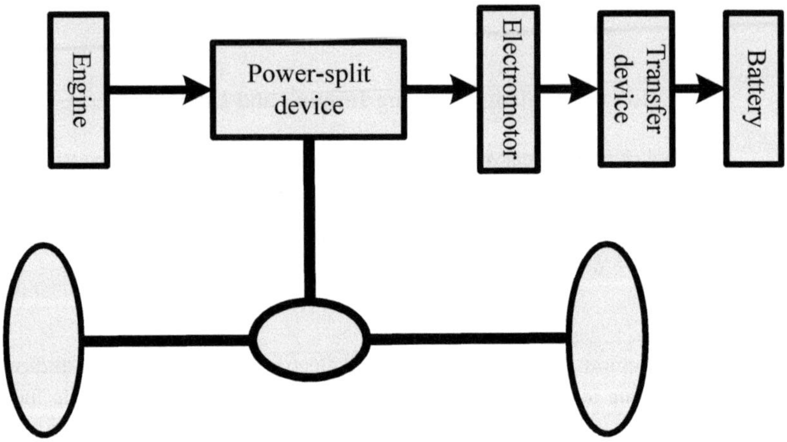

Fig. 1. The structure of HEV

Compared with single fuel vehicle, HEV,s structure is more complex, according to the above figure 1, power-split device、electromotor、transfer device and storage battery are added. HEV ,s changes of structure and key components are as follows.

2.2 Power System

(1) Change of dynamic output device. Single fuel vehicle,s driving depends on only the power provided by engine, but HEV depends on the power provided by both engine and electromotor, or driven by one of them separately.
(2) Added storage battery and energy conversion system. Storage battery can provide auxiliary power, and braking energy can be converted into electricity by generator, which may be stored in the storage battery.
(3) Change of transmission. Dynamic coupling device has been used in the HEV in order to synthesize or divert the power, which makes transmission more complex.

2.3 Auxiliary Power System

Auxiliary systems of single fuel vehicle are driven by the engine, however all of these auxiliary systems in the HEV such as air-conditioning system、steering force pump system、water pump system、cooling system, which are driven by electromotor, therefore electric compressor and electric pump are needed in order to keep the auxiliary systems work well.

2.4 Body Structure and Chassis System

Gross mass、load distribution and drive system have been adjusted according to the changes of power system and auxiliary system, in order to adapt to all these changes, HEV,s body structure and chassis system should have also been improved.

2.5 Characteristics of the Sources of HEV,s Vibration and Noise

Based on the analysis of HEV,s structure and its changes of key components, it is concluded that HEV,s sources of vibration and noise have its unique characteristics compared with single fuel vehicle as follows:

(1) Mass has a great influence on the characteristics of structure vibration and structure borne noise. High pressure nickel-hydrogen battery or lithium battery that consists of many battery modules is of large mass, which may reach tens of kilograms or even hundreds of kilograms; and its distribution and supporting way can affect the its structural modal, generally they reduces its modal frequency. In order to improve fuel efficiency, HEV uses low friction coefficient tire; and the mass of its body structure can also be greatly decreased; these factors will be a challenge for dealing with HEV,s structure vibration and structure borne noise.

(3) HEV,s acoustic feature has changed due to its changes of indoor and outdoor acoustic environment. HEV,s noise conditions has been changed because its driving power source are both low power engine and electromotor, or is either of them; noise caused by road excitation and tire noise still stay, therefore its noise level as a whole is quite low. With the decrease of noise level and the change of noise feature, the contribution ratio of its noise sources changes greatly, and it has a considerable influence on indoor acoustic environmental quality and outdoor noise level of HEV.

(3) Distribution of sources of HEV,s vibration and noise is more disperse. Vibration and noise sources of single fuel vehicle are mainly located near the engine; however electromotor/generator 、 storage battery and other high-power components are added into the HEV, which are all scattered in the automobile body or the trunk, thus the feature of dispersed multi-noise sources forms; resulted in the difficulty of analyzing the feature of vibration and noise sources.

(4) Transient impact vibration and high frequency noise are more prominent. HEV starts depending on auxiliary electromotor, whose torque fluctuation is quite small and torque response slope is big, suspend system of electromotor must have the ability to resist transient impact vibration. Start/Stop frequency of HEV,s engine increases greatly compared with single fuel vehicle, torque fluctuation of the engine,s dynamic process and torsional vibration of the transmission system and HEV,s longitudinal vibration are prominent. Apart from that, frequent start/stop of electromotor of auxiliary system results in transient vibration and noise at all levels. Driving motors and controllers of auxiliary system can cause high frequency electromagnetism noise, so it is a quite prominent problem that how to control high frequency electromagnetism noise in order to improve the riding comfort.

(5) Difficulty of controlling the sources of solid-borne and air-borne vibration and noise increases. Local impedance of body structure and the suspend and support of solid-borne vibration controlling are badly demanded due to scattered feature of sources of the HEV,s vibration and noise. High frequency electromagnetism noise of driving motors and controllers of the auxiliary system, plus the complexity of its circuit system and ventilating duct that pass through between body structure and auto inner trim structure make it more difficult to control the noise.

3 Analysis of Sources of HEV,s Vibration and Noise

HEV has its feature of vibration and noise compared with single fuel vehicle, this paper mainly studies the problems of vibration and noise which are related to HEV. From HEV,s structure and key parts, it is concluded that HEV,s main sources of vibration and noise are as follows: engine,s vibration and noise、 electromotor,s vibration and noise, vibration and noise of voltage boost circuit、 vibration and noise of power coupling device、 vibration and noise of accelerating and decelerating、 vibration and noise of full vehicle structure and braking,s vibration and noise.

3.1 Engine

Although the engine,s vibration and noise is decreased greatly, it is still the main source of HEV,s vibration and noise, which changes much in its feature. The characteriscs of HEV,s vibration and noise are as follows:

(1) Engine thermal efficiency is inefficient when HEV runs at a low speed in the city; to improve fuel economy, engine must start/stop frequently; switch of HEV,s frequent start/stop causes discontinuous vibration and noise of start/stop, which makes passengers feel quite uncomfortable and more easily arouse their notice.
(2) Transmission is at the driving shift when HEV starts, engine vibration and torque fluctuation can be transfered to power system, which causes the vibration and noise of HEV,s floor and seat.
(3) HEV may be affected by torque fluctuation of the engine when it drives at a low speed, causing buzz and body vibration, the main reason is that engine is directly connected with transmission system.

3.2 Electromotor/Generator

Electromotor/Generator of HEV works as the auxiliary power or generating electricity. When HEV speeds up, electromotor /generator works as auxiliary power, when HEV decelerates, electromotor /generator works as recycling braking energy that is stored in the storage battery. During the process of working as the auxiliary power or generating electricity, its main noise sources are electromagnetism noise、 mechanical noise and aerodynamic noise.

The main noise source of electromotor is electromagnetism noise, it transmits outside through field yoke. air gap magnetic wave acts on the stator core gear tooth, which results in radial and tangential magnetic force. Radial magnetic force causing the vibration of the stator core is the primary source of electromagnetism noise, the tangential magnetic force causing the local distortion of gear tooth is the secondary source of electromagnetism noise. It can resonate when radial magnetic force wave approachs the natural frequency of the stator, causing the increase of vibration and noise. Eccentricity of the stator/rotor or Asymmetry of the magnetic circuit can result in the phenomena of the uneven force on both sides of the stator/rotor or magnetic circuit, causing single side magnetic pull and vibration and noise.

It is inevitable that any moving components have mechanical noise, which is closely related to electromagnetic noise, because structural vibration can affect electromagnetic field; vibration frequency and amplitude characteristic of structural components can be changed by the electromagnetic force. Mechanical noise can increase with the increase of rotate speed and load, it is the main noise of electromotor when it is at the high speed, including the noise of bearing、 electric brush and structure resonance.

Vortex noise and whistling noise are the main aerodynamic noise of electromotor. Vortex noise is mainly caused by the alternate eddy current of rotating surfaces of the rotor and cooling fans, whose frequency spectrum varies widely; whistling noise is due to the compressed air or skin friction of the air passing through the surface of components, which is called whistling phenomena. Aerodynamic noise of the rotating motor is inevitable, which is related to surface peripheral speed、 roughness、 change of the duct cross section、 aerodynamic characteristic of fan and component shape, these noises are divided into wide frequency noise and discrete noise.

3.3 Voltage Booster Circuit System

To improve fuel economy and power performance, Voltage booster circuit system is added in the HEV; voltage booster circuit system will increase voltage to enhance the power and the efficiency of hybrid system. When the voltage that passes through the voltage booster circuit system increases, the voltage switching process of conversion circuit can cause the alternate magnetic field in the inductance coil, resulting in the magnetostriction、 magnetic core expansion or shrinkage and coil vibration; radiating vibration and noise of voltage booster circuit system is the main sources of internal and external high frequency vibration and noise of HEV. Radiating vibration and noise of voltage booster circuit system are proportional to the current, therefore vibration and noise of voltage booster circuit system becomes more prominent when electromotor produces high pressure current.

3.4 Dynamic Coupling Device

HEV is driven by both engine and electromotor when it accelerates or climbs, the equipment that is responsible for the synthesis of the two kinds of power is called dynamic coupling device. Performance of the dynamic coupling device has a great influence on efficiency and stationarity of dynamic coupling. If there is some problem during the dynamic coupling process, which can result in vibration、 shock、 noise of dynamic coupling device, unstationary output of coupling dynamic can cause vibration of driveline. In addition to that, dynamic coupling device is also responsible for power split and energy feedback; That is to say, energy produced by engine can be stored partially or entirely in the battery through the generator, resulting in unstationary problem of power switching, and it will cause vibration and noise during the process of switch if power system and driveline of HEV coordinate inconsistently. So it is a significant problem that how to control vibration and noise of dynamic coupling device.

3.5 Other Vibration and Noise

Electromotor increases or decreases output torque accordingly when HEV accelerates or decelerates, engine,s torque may be bigger or smaller than electromotor,s torque, so the synthesis of both of them may be unstationary, causing torsional vibration of driveline.

The added nickel-hydrogen battery in the HEV that are tens of kilograms and even hundreds of kilograms can affect its structural modal due to the distribution and support of storage cell, resulting in the decrease of modal frequency.

HEV,s circuit system and ventilating duct compared with single fuel vehicle is more complex because of applying advanced electric circuit technology; and they pass through between vehicle body and auto inner trim structure, resulting in cavity vibration and noise.

4 Controlling Measures of Sources of HEV,s Vibration and Noise

4.1 Development Status of HEV,s Vibration and Noise

HEV is driven by both low power engine and electromotor, therefore its total level of vibration and noise is much lower than single fuel vehicle, which is not paid much attention by automakers, so it is seldom studied on the field of HEV,s vibration and noise. According to the current research reference materials, the following organizations and universities have studied on HEV,s vibration and noise.

Toyota is in a leading position on the analysis and control of HEV,s vibration and noise, it has studied the following aspect: vibration and noise of engine, electromagnetic noise of voltage booster circuit system、 buzz at the low speed, electromagnetic noise of recycling the braking energy. These technology has already been applied on vibration and noise controlling of new type Prius; in addition, Toyota have as well developed new type THS (Toyota hybrid system), which optimizes the performance of dynamic coupling device and reduces shock noise and vibration. Ford has studied entire vehicle NVH performance of HEV of the hydrogen-fueled engine, and also has analyzed the change of mass and structural layout and power system, and its influence on its vibration and noise, controlling measures are proposed accordingly as well; Ford has adopted the PTH (parallel transfer hybrid) dynamic coupling device on its PHEV (parallel hybrid electric vehicle) and SPHEV (series-parallel hybrid electric vehicle). Warsaw University of Technology has developed dynamic coupling device of single-row planet gear mechanism, which works more stable, and its shock noise and vibration are greatly decreased. Hunan University has studied new type dynamic coupling device and HEV,s power stationary switching. Beijing University of Technology has studied noise controlling of HEV,s auxiliary system; Wuhan University of Technology has studied vibration and noise of electromotor.

4.2 Controlling Technology of HEV,s Vibration and Noise

Engine.Engine,s vibration and noise can be controlled by taking the following measures:

(1) Optimization of switching control between engine and hybrid power system can reduce additional vibration force of engine when it starts;
(2) Optimization of bracket can improve transfer properties of vibration;

(3) Adopting low noise structure pattern and transmission system;
(4) Improving quality of key parts and assembly accuracy;
(5) Adding viscoelastic material on the surface of vibrating components to absorb vibrating energy.

4.3 Electromotor

To reduce noise of HEV,s electromotor system, firstly we must decrease electromagnetic noise; and next mechanical noise. Methods of decreasing electromagnetic noise are as follows

(1) Optimization of claw pole, forming sinusoidal excitation magnetic field, reducing harmonic wave;
(2) Choosing appropriate air-gap magnetic density;
(3) Choosing proper slot combination, avoiding low magnetic force wave;
(4) Improving symmetry of stator and rotor, fastening between stator and its coil;
(5) Improving roundness and coaxiality of machining and assembly of stator and rotor.

Method of decreasing mechanical structure noise is maintaining dynamic balance of rotor, including structural components as many as possible during applying weight-removing method to keep dynamic balance.

Approaches of reducing and isolating mechanical structural vibration are as follows:

(1) Making natural frequency of main structural components deviate from primary exciting force frequency, especially making resonance frequency of structural components deviate from vibration frequency of high order electro-magnetic exciting force;
(2) Installing elastic link structure between motor cover and stator core;
(3) Adding structural damping in the electromotor to dissipate internal energy and decrease vibration response of electromotor.

4.4 Voltage Booster Circuit System

HEV,s noise is more prominent when it accelerates in high load or recycles braking energy, in order to reduce airborne noise and structure-borne noise of voltage booster circuit system, the following measures are generally taken.

(1) Replacing electromagnetic coil material termly to decrease applied compelling force caused by magnetostriction and magnetic core expansion;
(2) Improving vibration characteristics of shell and bracket of voltage booster circuit system;
(3) Improving performance of absorption and isolation of vibration and noise of vehicle body.

4.5 Dynamic Coupling Device

Performance of dynamic coupling device has a great impact on stationarity during the process of power conversion. On the one hand we can optimize control algorithm of dynamic coupling device; on the other hand we can increase its dynamic performance by using simulation software to optimize mechanical structural components.

4.6 Controlling of other Vibration and Noise

When HEV accelerates or decelerates, if torque between auxiliary power system and primary power system are not matching, it can cause shock and unstationarity. In order to solve this problem, firstly we can figure out output torque of primary power system by controller, and then we can get auxiliary power value demand according to overall power need, making torque between primary power system and auxiliary power system match.

Distribution and support of High pressure nickel-hydrogen battery or lithium battery can affect HEV,s structural modal, in order to reduce structural vibration and noise, nickel-hydrogen or lithium battery should be near the node of entire vehicle modal vibration shape, for example they can be installed under the second row seat.

The battery used in the HEV, due to its big mass, can reduce entire automobile modal frequency, the following measures can be taken to decrease low frequency vibration and noise.

(1) Avoiding the coupling between suspend and frame by optimization;
(2) Avoiding the coupling among engine and vehicle body and its connected systems;
(3) Avoiding the coupling between structural modal of vehicle body panel and acoustic modal of carriage;
(4) Improving structure strength.

HEV,s cavity vibration and noise caused by circuit system and ventilating duck can be improved on the basis of aerodynamics study, we can optimize structure according to aerodynamics experiment to lower level of vibration and noise.

5 Conclusions

We can draw the following conclusions of difficulty and development trend from the characteristics of HEV,s vibration and noise and the changes of structure.

(1) Comprehensive analysis of source-path-acceptor of entire vehicle vibration and noise;
(2) Design of vibration and noise of developing process of entire auto;
(3) Subjective and objective evaluation of vibration and noise and analysis of acoustic quality;
(4) Modal analysis and frequency planning of entire vehicle;
(5) Controlling technology of torque fluctuation and electromagnetic noise of auxiliary power system;
(6) Transient vibration analysis and controlling of engine,s frequent start/stop;
(7) Dynamic analysis and controlling technology of dynamic coupling device during the power coupling process;
(8) Analysis of vibration and noise during the switching process of different power mode;
(9) Influence analysis of HEV,s vibration and noise due to the changes of mass and layout.

To study characteristics of vibration and noise of HEV and cope with the new problems of vibration and noise caused by HEV is of great importance for its development. We can take old measures to control vibration and noise whose reasons are the same as single fuel vehicle; however as for the arising new problems of vibration and noise of HEV, we must study and take new controlling measures to lower HEV,s level of vibration and noise and to improve its comfort.

References

1. Ma, Y.-L., Chen, Q.-S.: The analysis of the structure and characteristic of HEV. Automobile Research & Development (4), 20–22 (2000)
2. Pang, J., Chen, G., He, H.: Automobile vibration and noise:Theory and Application, pp. 444–445. Beijing Institute of Technology Press, Beijing (2006)
3. Ito, Y., Tomura, S., Moriya, K.: Vibration-reducing Motor Control for Hybrid Vehicles. R&D Review of Toyota CRDL 40(2), 37–43 (2005)
4. Zhu, Z.-L., Yin, C.-L., Zhang, J.-W.: Application of CAN bus system in hybrid electric vehicle. Journal of Traffic and Transportation Engineering 4(3), 90–94 (2004)
5. Zhang, L.-J., Mou, W.-J.: Research Prospect of Electric Vehicle Vibration and Noise. Trading Web of Chinese Electric Vehicle, http://www.9ebike.com
6. Syed, F.U., Kuang, M.L., McGee, R.A., Gartner, P.M.: Method of operating a hybrid electric vehicle to limit noise, vibration, and harshness. US Patent, 6907325,2005-6-14
7. Lee, H.-D., Sul, S.-K.: Diesel engine ripple torque minimization for parallel type hybrid electric vehicle. In: IEEE Industry Applications Society Annual Meeting, New Orleans, Louisiana, October 5-9 (1997)
8. Wu, X.-J., Li, Z.-M.: Automobile Dynamo,s Noise Generation Principle and Some Ways to Cut Noise Down. Small & Special Electrical Machines 1, 14–16 (2003)
9. Lin, J.: Application of Vibration and Noise Reduction In New Generation Hybrid Electric Vehicle. Automobile and Fitting (14), 30–33 (2006)
10. Zhong, Y., Zhong, Z.-H., Jin, Q.-T., Zeng, Z.-W., Hu, C.-H.: A Simulation on a New-type of Power Coupler in HEV with ADAMS. Automobile Engineering 28(10), 877–880 (2006)
11. Yu, Q.-M., Zhou, J., Huang, W., Yu, W.-X., Wang, Y.-N.: A Study on the Smoothness of Power Switching in a Parallel-series Hybrid Electric Vehicle with CVT. Automobile Engineering 29(3), 216–219 (2007)
12. Zhang, B.-Y., Lei, Y., Zhou, D.-S.: Noise Control of the Auxiliary Power Unit of HEV. Chinese Internal Combustion Engine Engineering 26(6), 6–10 (2005)
13. Masashi, K., Takayoshi, Y.: Noise and Vibration Reduction Technology in New Generation Hybrid Vehicle Development. SAE Paper 114(6), 2648–2655 (2005)
14. Kawabata, N., Komada, M., Yoshioka, T.: Noise and Vibration Reduction Technology in the Development of Hybrid Luxury Sedan with Series/Parallel System. SAE Paper 116(6), 1747–1753 (2007)
15. Takayoshi, Y.: Development of Noise and Vibration Reduction Technology on Hybrid Vehicles. J. SAE 57(7), 34–40 (2003)
16. Kazunori, M., Masashi, K., Takayoshi, Y., Eiji, U.: Noise and Vibration Reduction Technology on New Generation Hybrid Vehicles[C]. ToyotaTech. Rev. 54(1), 68–75 (2005)
17. Naoto, K., Masashi, K., Kojii, T., Takayoshi, Y.: Noise and Vibration Reduction Technology in Rear-Wheel Drive Hybrid Vehicle Development. ToyotaTech. Rev. 55(1), 42–47 (2006)

18. Dhillon, J., Jih, E., Jaura, A.N.: Challenges and Improvements Implemented in Ford H2RV:An HEV Propelled with a Supercharged Hydrogen Internal Combustion Engine.SAE Paper,2004-01-0061
19. Yoshioka, T.: Noise and Vibration Reduction Technology in Hybrid Vehicle Development.SAE Paper,2001-01-1415

Brand Logo Design Symbol Research

Jiaqing Guan and Qian He

School of Urban Design, Wuhan University, China
ggjjqq9271@163.com, 540517693@qq.com

Abstract. Symbols themselves have no value until they are rooted in city characters. This paper sums up the avenues of city brand positioning and spreading via numerous researches on city brands and deep analysis of their formation, which leads to a conclusion that the designing of city brand not only serves as a creator of urban image but also an irreplaceable communication link in public relations.

Keywords: City brand, logo design, symbolization.

1 Introduction

"People gather in a city for living while live in the city for a better life", said Aristotle. A city logo, an integration of the urban history and culture, is a symbolic visual sign showing the image and spirit of a city. City logos originate from the marks and emblems used by the nobles and lords during European time,and became the predecessor of city logos when they extended to cities[1]. The demand of city brand promotion gave birth to modern city logos which help to shape unique city image, to create city identity and to bring economic benefits. With acceleration of urbanization and the recession of regional cultural diversity, now the importance of logos in promoting cities charm has been increasingly recognized.

2 City Symbol

2.1 The Connotation of City Symbol

Each region or each city has its unique history and distinctive culture which contain residents behavioral patterns and cultural heritage,the result of longterm residence. City, regarded as an interesting social space of high quality,earns its local style when be linked with its own special landscape and the environmental characteristics. Chances are that when people refer to a city, they always associate it with its corresponding city symbols, for instance, the west lake in Hangzhou, china made in Jingde town, Chongqing hot pot and so on. All these special city symbols, corresponding to each city, have become people's memory spot of the city.

Y. Zhang (Ed.): Future Communication, Computing, Control and Management, LNEE 141, pp. 383–388.
springerlink.com

2.2 The Category of City Brand

Different city symbols enjoy different content and expressing styles. "City symbol is an unity of the subjective purposes of the city designer and the subjective feelings of the cognition objects,and it also manifests the city's natural environment, history resources and economic conditions externally ", said Hongli Zhao, deputy dean of China city science research institute.

Haitian Xiao, a committee member of CPPCC (Chinese People's Political Consultative Conference), has synthesized city symbol systematically. First, the designing elements of city symbols including the city emblem and flag, standard font and color, maps in various specifications and similar designs in each region. Second, the setting elements of city symbols including the city flower and tree, the city bird and mascot. Third, the architectural elements of city symbols including symbolic architectures and sculptures. If the conditions permit, each district and each street should has its own symbolic buildings and statues. Fourth, the natural elements of city symbols including some famous mountains, rivers, lakes, or forest, all these famous landscape, and parts of certain city area. Fifth, the indexical elements of city symbols including guideboards, street signs, community and building boards, etc. Sixth, the advertising elements of city symbols including various public service and commercial signals, such as, warnings, reminders, instructions, advertisements, and directions[2].

3 City Brand Symbol

3.1 The Connotation of City Brand Symbol

The main connotation of a brand is its core interest sending to its consumers, namely, what on earth does the brand bring to its customers,and brands also stands for the promise of enterprises to their customers. As to city brand symbol, it also is such a wide-accepted core concept sending to the public according to the orientation in its development strategic during the promotion procedure of city image.

3.2 The Properties of City Brand Symbol

The design and implement of city brand symbol are tied to the enforcement of specific plans and strategies for city development. Urban development strategies lay down the foundation for the realization of brand tactic, which focus on developing and taking advantage of tangible assets and turn the city brand value into reality through a series of human resource arrangements and sophisticated financing

3.3 Representativeness

A city brand symbol has to have enough power to stand for the city. Therefore, the orientation of city image and the connotation of city brand require be clear and definite before the determination of a city brand, then collect and select an appropriate brand to represent the city image[3].

3.4 Identifiability

A city brand with high identifiability not only can be recognized and remembered easily, but also cut the promotion budge, facilitating the city image communication.

3.5 Popularity

A city brand enjoying great popularity has more understandability, which shorten the target audiences to accept it and to perceive the beauty and meaning of city brand symbol.

3.6 Regionalization

City is a space dimensional stuff, therefore, the brand symbol standing for city image is typical and is an unique regional symbol of which the city can be proud. As the result of regional characteristic, the brand symbol of a city cannot be imitative.

Designing plays a role of 'translation' in the process of symbol creation, which can precisely translate people's diversified sensorial information into designed symbol and be scrambled as a kind of impression, cognition, which satisfies people's functional, spiritual and psychological requirement, reflecting futures and conception of a era. Designing symbol focuses on the masses, so it must be considered on the bases of angles of the public and standpoint, meanwhile, it has to clearly define the symbol and to make it readable[4].

4 City Symbol and City's Brand Symbol

City symbol is one of the factors on deciding character of a city, however, the city's brand symbol is a core concept which is according to city's development strategic position, passing on this concept to the social public, and it is also a dense epitome in terms of city's strategic position, which is involved in city symbol. Cit symbol is the foundation of city's brand symbol, at the same time, city's brand symbol is the extracting of city symbol.

4.1 Current Situation Analysis about City Logo

Current Situation of City Logo in China

Table 1.

4.2 Predicament of City's Brand Logo in China

With the development of prompt urbanization in China, every city require an increasing number of city logos. However, for the majority of cities, the logos corresponding with its image are difficult to been found. The problems of being repeated, similar meaning and old fashion on displaying etc occur in many logos. Owing to the selling point is places of historical and interest, a lot of cities' logos have some visual elements such as landscape, architectures and so on. It's very easy for people to recognize these logos because of the repetition: the expression of some logos uses repetition. In order to present traditional cultures and temperament, many of logos use the style of drawing or writing, which is prominent. Nevertheless, some logos lack formally breakthrough, directly using the form of bold circle seal and putting headings on to the seal; the technique of expression of many logos is too simple, which directly applies city flower, city bird or city's Chinese or English names to the logos. The issues mentioned above reflect most cities are short of systematic mapping out, novel patterns and unique graphic originality, which only pay attention to single request without considering overall logo image's transmitting construction, leading to difficulties in precisely reflecting and extensively making use of its peculiar cultural features and connotation of the concept.

Currently in China, because city logo doesn't get unified using method or lacks of a kind of systematic applying rules, it cannot reach the predicting effect on its function and usability. City logos eagerly need to play the most functionary role in visual transmission systems continuously instead of putting it into an isolated situation which is unilaterally publicized after it's used ,or ending its function as a city logo after short-lived publicized period. Meanwhile, with regard of lacking standard application, the using of novel transmission patterns and merely using planar designing, city logo's using cycle and publicized effects are shortened and limited in a large scale[5].

Borrowing Ideas from Foreign Cities.

Beyond the sea, particular in western developed countries, city marketing concept develops in an early period, therefore, city logo's designing application is more mature and normative. A great number of city logos have jumped out of using frame work simply. By using logos' characteristic which forms change but its meanings don't, satisfying different using environment and space, to strengthen its using flexibility each other, to improve centralized planar visual elements to a multivariate, dynamic, visual application system with various sides, levels, thus make it have a bigger visual transmission expanding space. Melbourne released a unique city logo in 2009. The designer creates this logo by using a series of changeable, combinative and colorful "M" to present colorful, exquisite, and orderly Melbourne. The logo's fundamental visual elements are based on rigorous triangle network, though different colors' different mechanism assemble and apply, showing various changes. Not only does it keep logo integration, but avoid simple, straitlaced, disordered and complex visual effects so as to make this logo adapt to different fields and demands[6].

CITY OF MELBOURNE

The more important reason is that these designing are not only limited in a simple symbol, but focus more on derivative and expanding of symbols. One of the most visual symbols can compose with various elements, then present towards all kinds of people in each place in a city. As a city logo is designed, logos in branches of the city also come out. These logos have different visual changes and the fundamental connection. They not only adapt every aspect demand ,but coordinate with the rule of transmitted unify.

5 Summary

The designing of City logo cannot merely focus on designing brand visual symbol, but pay more attention to considering how to let symbol design closely combine with application systems. In terms of the city publicity system, city logo is a single visual symbol, so we are not ought to avoid the importance points and dwell on the trivial and ignore its connection with visual recognition system, the expanding and extending of symbol system as well.

References

1. Chou, R.: City Logo design. Tongji University Press, Shanghai (2004)
2. Bart, R.: Elements of Semiology. The Joint Press, Shanghai (1998)
3. Cheung, L.: Visual Symbol Manual. Shanghai Bookstore Publishing House, Shanghai (2008)
4. Lv, W.Q.: Designing on City Image. Southeast university press, Nanjing (2002)
5. Jin, D.H., Nie, Y.M., Wu, B.B.: Thinking of the Current Status of China's Urban Symbols and City Brand Symbol Spread. Advertisement Grand Sight (9), 155–170 (2009)

A CBR-Based CAD System for Subframe Design of Aerial Work Trucks

Li Qi, Liu Xinhua, Liu Youhui, and Yin Jilin

School of Mechanical and Electrical Engineering,
China University of Mining and Technology, XuZhou, China
l_xinhua_2006@126.com

Abstract. On the analysis of the case-based reasoning technology and the design process of subframe for the aerial work truck, the system framework and implementation method for subframe CAD system based on CBR are proposed. Then with analyzing the similarity between a new part and the typical part in instance base, the problems of selecting similarity system and similarity element are discussed. Meanwhile, the method of similarity analysis is proposed in detail. At last, an example is given to illustrate the effect of the proposed system.

Keywords: computer-aided design, case-based reasoning, similarity calculation, subframe design.

1 Introduction

Currently the design of subframe for the aerial work trucks is mainly used interactive CAD system, such as UG and Pro/E etc. The design mainly relies on the experience of the designers, and just some matching rules and computation works have been done by computers. For different vehicles, the subframe has to be redesigned. The designed subframe model cannot be reused effectively, which makes the design spending more time, high cost and low efficiency. This design method cannot meet the rapid development of modern enterprise. Meanwhile, the design experience and design knowledge cannot be expressed regularly and clearly because of the differences of the structure and size of subframe among different vehicles. So the expert system based on rule cannot be developed easily.

The CBR appeared in recent decades, which distinguishes from the Rule-Based Reasoning (RBR) is a kind of inference mode [1, 2]. The core of CBR is to use similar problems that was resolved previously to solve the current problem, which avoids the repeatability of design work to a certain degree. In view of the current situation of subframe design, the subframe CAD system based on CBR is proposed. In the system, the design experience of designers and design knowledge are stored in the case base. When a subframe is designed, one or more cases are obtained from the case base, then find out the most similar one or several cases comparing with current design to modify to meet the new design requirement. The CAD system is based on UG (Unigraphics NX) 3D graphics environment. On the base of intelligent decision-making function of CBR, the system realizes rapid design of subframe,

Y. Zhang (Ed.): Future Communication, Computing, Control and Management, LNEE 141, pp. 389–396.
springerlink.com

improves product design efficiency, shortens the development cycle and reduces the cost of product development.

2 Design Principle of the Subframe CAD System

2.1 The Structure Analysis of Subframe

The aerial work truck subframe consists of double-longerons, former fixed leg, latter fixed leg, rotary table bearing, rotary table and reinforcing rib etc [3]. To avoid the stress concentration on the main frame which caused by the sudden change of the stiffness of the subframe, the front shape of it adopts a gradual transition structure, usually, triangular shape, U-shape and L-shape are most commonly applied in the engineering practices. All kinds of specific subframe exist in the applications, but there exist obvious similarities among their composition and structure of each component. Therefore, the components and parts designed previously can be applied to the current design. In order to make full use of excellent examples designed previously, the CBR theory has been introduced to the design of subframe and the CAD system based on CBR is developed.

System structureThe subframe CAD system, as an intelligent design module, mainly consists of three parts: design function module, communication interface module and information sharing module [4, 5], as shown in fig.1.

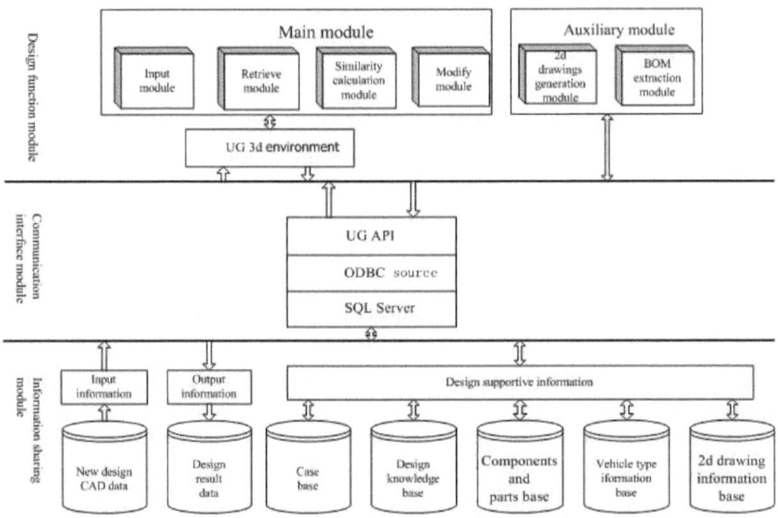

Fig. 1. Frame structure of the CAD system

Design function module consists of two parts: the main module and auxiliary module. The information for design function module and the design result information are shared between the design function module and the information sharing module, and the communication interface module is the channel of communication for them.

The communication interface module, which consists of three parts: UG API, ODBC data source and SQL Server, is the communication interface between the design function module and the information sharing module. In the system, all the informations are stored in data base, which is convenient for information sharing. The information in the database exchange external data using ODBC data source. UG API is an interface between UG and external environment, by which the data and information exchange between the UG and external environment.

Sharing information, which contains the sharing data between the CAD system and the other external system and the internal sharing information, is set of information. The information sharing module stores all kinds of information and provides data for other modules.

2.2 Design Procedure for Subframe Based on CBR

Subframe CAD system based on CBR is an expert system using CBR theory to realize the design of subframe. To realize the system, three aspects of things must be done: representing the case, establishing case-retrieval mechanism and modifying the case.

(1) Representing the case. Representing the case is the key for the CAD system. The complexity of similarity calculation could be increased and correctness could be affected if the case be represented unreasonably, subsequently the difficult of retrievaling case will be increased, which eventually case the low system efficiency. To the subframe CAD system, the distinctions are obvious among different trucks in different height. Now, according to the height, here are 6m, 10m, 14m, 16m, 18m, 22m, 25m, 30m, 50m and above 80 meters, highest level in foreign country. To the specific subframe, the front shapes are obviously different. In order to retrievaling case quickly and combined with the characteristics of subframe, the hierarchical structure is used to present the case [6, 7], which is not only beneficial to the case store, but also beneficial to similarity calculation. The hierarchical structure for the case is shown in fig.2.
(2) Similarity calculation. In the CAD system. The cases are stored in hierarchical structure. Therefore, the system matches the height firstly when retrievaling, and then the front shape. After the match of the two aspects being successful, the retrieval will go on, otherwise pause. Due to different cases having the similar geometric topology, the match of components or parts is vital [8]. The nearest neighbor algorithm is used for the similarity calculation. The distance among the cases is defined as weighted sum of the characteristics of the cases.

Different components or parts have the different importance for the whole subframe. If the whole subframe is treated as one similarity element when calculating similarity, the match of components and parts could be difficult. So the whole subframe be treated as a system [9], series of nominal similarity elements are defined, after which here are varies of Similarity characteristics.

Two systems compared are noted for Pa and Pb respectively, the same similarity units constitute similarity elements correspondly. For the subframe, there are mainly 6 similarity elements: vehicle type u1, front shape u2, former leg u3, latter leg u4, rotary table bearing u5, rotary table u6.

Taking the rotary table as an example, the way of calculating similarity grade is introduced. According to the design rule and the design experience, rotary table element has three characteristics: inner radius ra external diameter rb and height h. It is assumed that there is a case in the case base: ra =25cm, rb=22cm and h=30cm. And the new design: ra1=20cm, rb1=18cm and h1=25cm. The element similarity grade decided by the number of the similarity elements is calculated as follows:

$$q(u_6)_n = \frac{m}{k+n-m} = \frac{3}{3+3-3} = 1$$

Where: k is the number of characteristics of the new rotary table, n is the number of characteristics of the rotary table in case base and m is the number of characteristics of similarity units

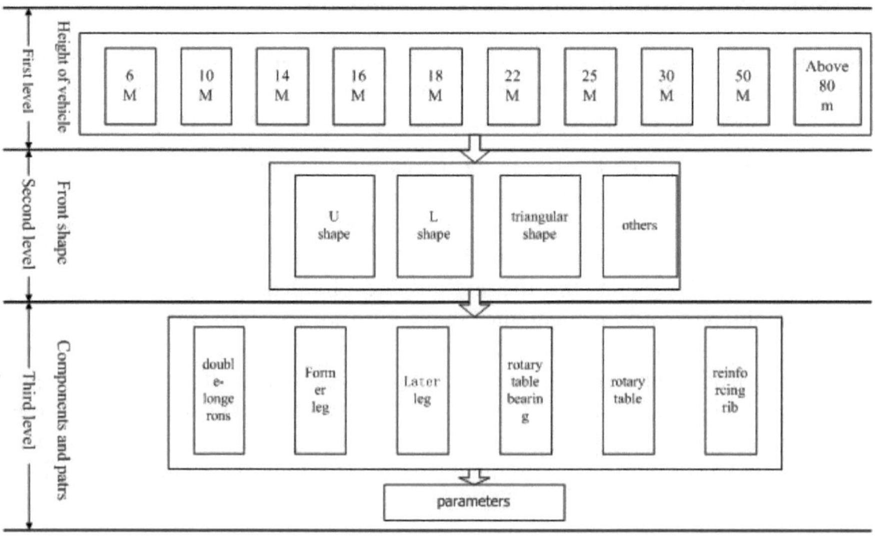

Fig. 2. Hierarchical structure of the case

The similarity ratio of the three characteristics is calculated as follows:

$$\begin{cases} R_1 = \dfrac{r_{a1}}{r_a} = \dfrac{20}{25} = 0.800 \\ R_2 = \dfrac{r_{b1}}{r_b} = \dfrac{18}{22} = 0.818 \\ R_3 = \dfrac{h_1}{h} = \dfrac{25}{30} = 0.833 \end{cases}$$

For the rotary table, the three characteristics have the same significance, so the three characteristics were given the same weighting coefficient, means 1/3 for each. The similarity grade of the similarity unit is calculated as follows:

$$q(u_6)_s = \sum_{i=1}^{3} d_i R_i = \frac{1}{3}\times 0.800 + \frac{1}{3}\times 0.818 + \frac{1}{3}\times 0.833 = 0.817$$

So the similarity grade of the rotary table is calculated as follows:

$$q(u_6) = q(u_6)_n \bullet q(u_6)_s = 1\times 0.817 = 0.817$$

According to the above method, the remaining five similarity elements: vehicle type, front shape , former leg , latter leg and rotary table bearing are calculated, they are 0.942, 0.964, 0.875, 0.715,0.831 respectively. According to the importance of each element for the whole subframe, the order ranged from high to low of the 6 elements is vehicle type, front shape, rotary table, rotary table bearing, former leg, latter leg. The judgment matrix is given as follows:

$$P = \begin{bmatrix} 1 & 2 & 3 & 4 & 5 & 6 \\ 1/2 & 1 & 3/2 & 2 & 5/2 & 3 \\ 1/3 & 2/3 & 1 & 4/3 & 5/3 & 2 \\ 1/4 & 1/2 & 3/4 & 1 & 5/4 & 3/2 \\ 1/5 & 2/5 & 3/5 & 4/5 & 1 & 5/6 \\ 1/6 & 1/3 & 1/2 & 2/3 & 5/6 & 1 \end{bmatrix}$$

The weighting coefficient for each element is calculated by the way of normalizing the judgment matrix, the result is given as follows:

$$\beta = (0.41 \quad 0.20 \quad 0.14 \quad 0.10 \quad 0.08 \quad 0.07)$$

The similarity grade decided by the number of the similarity elements for the whole subframe is calculated as follows:

$$Q_n = \frac{M}{K+N-M} = \frac{6}{6+6-6} = 1$$

The similarity grade decided by the value of each similarity element is calculated as follows:

$$\begin{aligned} Q_u &= \sum_{i=1}^{6} q(u_i)\beta_i \\ &= 0.942\times 0.41 + 0.964\times 0.2 + 0.817\times 0.14 \\ &\quad 0.831\times 0.1 + 0.875\times 0.08 + 0.715\times 0.07 \\ &= 0.897 \end{aligned}$$

According to the above, the similarity grade of the whole subframe case is calculated as follows:

$$Q = Q_n \bullet Q_u = 1\times 0.897 = 0.897$$

2.3 The Modifying Strategy of the Case Base

The case retrieved from the case base could have the difference with the new design commonly, which makes the designer to modify the retrieved case. In order to modify the case, first of all the difference between the new design and the retrieved case has to be assured, then modify the case based on the difference. Modifying the case means modifying one case or restructuring several candidate cases.

To the CAD system, retrieving case is based on the similarity grade. The similarity grade for the whole subframe is the highest, but for each component or part, the similarity grade cannot be guaranteed. So the modifying of case is essential to guarantee the similarity grade of the components and parts. Modifying the case is also based on the similarity theory. The similarity grade of components and parts can be obtained by comparing the retrieved case and the other cases in case base, then the designer can modify the design refer to the similarity grade of components and parts.

3 System Implementation

The CAD system based on CBR is a module for rapid design of subframe, the work procedure of it is shown in fig.3.

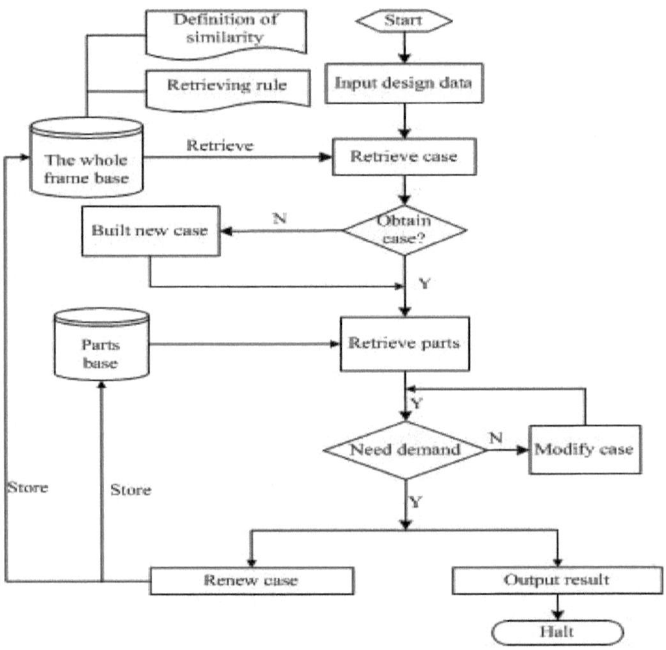

Fig. 3. The work procedure of the system

The main procedure of the system is shown as follows:

Step1: Input the information of the new design and retrieve the similar model in the case base.

Step2: Retrieve the components and parts to get the more perfect model when retrieving the needed case, otherwise retrieve the case again or modify the case.

Step3: The designer build a new module when there exists no similar module in the case base to match the new design.
Step4: Renew the case base, the new design has to be stored into the case base as a new case, and the design result is to be output.

According to the above and combined with the computer technology, database technology and the UG second development, the system can be realized. The interface for inputting new design information is shown in fig.4.

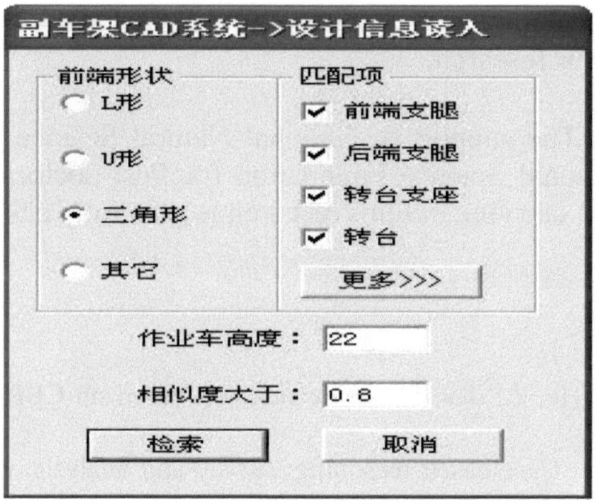

Fig. 4. Interface for inputting information

According to the information shown in fig.4, two cases can be retrieved from the case base automatically by the CAD system, as shown in fig.5.

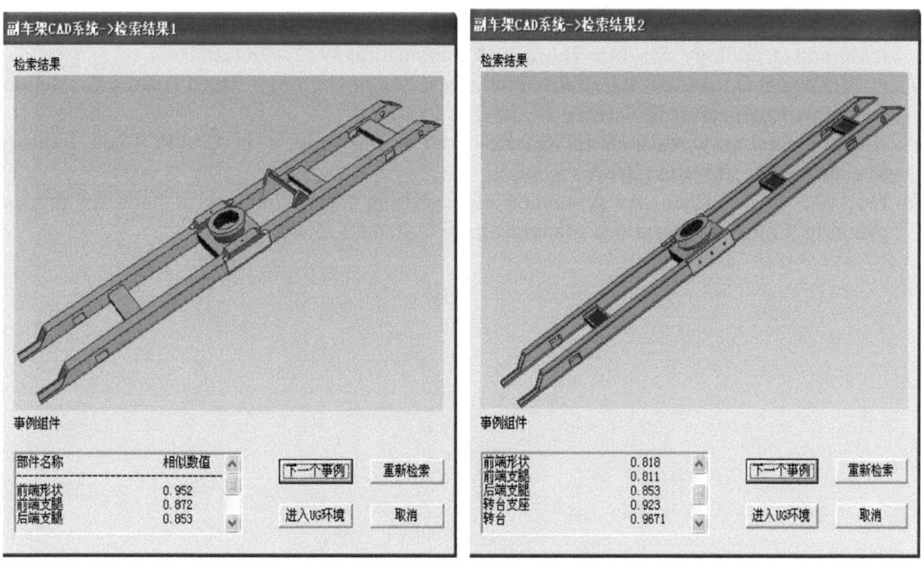

Fig. 5. The retrieved cases

4 Conclusions and Future Work

Based on the case-based reasoning technology, the subframe CAD system has been realized. The consumed time for designing a subframe is about one day, at least 10 days by traditional design method. A new subframe can be designed with shorten time, high quality and low cost through the proposed CAD system. Moreover, the advantages of the CAD system will be more evident with the increase of the number of cases in the case base.

In the next phase, realizing the CAD system based on CBR for the whole aerial work truck is the focus of our research.

Acknowledgements. The support of National Natural Science Foundation of China (No. 51005231), National Science Foundation for Post-doctoral Scientists of China (No. 20100471408) in carrying out this research is gratefully acknowledged.

References

1. Geng, W., Pan, Y., He, Z.: The reasoning research based on CBR. Computer Science 20, 36–40 (1993)
2. Yuan, X., Wang, Y.: Case-based reasoning: survey and analysis. Pattern Recognition and Artificial Intelligence 8, 18–31 (1995)
3. Si, J.: The design and analysis of subframe of special automobile. Highways & Transportation in Inner Mongolia 1, 28–29 (1997)
4. Kolodner, J.L.: Maintaining organization in a dynamic long-term memory. Cognitive Science 7, 243–280 (1983)
5. Wang, X., Sun, H.: A stamping die CAD system based on CBR. Journal of Xi'an Institute of Technology 24, 36–40 (2004)
6. Jiang, Y.: 3D Design of special-shaped slanted ejecting and new slide based on case reasoning technology. Machine Building & Automation 39, 27–28 (2010)
7. Li, Y., Wang, G.: A Case-Based Reasoning Research of Ontology-Based Feature Extraction. Computer Engineering & Science 29, 74–76 (2007)
8. Xie, Z.: Similarity research on feature-based whole process in CAPP. Light Industry Machinery 28, 100–102 (2010)
9. Liu, W., Xu, Z., Chen, Z.: A method of measuring form similarity of parts for process planning. Computer Integrated Manufacturing Systems 1, 58–62 (2009)

Assembly Sequence Planning Based on Ant Colony Algorithm

Liu Youhui, Liu Xinhua, and Li Qi

School of Mechanical and Electrical Engineering,
China University of Mining and Technology, XuZhou, China
l_xinhua_2006@126.com

Abstract. Assembly sequence planning (ASP) is the core content of assembly process planning, an improved ant colony algorithm (ACA) is presented to solve the ASP problem. This paper defines the concept of product's feasible assembly sequence, introduces the evaluation mechanism of assembly sequence, and establishes the mathematical model of assembly sequence optimizing in which the ASP problem was transformed into the problem of searching optimal assembly sequence in all of the feasible assembly sequences. Finally, the validity and feasibility of the proposed algorithm have been verified through an exemplification of vibration generator's ASP.

Keywords: Assembly sequence planning, ant colony algorithm, feasible assembly sequence, optimal assembly sequence.

1 Introduction

Assembly is the final links in product's manufacturing process. Product's assembly quality has influence on its performance, reliability and life to a great extent. With the unceasing development of product design and manufacturing technology, the proportion of assembly cost increases constantly. Thus, Assembly process planning can not only improve the quality and efficiency of assembly, but also shorten the development cycle of product and reduce the production cost [1].

Assembly is the final links in product's manufacturing process. Product's assembly quality has influence on its performance, reliability and life to a great extent. With the unceasing development of product design and manufacturing technology, the proportion of assembly cost increases constantly. Thus, Assembly process planning can not only improve the quality and efficiency of assembly, but also shorten the development cycle of product and reduce the production cost [2].

2 ASP Based on ACA

2.1 The Mathematics Description of the ASP Problem

Product's ASP problems can be described as the under definitions: Take a product which includes n parts, for example, the set of all of its parts is $P=\{P_i|i=1,2,\ldots,n\}$.These

Y. Zhang (Ed.): Future Communication, Computing, Control and Management, LNEE 141, pp. 397–404.
springerlink.com

parts constitute a permutation and combination $S=\{P_{i1}, P_{i2},P_{i3} ,...,P_{in} \}$, exist below conditions:

(1) Any one of the parts in S exists assembly constraints contact with at least one of all previous parts;
(2) The order between any two parts in S must meet the given set C which is composed of assembly constraints;
(3) For a given evaluation function F, S makes the value of F minimum.

If S only satisfies conditions (1) and (2), it is feasible assembly sequence, if S satisfies the conditions (1), (2) and (3) at the same time, it is optimal assembly sequences [3].

The assembly sequence between any two parts must accord with the given set C which consists of assembly constraints. This kind of constraint relation can be expressed by product's integrated interference matrix [4]. Matrix elements represent the corresponding parts' interference status with other parts along $+X$, $+Y$ and $+Z$ directions. Integrated interference matrix is defined as follows:

$$IIM = \begin{array}{c} \\ P_1 \\ P_2 \\ \cdots \\ P_n \end{array} \begin{array}{c} \begin{array}{cccc} P_1 & P_2 & \cdots & P_n \end{array} \\ \begin{bmatrix} I_{11} & I_{12} & \cdots & I_{1n} \\ I_{21} & I_{22} & \cdots & I_{2n} \\ \cdots & \cdots & \cdots & \cdots \\ I_{n1} & I_{n2} & \cdots & I_{nn} \end{bmatrix} \end{array}$$

Where I_{ij} represent the part P_j interference status with the part P_i along $+X$, $+Y$ and $+Z$ directions. While I_{ijX}, I_{ijY} and I_{ijZ} represent respectively the part P_j interference status with the part P_i along $+X$, $+Y$ and $+Z$ directions. I_{ij} is given as follows:

$$I_{ij} = \begin{cases} 0 & \text{if } I_{ijX}=0, I_{ijY}=0, I_{ijZ}=0 \\ 1 & \text{if } I_{ijX}=0, I_{ijY}=0, I_{ijZ}=1 \\ 2 & \text{if } I_{ijX}=0, I_{ijY}=1, I_{ijZ}=0 \\ 3 & \text{if } I_{ijX}=0, I_{ijY}=1, I_{ijZ}=1 \\ 4 & \text{if } I_{ijX}=1, I_{ijY}=0, I_{ijZ}=0 \\ 5 & \text{if } I_{ijX}=1, I_{ijY}=0, I_{ijZ}=1 \\ 6 & \text{if } I_{ijX}=1, I_{ijY}=1, I_{ijZ}=0 \\ 7 & \text{if } I_{ijX}=1, I_{ijY}=1, I_{ijZ}=1 \end{cases}$$

Where

$$I_{ijD} = \begin{cases} 1 & \textit{If interference occurs along} + D \textit{ direction} \\ 0 & \textit{otherwise} \end{cases} \qquad D = X, Y \textit{ or } Z$$

In order to illustrate the integrated interference matrix, a simple assembly which consists of ten parts is proposed in fig.1.

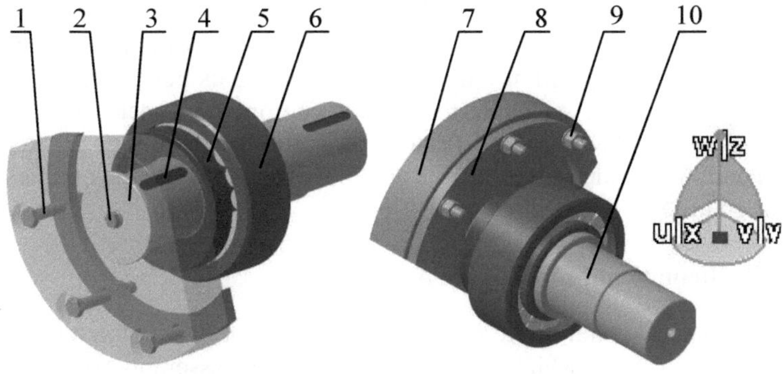

Fig. 1. The CAD model of a simple assembly

In fig.1: 1-M20 bolts; 2-M16 bolts; 3-Shaft gland; 4-Key; 5-Shaft sleeve; 6-Bearing; 7-Main eccentric block; 8-Vice eccentric block; 9-M20 nuts; 10-Shaft.

The assembly constraints set C of the above assembly can be expressed by its integrated interference matrix which is shown as follows. Here the Y-axis is along the axis direction of the shaft (part 10) in fig.1.

$$IIM = \begin{array}{c} \\ p_1 \\ p_2 \\ p_3 \\ p_4 \\ p_5 \\ p_6 \\ p_7 \\ p_8 \\ p_9 \\ p_{10} \end{array} \begin{array}{c} \begin{array}{cccccccccc} p_1 & p_2 & p_3 & p_4 & p_5 & p_6 & p_7 & p_8 & p_9 & p_{10} \end{array} \\ \begin{bmatrix} 0 & 1 & 1 & 1 & 7 & 0 & 7 & 7 & 7 & 1 \\ 0 & 0 & 7 & 0 & 0 & 0 & 5 & 0 & 0 & 7 \\ 0 & 5 & 0 & 2 & 2 & 2 & 7 & 0 & 0 & 2 \\ 0 & 0 & 0 & 0 & 3 & 2 & 7 & 0 & 0 & 7 \\ 0 & 0 & 0 & 5 & 0 & 2 & 5 & 0 & 0 & 7 \\ 0 & 0 & 0 & 0 & 0 & 0 & 0 & 0 & 0 & 7 \\ 5 & 5 & 5 & 7 & 7 & 2 & 0 & 2 & 2 & 7 \\ 5 & 0 & 0 & 1 & 1 & 0 & 1 & 0 & 2 & 1 \\ 5 & 0 & 0 & 1 & 1 & 0 & 1 & 0 & 0 & 1 \\ 0 & 0 & 0 & 7 & 5 & 5 & 5 & 0 & 0 & 0 \end{bmatrix} \end{array}$$

2.2 The Basic Principle of ACA

The main idea of ACA is to construct the pheromone used by real ants as a medium for communication and feedback. ACA is a population-based, cooperative search procedure that is derived from the behavior of real ants. ACA uses artificial ants to construct solutions to combinatorial optimization problems iteratively [5]. Artificial pheromone and problem-specific heuristic information guide ants to generate solution. ACA is often applied to solve combinatorial optimization problems.

The ASP problem belongs to the typical combinatorial optimization problem. Each product's feasible assembly sequence corresponds to the path from the nest of ant colony to the food. The process of ASP based on ACA can be described as follow:

First a certain number of ants must be generated. Then make the starting points of all the ants are located on the basal parts. Each ant chooses the next part according to the maximum probability and executes the local pheromone update after each step. A feasible assembly sequence will be obtained when all the parts are assembled. ACA completes an iterative algorithm only when all of the ants get the feasible assembly sequence. Then each ant executes the global pheromone update according to the evaluation results of the generated feasible assembly sequence [6, 7]. Repeat these procedures until all the ants gather in one assembly sequence and the optimal assembly sequence is acquired.

2.3 The Mathematical Model of ACA for ASP

In this mathematical model of ACA for ASP, the alterant times of assembly direction and tool are set to weights which are added to each of the assembly path as pheromone factors. Ants choose the next part according to the size of the pheromone amount [8].

The probability that ant from a assembling operation to the next assembling operation is called transition probability. Assume the trajectory that ants from the previous node to the current node is (i,j), the transition probability Pijk(t) is decided by the pheromone strength τij (t), direction guidance factor dij (t) and tools guidance factor tij (t) together, and it is defined as follows:

$$P_{ij}^{k}(t)=\begin{cases}\dfrac{[\tau_{ij}(t)]^{\alpha}\cdot[d_{ij}(t)]^{\beta}\cdot[t_{ij}(t)]^{\mu}}{\sum_{K\in allow_k}[\tau_{ij}(t)]^{\alpha}\cdot[d_{ij}(t)]^{\beta}\cdot[t_{ij}(t)]^{\mu}} & j\in allow_k \\ 0 & otherwise\end{cases}$$

Where: τij (t) is the pheromone strength on the path (i,j) in t-moment, dij (t) is the enlightening information which is related with the alterant times of assembly direction, and it is calculated as follows:

$$d_{ij}=\begin{cases}0.2 & \textit{if the direction is changed}\\ 1 & \textit{otherwise}\end{cases}$$

tij (t) is the enlightening information which is related with the alterant times of assembly tool, and it is calculated as follows:

$$t_{ij}=\begin{cases}0.2 & \textit{if the tool is changed}\\ 1 & \textit{otherwise}\end{cases}$$

allowk is a set that is make up of the number of nodes k and each node in which is visited possibly by ants. a, β and μ are respectively the weights of pheromone strength, direction guidance factor and tool guidance factor, and they are all nonnegative number.

With the passage of time, ants will leave pheromones on the passed path. The pheromone will influence subsequent ants, therefore, it must be updated regularly [9, 10]. This paper adopts two kinds of pheromone update, namely local pheromone update and global pheromone update.

Each ant executes the pheromone update after each step, applying pheromone to the last path traversed according to the following update function:

$$\tau_{ij}(t) = (1-\rho)\cdot\tau_{ij}(t) + \rho\cdot\tau_0$$

Where ρ∈(0,1]is the local pheromone decay coefficient, and τ0 is the initial value of the pheromone.

It is necessary to undertake global pheromone update when all the ants find the end node. The update function is shown as follows:

$$\tau_{ij}(t+1) = (1-\gamma)\cdot\tau_{ij}(t) + \sum_{1}^{m}\Delta\tau^{k}{}_{ij}(t)$$

Where γ∈(0,1]is the global pheromone decay coefficient, and m is the number of ants, Δτkij (t) is the increased quantity of pheromone, and is given as follows:

$$\Delta\tau^{k}{}_{ij}(t) = \begin{cases} Q/S & \textit{if the path in the k-th sequence} \\ 0 & \textit{otherwise} \end{cases}$$

Where Q is a regulation parameter, S is a quality evaluation function of assembly sequence, and is given as follows:

$$S = \omega_1 \cdot D + \omega_2 \cdot T + 1$$

Where D is the alterant times of assembly direction, T is the alterant times of assembly tool, ω1∈(0,1] is the weight concerning the alterant times of assembly direction, ω2∈(0,1] is the weight concerning the alterant times of assembly tool.

2.4 The Pseudo Code of ACA

The above algorithm is easy to realize by computer, and its implementation process can be expressed by the pseudo code. The pseudo code of ACA is shown as follows:

```
Begin
Input IIM, P(T)
Initialize α, β, μ, ρ, γ, τ0, Q, ω1, ω2, m, N
Set  t :=0,Δτij = 0
While (t < N)
{
t : = t +1
For (k: =0; k<m; k++)
{
Set allowk (t):=Ø, Lk (t):= Ø, tabuk (t):= Ø
i :=0
While (ants have not finished the search work of all of the parts)
{
Each ant selects next part according to probability formula Pijk (t)
// Update the local pheromone
Update τij (t)
Update allowk (t), Lk (t), tabuk (t)
```

```
i :=j
}
}
// Calculate the current optimal assembly sequence according to the target function S
Update τij (t)
Update τij (t+1)
}
// Calculate the optimal assembly sequence according when iterative operation is completed
// Output the best assembly sequence Lbest
End
```

In the above pseudo code, m equal to the number of ants, N equal to the maximal iteration times, t equal to the current iteration times, i refer to part number, P(T) is a matrix in which the information of part's assembly tool can be expressed, L is a feasible assembly sequence, *tabu* is a set in which the next part can't be chosen.

3 An Exemplification of Vibration Generator's ASP

In order to validate the efficiency of the proposed algorithm, an exemplification of vibration generator's ASP will be introduced.

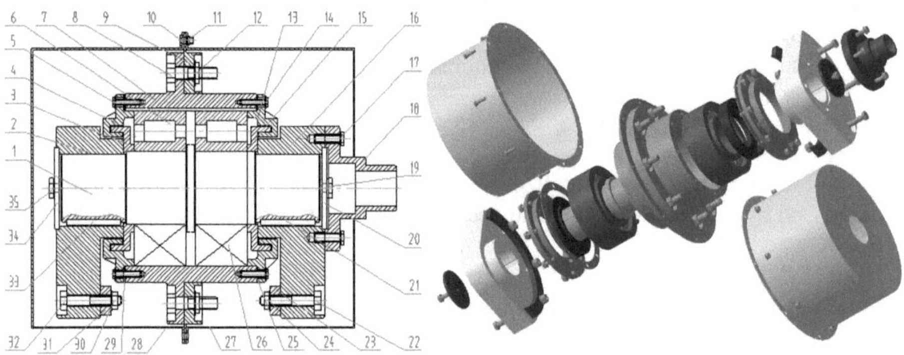

Fig. 2. The assembling drawing and exploded view of a vibration generator

Fig.2 is a vibration generator's assembling drawing and exploded view. The vibrator generator has 35 major parts. Here the washers and nuts and the washers and bolts are all considered as a part, which have the same function.

Assuming there are six kinds of assembly tools in the assembly process of the vibration generator shown in fig.2. The information of part's assembly tool can be expressed in the under matrix-P(T). The value of matrix element is 1 if the assembly tool is used in assembly process, otherwise, it is 0.

$$
P(T)=\begin{array}{c} \\ T_1 \\ T_2 \\ T_3 \\ T_4 \\ T_5 \\ T_6 \end{array}
\begin{array}{c}
\begin{array}{ccccccccccccccccc} P_5 & P_6 & P_8 & P_{10} & P_{11} & P_{12} & P_{14} & P_{17} & P_{19} & P_{21} & P_{22} & P_{24} & P_{26} & P_{30} & P_{32} & P_{33} & P_{35} \end{array} \\
\left[\begin{array}{ccccccccccccccccc}
0 & 0 & 0 & 1 & 1 & 0 & 0 & 0 & 0 & 0 & 0 & 0 & 0 & 0 & 0 & 0 & 0 \\
1 & 0 & 0 & 0 & 0 & 0 & 1 & 1 & 1 & 0 & 0 & 0 & 0 & 0 & 0 & 0 & 1 \\
0 & 0 & 0 & 0 & 0 & 0 & 0 & 0 & 0 & 0 & 1 & 1 & 0 & 1 & 1 & 0 & 0 \\
0 & 0 & 1 & 0 & 0 & 1 & 0 & 0 & 0 & 0 & 0 & 0 & 0 & 0 & 0 & 0 & 0 \\
0 & 1 & 0 & 0 & 0 & 0 & 0 & 0 & 0 & 1 & 0 & 0 & 1 & 0 & 0 & 1 & 0 \\
0 & 1 & 0 & 0 & 0 & 0 & 0 & 0 & 0 & 0 & 0 & 0 & 1 & 0 & 0 & 0 & 0
\end{array}\right]
\end{array}
$$

In the above matrix: T1-M12 Wrench; T2-M16 Wrench; T3-M20 Wrench; T4-M24 Wrench; T5-Hammer; T6-Sleeve; PI (I=5,6,8,…,35) is the part In fig.2 whose number is I.

Vibration generator's integrated interference matrix can be obtained through the same method which is introduced in detail in 2.1. The parameters of ACA are set as follows:

$$\alpha=1.0,\ \beta=0.8,\ \mu=0.6,\ \rho=0.2,\ \gamma=0.2,\ \tau0=1.0,\ Q=1.5,\ \omega1=0.7,\ \omega2=0.3,\ m=40,\ N=1000$$

Through calculation, the optimal assembly sequence is generated and is shown as follows:

1→6→26→3→15→7→29→4→5→25→13→14→33→2→34→35→21→16→20→ 19→18→ 17→32→31→30→22→23→24→8→9→12→28→10→27→11

The optimal assembly sequence accords with the actual assembly situation,thus the validity and feasibility of the proposed algorithm have been verified.

4 Conclusions and Future Work

According to the characteristics of the ASP problem, a method based on ACA for ASP is presented. The evaluation mechanism of assembly sequence is introduced to the mathematical model of ACA, which is contributed to obtain optimal assembly sequence. Finally, the validity and feasibility of the proposed algorithm have been verified through an exemplification of vibration generator's ASP.In the next phase, research work should focus on solving more complex product's ASP problem, besides ,the more personnel and tools collaborative ASP problem also need to be solved.

Acknowledgements. The support of National Natural Science Foundation of China (No. 51005231), National Science Foundation for Post-doctoral Scientists of China (No. 20100471408) in carrying out this research is gratefully acknowledged.

References

1. Wu, W.-J.: Research and application on virtual assembly technologies of aircraft based on DELMIA. Nanjing University of Aeronautics and Astronautics 5, 7–8 (2008)
2. Yin, Z.-P., Ding, H., Li, H.-X.: A connector-based hierarchical approach to assembly sequence planning for mechanical assemblies. Computer-Aided Design 35, 37–56 (2003)

3. Sun, K., Miu, W.: Assemble Sequence Planning Based on the Self Adaptive ant Colony Algorithm in the Small Satellite. Machine Building & Automation 36, 24–27 (2007)
4. Peng, T., Li, S., Wang, J.: Integrated interference matrix based ant colony algorithm for assembly sequence planning. Computer Science 37, 179–204 (2010)
5. Mullen, R.J., Monekosso, D., Barman, S.: A review of ant algorithms. Expert Systems with Applications 36, 9608–9617 (2009)
6. Liu, X., Zhang, X., Liu, W.: Multi2process routes decision2making methodology based on improved max2min ant system. Computer Integrated Manufacturing Systems 14, 2414–2420 (2008)
7. Zhang, X., Li, S., Wang, J.: Single object selective disassembly sequence planning based on ant colony algorithm. Computer Integrated Manufacturing Systems 13, 1109–1114 (2007)
8. Wu, S., Cui, H.: Virtual assembly sequence planning based on graphics theory and ant colony optimization. Computer Engineering and Design 30, 2270–2273 (2009)
9. Shi, S., Li, R., Fu, Y.: Assembly sequence planning based on improved ant colony algorithm. Computer Integrated Manufacturing Systems 16, 1189–1194 (2010)
10. Xue, J., Qiu, C., Xiang, D.: Planning of selective disassembly sequence based on ant colony optimization algorithm. Journal of Computer-aided Design & Computer Graphics 19, 742–747 (2007)

The Feasible Algorithm Based on a Class of Multiple Affine Bi-Frames with Arbitrary Real Translation Number

Yongcai Hu

Department of Fundamentals, Henan Polytechnic Institute, Nanyang 473009, P.R. China
nysslt88@126.com

Abstract. Wavelet analysis is nowadays a widely used tool in applied mathematics. Frame theory has been the focus of active research for twenty years. In this paper, the bivariate generalized multiresolution analysis (BGMA) of space $L^2(R^2)$ is proposed. The traits of bivariate multiple pseudoframes with arbitrary real translation number for subspaces is studied. The feasible constructive algorithm of a GMS of Paley-Wiener subspace of $L^2(R^2)$ is provided. The pyramid decomposition scheme is obtained based on such a BGMA and a sufficient condition for its existence is provided. A constructive method for affine frames of $L^2(R^2)$ based on a BGMA is established.

Keywords: Affine multiple pseudoframes, bivariate, generalized multiresolution analysis, filter functions, affine bi-frames, multi-scale dilation factor, functional analysis method.

1 Introduction

Wavelet analysis is nowadays a widely used tool in applied mathematics. The frame theory has been one of powerful tools for researching into wavelets. The structured frames are much easier to be constructed than structured orthonormal bases. Duffin and Schaeffer introduced the notion of frames for a separable Hilbert space in 1952. Later, Daubechies, Grossmann, Meyer, Benedetto, Ron revived the study of frames in[1,2], and since then, frames have become the focus of active research, both in theory and in applications, such as signal processing, image processing and sampling theory. The rise of frame theory in applied mathematics is due to the flexibility and redundancy of frames, where robustness, error tolerance and noise suppression play a vital role [3,4]. The concept of frame multiresolution analysis (FMRA) as described in [2] generalizes the notion of MRA by allowing non-exact affine frames. However, subspaces at different resolutions in a FMRA are still generated by a frame formed by translates and dilates of a single function. This paper is motivated from the observation that standard methods in sampling theory provide examples of multiresolution structure which are not FMRAs. Inspired by [2] and [5], we introduce the notion of a bivariate generalized

Y. Zhang (Ed.): Future Communication, Computing, Control and Management, LNEE 141, pp. 405–414.
springerlink.com

multiresolution analysis (BGMA) of $L^2(R^2)$, which has a pyramid decomposition scheme. It also lead to new constructions of affine frames of $L^2(R^2)$.

Let U be a scparable Hilbert space and Λ is an index set. We recall that a sequence $\{f_v : v\in Z\}\subseteq U$ is a frame for H if there exist positive real numbers B, C such that

$$\forall\ \Upsilon\in U, \qquad B\|\Upsilon\|^2 \le \sum_{v\in\Lambda}|\langle\Upsilon, f_v\rangle|^2 \le C\|\Upsilon\|^2 \tag{1}$$

A sequence $\{f_v : v\in Z\}\subseteq U$ is a Bessel sequence if (only) the upper inequality of (1) holds. If only for all $\Upsilon\in\Omega\subset U$, the upper inequality of (1) holds, the sequence $\{f_v\}\subseteq U$ is a Bessel sequence with respect to (w.r.t.) Ω. If $\{f_v\}$ is a frame, there exists a dual frame $\{f_v^*\}$ such that

$$\forall\Upsilon\in W,\quad \Upsilon=\sum_{v\in\Lambda}\langle\Upsilon, f_v\rangle f_v^* = \sum_{v\in\Lambda}\langle\Upsilon, f_v^*\rangle f_v. \tag{2}$$

To state our results, the Fourier transform of an integrable function $f(x)\in L^1(R^2)$ is defined by

$$Ff(\omega)=\hat{f}(\omega)=\int_{R^2} f(x)e^{-2\pi ix\omega}dx,\quad \omega\in R^2, \tag{3}$$

which, as usual, can be naturally extended to functions in $L^2(R^2)$. For a sequence $c=\{c(v)\}\in\ell^2(Z)$, we define its discrete-time Fourier transform as the function in $L^2(0,1)^2$ by

$$Fc(\omega)=C(\omega)=\sum_{v\in Z^2} c(v)e^{-2\pi ix\omega}dx, \tag{4}$$

Note that the discrete-time Fourier transform is 1-periodic. Let $T_v\phi(x)$ stand for integer translates of a function $\phi(x)\in L^2(R^2)$, i.e., $(T_{va}\phi)(x)=\phi(x-va)$, and $\phi_{n,va}=m^n\phi(m^n x-va)$, where a is a positive real constant number. Let $\hbar(x)\in L^2(R^2)$ and let $V_0=\overline{span}\{T_v\hbar : v\in Z^2\}$ be a closed subspace of $L^2(R^2)$. Assume that $\mathrm{H}(\omega):=\sum_v|\hat{\hbar}(\omega+v)|^2\in L^\infty[0,1]^2$. In $[5]$, the sequence $\{T_v\hbar(x)\}_v$ is a frame for V_0 if and only if there exist positive constants L_1 and L_2 such that

$$L_1\le \mathrm{H}(\omega)\le L_2,\ \text{ a.e., }\omega\in[0,1]^2\setminus N=\{\omega\in[0,1]^2 : \mathrm{H}(\omega)=0\}. \tag{5}$$

2 Binary Generalized Multiresolution Analysis and Pseudoframes of Subspace

To characterize such a BGMS, we first introduce the concept of pseudoframes of translates.

Definition 1. Let $\{T_{va}\Upsilon,\ v\in Z^n\}$ and $\{T_{va}\widetilde{\Upsilon},\ v\in Z^n\}$ be two sequences in $L^2(R^n)$. Let U be a closed subspace of $L^2(R^n)$. We say $\{T_{va}\Upsilon,\ v\in Z^n\}$ forms an affine pseudoframe for U with respect to $\{T_{va}\widetilde{\Upsilon},\ v\in Z^n\}$ if

$$\forall\, f(x)\in U\,,\ f(x)=\sum\nolimits_{v\in Z^n}\left\langle f,T_{va}\widetilde{\Upsilon}\right\rangle T_{va}\Upsilon(x) \tag{6}$$

Define an operator $K:U\rightarrow \ell^2(Z^n)$ by

$$\forall\, f(x)\in U\,,\quad Kf=\{\langle f,T_{va}\Upsilon\rangle\}, \tag{7}$$

and define another operator $S:\ell^2(Z^n)\rightarrow U$ such that

$$\forall c=\{c(k)\}\in \ell^2(Z^2),\ Sc=\sum\nolimits_{v\in Z^2}c(v)T_{va}\widetilde{\Upsilon}\,. \tag{8}$$

Theorem 1. Let $\{T_{va}\lambda\}_{v\in Z^n}\subset L^2(R^n)$ be a Bessel sequence w.r.t. the subspace $U\subset L^2(R^n)$, and $\left\{T_{va}\widetilde{\Upsilon}\right\}_{v\in Z^2}$ is a Bessel sequence in $L^2(R^n)$. Assume K be defined by (7), and S be defined by(8).

Assume P is a projection from $L^2(R^n)$ onto U. Then $\{T_{va}\lambda\}_{v\in Z^n}$ is pseudoframes of translates for U with respect to $\{T_{va}\widetilde{\lambda}\}_{v\in Z^n}$ if and only if

$$KSP=P\,. \tag{9}$$

Proof. The convergence of all summations of (7) and (8) follows from the assumptions that $\{T_{va}\lambda\}_{v\in Z^n}$ is a Bessel sequence with respect to the subspace Ω, and $\{T_{va}\widetilde{\lambda}\}_{v\in Z^n}$ is a Bessel sequence in $L^2(R^n)$, with which the proof of the theorem is direct forward.

We say that a bivariate generalized multiresolution structure (BGMS) $\{V_n,\lambda(x),\widetilde{\lambda}(x)\}$ of $L^2(R^n)$ is a sequence of closed linear subspaces $\{V_n\}_{n\in Z}$ of $L^2(R^n)$ and two elements $\lambda(x)$, $\widetilde{\lambda}(x)\in L^2(R^n)$ such that (i) $V_n\subset V_{n+1}$, $n\in Z$; (ii) $\bigcap_{n\in Z}V_n=\{0\}$; $\bigcup_{n\in Z}V_n$ is dense in $L^2(R^n)$; (iii) $h(x)\in V_n$ if and only if $h(mx) \in V_{n+1}$ $\forall n \in Z$ (iv) $g(x) \in V_0$ implies $T_{va} g(x) \in V_0$, for $v \in Z$; (v) $\{T_{va}\lambda\}_{v \in Z^n}$ forms pseudoframes of translates for with respect to $\{T_{va}\tilde{\lambda}, v \in Z^n\}$.

Proposition 1[6]. Let $f \in L^2(R^n)$, $n = 2$ satisfy $|\hat{f}| > 0$ a.e. on a connected neighbourhood of 0 in $[-\frac{1}{2}, \frac{1}{2})^2$, and $|\hat{f}| = 0$ a.e. otherwise. Define $\Lambda \equiv \{\omega \in R^2 : |\hat{f}(\omega)| \geq C > 0\}$, and $V_0 = PW_\Lambda = \{\phi \in L^2(R^2) :$

$\text{supp}(\hat{\phi}) \subseteq \Lambda\}$. Then for $\tilde{f} \in L^2(R^2)$, $\{T_v f, v \in Z^2\}$ is pseudoframes of translates for V_0 with respect to $\{T_v \tilde{f}, v \in Z^2\}$ if and only if

$$\overline{\hat{f}(\omega)}\hat{\tilde{f}}(\omega)\chi_\Lambda(\omega) = \chi_\Lambda(\omega) \quad \text{a. e.,} \tag{11}$$

where χ_Λ is the characteristic function on Λ. Moreover, if $\tilde{f}(\omega)$ is the above conditions then $\{T_v f, v \in Z^2\}$ and $\{T_v \tilde{f}, v \in Z^2\}$ are a pair of commutative pseudoframes of translates for V_0, i. .e.,

$$\forall\, \Gamma(x) \in V_0, \quad \Gamma(x) = \sum_{k \in Z^2} \langle \Gamma, T_k \tilde{f} \rangle T_k f(x) = \sum_{k \in Z^2} \langle \Gamma, T_k f \rangle T_k \tilde{f}(x). \tag{12}$$

Proposition 2[5]. Let $\{T_{va} f\}$ be pseudoframes of translates for V_0 w. r. t. $\{T_{va} \tilde{f}\}$. Define V_n by

$$V_n \equiv \{\Upsilon(x) \in L^2(R^2) : \Upsilon(x/4^n) \in V_0\}, \quad n \in Z, \tag{13}$$

Then, $\{f_{n,va}\}_{v \in Z^2}$ is an affine pseudoframe for V_n with respect to $\{\tilde{f}_{n,va}\}_{v \in Z^2}$.

The filter functions associated with a BGMS are presented as follows. Define filter functions $D_0(\omega)$ and $\tilde{D}_0(\omega)$ by $D_0(\omega) = \sum_{s \in Z^2} d_0(s) e^{-2\pi i \omega}$ and $\tilde{B}_0(\omega) = \sum_{s \in Z^2} \tilde{b}_0(s) e^{-2\pi i \omega}$ of the sequences $d_0 = \{d_0(s)\}$ and $\tilde{d}_0 = \{\tilde{d}_0(s)\}$, respectively, wherever the sum is defined. Let $\{b_0(v)\}$ be such that $D_0(0) = 2$ and $B_0(\omega) \neq 0$ in a neighborhoood of 0. Assume also that $|D_0(\omega)| \leq 2$. Then there exists $f(x) \in L^2(R^2)$ (see ref.[3]) such that

$$f(x) = 2\sum_{s \in Z^2} d_0(s) f(4x - sa). \tag{14}$$

There exists a scaling relationship for $\widetilde{f}(x)$ under the same conditions as that of d_0 for a seq. $\widetilde{d_0}$, i.e.,

$$\widetilde{f}(x)=2\sum_{s\in Z^2}\widetilde{d_0}(v)\widetilde{f}(4x-sa). \tag{15}$$

3 Construction of Affine Frames for $L^2(R^2)$

In order to split a function $f(x)$ of V_1 into two functions (mostly) in V_0 and X_0, respectively, we will construct an affine pseudoframe for V_0 making use of the existing affine pseudoframe structure for V_0. Conventional symbols, $\Upsilon(x)$ and $\widetilde{\Upsilon}(x)$, will be used as generating functions for X_0. But they need not be contained in X_0.

Definition 2. Let $\{V_n, f(x), \widetilde{f}(x)\}$ be a given GMS, and let $\Upsilon_\iota(x)$ and $\widetilde{\Upsilon}_\iota(x)(\iota=1,2,3)$ be 4 band-passfunctions in $L^2(R^2)$. We say $\{T_{va}f(x), T_{va}\widetilde{\Upsilon}_l(x), l\in I\}$ form pseudoframes of translates V_1 with respect to $\{T_{va}\widetilde{f}(x), T_{va}\widetilde{\Upsilon}_l(x), l\in I\}$, if

$\forall \Gamma(x)\in V_1$,

$$\Gamma(x)=\sum_{v\in Z^2}\left\langle \Gamma, T_{sa}\widetilde{f}\right\rangle T_{sa}f(x)+\sum_{l=1,2,3}\sum_{s\in Z^2}\left\langle \Gamma, T_{sa}\widetilde{\Upsilon}_l\right\rangle T_{sa}\Upsilon_l(x) \ , \tag{16}$$

where the index set $I=\{1,2,3\}$. Accordingly, $\{T_{va}\widetilde{f}, T_{va}\widetilde{\Upsilon}\iota, \iota\in I\}$ is called a dual pseudoframe of translates to $\{T_{va}f, T_{va}\Upsilon_\iota, \iota\in I\}$ in the sense of (16).

Proposition 3[6]. Let $\{V_n, f, \widetilde{f}\}$ be a given BGMS, and let $\{T_{sa}f, T_{sa}\Upsilon_\iota,\ \iota\in I\}$ be an affine pseusoframe for V_1 with respect to $\{T_{sa}\widetilde{f}, T_{sa}\widetilde{\Upsilon}_\iota,\ \iota\in I\}$. Then, for each $n\in Z$, the family of functions

$\{f_{n,va}, \Upsilon_{l:n,va}, l\in I\}$ forms pseudoframes of translates for V_{n+1} w. r. t. $\{\widetilde{f}_{n,sa}, \widetilde{\Upsilon}_{l:n,sa}, l\in I\}$, i.e.,

$\forall\, \Psi(x)\in V_{n+1}$,

$$\Psi(x)=\sum_{s\in Z^2}\left\langle \Psi, \widetilde{f}_{n,sa}\right\rangle f_{n,sa}+\sum_{\iota\in I}\sum_{v\in Z^2}\left\langle \Psi, \widetilde{\Upsilon}_{\iota:n,sa}\right\rangle \Upsilon_{\iota:n,sa}.$$

To characterize the condition for which $\{T_{sa}f, T_{sa}\Upsilon_\iota,\ \iota\in I\}$ forms an affine pseudoframe for V_1 with respect to $\{T_{sa}\tilde{f}, T_{sa}\tilde{\Upsilon}_\iota,\ \iota\in I\}$, we begin with developing the wavelet equations associated with band-pass functions Υ_ι $(\iota\in\{1,2,3\})$ and $\tilde{\Upsilon}_\iota$ $(\iota\in\{1,2,3\})$ based on a GMS, namely,

$$\Upsilon_\iota(x) = 2\sum_{s\in Z} d_\iota f(4x - sa),\quad \iota = 1,2,3 \ \text{in}\, L^2(R) \tag{17}$$

$$\tilde{\Upsilon}_\iota(x) = 2\cdot\sum_{s\in Z^2} \tilde{d}_\iota \hbar(4x - sa), l = 1,2 \quad \text{in } L^2(R). \tag{18}$$

Let $\chi_\Lambda(\omega)$ be the characteristic function of the interval Λ defined in Proposition 1. We shall use the following 1-periodic function $\Gamma_\Lambda(\omega) \equiv \sum_k \chi_\Lambda(\omega + k)$.

Theorem 2. Let Λ be the bandwidth of the subspace V_0 defined in Proposition 1. $\{T_v f, T_v\Upsilon_l,\ v\in Z,\ l\in I\}$ forms pseudoframes of translates for V_1 with respect to $\{T_v\tilde{f}, T_v\tilde{\Upsilon}_\iota, l\in I\}$ if and only if there exist D_0 and D_l, $(l\in I)$ in $L^2([0,1])$) such that

$$\sum_{l=0}^{3} D_l(\omega)\overline{\tilde{B}_\iota(\omega)}\Gamma_\Lambda(\omega) = 4\Gamma_\Lambda(\omega) \tag{19-1}$$

$$\sum_{l=0}^{3} D_l(\omega + (1/4)I_2)\overline{\tilde{B}_\iota(\omega + (1/4)I_2)}\Gamma_\Lambda(\omega) = 0 \tag{19-2}$$

$$\sum_{l=0}^{3} D_l(\omega + (1/2)I_2)\overline{\tilde{B}_\iota(\omega + (1/2)I_2)}\,\Gamma_\Lambda(\omega) = 0. \tag{19-3}$$

$$\sum_{l=0}^{3} D_l(\omega + (3/4)I_2)\overline{\tilde{B}_\iota(\omega + (3/4)I_2)}\Gamma_\Lambda(\omega) = 0. \tag{19-4}$$

Proof. Since the family $\{\tilde{f}_{1,n}\}$ is complete when restricted to V, equation (17) holds if and only if

$\forall n\in Z$,

$$\langle\tilde{f}_{1,n}\rangle = \sum_{v\in Z}\langle\Phi, T_v\tilde{f}\rangle\langle T_v f, \tilde{f}_{1,n}\rangle + \sum_{l\in I}\sum_{v\in Z}\langle\Phi, T_v\tilde{\Upsilon}_\iota\rangle\langle T_v\Upsilon_l, \tilde{f}_{1,n}\rangle. \tag{20}$$

Define $c_0(v) = \langle\Phi, T_v\tilde{f}\rangle$, $c_1(v) = \langle\Phi, \tilde{f}_{1,v}\rangle$, $g_\iota(v) = \langle\Phi, T_v\tilde{\Upsilon}_\iota\rangle$, $\iota\in I$, , and denote by

$$b_0(n-4v) = \langle T_v f, \tilde{f}_{1,n}\rangle, b_v(n-4v) = \langle T_v\Upsilon_l, \tilde{f}_{1,n}\rangle,\quad l\in I, \tag{21}$$

where the $n-4v$ indexing is easily verificable. Then equation (21) becomes

$$\forall n\in Z \quad , \quad c_1(n)=\sum_{v\in Z^2}c_0(v)d_0(n-4v)+\sum_{v\in I}\sum_{v\in Z^2}g_v(\upsilon)d_v(n-4v).$$

Implementing the Fourier transform, we arrive at

$$C_1(\omega)=C_0(4\omega)D_0(\omega)+\sum_{l\in Z}G_l(4\omega)D_l(\omega), \tag{22}$$

where $C_i(\omega)=Fc_i(\omega)$, $i=0,1$ and $G_l(\omega)=Fg_l(\omega)(l\in I)$, are the discrete Fourier transform of $c_0=\{c_0(k)\}$, $c_1=\{c_1(k)\}$ and $g_l=\{g_l(k)\}(l\in I)$, respectively. Note that (15) and (19) also imply $c_0(v)=\sum_{n\in Z}\overline{\tilde{b}_0(n-3v)c_1(n)}$, And $g_l(v)=\sum_{n\in Z}\overline{\tilde{b}_0(n-3v)c_1(n)}$, their discrete Fourier transform are, respectively,

$$4C_0(4\omega)=\sum_{\sigma=0}^{3}C_1\left(\omega+\frac{\sigma}{4}\right)\overline{\tilde{B}_0\left(\omega+\frac{\sigma}{4}\right)} \quad ,$$

$$4G_l(4\omega)=\sum_{\sigma=0}^{3}C_1\left(\omega+\frac{\sigma}{4}\right)\overline{\tilde{B}_0\left(\omega+\frac{\sigma}{4}\right)}, l\in I,$$

Combining (22), (23) and (24), we get that

$$\begin{aligned}3C_1(\omega)&=D_0(\omega)\sum_{\sigma=0}^{3}C_1\left(\omega+\frac{\sigma}{4}\right)\overline{\tilde{B}_0\left(\omega+\frac{\sigma}{4}\right)}+\sum_{l\in I}D_l(\omega)\sum_{\sigma=0}^{3}C_1\left(\omega+\frac{\sigma}{4}\right)\overline{\tilde{B}_l\left(\omega+\frac{\sigma}{4}\right)}\\&=\left[D_0(\omega)\overline{\tilde{B}_0(\omega)}+\sum_{l\in I}D_l(\omega)\overline{\tilde{B}_v(\omega)}\right]C_1(\omega)+\sum_{l=0}^{3}D_l(\omega)\overline{\tilde{B}_l(\omega+1/4)}C_1(\omega+1/4)\\&+\sum_{l=0}^{3}D_l(\omega)\overline{\tilde{B}_l(\omega+1/2)}C_1(\omega+1/2)+\sum_{l=0}^{3}D_l(\omega)\overline{\tilde{B}_l(\omega+3/4)}C_1(\omega+3/4)\end{aligned}$$

This relationship is to hold for all $\Phi\in V_1$. Moreover, for any $\Phi\in V_1$ such that $C_1(\omega)C_1(\omega+1/4)\cdot$

$\cdot C_1(\omega+2/4)\cdot C_1(\omega+3/4)=0$, the relationship holds only if (19) follows. The if part is clear. This establishes the result.

Definition 3. Let $\{V_\ell,\lambda,\tilde{\lambda}\}$ be a given GMS. We say that the GMS has a pyramid decomposition scheme if there are band-pass functions $\Upsilon_l,\tilde{\Upsilon}_l\in L^2(R^2)$, $l\in I$ such that

$$\forall \lambda(x)\in L^2(R^2) \quad ,$$

$$\sum_{v\in Z^2}\left\langle\Phi,\tilde{\lambda}_{1,va}\right\rangle\lambda_{1,va}=\sum_{v\in Z^2}\left\langle\Phi,T_{va}\tilde{\lambda}\right\rangle T_v\lambda+\sum_{l\in I}\sum_{v\in Z^2}\left\langle\Phi,T_{va}\tilde{\Upsilon}_l\right\rangle T_{va}\Upsilon_l. \tag{23}$$

Theorem 3. Let $f(x), \widetilde{f}(x), \Upsilon_{\iota}(x)$ and $\widetilde{\Upsilon}_{\iota}(x), \iota \in I$ be functions in $L^2\left(R^2\right)$ defined by (14), (15), (17) and (18), respectively. Assume that conditions in Theorem 2 are satisfied. Then, for any function $\Gamma(x) \in L^2\left(R^2\right)$, and any integer n,

$$\sum_{k \in Z^2} \left\langle \Gamma, \widetilde{f}_{n,k} \right\rangle f_{n,k}(x) = \sum_{\iota=1}^{3} \sum_{v=-\infty}^{n-1} \sum_{k \in Z^2} \left\langle \Gamma, \widetilde{\Upsilon}_{\iota:v,k} \right\rangle \Upsilon_{\iota:v,k}(x).$$

Furthermore, for any $\Gamma(x) \in L^2\left(R^2\right)$, $\Gamma(x) = \sum_{\iota=1}^{3} \sum_{v=-\infty}^{\infty} \sum_{k \in Z^2} \left\langle \Gamma, \widetilde{\Upsilon}_{\iota:v,k} \right\rangle \Upsilon_{\iota:v,k}(x)$.

Consequently, if $\left\{\Upsilon_{\iota:v,k}\right\}$ and $\left\{\widetilde{\Upsilon}_{\iota:v,k}\right\}$, $(\iota \in \Lambda, v \in Z,\ k \in Z^2)$ are also Bessel sequences, they are a pair of affine frames for $L^2\left(R^2\right)$.

Proof. (i) Consider, for $\sigma \geq 0$, $\sigma \in Z$, the operator $E_\sigma : L^2\left(R^2\right) \to L^2\left(R^2\right)$ such that

$$E_\sigma \Gamma(x) \equiv \Gamma_\sigma(x) \equiv \sum_{k \in Z^2} \left\langle \Gamma, \widetilde{f}_{(-\sigma),k} \right\rangle f_{(-\sigma),k}(x).$$

Then the operator E_σ are well defined and uniformly bounded in the norm on $L^2\left(R^2\right)$. To show that $E_\sigma \to 0$ as $\sigma \to \infty$, it is sufficient to show that, for all $g(x)$ in any dense subspace of band-limited functions in $L^2\left(R^2\right)$, $\sum_{k \in Z^2} \left\langle g, \widetilde{f}_{(-\sigma),k} \right\rangle f_{(-\sigma),k} \to 0$ as $\sigma \to \infty$ .In particular, we may choose the dense set of functions $g(x)$, whose Fourier transform have compact support, is continuous, and vanishes in a neighborhood of 0.

$$\left\| \sum_k \left\langle g, \tilde{\phi}_{(-\sigma),k} \right\rangle \phi_{(-\sigma),k} \right\| \leq \sup_{\|h\|=1} \left(\sum_k \left| \left\langle g, \tilde{\phi}_{(-\sigma),k} \right\rangle \right|^2 \right)^{1/2} \left(\sum_k \left| \left\langle h, \phi_{(-\sigma),k} \right\rangle \right|^2 \right)^{1/2} \leq$$

$$B^{1/2} \left(\sum_k \left| \left\langle g, \tilde{\phi}_{(-\sigma),k} \right\rangle \right|^2 \right)^{1/2}$$

where B is the Bessel bound of $\left\{\phi_{-\sigma,k}\right\}_k$. Taking standard calculation of the above inequlity gives

$$\sum_k \left| \left\langle g, \tilde{\phi}_{(-\sigma),k} \right\rangle \right|^2 = \int \sum_u \hat{g}\left(\omega + 2^{-\sigma} u\right) \overline{\hat{\tilde{\phi}}_v \left(2^\sigma \omega + u\right)} \hat{\tilde{\phi}}_v \left(4^\sigma \omega\right) \overline{\hat{g}(\omega)} d\omega$$

$\leq \quad \int\left(\sum_u\left|\hat{g}\left(\omega+4^{-\sigma}u\right)\right|^2\right)^{1/2}\left(\sum_u\left|\hat{\tilde{\phi}}_v\left(4^{\sigma}\omega+u\right)\right|^2\right)^{1/2}\cdot\left|\hat{\tilde{\phi}}_v\left(4^{\sigma}\omega\right)\hat{g}(\omega)\right|d\omega \quad \leq$

$\left(B^*\right)^{1/2}\int\left(4^{-\sigma}\sum_u\left|\hat{g}\left(\omega+4^{-\sigma}u\right)^2\right|\right)^{1/2}\cdot 4^{\sigma}\left|\hat{\tilde{\phi}}_v\left(4^{\sigma}\omega\right)\hat{g}(\omega)\right|d\omega$

where B^* is the Bessel bound of $\left\{\tilde{\phi}_{-\sigma,k}\right\}_k$. Following the lead of [8] and since $\hat{f}(\omega)$ is continuous with compact support, the term $4^{-\sigma}\sum_n\left(\left|\hat{g}\left(\omega+4^{-\sigma}n\right)\right|^2\right)^{1/2}\leq C^2$ $<+\infty$, being a Riemann sum to the finite integral $\int\left|\hat{g}(\omega+t)\right|dt$. Moreover, since $\hat{g}(\omega)$ vanishes in a neighborhood of 0 for all $\|\omega\|<\delta_g$, we get that

$\sum_k\left|\left\langle g,\tilde{\phi}_{-\sigma,k}\right\rangle\right|^2 \quad \leq \quad \left(B^*\right)^{1/2}C\int\left|4^{\sigma}\hat{\tilde{\phi}}\left(4^{\sigma}\omega\right)\hat{g}(\omega)\right|d\omega \quad \leq$

$\left(B^*\right)^{1/2}C\left\|g\right\|_2\left(\int_{\|\omega\|\geq\delta_g}\left|4^{\sigma}\hat{\tilde{\phi}}\left(4^{\sigma}\omega\right)\right|^2d\omega\right)^{1/2}$

Note that the last integral at the right-hand side tends to 0as $\sigma\to+\infty$. This proves the first part of the theorem since, by using (18) recursively, we have

$$\hbar_{\sigma}(x)=\sum_{k\in Z^2}\left\langle\hbar,\tilde{\phi}_{n,k}\right\rangle\phi_{n,k}(x)-\sum_{\iota=1}^{3}\sum_{v=-\infty}^{n-1}\sum_{k\in Z^2}\left\langle\hbar,\tilde{\psi}_{\iota:v,k}\right\rangle\psi_{\iota:v,k}(x).$$

(ii) Since $\overline{\cup V_{\ell}}=L^2\left(R^2\right)$, for any $\hbar\in L^2\left(R^2\right)$ and any $\varepsilon>0$ there exists $n_0=n_0(\varepsilon)>0$, and for any $n>n_0$ there exists $g\in V_{n_0}\subset V_n$ such that $g(s)=\sum_{k\in Z^2}\left\langle g,\tilde{\phi}_{v,k}\right\rangle\phi_{v,k}(t)$ .Moreover, for $C=\sqrt{BB^*}$, $\left\|\hbar-g\right\|_2<(1+C)^{-1}\varepsilon$. Now, by (18), for all $n>n_0$, we have

$\left\|\hbar-\sum_{\iota=1}^{3}\sum_{v=-\infty}^{n-1}\sum_{k\in Z^2}\left\langle\hbar,\tilde{\psi}_{\iota:v,k}\right\rangle\psi_{\iota:v,k}\right\|_2 \quad \leq \quad \left\|\hbar-g\right\|_2+C\left\|\hbar-g\right\|_2 \quad =$

$\left\|\hbar-g\right\|_2(1+C)<\varepsilon$.

If $\left\{\phi_{v,k}\right\}$ and $\left\{\tilde{\phi}_{v,k}\right\}$ are Bessel sequences, then equation(19) implies that both $\left\{\phi_{v,k}\right\}$ and $\left\{\tilde{\phi}_{v,k}\right\}$ will be affineframes. In fact, the lower frame bound of $\left\{\phi_{v,k}\right\}$ or $\left\{\tilde{\phi}_{v,k}\right\}$ is implied by the upper Bessel bound of the other. Therefore,this completes the proof of the second part of Theorem 3.

4 Conclusion

We characterize the pseudoframes of translates for the subspaces of $L^2\left(R^2\right)$. The pyramid decomposition scheme is derived based on such a BGMA. As a major new constribution the construction of affine frames for $L^2\left(R^2\right)$ based on a BGMA is presented.

References

1. Daubechies, I., Grossmann, A., Meyer, A.: Painless nonorthogonal expansions. J. Math. Phys. 27, 1271–1283 (1986)
2. Benedetto, J.J., Li, S.: The theory of multiresolution analysis frames and applications to filter banks. Appl. Comput. Harmon. Anal. 5, 389–427 (1998)
3. Ron, A., Shen, Z.: Affine systems in L2(Rd) (II) Dual systems. J. Fourier Anal. Appl. 4, 617–637 (1997)
4. Daubechies, I.: Ten Lectures on Wavelets. SIAM, Philadelphia (1992)
5. Li, S., et al.: Pseudoframes for Subspaces with Applications. J. Four. Anal. Appl. 10, 409–431 (2004)
6. Li, S.: A Theory of Geeneralized Multiresolution Structure and Pseudoframes of Translates. J. Fourier Anal. Appl. 6(1), 23–40 (2001)

Dense Sub Graph Based Video Semantic Mining

Shilin Zhang[1,2], Heping Li[2], and Shuwu Zhang[2]

[1] Computer Faculty, North China University of Technology, Beijing, China
[2] High Technology and Innovation Center, Institute of Technology, Chinese Academy of Science, Beijing, China
zhangshilin@126.com

Abstract. In this paper, we bring forward a method which integrating continuous speech recognition, video caption text recognition and object recognition. The video is firstly segmented to a serial of shots by shot detection. Then the caption text and speech recognition results are treated as two paragraphs of text. The object recognition results are presented by bag of words. The above three aspects of texts are processed by part of speech and stemming. Then only the noun words are kept. At last a video is represented by three bags of words. The words are further depicted as a graph. The graph vertices stand for the words and the edges denote the semantic distance between two neighboring words. In the last step, we apply the dense sub graph finding method to mine the video semantic meaning. Experiments show that our video semantic mining method is efficient.

Keywords: Video Semantic Mining, Auto Speech Recognition, Information Fusion, Object Recognition.

1 Introduction

Over the past years, there has been an explosive growth of available videos on the internet. Good examples are the video sharing website “youtube” in America and “youku” in China, hosting billions of videos with thousands of uploads each minute. The advent of such collections has sparked research on video semantic mining within these collections to implement a system processing videos without human intervention; see Snoek and Worring for a recent overview [1].One active line of research is on using only the visual contents for concept-based management tasks. But in this paper, we intend to integrate three aspects of information, namely the visual content, the speech signal and the video caption text. The Bag-of-Words method [2-3] has proven to be the most efficient strategy as a generic classification scheme for individual concepts. This is proven by their top performance in various major benchmarks over the past few years such as the TRECVID high-level feature extraction task [4] and the Pascal VOC Challenge [5]. The same is true for video semantic mining. Multi-modal information fusion based video semantic mining is the frontier research direction, as is reported in [6-7].

This paper is organized as follows. Section 2 presents the speech, object and caption text recognition procedures. Section 3 shows the text processing and the graph construction. Section 4 depicts dense sub graph finding procedure and the video

Y. Zhang (Ed.): Future Communication, Computing, Control and Management, LNEE 141, pp. 415–419.
springerlink.com

semantic mining method. Experiments on realworld videos are shown in Section 5. We end the paper with a conclusion and some prospective future work in Section 6.

2 Multi-modal Video Content Analysis

2.1 Overall Workflow

The workflow of our system is illustrated in Fig.1.Four layers constitute our whole system, that are user layer, semantic video mining layer, multi-modal video analysis fusion layer and the video resource layer. The user layer provides video services. The second layer implements the video mining tasks. The third layer fusions the multi-modal video recognition results. And the bottom layer stores the videos resources. The details of every layer will be discussed in the following sections.

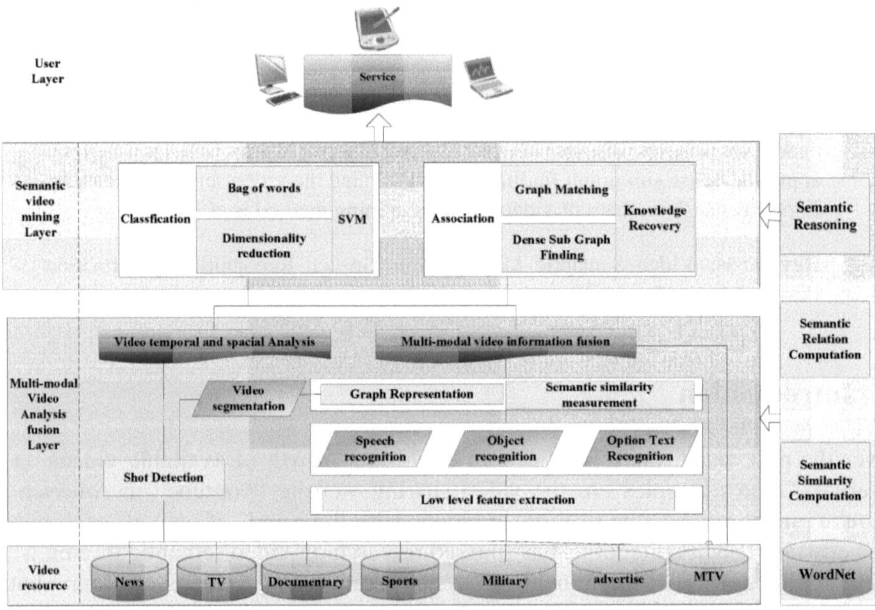

Fig. 1. Workflow of video semantic mining

2.2 Video Preprocessing

The videos are firstly segmented by silence detection and other shot detection methods [8]. The shots are the smallest unit in the following processing.

2.3 Continuous Speech Recognition

For every shot of videos, we implement an automatic speech recognition procedure to get the spoken words of the video. Speech recognition has achieved great improvements recently. So we can use the technique to get the speech content of the videos. However, speaker variability and the physical characteristics of speaker, especially accent, cause the speech recognition not robust.

3 Dense Sub Graph Construction and Mining

Through above analysis, the video is represented by three high level semantic features, namely the $\mathrm{Text}_{\mathrm{ASR}}$, $\mathrm{Text}_{\mathrm{VOCR}}$ and $\mathrm{Text}_{\mathrm{OBJECT}}$. They can be seen as three sub space of the video high level feature space. The three aspects of texts can be processed by part of speech and stemming, kept only the noun words to form three bags of words. So the video feature space can be seen as a tensor which constitutes three vectors. We can formally denote it as:

$$\mathrm{V} \in \Re^{\mathrm{T,A,I}} \tag{1}$$

Here, the V denotes the video's high level feature and is presented as a tensor. The T, A and I stand for the speech words vector, caption text words vector and the object words vector respectively. As the three aspects of words are highly correlative, the semantic similarity distance between them should be as short as possible. This can be seen as an optimization problem.

$$\min_{i,j} \sum_{\mathrm{m},n,m\neq n} \sum_{i,j,i\neq j} f(\mathrm{w}_i^m, \mathrm{w}_j^n) \tag{2}$$

$f(\bullet)$ denotes the similarity measure of two words, which is introduced in the next paragraph. "m" and "n" represent different modal text, such as $\mathrm{Text}_{\mathrm{ASR}}$, $\mathrm{Text}_{\mathrm{VOCR}}$ and $\mathrm{Text}_{\mathrm{OBJECT}}$. "i" and "j" denote different words. w_i^m stands for a word in modal m with sequence number i. The same is true with w_j^n.In the optimization procedure; we check every word in every modal. If a word is far away from the other two modal's words in semantic measure, in other words, there are no correspondences words in the other two modal, the isolated word will be dropped off. Though the above procedure is simple and comprehensible, the computation complexity is terrible.

4 Experiment

We evaluate the proposed method on TRECVID 07 dataset [11]. TRECVID is an annual video retrieval evaluation event organized by National Institute of Standards and Technology (NIST) to promote progress in content-based retrieval from digital video via open, metrics-based evaluation. The TRECVID 2007 dataset comprises 100 hours of documentary video with ~40, 000 key frames. The dataset is equally divided into development and test sets, with each containing approximately 50 hours and around 20,000 key frames. In our experiments, we use the 30 semantic concepts listed in Figure 2 used in the TRECVID 07 valuation. We used the ground truth annotations as baseline from MediaMill with a lexicon of 101 semantic concepts but we used only the most interesting 30 ones. Performance was then evaluated on the test set. We used the ground truth annotations as baseline from MediaMill with a lexicon of 101 semantic concepts but we used only the most interesting 30 ones. Performance was then evaluated on the test set.

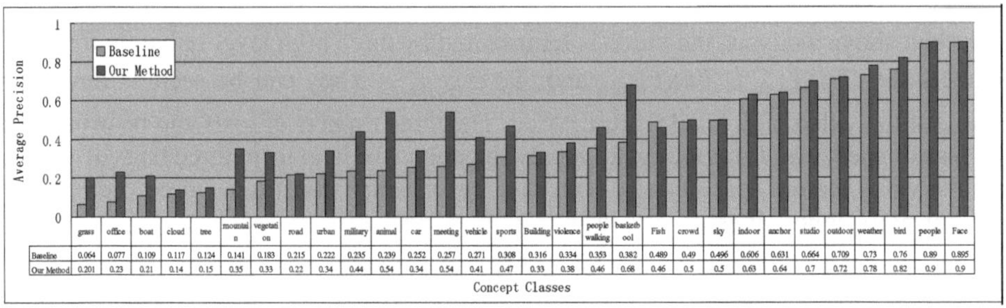

Fig. 2. AP of each concept by baseline and our method respectively

In Table 1, we illustrated some multi-modal recognition fusion results. The first column denotes the key frame of a video shot. The second column lists the annotation of our system.

Shot image	**annotation**	**Shot image**	**annotation**
	✓ Urban ✓ Car ✓ Building ✓ Road ✓ Outdoor		✓ Military ✓ Training ✓ Building ✓ Outdoor ✓ Troop
	✓ Anchor person ✓ Report ✓ News		✓ Meeting ✓ Crowd ✓ News ✓ Plan ✓ Indoor

5 Conclusion

In this paper, we have proposed a video semantic mining method which exploits the three aspects of modals in videos. By a graph based late fusion method, the text, audio and visual information can be unified to annotate a video. The experiments on TRECVID 07 showed that the graph based fusion can outperform other fusion method, such as baseline method, with statistically significant improvements. Several issues remain open. Our method depends on the accuracy of caption text recognition, speech recognition and object recognition. So the achievements in the above three aspects will boost out method. Secondly an effective semantic similarity and correlation computation method will make our method more accuracy.

References

1. Snoek, C.G.M., Worring, M.: Concept-based video retrieval. Trends Inf. Retriev. 4(2), 215 (2009)
2. Csurka, G., Dance, C., Fan, L., Willamowski, J., Bray, C.: Visual categorization with bags of keypoints. In: Proc. ECCV Int. Workshop Statistical Learning in Computer Vision, Prague, Czech Republic (2004)
3. Sivic, J., Zisserman, A.: Video Google: A text retrieval approach to object matching in videos. In: Proc. IEEE Int. Conf. Computer Vision (2003)
4. Smeaton, P.O., Kraaij, W.: Evaluation campaigns and TRECVID. In: Proc. ACM SIGMM Int. Workshop Multimedia Information Retrieval (2006)
5. Winn, J.: The PASCAL Visual Object Classes Challeng (VOC 2010) Development Kit, Tech. Rep., University of Leeds (2010)
6. Zhang, X., Liu, Y., Liang, C., Xu, C.: A visualized communication system using cross-media semantic association. In: Proceedings of the 17th International Conference on Advances in Multimedia Modelling, Taipei, Taiwan, January 5-7 (2011)
7. Hanbury, A., Müller, H.: Automated component-level evaluation, present and future. In: Proceedings of the 2010 International Conference on Multilingual and Multimodal Information Access Evaluation: Cross-Language Evaluation Forum, Padua, Italy, September 20-23 (2010)
8. Lee, H., Yu, J., Im, Y., et al.: A unified scheme of shot boundary detection and anchor shot detection in news video story parsing. Multimedia Tools And Applications, 1127 (2011)
9. Open Source Toolkit For Speech Recognition, `http://cmusphinx.sourceforge.net/`
10. RLeon, M., Mallo, S., Gasull, A.: A tree structured-based caption text detection approach. In: Proceedings of the Fifth IASTED International Conference on Visualization, Imaging, and Image Processing, p. 220 (2005)
11. Hoch, A.: On Finding Densest Subgraphs and their Applications. Technical report (2009)

CAM Software Products for Creation of Programs for CNC Machining

Peter Michalik[1], Jozef Zajac[1], Ján Dupľák[1], and Alexander Pivovarnik[2]

[1] Faculty of Manufacturing Technologies Technical University in Košice with seat in Prešov, Department of Production Technology, Slovak Republic
[2] INGMETAL, Prešov, Slovak Republic
{peter.michalik,jozef.zajac,jan.duplak}@tuke.sk, info@ingmetal.sk

Abstract. The paper deals with CAM software products that are used for creation of programs for CNC machining. Proceeding of programming itself depends on geometrical shape, requested accuracy and also on possibilities of software. On the base of accessibility there were software products selected: WorkNC, NX (Unigraphics) 7.5, Pro/Engineer Wildfire 5.

Keywords: CAM software, CNC program, milling.

1 Introduction

Computer-aided manufacturing (CAM) is use of computer software to control machine tools and related machinery in the manufacturing of workpieces. Most machining progresses through four stages, each of which is implemented by a variety of basic and sophisticated strategies, depending on the material and the software available [1,4,8,11].The stages are:

Roughing: This process begins with raw stock, known as billet, and cuts it very roughly to shape of the final model. In milling, the result often gives the appearance of terraces, because the strategy has taken advantage of the ability to cut the model horizontally. Common strategies are zig-zag clearing, offset clearing, plunge roughing, rest-roughing.

Semi-finishing: This process begins with a roughed part that unevenly approximates the model and cuts to within a fixed offset distance from the model. The semi-finishing pass must leave a small amount of material so the tool can cut accurately while finishing, but not so little that the tool and material deflect instead of shearing. Common strategies are raster passes, waterline passes, constant step-over passes, pencil milling.

Finishing: Finishing involves a slow pass across the material in very fine steps to produce the finished part. In finishing, the step between one pass and another is minimal. Feed rates are low and spindle speeds are raised to produce an accurate surface.

Contour milling: In milling applications on hardware with five or more axes, a separate finishing process called contouring can be performed. Instead of stepping

Y. Zhang (Ed.): Future Communication, Computing, Control and Management, LNEE 141, pp. 421–425.
springerlink.com

down in fine-grained increments to approximate a surface, the workpiece is rotated to make the cutting surfaces of the tool tangent to the ideal part features. This produces an excellent surface finish with high dimensional accuracy.

The first commercial applications of CAM was realized decades before. It was in large companies in the automotive and aerospace industries, for example UNISURF in 1971 at Renault for car body design and tooling. The output from the CAM software is usually a simple text file of G-code, sometimes few thousands of commands long, that is then transferred to a machine tool using a direct numerical control (DNC) program [3].

The 8 largest CAM software products and companies, by vendor revenues in year 2009 are, sorted alphabetically: BOBCAD-CAM from BobCAD-CAM, CATIA from Dassault Systèmes, CAM-Tool from C & G Systems, Cimatron from Cimatron group, Dynavista from Nihon Unisys, Edgecam from Planit, Esprit from DP Technoogy, HyperMill from Open Mind, Mastercam from CNC Software, NX (Unigraphics) from Siemens PLM Software, Powermill from Delcam, Pro/E from PTC, SolidCAM from SolidCAM, Space E from NTTD, SurfCAM from Surfware, TopCAM from Missler, Tebis from Tebis AG, VisiCAM from Vero, VisualMILL from MecSoft, Vericut from CGtech, WorkNC from Sescoi.

2 Component Modeling in CAD Software Products

CAD model of part was created using software ProEngineer WildFire 5, with use of features like Extrude, Removal Extrude, Round, Edge Chamfer (Fig.1). Modeling in NX 7.5 would be realized with use of same features and operations [2]. Philosophy of modeling in WorkNC is bit more complicated than in previous two cases.

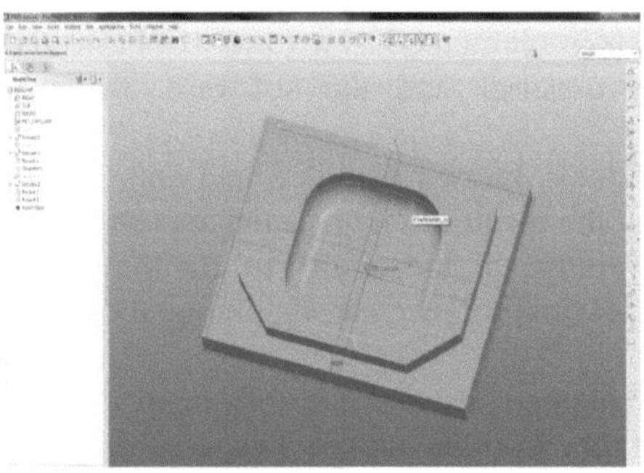

Fig. 1. Modeling in graphical environment ProE

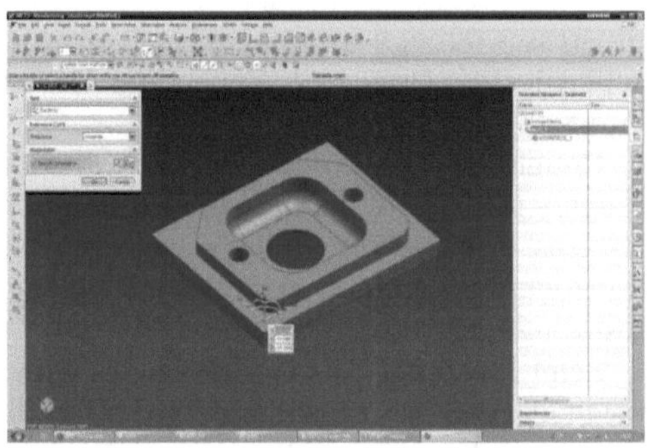

Fig. 2. Choice machine origin in NX 7.5

CAD model of part was saved in STEP (*.stp) format. Environments of different CAM software solutions are showed on further figures. They present Work NC (Fig.2), NX 7.5 and ProE Wildfire5.

3 Comparison and Evaluation of Creating of CNC Programs with Use of Different Cam Software

Prior to comparison of creation of CNC programs in environment of different software solution it is necessary to determine the categories and values that will be observed, measured and compared in order to make the final result objective and true. With this paper following values were selected for evaluation purposes [6,7]:

Number of mouse clicks while programming the particular operations [2,5,12].,

Time necessary for programs creation and completion of particular technological operations [9].

While setting particular technological operations and cutting conditions it is necessary to use the mouse clicks many times for selecting of actions, orders. For measuring of click count there were the same conditions set in all programs, for example also the initial clicks for software starting and environment getting ready were counted as well as clicks for creation and saving of part model [10]. Table 1 and Fig. 3 provides information on total click amount for selected technological operations.

Table 1. Number of clicks until the program was created

VALUE COMPARED	WORK NC	NX 7.5	PROE W5
Total click amount	334	185	285
Total time necessary for program creation	1169	686	907

Following figures shows the selection of technological parameters of manufacturing and also selection of tools in different CAM software and play path process with CL data: Work NC, NX 7.5, ProE Wildfire5.

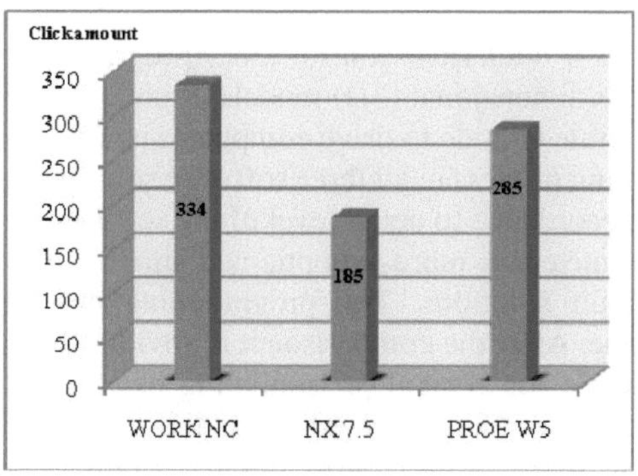

Fig. 3. Total click number for necessary

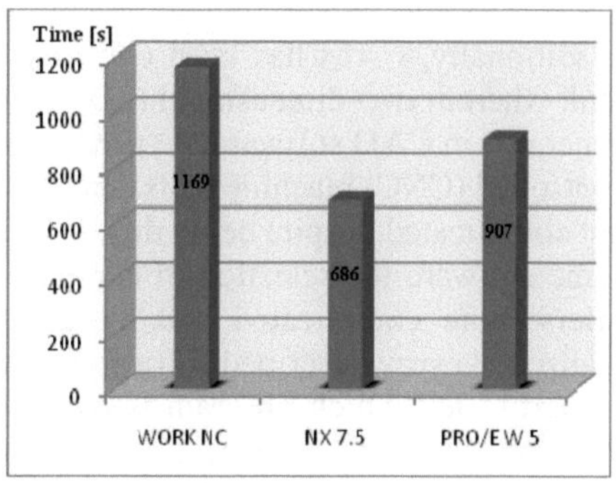

Fig. 4. Total programming time for creation of CNC program

4 Time of Model Cration

Measuring was started with first click on the icon of the program and stopped with complete and usable NC program created. Time measuring was realized between first and last mouse click. Table 1 and Fig. 4 provides information on total click amount for selected technological operations.

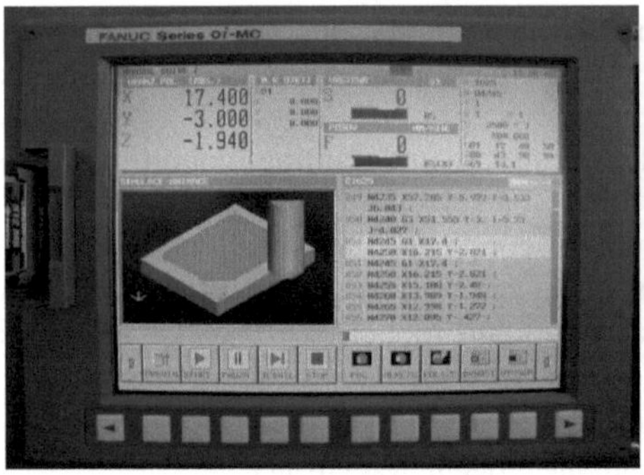

Fig. 5. Simulation on

Fig. 6. Part manufacturing on CNC machine

After generating of CL data and handling them with postprocessor they were inserted to control program of CNC machining center Pinnacle VMC 650-S, verified with use of simulation mode (Fig.5) and following manufacturing itself (Fig.6).

Fig.4 shows total time necessary for part programming in seconds. Shortest time of model creation was measured for NX 7.5 thanks to low number of necessary clicks. Second place took Pro/Engineer WildFire 5. Longest programming time was observed for WorkNC program, where the program creation took almost 20 minutes mainly because of complicated use of working windows.

5 Conclusion

Traditionally, CAM has been considered as a numerical control (NC) programming tool, wherein two-dimensional (2D) or three-dimensional (3D) models of components generated in CAD software are used to generate G-code to drive computer numerically controlled (CNC) machine tools. Programming process in all three software solutions is not complicated despite being different in accordance to detail level of the software. In some software the selection of certain parameters is more synoptic and simple, other offers more complicated and more difficult solutions. The programming process realized in every described software is unique. After the comparison it is obvious that it is hard to tell which program is most suitable for general programming of all parts. Program presenting the best solution here can change according to the change of criterion – time, working style, geometrical shape of part. Most suitable seems to be NX 7.5 mainly because of shortest programming time. This is not the only definition for CAM, but it seems to be the most common. CAM may also refer to the use of a computer to assist in all operations of a manufacturing plant, including planning, management, transportation and storage. CAM is a subsequent computer-aided process after computer-aided design (CAD) and sometimes computer-aided engineering (CAE), as the model generated in CAD and verified in CAE can present an input into CAM software, which then controls the machine tool.

Acknowledgement. The authors would like to thank in words the grant agency for supporting research work and cofinancing the projects: KEGA #3/7166/09 and VEGA #1/0048/10.

References

1. Murčinko, J.: The application of manufacturing templates in CADCAM systems -2008. Scientific Bulletin serie C 22, 1224–3264 (2008), ISSN 1224-3264
2. Jurko, J., Džupon, M., Panda, A., Gajdoš, M., Pandová, I.: Deformation of Material Under the Machined Surface in the Manufacture of Drilling Holes in Austenitic Stainless Steel. Chemickélisty 105(S) (2011), ISSN 0009-2770
3. Panda, J.J., Džupon, M., Pandová, I.: Optimalization of Heat Treatment Bearings rings With Goal to Eliminate Deformation of Material. Chemickélisty 105(S), 459–461 (2011), ISSN: 0009-2770
4. Jurko, J., Panda, A., Gajdoš, M.: Study of changes under the machined surface and accompanying phenomena in the cutting zone during drilling of stainless steels with low carbon content. Metalurgia Journal 50(2), 113–117 (2011), ISSN 0543-5846
5. Džupon, M., Gajdoš, M., Jurko, J., Ferdinandy, M., Jakubeczyová, D.: Plastic Deformation Around Holes Drilled in Austenitic Steel 1.4301. Chemickélisty 105(S), 606–608 (2011), ISSN 0009-2770
6. Jurko, J., Panda, A., Gajdoš, M.: Accompanying phenomena in the cutting zone machinability during turning of stainless steels. International Journal Machining and Machinability of Materials 5(4), 383–400 (2009), ISSN 1748-5711
7. Čilliková, M., Salaj, J., Mičietová, A., Pilc, J.: Influence of Cutting Fluids on Cutting Forces when Grinding Bearing Steel. Journal of Machine Manufacturing Design and Manufacturing, The Scientific Society for Mechanical Engineering XLIX(E3-E5), 101–103 (2009), INDEX: 25344 HU ISSN 0016-8580
8. Peterka, J., Pokorný, P., Václav, Š.: CAM strategies and surfaces accuracy. - registrovaný v ISI Proceedings. Annals of DAAAM and Proceedings of DAAAM Symposium 19(1), 1061–1062, ISSN 1726-9679
9. Žitňanský, M., Zrník, J., Martinkovič, M.: Developing processes of property improvement of nickel base superalloys. Journal of Materials Processing Technology 78, 204–209 (1998), ISSN 0924-0136
10. Mrkvica, V., Ochodek, R.: Influence of a Tool Geometry on Surface Tension of a Workpiece. International Journal Manufacturing Technology 8, 76–79 (2009), ISSN 1213-2489
11. Jurko, J., Lukovics, I.: Obrábateľnosť materiálov. 1. vyd - Zlín, p. 144. Univerzita Tomáše, Bati (2008), ISBN 978-80-7318-736-1
12. Panda, A., Jurko, J., Pandová, I.: Design and Evolution of New Article in the Earth. World Academy of Science. Engineering and Technology 41, 1193–1196 (2009), ISSN 2070-3740

Research on the Information Security Management in E-commerce System

Qin Qin[1] and Ge Li[2]

[1] School of Economic and Management,
Southwest University of Science and Technology Mianyang, 621010, China
[2] School of Economic and Management, Shanghai Ocean University
Shanghai 200090, China
qqxjm@163.com, 42324563@qq.com

Abstract. The issue of Information Resources security in E-commerce system is the most important core that E-commerce market faces with. It is also the important reason that constraints development of E-commerce. From the paper it starts with the current situation of information security management issues in E-commerce system, it introduces the related. Theory the existing problems of information security management issues in E-commerce system. Research on the data encryption and authentication techniques, and puts forward my own proposals of how to improve the information security management in E-commerce system.

Keywords: E-commerce, Information Security Management, Data Encryption, Authentication.

1 Introduction

Information security management is the basis for the development of E-commerce. With the rapid development of E-commerce, E-commerce is famous everywhere and each big enterprise, the means of the network exchange have become more diversified. Because the openness of the network, information security management may also become more outstanding. In order to solve this problem, the network must have security technology do guarantee. At present, the firewall technology, identity authentication technologies, and data encryption technology and calculation system security technology plays an important role. In addition, we still need to perfect the legal system, management system and credit system, so that we can guarantee the E-commerce information security management, to speed up the development of E-commerce.

Each country attaches importance to the development of the E-commerce safety problems research, especially the E-commerce information security standards of widespread advance. Such as the United States government studied and made it the password techniques standardization long ago. From the data encryption standard DES was published in1977, a series of relevant code of the federal information processing

Y. Zhang (Ed.): Future Communication, Computing, Control and Management, LNEE 141, pp. 427–431.
springerlink.com

technology standard (FIPS) which formulated by the American national standards institute of technology (NIST), Under the premise of the technical specification, it tested the password products strictly。 On July 1, 1998, the United States in the outline of the E-commerce put forward to establish some common standards, to ensure that through the network shopping consumers enjoy shopping consumers in the shop and have the same rights. South Korea established a union, signed a joint agreement, setted in 2000 of the E-commerce out the whole standard.

Although China's network information security started quite late, investment and attention more less than Japan, the United States and other developed countries. But in recent years, the state pay more attention to the network and information security, the industry has already developed a national information security management, policies, laws and regulations. Establishment of common standards in accordance with China National Information Security Testing Evaluation and Certification Center and industry center, a support network and information security research laboratory and the State Key Laboratory of departments. In short, while China's network information security business got great development, but the difficulties and problems need to be solved is still a lot. system.

2 The Importance of E-commerce Security Management

2.1 Realize the Problems from the System

Any information system engineering has a very complicated structure with many parts, its security measures are not hidden danger of integrity and security hole is likely to permeate all parts of the system, some hidden danger and may have even system designer, the user is not necessarily know very well that it's there. Therefore, the system safety factor is always present, so there is no absolute safety, only relative safety.

Since the security of the system is always present and so we should focus on how to prevent it. First of all, information security management should not to be perfect, but should make the system safety according to the characteristics of the system . If only emphasize security itself and many of the restrictions are not reality, it will only lead the staff feeling more antipathy.

Second, on to a certain conditions to develop real suitable standard, it is a must to assess their risk that can assume, will attack of the increase in the cost of system to let want to crack the people regard unsuited to costs and not willing to take time to attack your system.

At the same time, from another level, the information security management should be strengthened. Information security management is the social system project which contains the technical level, management level, legal aspects: it is not a product,it is a complete process which composed by the people, the technology, the process. These components matching better the progress is more smoothly, to protect the safety of information systems play a more important role.

2.2 The Importance of the Information System Security Form the Current Government and the Enterprise

In recent years, our country continuous appears different degrees of information system safety accidents. The accident is not only simple information system, the direct consequence is the huge economic losses, it also caused the bad influence in society. If the economic loss also can make up, then the public to information system project credibility crisis which caused by the information network security system in the aspects of management and reflects the vulnerability can not be restored in short period.

In generally ,the industry thinks that the government and enterprise must hand in hand to faced with the information security. The government and enterprise management have the responsibility to ensure that all users to provide a safe environment information system.

Our government departments and all industries have been aware of the importance of information security. The government began a series of related strategy, push on the application and development of information security. The government-led related industry, also began to come to the application of information security technology product standards and regulations. The people's bank of China in the hustling to online banking system safety assessment guide, and clearly put on information security investment to reach a total investment of more than 10%, and in some other key industry, information security investment has even more than the 30%-50% of the total budget.

At present, the government and every trade to the importance of information security have understanding, the provisions of the relevant standards are formed, the investment is increased,safety technology, products, and the market are developed. Most enterprise organization is making the comprehensive security strategy which according to the different business information system and network security level need. Then, it need what kind of method or mechanism to manage or control information safety? In the longer term, the key is to establish a set of information security organization able to cover systems, including the management measures and system, which is about how to information system security from mechanism and the structure of the ascend to a new level. How to set up and maintenance through a framework to ensure that information security strategy and the organization's business objectives.

3 The Problems of E-commerce System Information Security Management

E-commerce is an application of the network in business activity, the network is the important carrier of information flow in electronic business affairs . With the rapid growth of Internet users, the foundation of the E-commerce work completed. The above characteristic of electronic business information will make the E-commerce has more and more influence. And now participate in the E-commerce of all parties to users also difficult to adapt to the characteristics and satisfy these requirements of E-commerce, so that the security and integrity are highly threat.

3.1 Electronic Business Information Storage Security Risk

Information storage security is refers to the electronic business information stored in the static in the process of safety. The information security hidden danger basically has:

to unauthorized call all kinds of company information and manipulate the enterprise information content. When the Intranet of the enterprise electronic commerce system was connected with Internet, the information storage security faces with the internal and external hidden trouble.

The internal hidden danger is mainly the user of one enterprise consciously and unconsciously unauthorized calls the enterprise electronic commerce information. Or the user without permission to add、modify、delete the E-commerce information.

The external hidden danger is mainly the unsuitable software configuration, which caused other organizations or individuals who broke into the enterprise, and one of the enterprise information intentionally or unintentionally to unauthorized calls or delete, add, modify. The main sources of the hidden troubles are the competitors enter, the information of malicious spy ware illegally broken into played about the curiosity of the users, who ran the hackers and harassment to break in.

3.2 Information Flow Security Hidden Danger of the E-commerce

Information flow security is refers to the process of dynamic transmission security in the electronic business operation process which are merged with logistics, cash flow. The main security hidden dangers include:

1. Stealing trade secrets

Because the information flow of E-commerce is basically transferred by clear text, the malicious destroyer which include information Spy and competitors can be easier to intercept and monitoring, and steal trade secrets.

2. Attack e-commerce site

Competitive enterprise or hackers spread through computer virus, send E-mail bombs, breached e-commerce sites firewall, damage, delete, modify, computer software and hardware, and to increase the victims of business information, making them unable to normal operation.

3. Implement business fraud

Now the outlaws steal deposits by cracking savers password ; play the stock market by infringing upon the people account draws ; swindle money through the Internet publishing false business advertising information ; malignant overdraw by stealing credit card Numbers. That entire make users and consumers feel distrust of e-commerce transactions which seriously hinder the development of electronic commerce.

4. Invade other's rights

Illegal molecules steal business order through the Internet, collect others information and sell products, encroach consumer privacy. Illegal enterprise destroy competition image, demeaning the competitor's products, infringe upon the enterprise's reputation and so on, that disturb the E-commerce market order.

5. Spread bad information

Outlaws proclaim violence, superstitions, pornography and other bad information in E-commerce to sell themselves.

6. Deny to send information

Based on all kinds of reasons, the businessman and the user may make e-commerce information or malicious helpless denied.

4 Conclusion

With the rapid development of Internet, people pay more attention to the E-commerce, but security problems affect the electronic commerce system development which is one of the key factors.How to protect the safety of the information resources become the hot topic of research at home and abroad. With the development of e-commerce, information security is the foundation of the development of electronic commerce, through the various network of business means will be a more diverse, safety problems become more prominent. In order to solve this problem, people must have the guarantee safety technology.

Therefore this paper researches on the information security management in E-commerce system and both researches on information security management in the importance of electronic commerce in detail. This paper calls for national, enterprise and individual to maintain the safety management with the electronic commerce system.

References

1. Feitzing, E., Lee, H.: Mass customization at Hewlett-Pack-ardt he power of postponement. Harvard Business Review 75(1), 116–121 (1997)
2. Qi, G., Gu, X., Yang, Q., et al.: Principles and key technologies of mass customization. Computer Integrated Manufacturing Systems 9(9), 776–783 (2003) (in Chinese)
3. Kraslawski, A., Koiranen, T., Nystrom, L.: Concurrent engineering: robust in fuzzy environment. Computers and Chemical Engineering 17(S1), 447–452 (1993)
4. Deng, C., Ma, X., Wu, J., et al.: Analysis technique of customers requirements in HOQ based on rough set theory. Computer Integrated Manufacturing Systems 13(6), 1191–1195 (2007) (in Chinese)
5. Pawlak, Z.: Rough set. International Journal of Computer and Information Sciences 11, 341–356 (1982)
6. Likert, R.: A Technique for the Measurement of Attitudes. Archives of Psychology (140), 1–55 (1932)
7. Zadeh, L.: Fuzzy Sets. Information and Control (8), 338–353 (1965)
8. Ertay, T., Kahraman, C.: Evaluation of design requirements using fuzzy outranking methods. International Journal of Intelligent Systems 22(12), 1229–1250 (2007)

Multi-objective Optimization of Coal-Fired Boiler Efficiency and NOx Emission under Different Ecological Environment

Wang Weiqing

Longfeng Thermal Power Plant, Daqing Oilfield Electric Power Group, Daqing 163711, Heilongjiang, China
wangwq163711@126.com

Abstract. Energy-saving and emission reduction technologies for reducing pollution and efficiency-rising for reducing loss are new task for thermoelectric power station. Least Square Support Vector Machine (LS-SVM) was used to train 12 main operating parameters which influence boiler efficiency and NOx emission. A mixed model for boiler efficiency and NOx emission as the outputs was put forward in this paper. In relation to multi-objective optimization problems, Pareto front was obtained though NSGA-Ⅱ. Mathematical model for optimal Pareto front under two conditions was established. The comparative analyses of the operation parameters before and after optimization can ensure the adjusting actions provided to the boiler operation staff as reference is useful and reliable.

Keywords: Optimization, Coal-fired boiler efficiency, NOx emission, LS-SVM, NSGA-Ⅱ.

1 Introduction

The efficiency promotion and the fuel consumption reduction can largely contribute to the economy of the coal-fired boiler on the grounds that the coal price as well as the transportation cost has been considerably ascended over the recent years. In spite of the fact that there have been diverse fully developed approaches — the low NOx combustion technology, the selective catalytic reduction technology and the advanced reburning technology, which involves abundant human and material resources— to control the NOx emission, the common most economical technique to lower the NOx emission is adjusting the operating parameters which affect the NOx generation amount. The key control technology of the NOx emission and the boiler efficiency rest with the coordination of the temperature and the coal powder density in the combustion area, which requires the combustion process to be optimized.

In this paper, the least square support vector machine and the NSGA-Ⅱ are jointly applied to present a combustion optimization control model. The least square support vector machine is used to train the combustion samples, formatting a combustion multi-objective optimization model choosing the coal-fired boiler efficiency and NOx emission as output. On the basis of this, the NSGA-Ⅱ is utilized to develop a mathematical model for picking out the optimal Pareto solution under two different

Y. Zhang (Ed.): Future Communication, Computing, Control and Management, LNEE 141, pp. 433–439.
springerlink.com

ecological environments. Comparing with the operating conditions before and after optimization, the combustion optimization adjustment mode is acquired to direct the reasonable economical and environmental course of the power boiler.

2 Sample Training Based on the LS-SVM

2.1 Inputs and Outputs of the Black Box Model

There are many factors that influence the boiler efficiency and NOx emission, such as the furnace configuration, the burner type, the load, the furnace temperature and the pressure, the coal characteristics and the like. All the inputs of the LS-SVM model are important combustion operating parameters, namely the total air supply, the primary air velocity, the furnace outlet oxygen volume, the burner tilting, the secondary air valve(6 inputs) and the over-fire air valve(2 inputs), excluding some flexible parameters such as the unit load and the coal characteristics, some local parameters such as the furnace temperature and the exhaust gas temperature, and some fixed parameters such as the burner type and the furnace configuration. The requirements of other parameters are stable coal characteristics, shrinked-middle wind of the secondary air, horizontal bias burner and one coal pulverizing system operated.

The outputs of the LS-SVM model are the boiler efficiency and NOx emission. The black box model is shown in Fig. 1.

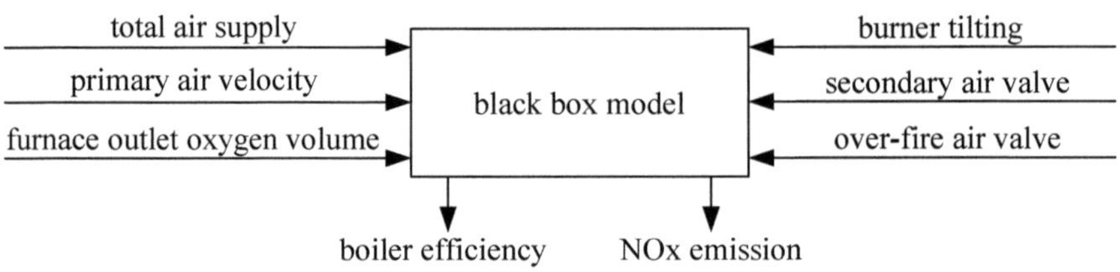

Fig. 1. Black Box model based on LS-SVM

2.2 Training Results and Analysis

Recent studies have been shown that the LS-SVM is of better prediction accuracy, faster convergence rate and better generalization, comparing with the neural network model [1,2,3], which indicates that the LS-SVM is suitable for establishing the of mathematics response model the boiler efficiency and NOx emission.

The training data that have received a significant amount of correction and processing are shown in Table 1 and Table 2. In all the samples, the former 10 samples are provided to train the model and the eleventh sample is used to test the accuracy of the model. The best combination of the regularized set of parameters and kernel parameters, namely (18,1), are selected through network search and cross-validation. The training results through the LS-SVM are shown in Table 3, from which the training results show that the absolute relative error of the boiler efficiency training group is 0.408% whereas the testing group is 1.062%, and the absolute relative error of the NOx emission training group is 0.376% whereas the testing group is 0.890%. The results reveal that the fitting accuracy of the training samples through the LS-SVM is very high and the model established is of good generalization ability.

Table 1. Operating parameters A for training

	total air supply[t/h]	primary air velocity[m/s]	furnace outlet oxygen volume[%]	burner tilting[%]
1	2290	26.8	3.10	0.5
2	2105	24.8	3.32	0.5
3	2308	22.5	2.84	0.5
4	2001	24.8	3.29	0.5
5	1862	25.0	3.41	0.6
6	2408	21.5	5.12	0.5
7	1908	20.3	3.25	0.4
8	2254	21.7	3.94	0.5
9	2097	23.7	4.15	0.4
10	1923	24.7	3.58	0.5
11	2549	26.9	3.97	0.6

Table 2. Operating parameters B for training

	secondary air valve[%]						over-fire air valve[%]		boiler efficiency	NOx emission
	AA	AB	BC	CD	DE	EF	OFA	OFB	[%]	[mg•m^{-3}]
1	65	65	65	65	65	65	60	40	93.77	730.48
2	65	50	65	65	65	65	30	0	92.31	652.58
3	65	65	65	65	65	65	60	0	93.55	774.61
4	65	45	45	25	45	45	60	40	92.61	713.32
5	65	65	65	65	65	65	60	40	90.07	647.98
6	65	40	40	80	80	85	60	40	91.12	631.62
7	65	65	65	65	65	65	60	40	92.63	661.39
8	45	45	45	45	25	45	48	0	91.57	629.84
9	65	40	50	65	50	50	60	40	92.77	632.54
10	50	65	65	40	80	50	48	40	90.32	674.85
11	65	40	80	40	50	50	60	40	91.33	654.03

Table 3. Training results of boiler efficiency and NOx emission

	boiler efficiency			NOx emission		
	actual value[%]	calculated value[%]	absolute relative error[%]	actual value[%]	calculated value[%]	absolute relative error[%]
1	93.77	93.71	0.064	730.48	732.24	0.241
2	92.21	92.22	0.011	652.58	650.87	0.262
3	93.55	93.47	0.086	774.61	771.20	0.440
4	92.61	92.54	0.756	713.32	716.30	0.418
5	90.07	90.10	0.033	647.98	649.38	0.216
6	91.12	92.03	0.998	631.62	634.32	0.427
7	92.63	91.64	1.069	661.39	657.87	0.532
8	91.57	90.54	0.033	629.84	628.27	0.249
9	92.77	93.00	0.248	632.54	630.03	0.396
10	90.32	91.03	0.786	674.85	670.94	0.579
11	91.33	90.36	1.062	654.03	659.85	0.890

2.3 Multi-objective Combustion Optimization

There are a lot of traditional multi-objective optimization methods, such as the weighted sum method, the multi-objective programming approach, the maximum-minimum method, the evaluation function method and the like, which contain limited non-inferior solutions and depend on the experience of decision makers. For this constrained double-objective optimization problem in this paper, genetic algorithm is a good method in multi-objective processing [4,5].

Multi-objective optimizations synchronously optimize multi-objective functions which possibly conflict reciprocally, obtaining a Pareto non-dominated solution (Pareto front), leading the group to access Pareto fronts and sustaining diverse solutions. Deb et al [6] considered the elitist non-dominated sorting genetic algorithm is more close to the Pareto front solutions rather than other multi-objective genetic algorithms. Therefore, the elitist non-dominated sorting genetic algorithm is used to optimize the model established in this paper.

2.4 Elitist Non-dominated Sorting Genetic Algorithm

To deal with a problem of potential solutions loss in the process of multi-objective optimization, Deb and Srinivas presented the elitist strategy, improving some conspicuous defects of the non-dominated sorting genetic algorithm such as complex computation, elitist retainment insufficiency and parametric variables sharing dependence [6]. NSGA- II is described with two novel operators, one is non-dominated solution series which are not controlled by other solutions and are used for the following iterative calculation, and the other is dense compared operators which provided a selection criteria.

The NSGA- II is shown in Fig. 2. The parent population derives a progeny population of equal size in the process of choosing, intersection, recombination and variation, and then the parent population and the progeny population are combined to become a novel doubled population for the sake of reserving excellent individuals. The non-dominated solution series and dense compared operators are calculated to non-dominantly sort the new population. Then sorting with distance of density and the former 50% individuals of the new population are selected and substitute for the primary parent population. The cyclic iteration enables the elite reservation of each population and the variety of the Pareto front.

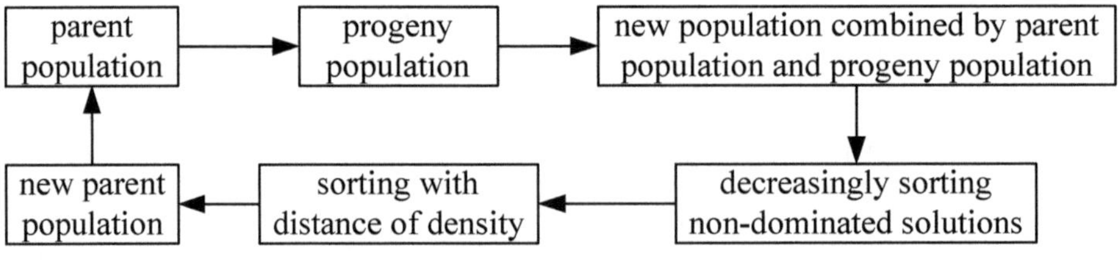

Fig. 2. NSGA- II

2.5 Typical Cases and Analysis

Researchers often list the Pareto optimal solution set rather than present the principle of picking out the optimal solution set, and are unable to make the most notable analysis and comparison. In this paper, the optimization results of the most representative second data is analyzed and referenced for the principle of picking out the optimal solution set. The optimization results are shown in Fig. 3.

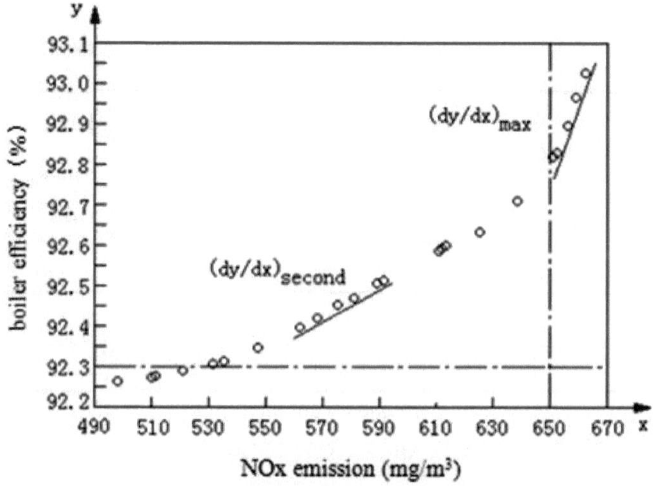

Fig. 3. Pareto front of NO.2 training data

For the second training group, the boiler efficiency is 92.31% and the NOx emission is 652.58 mg/m3. From Fig. 3, the boiler efficiency of the Pareto front moves upward but the restricted area M (the NOx emission concentration is ruled to be lower than 650mg/m3 in China) exists in the whole area. So the optimal points of which the NOx emission exceeds 652.58 mg/m3 are refused to take into consideration. The following analysis can be divided into two categories.

For the areas with greater ecological environment pressure, such as cities, attractions and densely populated areas, the optimization approach is reducing NOx emission as possible on the premise of that the boiler efficiency is higher than the original figure. As far as the second training group is concerned, the optimal points of which the boiler efficiency is less than 92.31% are refused to take into consideration. This is the restricted area N. The first situation is mathematically described as

$$\begin{cases} (NO_X)_{\min} \\ s.t.1: (NO_X) < (NO_X)_M \\ s.t.2: (\eta) \geq (\eta)_N \end{cases} . \quad (1)$$

For the areas with minor ecological environment pressure, the optimization approach is selecting the optimal points whose differentiation of the boiler efficiency with respect to the NOx emission is most notable, the significance of which is the boiler efficiency increases considerably while the NOx emission raises marginally. The mathematical description is expressed as

$$\begin{cases}(d\eta / d(NO_X))_{\max} \\ s.t.1:(NO_X)<(NO_X)_M \\ s.t.2:(\eta)\geq(\eta)_N\end{cases} \cdot \tag{2}$$

Table 4. Comparision A of NO.2 training optimal data and former data

second training group		total air supply (t/h)	primary air velocity (m/s)	furnace outlet oxygen volume (%)	burner tilting (%)	over-fire air valve(%) OF A	OF B
pre-optimization		2105	26.8	3.32	0.5	30	0
post-optimization	situation 1	2151	24.8	3.54	0.447	60.7	15.8
	situation 2	2142	23.7	4.77	0.449	58.2	6.8

Table 5. Comparision B of NO.2 training optimal data and former data

second training group		secondary air valve (%) AA	AB	BC	CD	DE	EF	boiler efficiency (%)	NOx emission (mg•m^{-3})
pre-optimization		65	50	65	65	65	65	92.31	652.58
post-optimization	situation 1	50.2	30.4	35.8	60.2	70.5	60.1	92.35	547.52
	situation 2	55.9	35.4	30.2	40.8	71.8	45.4	92.45	578.08

According to the two situations above, the appropriate solutions of the second training group Pareto front are respectively extracted to be compared before and after optimization (see Table 4 and Table 5). The operating parameters comparison of pre-optimization and post-optimization has prevalent meaning of guidance for the optimization of the boiler efficiency and NOx emission.

3 Summary

With consideration of the multi-objective combustion optimization, 12 important operating parameters are extracted as inputs to establish the LS-SVM model, excluding some flexible, local and fixed parameters, which has greater guiding significance for boiler operating personnel. The NSGA- II is applied to present a combustion optimization control model for picking out the optimal Pareto solution under two different ecological environments. The combustion optimization adjustment mode is acquired to direct the reasonable economical and environmental course of the power boiler.

References

1. Chen, M.S., Liu, D.P.: East China Electric Power 34, 50 (2006)
2. Suykens, J.A., Vandewalle, J.: Neural Network Letters 19, 293 (1999)
3. Wang, H., Li, E.Y., Li, G.Y.: Materials & Design 30, 1468 (2009)
4. Furtuna, R., Curteanu, S., Leon, F.: Engineering Applications of Artificial Intelligence 24, 772 (2011)
5. Cabello, J.M., Cejudo, J.M., Luque, M., Ruiz, F., Deb, K., Tewari, R.: Renewable Energy 36, 3146 (2011)
6. Deb, K., Agarwal, S., Megarivan, T.: IEEE Transactions on Evolutionary Computation 6, 182 (2002)

Ranking the Importance of Classes via Software Structural Analysis

Li Ding-wei[1], Li Bing[1,2,*], He Peng[1], and Pan Wei-feng[3]

[1] State Key Laboratory of Software Engineering, Wuhan University, Wuhan, 430072, China
[2] Complex Networks Research Center, Wuhan University, Wuhan 430072, China
[3] School of Computer Science and Information Engineering, Zhejiang Gongshan University, Hangzhou 310018, China
lidingwei2006@163.com, bingli@whu.edu.cn, jxgaych@163.com, panweifeng1982@gmail.com

Abstract. How to identify important classes in an object-oriented system is a significant task for resources allocation and decision making. However, the work on ranking the importance of classes is scarce. This paper introduces a weighted software network model to represent the structure of object-oriented software, and based on which an indicator, IC, is defined to measure the importance of classes. By ranking the ICs of all classes we can obtain the important classes we should focus on. Manually inspection of the source code verifies the meaningfulness of the classes obtained and also validates the feasibility of our approach.

Keywords: Importance of classes, software, software networks, complex networks.

1 Introduction

For the task of resources allocation and decisions making in software development practices, it is inevitable to consider the importance of classes in object-oriented (OO) software systems. For example, when designing and testing a software system, properly allocating the limited resources can be a main factor influencing whether you can successfully complete the work or not in a given period of time. At present, people usually identify the dangerous classes (those may contain defects) by recognizing the abnormal values of some metrics such as cohesion, coupling and complexity. Identifying the classes containing defects is not still sufficient to make a decision or allocate resources. We should also identify important classes [1].

To solve the above mentioned problem, this paper introduces a weighted software network model to represent the structure of object-oriented software, and based on which an indicator, IC, is defined to measure the importance of classes. And it uses a Simulation Algorithm (SA) to calculate the value of the ICs. By ranking the ICs of all classes we can obtain the important classes we should focus on. We test our approach

* Corresponding author.

Y. Zhang (Ed.): Future Communication, Computing, Control and Management, LNEE 141, pp. 441–449.
springerlink.com

against one case study on an OO software system, S2R2. The results of the case studies validate the feasibility of the method. The approach is fully automated by a tool written in Java.

2 Related Work

Our approach uses a weighted software network to represent the software structure. So first let us brief the work on software networks.

In recent years, some researchers have introduced the theories of complex networks to software engineering, building software network model to analyze structural features of OO software. In 2002, Valverde S, etc. introduced the methods of complex network to software structure analysis [2]. They used undirected graph to abstract the system classes, namely: the network nodes represent the classes, and the edges between nodes represent the relationships between classes, such as inheritance, association, etc. The relationships (collaboration, call, etc.) between system modules (class, method, etc.) reflect the system's control flow. Therefore, the direction of edge in the software network is significant [3]. In 2003, Myers C R used directed network to study software structure, analyzed three class collaboration graphs of OO software and three static procedure call graphs of process-oriented software.

Then, people started to study the structure of a large number of open source software from different particle size (methods, classes, packages, files, etc.). They found that such networks have "small world" and "scale-free" features. D Li, B Li, et al also analyzed a number of software systems written in Java, and found "small world" and "scale-free" features [4-8].

3 The Approach

In this section, we will use a weighted software networks model at class level to model OO software systems. The model has been already proposed in [5]. And the main steps used here are also adopted from that in [5]. Fig.1 gives a short overview of the workflow of the proposed approach.

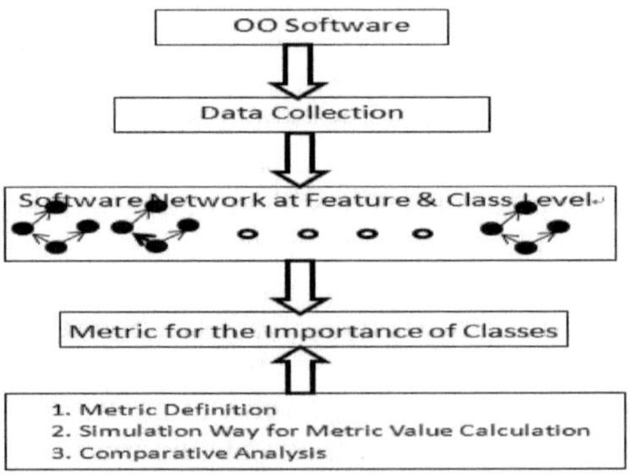

Fig. 1. Workflow of the proposed approach

3.1 OO Software [5]

OO has been used widely as the development paradigm in recent twenty years. So we will focus on the OO domains in this paper. People can easily get a lot open source object-oriented (OSOO) software systems which are developed under the OO paradigm on the Web. OSOO software systems have clear structures and the software components such as attributes, methods, classes, etc. And their dependencies are agreeable to extraction and analysis. It can easily use many software design principles to improve the quality of OO software systems. And the interrelationships between design principles and software quality can be revealed by the research on OO designs.

3.2 Data Collection [5]

Data collection is the process to extract software components such as attributes, methods, classes and their dependencies. A tool has been developed which can be used to analyze compiled Java codes to get needed data. In this paper, we only take into consideration two kinds of dependencies which are method accessing attribute dependency and method call dependency. We obtain the dependencies between classes according to these two kinds of dependencies.

3.3 Software Network at Feature and Class Level [5]

In this subsection, we built two kinds of WSNs, weighted feature dependency network (WFDN) and weighted class dependency network (WCDN). And WCDN is mainly built with the use of WFDN. The formal description of WFDN and WCDN will be first given[5].

3.4 Definition 1 [5] Weighted Feature Dependency Network (WFDN)

In WFDN the nodes represent features (attributes or methods) of a specific OO software system. And each feature is represented by only one node. Directed edge between two nodes denotes one feature uses another feature, i.e., if feature A uses feature B, there will be an edge from the node denoting A to the node denoting B. And here we only consider the presence of dependency and neglect the multiplicity of dependencies such as A depends three times on B. And the weight of each directed edge denotes the probability that a bug in B will propagate to A. See Fig.2 for example. Therefore WFDN can be described as:

$$WFDN = (N^f, E^f, M^f), \tag{1}$$

where Nf is the set of nodes which are all features (methods and properties) of the OO software system; Ef is the set of edges denoting all dependencies between features; Mf is a matrix, if there is one directed edge in WFDN, <j,i> (<> represents an ordered pair, the same below) denoting that node j links to node i, and the weight of Mf (i, j) is equal to 1.

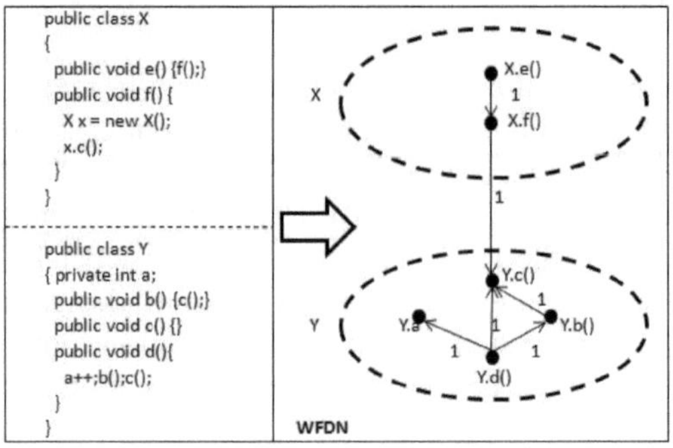

Fig. 2. Illustration of WFDN

3.5 Definition 2 [5] Weighted Class Dependency Network (WCDN)

In WCDN the nodes represent the classes of the specific OO software systems. And each class is represented by only one node in the whole WCDN. Directed edges between two nodes denote the interactions between the corresponding classes. And such interactions are obtained from the corresponding WFDN, i.e., if the features in class X use the features in class Y, there will be an edge from the node denoting class X to the node denoting class Y, and vice versa. Here we only consider the presence of dependency and neglect the multiplicity of dependencies such as class X depends three times on class Y. And the weight of each directed edge denotes the probability that a bug in class Y will propagate to class X. See Fig.3 for example. Therefore WCDN can be described as:

$$WCDN = (N^c, E^c, M^c), \tag{2}$$

The meanings of the notations of Nc, Ec and Mc are very similar to that we talked in WFDN. Only are they now applied to WCDN. So we will not go further into details about these notations here. Please refer to those in WFDN.

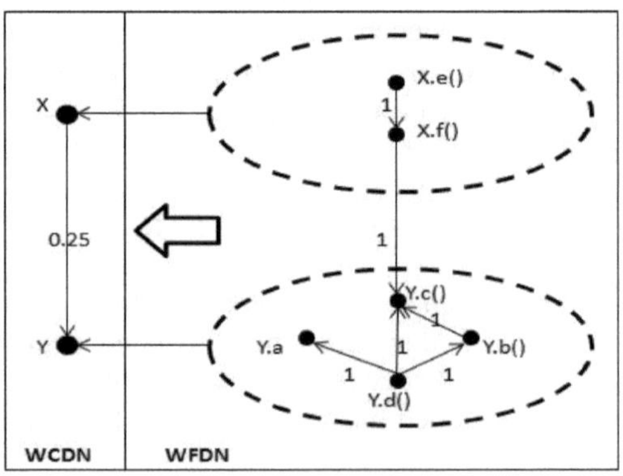

Fig. 3. Illustration of WCDN

In the following, the way to calculate the probability matrix Mc will be detailed clearly. But before that, we should give some primary definitions.

3.6 Definition 3 [5] Extra-Class Method Reachability Set (ECMRS)

Suppose there are two different classes, Ci and Cj (Ci, Cj ∈ Nc). And Ci depends on Cj directly (i.e., there is an edge from Ci to Cj in WCDN). Then we define the extra-class method reachability set of each method Cik in Ci, ECMRS(Cik , Cj), as the set of features in Cj reachable from Cik . A feature w in Cj is reachable from Cik if in WFDN there is a directed path from Cik to w. And the path is only composed of nodes denoting features in Ci or Cj , and it travels each node only once.

Clearly, the ECMRS of a specific method in one class is the set of features in another class that will have direct and indirect impact on this method. In order to obtain the ECMRS of each method in every pair of classes which are directly linked, we propose a queue-based level order traversal algorithm which is shown in Algorithm 1, where Fi is the set of features in class i, |Ω| denotes the counting of the elements in set Ω (the same below).

Algorithm 1 Queue-Based Level Order Traversal Algorithm [5]

Input: WFDN and the edge set Ec of the corresponding WCDN.

Output: ECMRS(Cik, Cj), i ≠ j, i, j = 1, 2, . . . , |Nc|; k = 1, 2, . . . , |Fi|.

Step 1. Prepare a queue.

Step 2. Select a directed edge from Ec, and delete it from Ec.

Step 3. Find the classes i and j on its two sides, and suppose the direction is from i to j.

Step 4. Travel through the nodes in WFDN one by one, if (node k ∈ Fi) then add node k to queue.

Step 5. Get a node from the queue and store it in a set setF, then add other nodes that are dependent by this node and belong to Fi or Fj to queue.

Step 6. Go to step 5 until node is NULL.

Step 7. Let ECMRS(Cik, Cj) = setF ∩ Fj, and setF = NULL.

Step 8. Go to step 2 until edge is NULL.

3.7 Definition 4 [5] Inter-Class Ripple Basin (ICRB)

The inter-class ripple basin of class i on class j, ICRB(i, j), describes the total set of features in class j that are directly or indirectly used by all methods in class i. So, in fact, it is the union of ECMRS of all methods in classes i. Therefore ICRB(i, j) can be described as:

$$ICRB\ (i,\ j) = \sum_{C_k^i \in F^i} \cup\ ECMRS\ \ (C_k^i, C^j) \tag{3}$$

Considering that the ECMRS of the methods in class i, all should have indirect relationships with the features in another class via the critical methods (those directly linked to features of another class), we only calculate the ECMRS of the critical nodes. See Fig.3 for example. X.f() is a critical method. X.e() is an ordinary method. And X.e() uses X.f(). So we only need to calculate ECMRS(X.f(),Y) to get ECMRS(X.e(),Y). By the same token, ICRB(i, j) is the union of ECMRS of all critical methods in class i. Therefore we can take a more simple formula to calculate ICRB(i, j) which can be described as:

$$ICRB\ (i,\ j) = \sum_{C_k^i \in S_c} \cup\ ECMRS\ \ (C_k^i, C^j) \tag{4}$$

Where Sc is the set of critical features in class i. We can find that ICRB(i, j) in fact is the set of features in class j whose bug will propagate to the methods in class i, finally making class i faulty. In Fig.3, ICRB(X,Y) = {y.c()}.

So given class i and class j of an OO software system, the weight from class i to class j is denoted by Mc(i, j) and defined in this paper as the following formula:

$$M^{\,c}(\,j,i) = \frac{|\ ICRB\ (i,\ j)\ |}{|\ F^{\,j}\ |} \tag{5}$$

In Fig.3 Mc(Y, X) = |{Y.c()}|/|{Y.a, Y.b(),Y.c(),Y.d()}| = 1/4 = 0.25.

3.8 Metric for Importance of Classes

In this subsection we will use a simulation way to calculate the metric for IC. So the definitions of the corresponding metrics all take into consideration of a specific running of the simulation algorithm.

Definition 5. Propagation Filed of Node (PFN).
The propagation filed of a node i, PFN (i), is defined as the set of nodes that have been affected by the faulty node in a running of the simulation algorithm

Definition 6. Importance of Classes (IC).
The importance of a specific class is defined as the average number of nodes that have been affected by the faulty node in N running of the simulation algorithm. Numerically it can be calculated by:

$$IC = \frac{1}{N}\sum_{j=1}^{N} |\ PFN(i)\ |_j \tag{6}$$

In order to obtain the IC, we propose a queue-based level order traversal algorithm which is shown in Algorithm 2.

3.9 Algorithm 2 Simulation Algorithm [8]

Input: WFDN, Mc, N.
Output: PFN[i] (i = 1, 2, …, |N|).

Step 1. Set simT = N. Set bChanged[i] = 0 (i = 1, 2, ..., |N|), and t = 1. Set cChgNN = 0. Prepare a queue cQueue.

Step 2. If (cChgNN <= |N|) then select a node i one by one which satisfies bChanged[i] = true, bChangedBak[i] = true, bSelected[i] = true, cChgNN++, and go to Step 3, else go to Step 6.

Step 3. If (cQueue is Null) then go to Step 6, else go to Step 4.

Step 4. Pop one node from cQueue, denoted as Ni. Travel through the nodes in N one by one (each node denoted as Nj), if (Mc[i][j] = p) then add node j to a temporary set tempSet.

Step 5. For each node Nk in tempSet, randomly generate a decimal deck. If (deck >= p), then (1) add Nk into cQueue (i.e., the change in Ni will propagate to Nk), (2) delete it from tempSet, (3) set the corresponding bChanged[k] and bChangedBak[k] of Nk to be true, and (4) go to Step 2, else delete it from tempSet and go to Step 2.

Step 6. Set PFN[t] to be the number of non-zero elements of array bChangedBak[l] (l = 1, 2, …, |N|). Set t = t + 1. If (t < simT), then set bChanged[i] = false and bChangedBak[i] = false (i = 1, 2, …, |N|) and go to Step 2, else go to Step 7.

Step 7. Finish the simulation algorithm and output PFN[i] (i = 1, 2, …, |N|).

4 Case Analysis

In this section, the method mentioned above will be tested by an example to verify the validity of the method. We have developed a web service registry and repository system (S2R2[1]). Here we will used it as a case study.

4.1 Service Registry and Repository System (S2R2)

In Wuhan University, Professor Li Bing and his research team have developed a web service registry and repository system (S2R2). Through the study of basic theories methods and technical standards of software services on the Internet, the system provides a unified web service registry and management platform which manages all registered software services and is convenient for users to find the service.

Based on the platform, the infrastructure of software services development is built and the relevant international and national standards are developed, promoting the development and application of service-oriented software and the sharing of services resources. Till now S2R2 has many versions. For limitation of space, here we only analyze S2R2 on March 17, 2010.

4.2 Numerical Results and Analysis

We calculate the ICs of all classes in S2R2, and rank them by their IC from large to small. There are total 153 classes in S2R2. For a clear demonstration, here we only list the 8 classes with largest IC values. They are shown in table 1.

[1] http://www.s2r2.org

Table 1. ICs of the rank first 8 classes in S2R2

Classes Name	IC
org.sklse.wsrr.database.DBConnector	23.9649
org.sklse.wsrr.util.WSDLInfo	6.351
org.sklse.wsrr.registry.WSRegister	5.4287
org.sklse.wsrr.util.WSDLAnalyzer	5.3442
org.sklse.wsrr.dao.WebServiceDAOFactory	5.2546
org.sklse.wsrr.bean.WebService	4.7061
org.sklse.wsrr.dao.WebServiceDAO	4.109
org.sklse.wsrr.test.TestWSDLAnalyzer	1

From table 1, the IC of class DBConnector is maximum, TestWSDLAnalyzer minimum, and WebServiceDAOFactory is normal. That is to say the class DBConnector is most important, WebServiceDAOFactory less important, and TestWSDLAnalyzer is least important. Obviously, the results are consistent with the actual situation. The operation of database connection is much important in S2R2 because other operations (web service registry and management) always use it inevitably. Of course, in web service registry system, it is no doubt that the class WebServiceDAOFactory is important. But the class TestWSDLAnalyzer is used least, so its importance is little.

5 Conclusion

This paper presents a method to rank the importance of classes in software system. This method is designed based on the weighted software network, weighted feature dependency network (WFDN) and weighted class dependency network (WCDN). We got the metric for the relationship between each two classes. And then we use the simulation way to calculate the metric for importance of classes. The results show that it's credible for IC via software structural analysis. Therefore, the method based on software structure is useful for the research in the future.

Acknowledgments. This work is supported by National Basic Research Program (973) of China under grant No.2007CB310801, National Natural Science Foundation of China under grant No.60873083, 60803025, 60703009, 60303018, 60970017 and 60903034.

References

1. Zhou, Y.M., Xu, B.W.: Dependence Structure Analysis-Based Approach for Measuring Impotance of Classes. Journal of Southeast University (Natural Science Edition) 38(3), 380–384 (2008)
2. Valverde, S., Cancho, R.F., Sole, R.: Scale Free Networks from Optimal Design. Europhysics Letter 60, 512–517 (2002)

3. Myers, C.R.: Software Systems as Complex Networks: Structure Function, and Evolvability of Software Collaboration Graphs. Physical Review E 68(4), 046116 (2003)
4. Han, M.C., Li, D.Y., Liu, C., et al.: Networks Characteristics in Software and Its Contribution to Software Quality. Computer Engineering and Applications 42(20), 29–31 (2006)
5. Pan, W.F., Li, B., Ma, Y.T., et al.: Measuring Structural Quality of Object-Oriented Softwares via Bug Propagation Analysis on Weighted Software Networks. Journal of Computer Science and Technology 25(6), 1202–1213
6. Liu, J., He, K., Ma, Y.T., et al.: Scale Free in Software Metrics. In: Proc. of the 30th Annual International Computer Software and Application Conference, USA, pp. 229–235 (2004)
7. Pan, W.F., Li, B., Ma, Y.T., et al.: Class Structure Refactoring of Object-Oriented Softwares using Community Detection in Dependence Networks. Frontier of Computer in China 3(3), 396–404 (2009)
8. Pan, W.F.: Software Networks-Based Software Static Structure Analysis and Its Applications. Wuhan University, Wuhan (2011)

Measurement Study of Self-concept and Academic Help-Seeking on Junior Middle School Students

Changxiu Shi[1], Jiajun Yao[2,*], Shuoqiu Gan[1], and Fan Yu[1]

[1] School of Education, Anqing Normal College, P.R. China
[2] School of Education, Fujian Normal University, P.R. China
shichx@aqtc.edu.cn, zhaoxj@aqtc.edu.cn

Abstract. 280 junior middle school students from Anhui were measured formally by scales. The study used 3(Grade)×2(sex)bivariate analysis of variance, made path analysis about self-concept on junior middle school students in the Anhui province China and academic help-seeking. The results showed that there were 2 significant paths in help for instrumental - students, 6 significant paths in help for instrumental - teacher and 8 significant paths in help for none instrumental behavior. Self-concept affected academic help-seeking behaviour by academic help-seeking attitude.

Keywords: Junior Middle School Students in Anhui Province, Self-concept, Academic Help-seeking Attitude, Academic Help-seeking Behaviour.

1 Introduction

Academic help-seeking refer to the students who meet difficulties in the learning process request help behavior from others. American scholar Nelosn - LeGall (1985) according to the help-seeking men' purpose divided help behavior into executive help-seeking and instrumental help-seeking. Executive help-seeking have been also called none adaptability for help, which indicated the students in the face of the questions should solve those themselves but ask others instead. Instrumental help-seeking have been also called adaptability for help, which indicated the students with the help of others in order to achieve their goal or solve the problems on purpose. The students mastered instrumental help-seeking would like to refuse help when they can solve independently, while get help when they needed. Therefore, the instrumental help-seeking is representative of ability. It is a kind of problem-solving behavior. It can help students solve the problems and realize the objectives and adapt to the situation. Academic help-seeking behavior is not only a kind of important cognitive strategies, but also a kind of interaction skills, relevant research for the actual teaching practice is of great significance.

The frailty hypothesis have pointed out that someone of low self- esteem would be less willing to help, while the consistency hypothesis pointed out that someone of high self-esteem would be more reluctant to request help. The supports are obtained in some studies for these two hypotheses. For example, Tessler and Schwartz, found that the

* Corresponding author.

Y. Zhang (Ed.): Future Communication, Computing, Control and Management, LNEE 141, pp. 451–456.
springerlink.com

subjects of low self-esteem request help faster and more often. So it supported consistency hypothesis. However, the results of the study of college students from Karabenick and Knapp support frailty hypothesis. Later, Newman's, Ryan and Pintrich's research also support frailty hypothesis. Many scholars believe that the relationship between esteem and academic help-seeking corresponds with frailty hypothesis. This study aimed to explore the relationship between each component of self-concept and academic help-seeking, which aspects of self-concept exactly have an impact on academic help-seeking.

The researchers through the observation of the middle school students' current situation in Lu An and An Qing city learn about parts of students always request help from students and teachers, instead, some other students never request help from them. What would be the cause? The researchers from the analysis of the data of 16PF,SCL -90 measurements and the information about relevant part of students in psychological counseling, found that family, school, individual and any other elements could be the factors, but the individual play a crucial role. Whether family or school, they were just a kind of external influence variables for individual transformation, nevertheless the individual would be more important, the change of self-awareness such as ability, specialty, socially acceptable and so on often affect the behavior of individual reactions. Therefore, this research attempted to explore the relationship between self-concept and academic help-seeking.

The teacher Liu An Ying consider that paying attention to the cultural background of the selfconcept is the new trend. Different cultural background affect Self-concept, as individualism society emphasize the Private self, group society emphasize the Public self. So in different cultures, Self nature and importance is different. To establish the self-concept mode adapted to their own nationalcharacteristics and the empirical research are Chinese ego psychology researcher's historical mission and also the embodiment of the localization trend. The study has the following realistic significance: 1. In order to improve the frequency of the academic help-seeking behavior of the juniors in AnHui, whether it is necessary to intervene self-concept or not? 2. If necessary, what components of the self-concept could be started? 3. What kind of intermediary variable the different components of self-concept via influence the academic help-seeking behavior.

2 Method

Subject 300 junior middle school students from Shouxian, Lu'an and Aanqing in Anhui were measured formally by Junior high school students self-concept scale , Junior high school student academic help-seeking attitude scale, Junior high school student academic help-seeking behavior scale.

Tool The Junior high school students self-concept scale is referred to SDQ(marsh) and Tennessee self-concept scale which has exhibited successful applications, the scholar Zhao xiaojun compiled the one which is more appropriate for juniors in china. The other two scales was compiled by Li donghai, revised by Zhao Xiaojun.

Design In the followings, the researchers made the 3 (grade: grade one, grade two, grade three) x 2 (gender: male, female) between groups design. Grade and gender are between group variables, the factors in the scales are the dependent factors. Then the study used 3(Grade)×2(gender)bivariate analysis of variance and made path analysis.

Measurement In this section, 280 effective scales have been selected, 105 subjects come from grade1, 90 ones come from grade 2, another 85 ones come from grade3, including155 boys and 125 girls.

2.1 The Result and Analysis

There are 2 paths in Instrumental help-seeking- influence of schoolmate being significant. One is gender ---instrumental help-seeking---schoolmates. The other one is friend self---instrumental help-seeking---schoolmates. Gender, friend self have a directly significant effect on students. These effects on instrumental help-seeking---schoolmates are called "direct path." It can be seen from standardized regression coefficient, girls in Anhui junior high school of instrumental help-seeking---schoolmates behavior have higher frequencies than boys (beta = 0.075 *). So the higher the friend self-concept, the higher the frequency of instrumental help-seeking---schoolmates behavior will be (beta =0.190 * * * *).

There are 6 significant paths in instrumental help-seeking---influence of teachers. Specifically as follows: the first one is the benefits of help-seeking---instrumental help-seeking---teachers; the second one is happiness---instrumental help-seeking----teachers; the third one is grade ------instrumental help-seeking-----teachers; the fourth one is art self---the benefits of help-seeking---instrumental help-seeking---teachers; the fifth one is the family self-concept the benefits of help-seeking--- instrumental help-seeking---teachers; the last path is honest---the benefits of help-seeking---instrumental help-seeking---teachers.

Among them, the family self-concept, honest self - concept have an indirect effect on instrumental help-seeking---teachers, while the attitude toward the benefits of help-seeking, happiness and grade affected instrumental help-seeking---teachers directly. It can be seen from standardized regression coefficient: the higher happiness self-concept, the higher frequency of instrumental help-seeking ---teachers is(beta = 0.086 * *);the higher grade, the lower the frequency of instrumental help-seeking---teachers will be (beta = - 0.186* * * *). What's more, the higher the family self-concept, the stronger the attitude toward the benefits of help-seeking (beta =0.078 *). The honest self-concept (beta = 0.167 * * * *) and art self-concept (beta = attitude 0.176 *) is the same. And for the stronger the attitude toward the benefits of help-seeking, the higher the frequency would be (beta = 0.214 * * * *)

At the same time, there are 8 significant paths in academic help-seeking, specifically as follows: the first one is the price of help-seeking--- academic help-seeking; the second one is grade--- academic help-seeking; the third one is happiness---academic help-seeking; the fourth one is self value---academic help-seeking; the fifth one is friend self--- academic help-seeking; the sixth one is appearance self-concept---the price of help-seeking---academic help-seeking; the seventh one is science self concept---the price of help-seeking---academic help-seeking; the last path is grade---the price of help-seeking---academic help-seeking.

Among them, the effect of appearance self-concept, science self-concept, grade is indirectly, then the attitude toward the price of help-seeking, grade, regions, happiness, self value, and friend self concept have a direct effects. Standardized

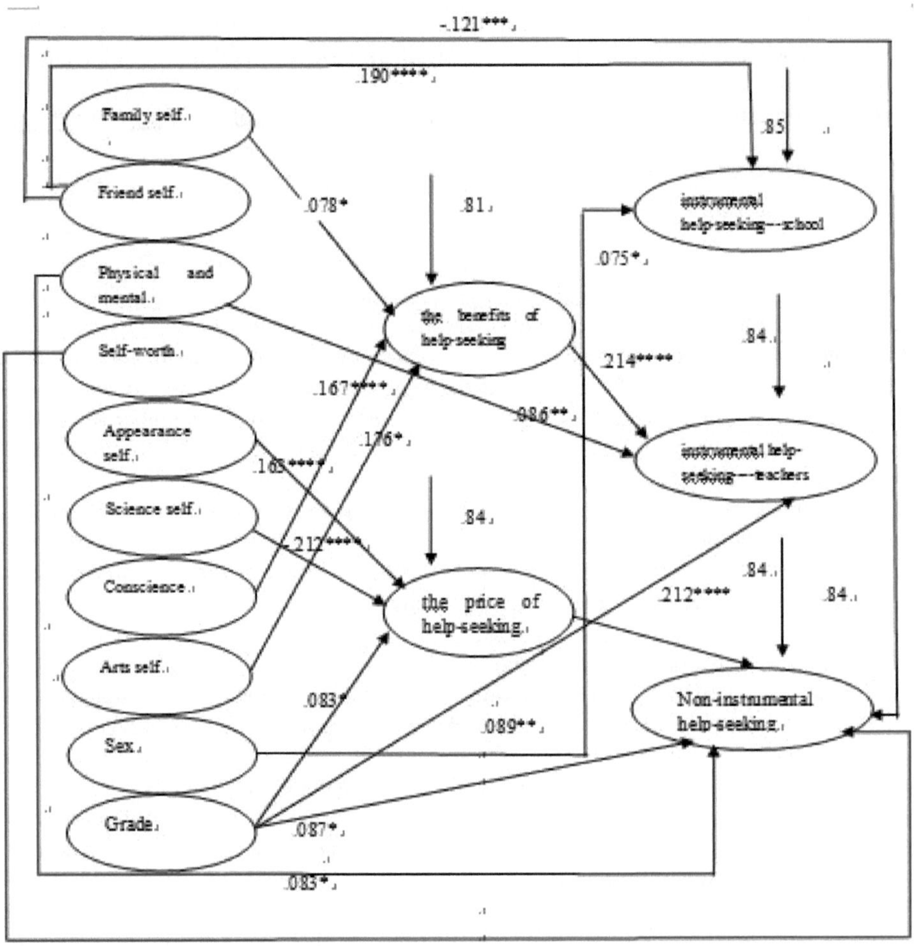

Chart 1 Path analysis on junior middle school students' self -concept influence to academic help-seeking

regression coefficient indicates that, the higher the grade, the higher frequency of academic help-seeking behavior is (beta = 0.089 * *). The higher happiness self-concept, the higher frequency of academic help-seeking behavior is(beta = 0.087 *). The common self value did the same (beta = 0.083 *). Oppositely, if the Friend self concept is higher, the frequency would be lower (beta = - 0.121 * * *). If the science self concept is higher, the attitude toward the price of help-seeking become weaker. But when Appearance self concept (beta = attitude 0.163 * * * *) and grade (beta =0.083 *) is higher, the attitude is stronger. (beta = - 0.212 * * * *). And then the frequency of academic help-seeking behavior improved higher(beta = 0.212 * * * *).

2.2 Discussion

Teachers in China such as Wang Zhenhong have the thought of that, the self-concept is an important aspect in individuality development of teenagers, the formation and development of the self-concept would have more extensive impact on teenagers' social development. This study found that, gender affect instrumental help-seeking – schoolmate behavior in mathematics learning, and girls have more instrumental help-seeking – schoolmate behavior than boys. Furthermore, Grades through the price of academic help-seeking attitude affect the academic help-seeking behavior in mathematics. If the grade is higher, the price of help-seeking attitude leads to the more academic help-seeking behavior is higher too. When pupils get into junior high school, because of the new environment, many of them try to get good results, so they are more willing to comprehend, the price of academic help-seeking behavior goes down, but the grade become higher, due to the adverse adaptation to the new environment or the depression in their study, the polarization appear, the idea of the price of help-seeking become more and more, the academic help-seeking behavior increases. The teacher Li XiaoDong also thinks, the influence of cultural factor on the academic help-seeking of different grade students in different modes. The researchers pointed, that was related with the culture of each region, education valued degree, education investment, education environment, family income, etc, then concrete analysis needs to be further discussed. Similarly, Zhang ChunXing of Taiwan normal university thinks, the influence of cultural factor on psychology and behavior is more significant than region factor.

Honest self-concept through the benefits of help-seeking affects mathematical instrumental help-seeking behavior -- the teacher behavior. If Honest self-concept is higher, the benefits attitude is stronger, instrumental help-seeking behavior -- the teacher behavior become more and more. The juniors who have higher honest self-concept would like to reflect the current situation of mathematics learning, to define their strengths and weaknesses, to strengthen the idea benefits of help-seeking, and to strengthen the confidence when ask teachers for help. At the same time, it is also proved the proverb-- don't take the MoMo behind, that means don't be self-deception.

Family self-concept through the attitude towards benefits of help-seeking affects mathematical instrumental help-seeking behavior -- the teacher behavior. Family self-concept through the attitude towards the price of help-seeking affects to academic help-seeking behavior. The higher the family self-concept is, the stronger the benefits of help-seeking in mathematics will be, and the more instrumental help-seeking behavior-the teachers behavior appears. Why? The students with high family self-concept consider the good treatment of parents, brothers and sisters, the idea is generalized to their teachers. The further investigation found that, the junior high school students in Anhui province, if not be in the single-child family, they always ranks the last-borns or penultimate, therefore, students with high self-concept in school, part of them will take teachers as parents or elder brother and elder sister, willing to ask them questions. An example of this is the children from special family. The results based on the nearly four years' research of the children from special family showed that, the family's the incomplete, the tension of family relations, atavism raising can seriously affect children's cognitive, then the deviation of cognitive can lead the attitude to changes (enhancement of academic help-seeking attitude),and more academic help-seeking behavior.

Appearance self-concept though indirectly the price of help-seeking affects the academic help-seeking. Oppositely happiness self-concept have a directly effects on instrumental help -seeking- the teacher behavior. Science self-concept through the attitude toward the price of academic help-seeking affect students' the academic help-seeking behavior in mathematics.The lower science self-concept is , the stronger price of academic help-seeking and the more academic help-seeking behavior will be. The reduction of science self-concept will make the student lose interest in learning mathematics, finally, cause of a pernicious circle, the worse math is, the lower science self-concept will be, and more reluctant to go and ask the teacher and classmate for math problems. Domestic teacher Li Xiaodong studies suggest that students have more interest in study, the more important and meaningful they consider, the less likely to adopt the method and means against their study.

Acknowledgements. This paper supported by 2010 Anhui province humanities and social science project Foundation of China, 2010 sk359.

References

1. Karabenick, S.A., Knapp, J.R.: Relationship of Academic Help-Seeking to the Use of Learning Strategies and other Instrumental Acheievement Behavior in College Students. Journal of Educational Psychology 83, 221–230 (1991)
2. Newman, R.S.: Children'Help-Seeking in the Classroom:The Role of Motivational Factors and Attitudes. Journal of Educational Psychology 82, 71–80 (1990)
3. Zhao, X.: Statistical Consciousness in the Instructional Design of Psychological Statistics. Journal of Psychological Science 27, 1226–1227 (2004)
4. Zhao, X.: Measurement Study of Self-Concept and Academic Help-Seeking of Junior Middle School Students in the Northwest China. Journal of Hexi University 23, 86–90 (2007)
5. Zhang, X., Wang, X.: Domestic and foreign academic help-seeking behavior research about the. Journal of Zhangzhou Normal University 20, 101–104 (2003)
6. Liu, A.: Self-concept of the theory to review and development trend [. Chinese Journal of Psychological Science 27, 248–249 (2004)
7. Li, X., Lin, C.: Personal goal orientation, classroom goal structure and cultural factors and the relationship of academic help-seeking strategy. Chinese Journal of Psychological Development and Education 17, 1–6, 18 (2001)
8. Li, X., Lin, C., Nie, Y., et al.: Class target structure, personal goal orientation, self-efficacy and value and academic self hinder. Chinese Journal of Psychological Science 26, 590–594 (2003)

The Description of Characters of Matrix-Block Multielement Wavelet Wraps according to a Dilation Matrix

Ruihua Hu[1] and Xianghai Li[2]

[1] Dept. of Electron & Electricity, Nanyang Institute of Technology, Nanyang 473004, China
[2] Dept. of Computer Science, Nanyang Institute of Technology, Nanyang 473004, China
asdfgh84sxxa@126.com, bjhnsx123@126.com

Abstract. Wavelet analysis has become a developing branch of mathematics for over twenty years.

In this paper, the notion of matrix-valued multiresolution analysis of space $L^2(R^n, C^{r\times r})$ is introduced. A method for constructing biorthogonal matrix–valued trivariate wavelet packets is developed and their properties are discussed by means of time-frequency analysis method, matrix theory and functional analysis method. Three biorthogonality formulas concerning these wavelet packets are provided. Finally, new Riesz bases of space $L^2(R^n, C^{r\times r})$ is obtained by constructing a series of subspaces of biorthogonal matrix-valued wavelet packets.

Keywords: Matrix-block scaling function, matrix-block multielement wavelet packets , Riesz bases, Bessel sequence, Bracket product, filter banks, time-frequency analysis, frame theory.

1 Introduction

Wavelet analysis, as a very active mathematical branch of the nonlinear science, has drawn many attentions since it appeared, which have led to exciting applications in signal analysis, fractal theory, image processing and so on. The rise of wavelet analysis in applied mathematics is due to its applications and the flexibility. Multiwavelets can simultaneously possess many desired properties such as short support, orthogonality, symmetry, and vanishing moments, which a single wavelet cannot possess simultaneously. This suggests that multiwavelet systems can provide perfect reconstruction, good performance at the boundaries (symmetry), and high approximation order (vanishing moments). Already they have led to exciting applications in signal analysis [1], fractals [2] and image processing [3], and so on. Matrix-block wavelets are a class of special multiwavelets. Chen [4] introduced the notion of orthogonal vector-valued wavelets [5]. Wavelet packets, due to their nice characteristics, have been widely applied to signal processing, image compression, and pattern recognition and so on. Coifman and Meyer are those who firstly introduced the notion of univariate orthogonal wavelet packets. The introduction for biorthogonal wavelet packets attributes to Cohen and Daubechies [6]. Yang and Cheng constructed

Y. Zhang (Ed.): Future Communication, Computing, Control and Management, LNEE 141, pp. 457–465.
springerlink.com

a-scale orthogonal multiwavelet packets which were more flexible in applications. Vector-valued wavelets are a class of special multi- wavelets. Chen [7] introduced the notion for orthogonal vector-valued wavelets. Moreover, he investigated the existence and constructive method for vector-valued wavelets. However, matrix-block wavelets and multiwavelets are different in the following sense. For example, prefiltering is usually required for discrete multiwavelet transforms but not necessary for discrete matrix-block wavelet transforms. Examples of matrix-block signals are video images. Therefore, studying matrix-block wavelets is useful in multiwavelet theory and represention of signals. It is known that the majority of information is multi-dimensional information. Shen [8] have generalized the univariate orthogonal wavelet wraps to the case of multivariate orthogonal wavelets such that they may be used in a wider field. The goal of the paper is to give the definition and the construction of matrix-block multielement wavelet wraps and design some new Riesz bases of space $L^2(R^n,C^{r\times r})$.

2 Notations and Matrix-Block Multiresolution Analysis

Let R and C be all real and all complex numbers, respectively. Z and N denote the set of integers and positive integers, respectively. Set $Z_+=\{0\}\bigcup N$, and $n,\, r\in N$, $n\geq 2$, $Z^n=\{(k_1,k_2,\cdots,k_n):\quad n_1,n_2,n_3\in Z\}$, $Z_+^n=\{(k_1,k_2,\cdots,k_n):\ k_\iota\in Z, \iota=1,2\cdots,n\}$. Suppose that D is an $n\times n$ matrix whose all entries are integers and all eigenvalues is large than one in modulus. The absolute value of the determinant of matrix D is denoted by m, i.e., $|\det(D)|=m$. Order D^T stands for the transpose of matrix, and D^{-T} is the inverse of the transpose of matrix D. It is known that there exist m elements $\rho_1, \rho_2 \dots,\ \rho_{m-1}\in Z_+^n$,by the finite group theory such that

$$Z^n=\bigcup_{v\in\Lambda_0}(v+DZ^n)\quad (v_1+DZ^n)\bigcap(v_2+DZ^n)=\varnothing,$$

where $\Lambda_0=\{v_1, v_2 \dots, v_{a-1}\}$ denotes the set of all different representative elements in the quotient group Z^n/DZ^n and v_1, v_2 denote two arbitrary distinct elements in Λ_0. Set $v_0=\{0\}$,where $\{0\}$ is the null of the set Z_+^n. Let $\Lambda=\Lambda_0-\{0\}$ and Λ, Λ_0 be two index sets. For $X,X_1,X_2\subset R^n$, Set $DX=\{Dx:x\in X\}$, $X_1+X_2=\{x_1+x_2:\quad x_1\in X_1,\quad x_2\in X_2\}$, $X_1-X_2=\{x_1+x_2:x_1\in X_1, x_2\in X_2\}$.

The space $L^2(R^n,C^{r\times r})$ is defined to be the set of all multiple vector-valued functions $\Phi(x)$, i.e.,

$$L^2(R^n,C^{r\times r}):=\left\{\Phi(x)=\begin{pmatrix}\phi_{11}(x) & \phi_{12}(x) & \cdots & \phi_{1r}(x)\\ \phi_{21}(x) & \phi_{22}(x) & \cdots & \phi_{2r}(x)\\ \cdots & \cdots & \cdots & \cdots\\ \phi_{r1}(x) & \phi_{r2}(x) & \cdots & \phi_{rr}(x)\end{pmatrix}:\begin{array}{l}\phi_{l,j}(x)\\ \in L^2(R^n)\\ l,j=1,\\ 2,\cdots,r\end{array}\right\}.$$

Examples of matrix- block signals are video images where $\phi_{lj}(x)$ denotes the pixel on the l th row and the j th column at the point x. $\|\Phi\|$ stands for the norm of the multiple vector-valued function $\Phi(x)\in L^2(R^n,C^{r\times r})$,

$$\|\Phi\|:=(\sum_{l,j=1}^{r}\int_{R^n}|\phi_{l,j}(x)|^2\,dx)^{1/2}, \tag{1}$$

For $\Phi(x)\in L^2(R^n,C^{r\times r})$, its integration and Fourier transform are defined, respectively as follows,

$$\int_{R^n}\Phi(x)\,dx=(\int_{R^n}\phi_{l,\tau}(x)\,dx)_{l,\tau=1}^{r},\qquad \hat{\Phi}(\omega):=\int_{R^n}\Phi(x)\cdot\exp\{-ix\cdot\omega\}dx,$$

where $x\cdot\omega$ stands for the inner product of real n-dimensional vectors x and ω. For two matrix-valued functions $G,H\in L^2(R^n,C^{r\times r})$, their symbol inner product is defined to be

$$\langle G,H\rangle:=\int_{R^n}G(x)H(x)^*\,dx,$$

where * means the transpose and the complex conjugate.

Definition 1. We say that a pair of matrix-block functions $G(x)$, $\widetilde{G}(x)\in L^2(R^n,C^{r\times r})$ are biorthogonal ones, if their translations satisfy $\langle G(\cdot),\widetilde{G}(\cdot-v)\rangle=\delta_{0,v}I_r$, $v\in Z^3$, where I_r denotes the $r\times r$ identity matrix and $\delta_{0,v}$ is the Kronecker symbol.

In the following, we introduce the notion of matrix-block multiresolution analysis and give the definitionof biorthogonal matrix-block wavelets of space $L^2(R^n,C^{r\times r})$. A matrix-block multiresolution analysis of $L^2(R^n,C^{r\times r})$ is a nested sequence of closed subspaces $\{Y_j\}_{j\in z}$ such that (i) $\cdots\subset Y_{-1}\subset Y_0\subset Y_1\subset\cdots$; (ii) $\bigcap_{j\in Z}Y_j=\{O\}$ and $\bigcup_{j\in Z}Y_j$ is dense in $L^2(R^n,C^{r\times r})$, where O denotes an

$r\times r$ zero matrix; (iii) $H(x)\in Y_j \Leftrightarrow H(Dx)\in Y_{j+1}, \quad \forall j\in Z;$ (v) $H(x)\in Y_j \Leftrightarrow H(x-k)\in Y_j,$

$\forall\, k\in Z;$ (iv) $\exists\ \Upsilon(x)\in Y_0$, called a matrix-block scaling function, such that the set $\{\Upsilon_k(x):=\Upsilon(x-k),$

$k\in Z^n\}$ forms a Riesz basis for subspace Y_0.

Since $\Upsilon(x)\in Y_0\subset Y_1$, by definition 3 and (6), there exists a finitely supported sequence of constant $r\times r$ matrice $\{Q_k\}_{k\in Z^n}\in l^2(Z^n)^{r\times r}$ such that

$$\Upsilon(x)=\sum_{k\in Z^n} Q_k\Upsilon(Dx-k). \tag{2}$$

Equation (8) is said to be a refinement equation. Let $mQ(\omega)=\sum_{k\in Z^n} Q_k\cdot\exp\{-ik\cdot\omega\}$, where $Q(\omega)$, is called a symbol of $\Upsilon(x)$, which is a $2\pi Z^n$ periodic function, Thus, equation (8) becomes

$$\widehat{\Upsilon}(\omega)=Q(D^{-T}\omega)\widehat{\Upsilon}(D^{-T}\omega), \quad \omega\in R^n. \tag{3}$$

Let $W_{j,}\ j\in Z$ be the direct complementary subspace of Y_j in Y_{j+1}. Assume that there exist $a-1$ matrix-valued functions $F_\rho(x)\in L^2(R^n,C^{r\times r}),\quad \rho\in\Lambda$, such that their thanslates and dilations form a Riesz basis of W_j, i.e.,

$$W_j=\mathrm{clos}_{L^2(R^n,C^{r\times r})}(\mathrm{Span}\{F_v(D^j\cdot-k):\ k\in Z^n,\ v\in\Lambda\}), \tag{4}$$

Since $F_\rho(x)\in W_0\subset Y_1, \rho\in\Lambda$, there exist $a-1$ finite supported sequences of constant $r\times r$ matrice $\{B_v^{(\rho)}\}_{v\in Z^3}$ such that

$$F_\rho(x)=\sum_{k\in Z^n} B_k^{(\rho)}\Upsilon(Dx-k), \quad \rho\in\Lambda. \tag{5}$$

Implementing the Fourier transform for the both sides of (5) yields

$$\hat{F}_\rho(\omega)=\mathcal{B}^{(\rho)}(D^{-T}\omega)\widehat{\Upsilon}(D^{-T}\omega), \quad \omega\in R^n, \rho\in\Lambda. \tag{6}$$

where
$$\mathcal{B}^{(\rho)}(\omega)=\frac{1}{m}\sum_{k\in Z^n} B_k^{(\rho)}\cdot \mathrm{e}^{-ik\cdot\omega}, \quad \rho\in\Lambda. \tag{7}$$

If $\Upsilon(x),\ \widetilde{\Upsilon}(x)\in L^2(R^n,C^{r\times r})$ are a pair of biorthogonal matrix-block scaling functions, then it follows that

$$\left\langle \Upsilon(\cdot), \widetilde{\Upsilon}(\cdot - u) \right\rangle = \delta_{0,u} I_r, \quad u \in Z^n. \tag{8}$$

We say that $F_\rho(x), \tilde{F}_\rho(x) \in L^2(R^n, C^{r\times r})$, $\rho \in \Lambda$ are pairs of biorthogonal matrix-block wavelets associated with a pair of biorthogonal matrix-block scaling functions $\Upsilon(x)$ and $\widetilde{\Upsilon}(x)$ if the family $\{F_v(x-k),\ k \in Z^n,\ \rho \in \Lambda\}$ is a Riesz basis of subspace W_0, and

$$\left\langle \Upsilon(\cdot), \tilde{F}_v(\cdot - k) \right\rangle = \left\langle \widetilde{\Upsilon}(\cdot), F_v(\cdot - k) \right\rangle = 0, \quad v \in \Lambda, \quad k \in Z^n, \tag{9}$$

$$\left\langle F_v(\cdot), \tilde{F}_\mu(\cdot - u) \right\rangle = \delta_{v,\mu} \delta_{0,u} I_r, \quad v, \mu \in \Lambda, \ u \in Z^n. \tag{10}$$

Let us set $W_j^{(\rho)} = \mathrm{clos}_{L^2(R^n, C^{r\times r})} \left\langle F_\rho(D^j \cdot - k) :\ k \in Z^n \right\rangle$, where $\rho \in \Lambda, \quad j \in Z$.

Proposition 1. If $F_\rho(x), \tilde{F}_\rho(x) \in L^2(R^n, C^{r\times r}), \rho \in \Lambda$ are a pair of biorthogonal matrix-block wavelets associated with a pair of biorthogonal matrix-block $\Upsilon(x)$ and $\widetilde{\Upsilon}(x)$, then

$$L^2(R^n, C^{r\times r}) = \biguplus_{j\in Z} W_j = \biguplus_{j\in Z} \Big(\biguplus_{\rho\in\Lambda} W_j^{(\rho)} \Big), \tag{11}$$

where $\biguplus$ denotes the direct sum.

Similar to (8) and (12), there exist m finite supported sequences of $r \times r$ constant matrice $\{\widetilde{Q}_k\}_{k\in Z^n}$ and $\left\{\tilde{B}_k^{(\rho)}\right\}_{k\in Z^n}, \rho \in \Lambda$ such that $\widetilde{\Upsilon}(x)$ and $\tilde{F}_\rho(x)$ satisfy the following equation:

$$\widetilde{\Upsilon}(x) = \sum_{k\in Z^n} \tilde{Q}_k\, \widetilde{\Upsilon}(x)(Dx - k), \tag{12}$$

$$\tilde{F}_\rho(x) = \sum_{k\in Z^n} \tilde{B}_k^{(v)}\, \widetilde{\Upsilon}(Dx - k), \quad v \in \Lambda. \tag{13}$$

3 The Traits of a Sort of Matrix-Block Wavelet Wraps

Let us introduce some notations. $\Psi_0(x) = \Upsilon(x)$, $\Psi_v(x) = F_v(x)$, $\tilde{\Psi}_0(x) = \widetilde{\Upsilon}(x)$, $\tilde{\Psi}_\rho(x) = F_\rho(x)$,

$$M_k^{(0)} = Q_k,\ M_k^{(v)} = B_k^{(v)},\ \tilde{M}_k^{(0)} = \widetilde{Q}_k,\ \tilde{M}_k^{(v)} = \tilde{B}_k^{(v)},\ v \in \Lambda,\ k \in Z^n.$$

For any $\alpha \in Z_+^n$ and the given biorthogonal matrix-block scaling functions $\Psi_0(x)$ and $\tilde{\Psi}_0(x)$, iteratively define, respectively,

$$\Psi_\alpha(x) = \Psi_{D\beta+\lambda}(x) = \sum_{k \in Z^n} M_k^{(\lambda)} \Psi_\beta(Dx - k), \tag{14}$$

$$\tilde{\Psi}_\alpha(x) = \tilde{\Psi}_{D\beta+\lambda}(x) = \sum_{k \in Z^n} \tilde{M}_k^{(\lambda)} \tilde{\Psi}_\beta(Dx - k), \tag{15}$$

where $\beta \in Z_+^n$ is the unique element such that $\alpha = D\beta + \lambda,\ \lambda \in \Lambda_0$ follows.

Definition 3. Two families of multiple vector-valued function $\{\Psi_{D\beta+\lambda}(x), \beta \in Z_+^n, \lambda \in \Lambda_0\}$ and $\{\tilde{\Psi}_{A\beta+\lambda}(x),\ \beta \in Z_+^n, \lambda \in \Lambda_0\}$ are said to be wavelet packets with respect to a pair of biorthogonal multiple vector-valued scaling functions $\Upsilon(x)$, and $\widetilde{\Upsilon}(x)$, respectively, where $\Psi_{D\beta+\lambda}(x)$ and $\tilde{\Psi}_{D\beta+\lambda}(x)$ are given by (14) and (15), respectively.

Applying the Fourier transform for the both sides of (14) and (15), respectively , yields,

$$\hat{\Psi}_{D\beta+\rho}(\omega) = \mathcal{M}^{(\rho)}(D^{-T}\omega)\hat{\Psi}_\beta(D^{-T}\omega), \quad \rho \in \Lambda_0, \tag{16}$$

$$\hat{\tilde{\Psi}}_{D\beta+\lambda}(\omega) = \mathcal{M}^{(\lambda)}(D^{-T}\omega)\hat{\tilde{\Psi}}_\beta(D^{-T}\omega), \quad \lambda \in \Lambda_0, \tag{17}$$

where

$$\mathcal{M}^{(\rho)}(\omega) = \frac{1}{m} \cdot \sum_{u \in Z^n} M_k^{(\rho)} \cdot \exp\{-iu \cdot \omega\}, \quad \rho \in \Lambda_0,$$

$$\tilde{\mathcal{M}}^{(\rho)}(\omega) = \frac{1}{m} \sum_{u \in Z^n} \tilde{M}_u^{(\rho)} \cdot \exp\{-iu \cdot \omega\}. \quad \rho \in \Lambda_0, \quad \omega \in R^n.$$

Lemma 1 [4]. Let $H(x), \widetilde{H}(x) \in L^2(R^n, C^{r\times r})$. Then they are biorthogonal ones if and only if

$$\sum_{k \in Z^n} \widehat{H}(\omega + 2k\pi)\widehat{\widetilde{H}}(\omega + 2k\pi)^* = I_r, \quad \omega \in R^n. \tag{18}$$

Lemma 2 [6]. Assume that are a pair of biorthogonal matrix-valued wavelets with a pair of biorthogonal scaling functions $\Psi_0(x)$ and $\tilde{\Psi}_0(x)$. Then, for $\mu, v \in \Lambda_0$, we have

$$\sum_{\sigma \in \Lambda_0} \mathcal{M}^{(\mu)}(D^{-T}(\omega + 2\sigma\pi))\mathcal{M}^{(v)}(D^{-T}(\omega + 2\sigma\pi))^* = \delta_{\mu,v} I_r,$$

Theorem 1. Suppose that $\{\Psi_\alpha(x), \alpha\in Z_+^n\}$ and $\{\tilde{\Psi}_\alpha(x), \alpha\in Z_+^n\}$ are wavelet packets with respect to a pair of biorthogonal matrix-valued scaling functions $\Psi_0(x)$ and $\tilde{\Psi}_0(x)$. Then, for $\alpha\in Z_+^n$, we have

$$\left\langle \Psi_\alpha(\cdot), \tilde{\Psi}_\alpha(\cdot-u)\right\rangle = \delta_{0,u} I_r, \quad k\in Z^n. \tag{19}$$

Proof. The result (29) follows from (15) as $\alpha=0$. Assume that (19) holds when $|\alpha|=\sum_{l=1}^n \alpha_l<\eta$, where η is a positive integer, and $\alpha=(\alpha_1,\alpha_2,\alpha_3)\in Z_+^n$. For the case of $\alpha\in Z_+^n$, $|\alpha|=\eta$, we will prove (29) holds. Order $\alpha=D\beta+\rho$ where $\beta\in Z_+^n, \rho\in\Lambda_0$. Since all eigenvalues of D is large than one in modulus, then $|\beta|<|\alpha|$. By induction assumption, we have:

$$\left\langle \Psi_\alpha(\cdot), \tilde{\Psi}_\alpha(\cdot-k)\right\rangle = \frac{1}{(2\pi)^n}\sum_{u\in Z^n}\int_{D^T([0,2\pi]^n+2u\pi)} \mathcal{M}^{(\rho)}(D^{-T}\omega)$$

$$\cdot\hat{\Psi}_\beta(D^{-T}\omega)\hat{\tilde{\Psi}}_\beta(D^{-T}\omega)^*\tilde{\mathcal{M}}^{(\rho)}(D^{-T}\omega)^* e^{ik\cdot\omega}d\omega$$

$$=\frac{1}{(2\pi)^n}\int_{[0,2\pi]^n}\sum_{\sigma\in\Lambda_0}\mathcal{M}^{(\rho)}(D^{-T}(\omega+2\pi\sigma))\cdot\tilde{\mathcal{M}}^{(\rho)}(D^{-T}(\omega+2\pi\sigma))^* e^{ik\cdot\omega}d\omega$$

$$=\frac{1}{(2\pi)^n}\int_{[0,2\pi]^n} I_r e^{ik\cdot\omega}d\omega=\delta_{0,k}I_r.$$

Therefore, the result is established.

Theorem 2. Suppose that $\{\Psi_\alpha(x), \alpha\in Z_+^n\}$ and $\{\tilde{\Psi}_\alpha(x), \alpha\in Z_+^n\}$ are wavelet packets with respect to a pair of biorthogonal matrix-valued functions $\Psi_0(x)$ and $\tilde{\Psi}_0(x)$, respectively. Then, for $\beta\in Z_+^n$, $n\in Z^n$, we have $\left\langle \Psi_{A\beta+\mu}(\cdot), \tilde{\Psi}_{A\beta+\rho}(\cdot-n)\right\rangle=\delta_{0,n}\delta_{\mu,\rho}I_r$, $\mu,\rho\in\Lambda_0$.

Proof. Since R^s has the following partition: $R^n=\bigcup_{k\in Z^n}([0,2\pi]^n+2k\pi)$, and $([0,2\pi]^n+2k_1\pi)\bigcap$

$([0,2\pi]^n+2\pi k_2)=\varnothing$, where $k_1\neq k_2, k_1,k_2\in Z^n$, then by Lemma 1, we have

$$(2\pi)^n\left\langle \Psi_{D\beta+\mu}(\cdot), \tilde{\Psi}_{D\beta+\rho}(\cdot-v)\right\rangle=\int_{R^n}\hat{\Psi}_{D\beta+\mu}(\omega)\hat{\tilde{\Psi}}_{D\beta+\rho}(\omega)^*\cdot\exp\{iv\cdot\omega\}d\omega$$

$$=m\int_{[0,2\pi]^n}\mathcal{M}^{(\mu)}(\omega)\{\sum_{u\in Z^n}\hat{\Psi}_\beta(\omega+2k\pi)\cdot\hat{\tilde{\Psi}}_\beta(\omega+2k\pi)^*\}\tilde{\mathcal{M}}^{(\rho)}(\omega)^* e^{iv\cdot D^T\omega}d\omega$$

$$= \int_{[0,2\pi]^n} \delta_{\mu,\rho} I_r e^{iv\cdot\omega} d\omega = \delta_{0,v}\delta_{\mu,\rho} I_r .$$

This completes the proof of the theorem.

Theorem 3. Suppose that $\{\Psi_\mu(x),\ \mu\in Z_+^n\}$ and $\{\tilde{\Psi}_\mu(x),\ \mu\in Z_+^n\}$ are wavelet wraps with respect to a pair of biorthogonal matrix-block scaling functions $\Psi_0(x)$ and $\tilde{\Psi}_0(x)$, respectively. Then,

$$\left\langle \Psi_\alpha(\cdot), \tilde{\Psi}_\sigma(\cdot - v) \right\rangle = \delta_{\alpha,\sigma}\delta_{\underline{0},v} I_r,\ v\in Z^n. \tag{20}$$

Proof. When $\alpha=\sigma$, (20) follows by Lemma 3. as $\alpha\neq\sigma$ and $\alpha,\sigma\in\Lambda_0$, it follows from Lemma 4 that (20) holds, too. Assuming that α is not equal to β, as well as at least one of $\{\alpha,\sigma\}$ doesn't belong to Λ_0, we rewrite α,σ as $\alpha = D\alpha_1+\rho_1$, $\sigma = D\sigma_1+\mu_1$, where $\rho_1,\mu_1\in\Lambda_0$. Case 1. If $\alpha_1=\sigma_1$, then $\rho_1\neq\mu_1$. (20) follows by virtue of (17), (18) aswell as Lemma 1 and Lemma 2,i.e.,

$$(2\pi)^n \left\langle \Psi_\alpha(\cdot), \tilde{\Psi}_\sigma(\cdot - k) \right\rangle = \int_{R^n} \widehat{\Psi}_{D\alpha_1+\rho_1}(\omega) \widehat{\tilde{\Psi}}_{D\sigma_1+\mu_1}(\omega)^* \exp\{ik\omega\} d\omega$$

$$= \int_{[0,2\pi]^n} \mathcal{M}^{(\rho_1)}(D^{-T}\omega) \{\sum_{u\in Z^n} \widehat{\Psi}_{\alpha_1}(D^{-T}\omega + 2u\pi)\cdot \widehat{\tilde{\Psi}}_{\alpha_1}(D^{-T}\omega+2u\pi)^*\} \widetilde{\mathcal{M}}^{(\mu_1)}(D^{-T}\omega)^* \cdot e^{ik\cdot\omega} d\omega$$

$$= \int_{[0,2\pi]^n} \delta_{\rho_1,\mu_1} I_r \cdot \exp\{ik\cdot\omega\} d\omega = O.$$

Case 2. If $\alpha_1\neq\sigma_1$, order $\alpha_1 = D\alpha_2+\rho_2$, $\sigma_1 = D\sigma_2+\mu_2$, where $\alpha_2,\sigma_2\in Z_+^n$, and $\rho_2,\mu_2\in\Lambda_0$. Provided that $\alpha_2=\sigma_2$, then $\rho_2\neq\mu_2$. Similar to Case 1, (20) can be established. When $\alpha_2\neq\sigma_2$,we order $\alpha_2 = D\alpha_3+\rho_3$, $\sigma_2 = D\sigma_3+\mu_3$, where $\alpha_3,\sigma_3\in Z_+^n$, $\rho_3,\mu_3\in\Gamma_0$. Thus, after taking finite steps (denoted by κ), we obtain $\alpha_\kappa\in\Lambda_0$, and $\rho_\kappa,\mu_\kappa\in\Lambda_0$. If $\alpha_\kappa=\sigma_\kappa$, then $\rho_\kappa\neq\mu_\kappa$. Similar to the Case 1, (20) follows. If $\alpha_\kappa\neq\sigma_\kappa$, then it gets from (12)-(15):

$$\left\langle \Psi_\alpha(\cdot), \tilde{\Psi}_\sigma(\cdot - k) \right\rangle = O \Leftrightarrow \sum_{u\in Z^n} \widehat{\Psi}_{\alpha_\kappa}(\omega+2u\pi) \widehat{\tilde{\Psi}}_{\beta_\kappa}(\omega+2u\pi)^* = O,\ k\in Z^n.$$

Furthermore, we obtain

$$(2\pi)^n \left\langle \Psi_\alpha(\cdot), \tilde{\Psi}_\sigma(\cdot - k) \right\rangle = \int_{R^n} \widehat{\Psi}_\alpha(\omega) \widehat{\tilde{\Psi}}_{\sigma_1}(\omega)^* \cdot e^{ik\cdot\omega} d\omega$$

$$= \int_{R^n} \widehat{\Psi}_{D\alpha_1+\lambda_1}(\omega) \widehat{\tilde{\Psi}}_{D\beta_1+\mu_1}(\omega)^* \cdot \exp\{ik\cdot\omega\} d\omega = \cdots\cdots\cdots\cdots\cdots\cdots\cdots\cdots$$

$$=\int_{[0,2D^{\kappa}\pi]^{3}}\{\prod_{l=1}^{\kappa}\mathcal{M}^{(\rho_l)}((D^{-T})^{l}\omega)\}\{\sum_{u\in Z^{n}}\hat{\Psi}_{\alpha_\kappa}((D^{-T})^{l}\omega)+2u\pi)\cdot\hat{\tilde{\Psi}}_{\sigma_\kappa}((D^{-T})^{l}\omega+2u\pi)^{*}\}\{\prod_{l=1}^{\kappa}\widetilde{\mathcal{M}}^{(\mu_l)}((D^{-T})^{l}\omega)\}^{*}\cdot e^{ik\cdot\omega}d\omega$$

$$=\int_{([0,2\cdot D^{\kappa}\pi]^{n}}\{\prod_{l=1}^{\kappa}\mathcal{M}^{(\rho_l)}((D^{-T})^{l}\omega)\}\cdot O\cdot\{\prod_{l=1}^{\kappa}\widetilde{\mathcal{M}}^{(\mu_l)}((D^{-T})^{l}\omega)\}^{*}\cdot\exp\{-ik\cdot\omega\}d\omega=O.$$

Therefore, for any $\alpha,\sigma\in Z_{+}^{n}$, result (20) is established.

4 Conclusion

The concept of biorthogonal matrix-valued wavelet packets of space $L^2(R^n,C^{r\times r})$ is introduced. Three biorthogonality formulas concerning these wavelet packets are presented.

References

1. Telesca, L., et al.: Multiresolution wavelet analysis of earthquakes. Chaos, Solitons & Fractals 22(3), 741–748 (2004)
2. Iovane, G., Giordano, P.: Wavelet and multiresolution analysis:Nature of ε^{∞}Cantorian space-time. Chaos, Solitons & Fractals 32(4), 896–910 (2007)
3. Zhang, N., Wu, X.: Lossless Compression of Color Mosaic Images. IEEE Trans. Image Processing 15(16), 1379–1388 (2006)
4. Chen, Q., et al.: A study on compactly supported orthogonal vector-valued wavelets and wavelet packets. Chaos, Solitons & Fractals 31(4), 1024–1034 (2007)
5. Shen, Z.: Nontensor product wavelet packets in $L_2(R^s)$. SIAM Math. Anal. 26(4), 1061–1074 (1995)
6. Chen, Q., Shi, Z.: Biorthogonal multiple vector-valued multivariate wavelet packets associated with a dilation matrix. Chaos, Solitons & Fractals 35(2), 323–332 (2008)

Mine Mechanical Equipments Fault Detection Based on Artificial Immune System

Xin Yue[a], Dongge Wen[b], Haifeng Ma[c], and Jingfeng Song[d]

Computer & Information Engineering College,
Heilongjiang Institute of Science and Technology, Harbin, China
[a] yuefour@gmail.com, [b] wdongge@126.com,
[c] mhf_1977@163.com, [d] Songjf_2000@hotmail.com

Abstract. The paper introduces the characteristics of coal mine mechanical equipment's maintenance. And according to their features, the thesis proposes current research condition and some of the research findings both home and abroad about artificial immune algorithm in fault diagnoses of coal mine machinery as well as the biotic mechanism of the artificial immune system, offering a new idea and approach to deal with fault diagnose. Besides, the paper also illustrates the artificial immune's application model and algorithm in fault diagnoses of coal mine machinery, makes an exploration about concurrent malfunctions, and analyzes the existing problems at current stage and the future prospects.

Keywords: Artificial immune system, Mine Mechanical Equipments, Fault Detection, Negative selection.

1 Introduction

As the largest source of energy in China, whether the safe, efficient production of coal can be guaranteed or not will directly affect the interests of the whole national economy. In coal enterprises, except the necessary conditions needed by coal production such as large quantities of coal-cutting workers, and coal mine mechanical equipment, with an increasing progress in science and technology, the level of mechanization will be raised day by day. Practices have proved that, during the production process, the frequent malfunctions caused by various mechanical equipment failures badly obstacle the effective production of coal.

As the coal industry develops, the functions of coal mine machineries become more and more complicated. The causes of malfunctions have become more complex due to the complicated structures and components of the machineries. Equipment diagnoses and maintenances are complex experimental jobs, which can be well done only by an application of information from various aspects. However, current diagnosing technologies of equipment are mostly the traditional approaches based on the detection data. This approach can find earlier and even predict stoppages, however, when the data of the diagnose object is large and complicated this simple data processing mean will fail to do the job.

Y. Zhang (Ed.): Future Communication, Computing, Control and Management, LNEE 141, pp. 467–472.
springerlink.com

Artificial immune system is an imitation of biological immune system and has a strong power in information processing. The biological immune system can accurately recognize various non-self substances under a shortage of priori knowledge, and continuously improve immune function during the process of self-directed learning, and conduct an automatic online malfunction detection and health monitoring toward organism, which provides the coal mine machinery malfunction diagnoses with new thoughts as well as new approaches. In order to improve the coal mine machinery performance, reliability, and safety, many researchers have focused their attention on the issues of fault diagnosis during the last two decades, as can be found from the related survey papers and books [1,2,3,4].

2 Research of Artificial Immune System

The earliest utilization of the immune system mechanism in solving engineering problems occurred in the 1990s. Before 1998, we had not attached great importance to researches of the immune system. Recently, various researchers have generated an increasing interest towards immune system, and solved many different engineering problems through its mechanism. In 1990, Japanese scholar Ishida solved the problem in network failure diagnoses of sensors. In 1994, Forrest first used the immune system in computer safety and virus detection. Hereafter, more and more scholars tried to establish a model of immune system, and the application fields of artificial immune system therefore are gradually enlarged, which now have covered information security, pattern recognition, machine learning, data mining, and fault diagnosis and automatic control etc. on December 1996, an international workshop based on immune system was launched in Japan, putting forward the "artificial immune system" concept for the first time. Then, the corresponding researches on artificial immune system emerged gradually, with the thesis and findings increase year by year. In 1997 and 1998, IEEE Conference organized related workshops, and established the artificial immune system and its application branches. Then, some of the famous international conferences in the field of artificial intelligence opened up a topic on artificial immune system one after another. From 2002, international conferences on artificial immune system had been launched on end in Britain, Italy, and Canada etc. In short, artificial immune is a learning technology enlightened by biological immune system and natural defense mechanism of the external substances. It offers an evolutionary learning mechanism, including noises endurance, study without a teacher, self-organization and memory etc. and combines a series of advantages of systems in classifiers, neural network, and machine reasoning. Therefore, it possesses the potential to solve complicated problems which has caused great attention of the mass researchers.

3 Characteristics of Coal Mine Mechanical Equipments

Less than 10% of our mine is open. Accordingly, the coal producing is basically borehole operation which renders far more distinctive characteristics of coal mine mechanical equipment that is totally different from the general equipment utilized in industry and agriculture :

Bad working environments: poor lighting conditions, ubiquitous dusts in the mine, water vapors, noises, harmful gases and liquids etc.

Demanding operation conditions: most mechanical equipments working under the conditions of high-speed, over-loading, vibration, impact, friction and bad lubrication.

Long working hours: most of the mechanical equipment working continuously day after night.

The above features result in the frequent malfunctions and accidents in coal mine machineries, whose repairing work is vital yet difficult.

In their developmental process, the machineries repair approaches mainly include: breakdown maintenance, preventive maintenance, predicting maintenance, and active preventive maintenance.

Breakdown Maintenance, it refers to an emergency maintenance without any preparation beforehand with a forced stoppage, adopted when the equipment fails or damages. Because there is no preparation in advance, the maintenance may take a long time which will cause a great loss due to production halts.

Preventive Maintenance, aiming at the obvious shortcomings of breakdown maintenance, is a maintenance system based on time. It takes a set time and adopts a maintenance approach to various degrees, therefore, to some extent, it can ensure the normal work and avoid the invalidation due to the extended life-cycle of the spare parts. However, its frequent maintenances may cause excessive unnecessary stoppages, which may increases the maintenances expenses, decreases the utilization efficiency, as well as posing a large quantity of maintenance costs.

Predicting Maintenance, it is adopted when the equipment is under working, to detect the working condition information, and judge whether its working is normal or not. The detection objects are generally initial failure such as violations and noises due to the wear and tear of materials and the function degradation. I f there is anything abnormal, they will judge the locations of fault points and instruct the maintenance staff to do the repair job timely to reduce unnecessary downtime and lower the maintenance fees. The shortcomings lie in that it can only lower the damage degree but cannot lower the accidents frequency and prolong the lifecycle of spare parts.

Active Preventive Maintenance, it and predicting maintenance are both based on operation condition monitoring of the system. The only difference is their monitoring targets. The former, as a preventive measure to avoid material invalidation, mainly detects the system original parameters which cause the material abrasion and functional degradation, such as the physical and chemical properties of oils, working temperature etc. For a long period of time, our coal mine enterprises basically adopt breakdown maintenance system and periodic maintenance (preventive maintenance) system; obviously, they cannot meet the need of modern production any longer. Currently, as a new strategy of equipment maintenance, active preventive maintenance needs to be promoted vigorously to greatly prolong the maintenance circle and efficient lifecycle of the equipment, lower the accidents ratio in coal mine machinery, substantially raise the working ratio of machinery in the process of coal producing, and create overall economic profits for the undertakings.

4 Negative Selection Algorithm

One of the major algorithms developed within emerging field of AIS is Negative Selection Algorithm, proposed by Forrest et al.. This algorithm is inspired by the mechanism of T-cell maturation and self tolerance in the immune system, and believed to have distinct process from alternative methods and be able to provide unique results with better quality [5]. It is based on principles of self/nonself discrimination in the immune system. There are two kinds of detector of Negative selection algorithms: Binary and Real-value. NSA in binary representation can hardly process many applications that are natural to be described in real-valued space, Real-valued representation provides some advantages such as increased expressiveness, the possibility of extracting high-level knowledge from the generated detectors. A negative selection algorithm basically consists of two phases. First, the detector set is generated in the training or generation phase. Then, the new sample is examined using the detector set during the detection phase[6], shown in the figure 1.

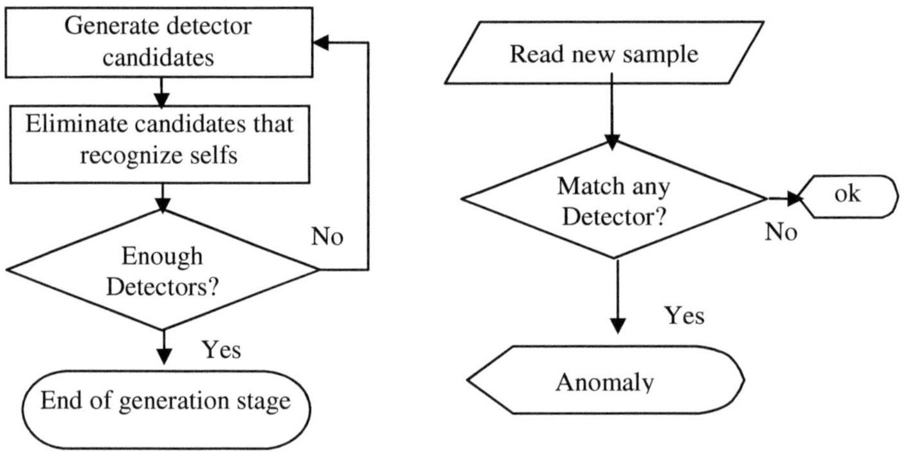

Fig. 1. Negative selection algorithm

Gonzalez et al.[6,7] proposed the real-value representation for the self/non-self space, which differs from the original binary representation. The real-valued self/detector is a hypersphere in n-dimensional real space; it can be represented by an n-dimensional point and a radius. Real-Valued Negative Selection Algorithm compared with the binary representation, real-valued representation of the most important feature is that self/non-self space corresponds to a subset of Rn, normalized to the super-rectangular space [0,1]n. Where, n is defined as the space dimension. The distance between two points uses Euclidean distance calculation. where every element is represented as a n-dimensional point or simply as a vector represented by a list of n real numbers. The Euclidean distance d is the (standard) distance between any two vectors x={x1,x2,…,xn}, y={y1,y2,…,yn}, defined as:

$$\text{Euclidean}(x, y) = \left(\sum_{i=1}^{n} | x_i - y_i |^2\right)^{1/2} \tag{1}$$

The volume of a n-dimensional hypersphere with radius r can be calculated as follows:

$$V(n,r) = r^n \bullet \frac{\pi^{n/2}}{\Gamma(n/2+1)} \tag{2}$$

Real-valued representation's search space is usually continuous, and it is suitable for network feature which can not use binary representation effectively .In the real-valued representation, it generally represents detector in the form of hypersphere or hypercube. The matching rule is expressed by the membership function of the detector, which is a function of the detector-antigen Euclidean distance and the radius of the detector.

5 Experiment and Analysis

The proposed approach is tested on the data collected from the process simulation for the Tennessee Eastman process (TEP). The TEP process was created by the Eastman Chemical Company to provide a realistic process for evaluating process control and monitoring methods. A detail description for TEP is available elsewhere. The data sets used in this paper are downloaded from http://brahms.scs.uiuc.edu. The data contains 41 measured and 12 manipulated variables and consists of one normal condition and 21 faulty operating conditions. Each data set contains 480 and 960 observations for training and testing respectively. The detection results for the normal operating data and the faults has better performance, where parameter rs is 0.1 and 0.05, the detection rate is 93.4% and 90.8 respectively.

It is crucial to determine the radius of self. If the radius is too large, the self will cover the non-self space and the non-self space covered by self will not generate detectors, which will finally result in the hole during fault detection; if the radius is too small, it can not ensure the full coverage to the self space, which will cause partial detectors to cover the self space and result in the false-alarm during detection. Both hole and false-alarm are avoidable and we must split the difference in the realization of algorithm. In the testing process, we can determine the radius of self by the way of testing, that is, adjust the radius of self from small to large and choose the self with radius that can best compromise the hole and false-alarm.

6 Conclusions

Because of the nature similarity of biological immune system and fault detection system, the AIS possesses promising potential in fault detection and diagnosis. by increasing a certain number of training samples, the accuracy of the method based on RNAS to fault diagnosis has made great changes. when the training samples are enough, the fault diagnosis method on the basis of artificial immunity proposed by the dissertation will have a good performance.

The results on the Tennessee Eastman process illustrate the effectiveness of the proposed method. However, much further work still remains to be done to use this method in some more complicated processes.

Acknowledgment. This work was supported by the Education Department Research Foundation of Heilongjiang Province (No. 11551434).

References

1. Frank, P.M.: On-line fault detection in uncertain nonlinear systems using diagnostic observers—a survey. Int. J. Syst. Sci. 25, 2129–2154 (1994)
2. Gertler, J.J.: Survey of model-based failure detection and isolation in complex plants. IEEE Control Syst. Mag. 8(6), 3–11 (1988)
3. Chen, J., Patton, R.J.: Robust Model-based Fault Diagnosis for Dynamic Systems. Kluwer Academic Publishers, Dordrecht (1999)
4. Venkatasubramanian, V., Rengaswamy, R., Yin, K., Kavuri, S.N.: A Review of Process Fault Detection and Diagnosis Part II: Quantitative Model-based Methods. Computers and Chemical Engineering 27, 293–311 (2003)
5. Garrett, S.M.: How do we evaluate artificial immune system? Evolutionary Computation 13(2), 145–177 (2005)
6. Gonzalez, F., Dasgupta, D., Kozema, D.: Combining negative and classification techniques for anomaly detection. In: Proc. of the 2002 Congress on Evolutionary Computation, pp. 705–710 (2002)
7. Ji, Z., Dasgupta, D.: Augmented Negative Selection Algorithm with Variable-Coverage Detectors. IEEE Congress on Evolutionary Computation, 1081–1088 (2004)

Study on Rotor-Side Converter of Doubly Fed Induction Generator*

Haihui Song[1,2] and Tian De[2]

[1] Shanghai Second Polytechnic University, Shanghai, China
[2] North China Electric Power University, Beijing, China
gzshh@163.com, gz3849@163.com

Abstract. The doubly fed induction generator has been widely used in wind power generation system. It can alter the rotational speed according to change of the wind speed, and contains the maximum tip speed ratio, in order to capture the maximum power from the wind energer. But the generator speed changes, the output frequency will chang. This paper studys the rotor-side converter of DFIG, compare the output frequency to the desired frequency input, spwm module sends 3-channel SPWM wave to dSPACE DS1103, and send out 3-channel PWM wave, then become 6-channel PWM wave to trigger the SKiiP after amplified and inverted, alter rotor excitation current, to maintain the generator output frequency stable.

Keywords: DFIG, VSCF, converter, PWM.

1 Introduction

The variable speed constant frequency (VSCF) doubly fed induction wind generator is currently the main unit today. It is commonly used a double-fed wind turbine controller, to achieve control of wind energy capture. It can alter the rotational speed according to change of the wind speed, and contains the maximum tip speed ratio, in order to capture the maximum power from the wind energer. But the generator speed changes, the output frequency will chang. Domestic and foreign experts and scholars have made a number of different control methods and strategies. But these studies is very complex, because of the uncertainty and aerodynamic complexity of the model power electronics. We construct a rotor-side converter of DFIG. It can avoid the drawbacks of traditional methods, follow automatically changes of unit performance and wind speed. Predict accurately the aerodynamic characteristics of wind turbines, adjust the 6-channel PWM wave frequency to trigger the SKiiP, alter rotor excitation current, to maintain the generator output frequency stable. It can not only give a constant frequency, but also enable the unit capturing maximum wind power.

---

* Supported by Leading Academic Discipline Project of Shanghai Municipal Education Commission (No.J51801).

Y. Zhang (Ed.): Future Communication, Computing, Control and Management, LNEE 141, pp. 473–478.
springerlink.com

2 Converter Structure

The rotor-side converter of doubly fed induction generator has been designed and simulated using both the Simulink and the dSPACE blocksets, the MATLAB to-DSP interface libraries, Real-Time Interface to Simulink, and Real-Time Workshop, all located on the workstation PC. The controller will be downloaded onto the Texas Instruments'TM320F240 DSP located on the DS1103 board.

A general block diagram for the converter is shown in figure 1 below.

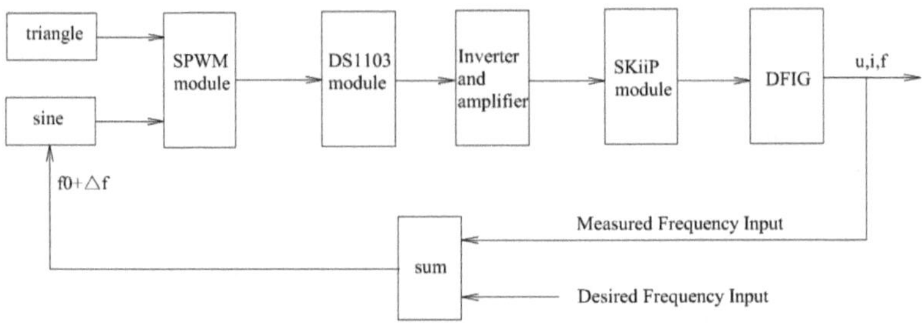

Fig. 1. Rotor-side converter block diagram

2.1 SPWM Module

SPWM module compares the output sine to the triangle wave, then sends 3-channel SPWM wave to dSPACE DS1103.

2.1.1 ControlDesk

ControlDesk serves multiple purposes. It provides the interface for downloading controller models onto the DSP and provides the ability to interface with the entire system so inputs, such as the desired motor speed, can be altered and output data can be monitored on the PC display in real-time.

2.1.2 DS1103 Module

The DS1103 serves as an interface between the dSPACE and all external hardware. The DS1103 contains connectors for twenty (20) Analog-to-Digital inputs, eight (8) Digital-to-Analog outputs, and several other connectors that can be used for Digital I/O, Slave/DSP I/O, Incremental Encoder Interfacing, CAN interfacing, and Serial Interfacing. Only the Slave I/O (for PWM output) and Incremental encoder interfaces are used in this tutorial.

2.1.3 Inverter and Amplifier Module

The inverter and amplifier module makes the 3-channel PWM wave out of DS1103 reversed and amplified, sends 6-channel PWM wave frequency to trigger the SKiiP.

2.1.4 SKiiP Module

SEMIKRON integrated intelligent Power is a power semiconductor subsystem, SKiiP integrates power semiconductor switches, heatsink and gate driver unit with protection and monitoringcircuit. Based on SKiiP pressure contact technology which allows a compact power module design with very low thermal resistances, high thermal cycling capability and low parasitic stray inductances. Equipped with closed loop current sensors, used for short circuit and over-current protection.

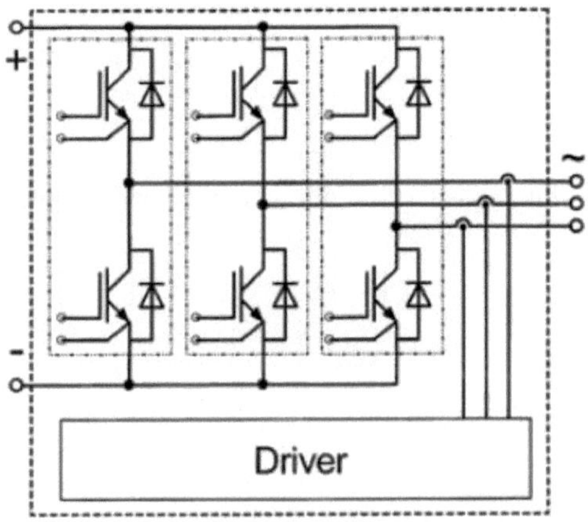

Fig. 2. SkiiP module block diagram

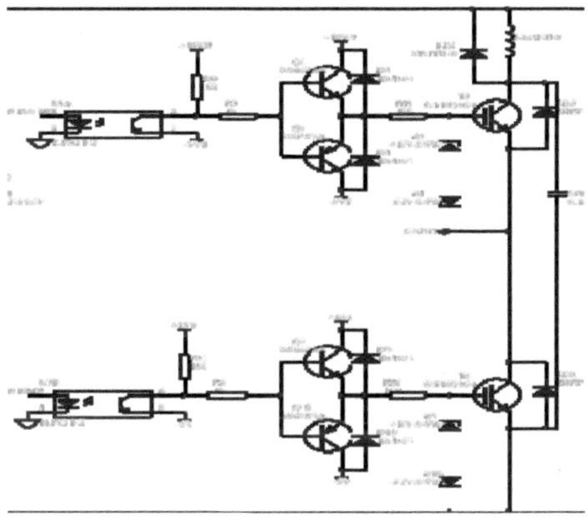

Fig. 3. The driving subsystem hardware schematic

3 The Interface between DS1103 and Skiip

Construct the driving subsystem shown in the figure 3 above, the 15 Volt PWM signal send to IGBT module after optical isolation and enlargeing.

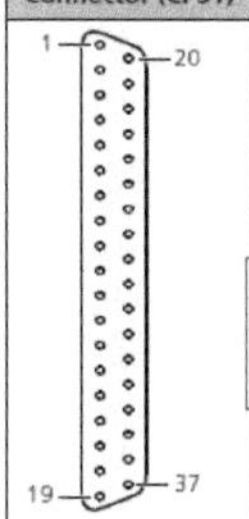

Slave I/O Connector (CP31)	Pin	Signal	Pin	Signal
	1	GND		
	2	SCAP1	20	GND
	3	SCAP3	21	SCAP2
	4	GND	22	SCAP4
	5	ST2PWM	23	ST1PWM
	6	GND	24	ST3PWM
	7	SPWM1	25	GND
	8	SPWM3	26	SPWM2
	9	SPWM5	27	SPWM4
	10	SPWM7	28	SPWM6
	11	SPWM9	29	SPWM8
	12	STMRCLK	30	GND
	13	GND	31	STMRDIR
	14	STINT1	32	SPDPINT
	15	GND	33	STINT2
	16	SSIMO	34	SSOMI
	17	SCLK	35	SSTE
	18	SXF	36	SBIO
	19	VCC (+5 V)	37	GND

Fig. 4. PWM connections on Slave I/O connector

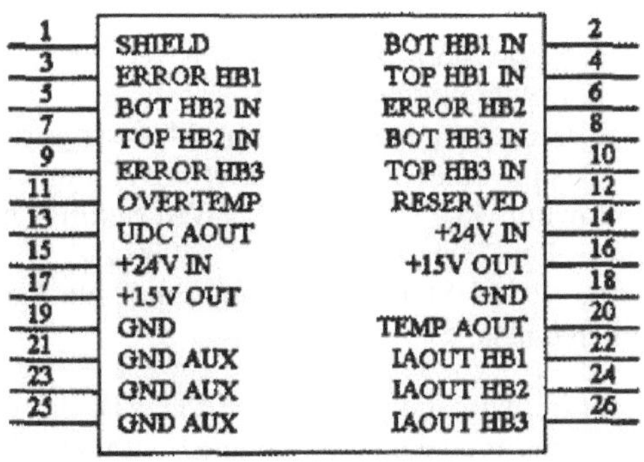

Fig. 5. The interface of skip

The slave I/O connector (CP31) is a 37-pin, female Sub-D connector located on the front of the connector panel. The following figure 4 shows the numbering scheme used(front view), and the required connections to the Slave I/O is shown in the figure.

The interface of skiip is shown in figure 5 above. From the figure, we can know that the 2#,5# and 8# Pin receives the each phase BOT HB trigger signal from the DS1103 DSP I/O, and the 4#,7# and 10# Pin receives the each phase TOP HB trigger signal. These signals are used to drive each phase IGBT.

4 Test

Begin the work with ControlDesk and Simulink, and start by following the instructions:

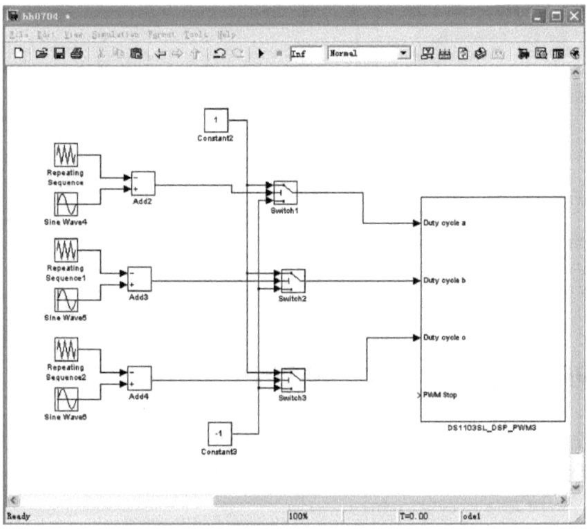

Fig. 6. DS1103SL-DSP-PWM

a. Setup/Startup

b. Detecting the DS1103

c. Creating a New Simulink Model(figure 6)

Change the Properties/Settings/Parameters of the PWM block. Change the following DS1103SL_DSP_PWM Settings in the dialog box:

a. “Unit” tab: Frequency = 12000 Hz

b. “PWM Stop and Termination” tab: Remove check from box for “set Ch” X for all channels.

Capture data and plot it in MATLAB, the value of any gain or constant block in the Simulink model can be changed without rebuilding and downloading code and while the system is running. The PWM signals of 3 phases TOP HB is shown in figure 7 below, and Volt waveform from Volt transformer is shown in figure 8.

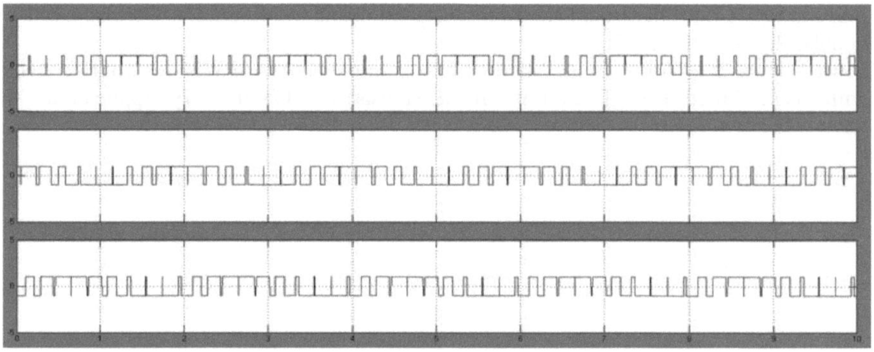

Fig. 7. PWM signals of 3 phases TOP HB

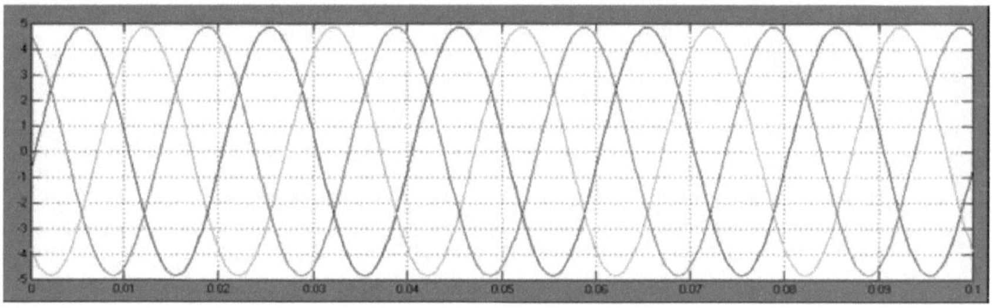

Fig. 8. Volt waveform from Volt transformer

5 Conclusion

The variable speed constant frequency (VSCF) doubly fed induction wind generator is currently the main unit today.This paper studys the rotor-side converter of DFIG, compare the output frequency to the desired frequency input, spwm module sends 3-channel SPWM wave to dSPACE DS1103, and send out 3-channel PWM wave, then become 6-channel PWM wave to trigger the SKiiP after amplified and inverted, alter rotor excitation current, to maintain the generator output frequency stable. The simulation results indicates that the wind power system has good wind compression and good capacity of capturing the maximum power. It shows that the converter is convenient and effective, and this offers a good platform for researching.

References

1. Semikron International. Power Electronics (2004)
2. Holtz, J.: Pulsewidth modulation for electric power conversion. Proceedings of the IEEE 82(8), 1194–1213 (1994)
3. DS1103 PPC Controller Board, Germany: dSPACE (July 2008)
4. Thomas, A.: dSPACE DS1103 Control Workstation Tutorial and DC Motor Speed Control. Senior Project Report, Bradley University ECE Department (May 2009)

5. DS1103 PPC Controller Board Hardware Installation and Configuration. Germany: dSPACE GmbH, Release 6.0 (2007)
6. Mwinyiwiwa, B., Wolanski, Z.: Microprocessor Implemented SPWM for Multi–converters with Phase-shifted Tnsllg]e Carriers. IEEE Transactions Oil Industry Applieationa 34(3), 487–494 (1998)

The Study on the Multi-object Fuzzy Decision-Making of the M&A of Resource –Typed Enterprises Based on Rough Sets

Tian Liang[1] and Ying Chu[2]

[1] Library, Huazhong University of Science & Technology, Wuhan 430074, China
[2] School of Management, Huazhong University of Science & Technology, Wuhan 430074, China
liangt1962@sina.com, Chuying_hust@tom.com

Abstract. The resource-typed enterprises have become the targets of M&A of global capital because of the shortage of non-renewable resources. The difficult problem that decides the success of enterprises' M&A is how to choose their targets. Through analyzing the characteristics of resource-typed enterprises and the selection decision-making system of the M&A' s targets, this paper establishes an index system which can be utilized to evaluate the M&A' s targets comprehensively and scientifically. Moreover, this paper uses the basic theory of rough sets and gray system to conduct a quantitative decision-making, presents a decision-making model to choose the optimum alternative for M&A' s targets. Finally this paper realizes the multi-object decision-making in mergers and acquisitions of the resource-typed enterprises.

Keywords: Resource, Mergers and acquisitions, Gray system, Rough sets.

1 Introduction

As to Chinese capital market, there is a fierce M&A competition in grabbling the high quality assets happened both in domestic and foreign capitals. Thus, a lot of opportunities have been provided to those bottlenecked energy and resource-typed enterprises whose price is still subjected to the government regulation. Because of the gradually decrease of non-renewable resources, the focus of the competition is to strategically get control of such kind of enterprises. The continuous rise of the price of natural gas, coal, copper, chemicals, rare metals, minerals and so on has resulted in that the global capital is aiming at the resource-typed enterprises [1].

With regard to resource-typed enterprises, no corresponding evaluation systems have been formed recently in the theoretical field. Most of the recent various evaluations are qualitative analyses while few quantitative ones, if any, they mainly evaluate a few financial indexes and have not formed an integrated system yet. As the imperfectness of the evaluation systems, the enterprises lack the systematic and scientific evidence to support their choices in M&A, particularly the systematic analysis in the long term [2]. Therefore, in the morass of a plethora of indexes about resource-typed enterprises, by adopting modern management theory and information technology, to establish a comprehensive and scientific index system in evaluating the M&A' s targets and then to conduct quantitative decision-making are of great significance in the modern society.

Y. Zhang (Ed.): Future Communication, Computing, Control and Management, LNEE 141, pp. 479–486.
springerlink.com

In terms of the characteristics of resource-typed enterprises and the selection decision-making system of the M&A' s targets, combining the gray system theory and rough sets theory, this paper establishes an index system which can be utilized to evaluate the M&A' s targets, presents a decision-making model to choose the optimum alternative for M&A' s targets, and realizes the multi-object decision-making in the mergers and acquisitions of the resource-typed enterprises. The purpose is to exclude interference of the artificial factors, and then rate the target enterprises comprehensively, concretely and objectively, finally choose the appropriate ones to strengthen the competitiveness of resource-typed enterprise and ultimately boost their competitiveness in the world market.

2 The M&A' s Target Selection Evaluation Index System of Resource-Typed Enterprises

In terms of the practical condition of resource-typed enterprises, we can divide their core competence into four parts: resources, technology, management and infrastructure, each of which can be divided further. See details in table1.

Table 1. The M&A' s Targets Selection Evaluation Index System

V1 The resources possession ability	V11 Resource	V111 Mineral output
		V 112 Mining difficulty
		V 113 Proved reserves
		V 114 Reserve-production ratio
	V12 Human resource	V121 Labor productivity
		V122 Technician ratio
		V123 Management stuff ratio
V2 The production technology ability	V21 Production technology level	V211 High-tech degree
		V212 Comprehensive technology service
		V213 The production efficiency of technicians
	V22 Science and technology Innovation ability	V221 Technology advancement contribution rate
		V222 The degree of science and technology investment
V3 The marketing and management ability	V31 Resources management	V311 Resource exploitation and management
		V312 The procedure of resource management and supervision
		V313 The accumulation of management experience
	V32 Financial evaluation	V321 Cost structure
		V322 Capital management ability
		V323 Ratio of liabilities to assets
	V33 Management efficiency	V331 Decision-making policy
		V332 Management efficiency
		V333 Management atmosphere
V4 The infrastructure match ability	V41 The Internal environment	V411 Enterprise culture
		V412 The sufficiency of production equipment
	V42 The external environment	V421 Legal and administrative stipulation
		V422 Government support
		V423 The relationship with social communities

The resources of enterprises reflect the concentration degree and accumulation level of each productive element, are the overall manifestation of enterprises' management condition and financial power, and are also the foundation and reliance on which enterprises can compete with others in the market. The strength indicating the resource-typed enterprises' scale is the historical accumulation about the past resource-exploitation activities and how to recognize and control resources, and is also the essential foundation on which enterprises can participate in market competition. Hence, this strength is the fundamental element that reflects the competitiveness of resource-typed enterprise.

3 The Multi-object Fuzzy Decision-Making Model Based on Rough Sets

The gray system theory can be applied to many fields such as economy prediction, decision-making and control etc. The multi-object gray relational degree fuzzy decision-making is to make sure the quality of each optional enterprise for through ascertaining the degree of similarity between the target enterprises' indexes and relative optimum enterprise indexes [3] .

3.1 The Establishment of Math Model

The resource-typed enterprise M&A' s target selection gray relational degree model:

$$\xi' = \xi \times W = (\xi_1, \xi_2, ..., \xi_m) \tag{1}$$

In the above formula, ξ' is the outcome matrix of m target enterprises; W is the weighted distribution matrix of n examined indexes(where n is the total sum of all radix in the above table, , n =30). $W = (W_1, W_2, ..., W_n)^T$; ξ is the evaluation matrix of each index, which can be defined as:

$$\xi = (\xi_{ij})_{m\times n} = \begin{bmatrix} \xi_{11} & \xi_{12} & \cdots & \xi_{1n} \\ \xi_{21} & \xi_{22} & \cdots & \xi_{2n} \\ \cdots & \cdots & & \cdots \\ \xi_{m1} & \xi_{m2} & \cdots & \xi_{mn} \end{bmatrix} \tag{2}$$

where ξ_{ij} is the relational degree, which denotes the relational between the j th decision-making index v_j of the target enterprise u_i for option and the j th index of the relative optimum target enterprise u^* .

$$\xi_i = \sum_{j=1}^{n} \xi_{ij} \times W_i \qquad i = 1,2,...,m \tag{3}$$

In the above formula, the larger the ξ_i, the closer the target enterprise u_i for option is to the relative optimum enterprise u^*, namely, the i th target enterprise is superior to all other optional targets and by which we can get the sequence of the optional targets. The optional target enterprise that corresponds to the largest ξ_i should be the optimum target enterprise, namely, if $\xi_i = \max(\xi_1, \xi_2, ..., \xi_m)$, the target enterprise is the best.

3.2 The Computation of the Optimum Alternative's Relational Degree

Suppose in the decision field, U is the optional target sets and:

$U = \{\text{target1}, \text{target2}, ..., \text{target}m\} = \{u_1, u_2, ..., u_m\}$;

V is the factor sets and

$V = \{\text{target1}, \text{target2}, ..., \text{target}n\} = \{v_1, v_2, ..., v_n\}$.

The index value of the target enterprise u_i to the factor index v_j is $r_{ij} (i = 1,2,..., m; j = 1,2,..., n)$, therefore, the target enterprise selection factor index matrix R from mn r_{ij} is:

$$R = (r_{ij})_{m \times n} \tag{4}$$

In order to eliminate the influence from factor indexes of different dimensions upon target selection, it is necessary to firstly convert the different values of evaluation indexes into values in range of with no dimension in the computation, in another word, conducting an initialization of the elements r_{ij} in matrix R, thus $R' = (r_{ij}')_{m \times n}$. The selection of target purpose measurement can be decided according to the practical situation [4].

In terms of gray system theory, the optimum selection relational degree of the target enterprise is the relational value between one target enterprise u_i in space V when taking the factor v_j into account and relative optimum target enterprise u^*, which can be measured by the formula below:

$$\xi_{ij} = \frac{0.5 \times \max_i \max_j \left| r_{ij}' - 1 \right|}{\left| r_{ij}' - 1 \right| + 0.5 \times \max_i \max_j \left| r_{ij}' - 1 \right|} \qquad i = 1,2,..., m; \qquad j = 1,2,..., n \tag{5}$$

3.3 The Determination of Evaluation Weights

Weights are the reflection of the physical attribute of indexes itself. The value of weights indicates the relative importance of each factor participating evaluation in the comprehensive judgment. Their values can directly influence the quality of evaluation result. Therefore, how to eliminate the problems of subjectivity and vagueness has become the focus of current researches. In order to judge precisely by introducing experts' knowledge and experience, to decrease the disturbance of artificial factors in the evaluation process and to guarantee the objectivity of evaluation result better, the emergence of rough sets theory has solved such problems. Through converting the problem of granting weights into the problem of evaluating the importance of rough sets' attributes, establishing the knowledge systems by building the relation data model and making attribute value into eigenvalue, and analyzing the support degree and importance of the targets participating in evaluation driven by data, rough set theory can work out the weight coefficients of comprehensive evaluation model. The rough set theory [5,6] and the fuzzy set theory [7] are two main tools used for processing these problems. The relevant definitions about rough sets are:

Definition 1: Suppose quaternary group $S=(U,A,V,f)$ is a knowledge expressing system. U is the no empty finite set of objects, called Universe, and $U=\{x_1,x_2,\Lambda,x_n\}$. $A=C\cup D$ is the no empty finite set of attributes and $C\cap D=\Phi$. Subsets C and D are called condition attribute set and result attribute set respectively. $V=\underset{a\in A}{\gamma} V_a$ is the attribute value set, where V_a denotes the value range of attribute $a\in A$, namely the range of attribute a. $f:U\times A\rightarrow V$ is an information function, which specifies the attribute value of every object x.

Definition 2: Every attribute subset $P\subseteq A$ determines a binary indiscernible relation $ind(P)=\{(x,y)\in U\times U|\forall a\in P, f(x,a)=f(y,a)\}$. $ind(P)$ is the equivalent relation in universe U. In relation $ind(P), P\in A$ constitutes a partition of U, expressed as $U/ind(P)$, in which every element is called equivalent class. Information system (U,A,V,f) is also called Knowledge A.

Definition 3: Suppose $S=(U,A,V,f)$ is an information system, and $P\subseteq A, U/ind(P)=\{x_1,x_2,\Lambda,x_n\}$. Thus the information capacity of Knowledge P can be defined as:

$$I(P)=\sum_{i=1}^{n}\frac{|X_i|}{|U|}\left[1-\frac{X_i}{|U|}\right]=1-\frac{1}{|U^2|}\sum_{i=1}^{n}|X_i|^2 \tag{6}$$

$|X|$ denotes the radix of X, and $|X_i|/|U|$ represents the probability of equivalent class X_i in U.

Definition 4: Suppose $S=(U,A,V,f)$ is an information system, and the importance of the attribute $a\in A$ in A can be defined as:

$Sig_{A-\{a\}}(a)=I(A)-A(A-\{a\})$. Especially, when $A=\{a\}$, $Sig(a)$ denotes $Sig_{\Phi}(a)$, and $Sig(a)=Sig_{\Phi}(a)=I(a)-I(\Phi)=I(\{a\})$, where $U/ind(\Phi)=\{U\},I(\Phi)=0$. Thus the importance of the attribute $a\in A$ in A is measured by the information capacity change after deleting a .

Definition 5: Suppose $S=(U,A,V,f)$ is an information system, so the weight of attribute $a_i\in A\{a_1,a_2,\Lambda,a_n\}$ is

$$w_i=\frac{Sig_{A-\{a_i\}}(a_i)}{\sum_{j=1}^{n}Sig_{A-\{a_j\}}(a_j)}=\frac{I(A)-I(A-\{a_i\})}{nI(A)-\sum_{j=1}^{n}I(A-\{a_j\})} \tag{7}$$

According to the definition 5, we can figure out the weight value of attribute a_i by computation.

Compared with the weights from experts' experience, the weights worked out by rough sets are based on measurement and experimental data and find out the internal laws among matters by digging the data themselves. Hence, this approach is more objective than Subjective Granting Weight and promotes the truthfulness of evaluation results.

Thus, we can get the weights of evaluation indexes:

$$W=(W_1,W_2,......,W_n)^T \tag{8}$$

4 Example

For the strategic purpose, one steel enterprise in China is considering to choose the M&A target from 10 optional mines. In order to simplify the process, the selection criteria will not take into account the synergic effect such as the complementation of advantages between this enterprise and the target enterprises for option, and we judge just from the target enterprises' strength. Thus:

$$U=\{t\arg et_1,t\arg et_1,......t\arg et_{10}\}=\{u_1,u_2,......u_{10}\}$$

To evaluate the 10 enterprises, we will adopt some indexes in the table 1. Subjected by the non-renewable resources' characteristics and the actual control right in resource exploitation, in the practical M&A decision-making, more attention is paid to two core competences: resource and technology. Likewise, for the convenience in citing examples, we just select 5 indexes in the first and second classes for demonstration, which are mineral output V111, mining difficulty V112, proved reserves V113, reserve-production ratio V114, science and technology contribution rate V221. Then the evaluation factor index set:

$$V = \{V_{111}, V_{112}, V_{113}, V_{114}, V_{221}\} = \{v_1, v_2, v_3, v_4, v_5\}$$

The relational data are in table 2.

Have a discretization to the attributes in table 2 (ascertaining the classification labels according to " high, middle and low"):

Mineral output: 1(v1>=10000),2(10000>v1>5000), 3(v1< =5000)
Mining difficulty: 1(v2<=65),2(80> v2 >65), 3(v2>=80)
Proved reserves: 1(v3>=2),2(2 > v3>0.5), 3(v3 <= 0.5)
Reserve/production ratio: 1(v4<=2),2(4.5> v4 > 2), 3(v4>=4.5)
Sci&Tech contribution rate: 1(v5>=40),2(40> v5 > 20), 3(v5 <=20)
Average of experts' evaluating points. 1(f<=11),2(11 > f >14), 3(f>=14)

Table 2. The Comprehensive Evaluation Results

	Mineral Output	Cost	Reserves	Reserve/ production Ratio	Sci&Tech Contribution rate
Weights	Thousand Tons	Yuan/ton	Mil. tons	Meter/thou. tons	%
Target 1	8000	100	800	6.8	24
Target 2	20000	98	1200	6.5	34
Target 3	3000	75	1500	3.4	20
Target 4	5000	72	200	3.6	30
Target 5	1000	65	50	4.0	40
Target 6	6000	80	400	4.5	27
Target 7	8500	77	600	5.9	37
Target 8	7000	82	350	3.8	28
Target 9	10000	79	480	4.4	35
Target 10	6500	84	180	4.6	31

Then, have normalization:

U/ind(c2,c3,c4,c5)={(1,2),(3),(4,9),(5),(6,10),(7),(8)}
U/ind(c1,c3,c4,c5)={(1,7),(2),(3),(4),(5),(6,10),(8),(9)}
U/ind(c1,c2,c4,c5)={(1,6,10),(2),(3),(4),(5),(7),(8),(9)}
U/ind(c1,c2,c3,c5)={(1),(2),(3),(4),(5),(6,8,10),(7),(9)}
U/ind(c1,c2,c3,c4)={(1),(2),(3,4),(5),(6,10),(7),(8),(9)}
U/f={(1,8,10),(2,5,7,9),(3,4,6)}
pos(c2,c3,c4,c5)(f)=4 1-Y=0.600 pos(c1,c3,c4,c5)(f)=6 1-Y=0.400
pos(c1,c2,c4,c5)(f)=7 1-Y=0.300 pos(c1,c2,c3,c5)(f)=7 1-Y=0.300
pos(c1,c2,c3,c4)(f)=8 1-Y=0.200
W1=0.333 W2=0.222 W3=0.167 W4=0.167 W5=0.111
Finally, the result of calculation is:

$$\xi' = \xi \times W = (0.5169, 0.7986, 0.5741, 0.5979, 0.6275, 0.5414, 0.5852, 0.5688, 0.6005, 0.5329)$$

From that above, we know that the target2 is the optimum target enterprise for option.

5 Several Interpretations

In order to research the core competence of resource-typed enterprises conveniently and comprehensively, this paper proposes some indexes for examination according to the concrete characteristics of resource industry. Particularly, by integrating the core competence indexes that can be calculated and commentarial indexes, this paper strives to handle the problem that the indexes in the international strategy research of enterprise development placed too much emphasis on financial indexes but paid few attention to technology innovation ability, management characteristics and talents. In the process of determining the values and weights of those indexes, it is necessary to calculate the two kinds of indexes respectively. Namely, for the indexes calculated by statistical data, their results are satisfied because no artificial factors exist; for the commentarial indexes, it converts the problem of granting weights into the problem of evaluating the importance of rough sets' attributes, establishes the knowledge systems by building the relation data model and making attribute value into eigenvalue, and works out the weight coefficients of comprehensive evaluation model by analyzing the support degree and importance of the targets participating in evaluation driven by data, which overcomes the problems of subjectivity and vagueness in Subjective Granting Weight and Objective Grant Weight and make the results more precise.

As one of the effective ways in target selection, gray relational degree decision-making has been applied extensively and internationally. The writer in this paper tries to apply it to the selection of M&A targets in resource industry. In the practical application, the M&A among enterprises is always out of strategic purpose, for example, to make up the weakness in competition or to gain the scale effect. Therefore, this model can be used as an auxiliary tool and can provide the proper quantitative criteria by combining the decision-makers' subjective intention.

References

1. Yu, C.: The global personal capital is targeted at resource-typed enterprises and media industry. Shanghai Negotiable Securities, April 22 (2006)
2. Liu, H.: The M&A activities of global resource-typed industries and their influence. Economic Analysis 1 (2001)
3. Deng, J.: The Basics of Gray System. Huazhong University of science and technology press, Wuhan (1987)
4. Cha, X., Huo, Z.: The fuzzy evaluation model and system analysis based on gray relational analysis. Mathematic Theory and Application 3, 38–41 (2005)
5. Pawlak, Z.: Rough sets. International Journal of Comput. Inform. Sci. 5, 341–356 (1982)
6. Pawlak, Z.: Rough Sets-theoretical Aspects of Reasoning About Data. Kluwer Academic Publishers, Dordrecht (1991)
7. Zadeh, L.A.: Fuzzy Sets. Information and Control 8, 338–353 (1965)

Algorithm of License Plate Location Based on Morphological Multi-Structural Elements*

Xiong Chun-rong, Chen Tie-jun, and Lv Jie

Department of Professional Technical Institute, Yulin Normal University, Yulin Guangxi 537000, China
xiongchunr@163.com, chentiejun2000@163.com, lvjie2227@126.com

Abstract. In traditional morphological algorithm of license plate location, they mostly use a single structure element on edge detection, edge information easy to miss, it is difficult to achieve better location performance, a new algorithm for license plate location based on morphological multi-structural elements is proposed. Firstly, we construct an anti-noise morphological edge detection operator. Secondly, construct structure element in different directions, and the final plate edge image is obtained with syntheticing the edge images which be geted by each structural elements, Finally, both the line scanning are used to locate the license plate area. The experimental results indicate that the algorithm is a better method for car license image edge detecting and noise filtering than the usually used, thus realizes the accurate localization of car license.

Keywords: License plate location, mathematical morphology, structure element, edge detection.

1 Introduction

Car license plates recognition (CLPR) system consists of license plates location, characters segmentation and characters identification. License plates location is critical to the operation CLPR system whose accurate positioning will have great influence on the identification performance of the system. The essence of license plates location is image segmentation aiming to segment the area of the license plate from the original image.

There are a variety of methods to image segmentation, among which edge detection is often applied. Proper edge detection serves image segmentation much better. As traditional edge detection is based on the use of space operations and space differential operator, and completes edge detection through the convolution of template operators and image. According to the size of template and the difference of element value, there are many different differential operators, such as Roberts, Sobel, Prewitt and so on. These edge detection operators are sensitive to noise, and can increase the noise while detecting edge, which brings difficulty to the edge detection of noise picture.

* Fund: Supported by Foundation of Guangxi Key Laboratory of Information and Communication(No:20901); Supported by Department of Education 2010 research projects in Guangxi(No:201010LX441); Suported by Guangxi, Yulin Normal University youth research projects in 2010(No:2010YJQN26).

Y. Zhang (Ed.): Future Communication, Computing, Control and Management, LNEE 141, pp. 487–493.
springerlink.com

Mathematical morphology is one kind of nonlinear image processing theory, whose basic idea is to use structure element with certain form to measure and detect the shape in targeted image, and to achieve the aim of analyzing and identifying the image [1]. In recent years, mathematical morphology has been widely applied to the area of image edge detection, and many scholars have studied some edge detection operators based on mathematical morphology [2,3]. However, most of they make use of single structure elements which contributes to the effects of edge detection but not good for keeping the informational geometrical characteristics. This article proposes a new license plate location algorithm based on morphological multi-structural elements. This new algorithm will be proved more effective on filtering noise, and it will get more accurate plate positioning as well as obtaining good performance of edge detection image.

2 Edge Detection Operators Based on Mathematical Morphology

Mathematical morphology is a mathematical method of analyzing geometrical shape and structure and the science of using set theory method to describe geometrical structure on the basis of set algebra.

The basic operations of mathematical morphology include: erosion, dilation, open and close [4]. The following is the introduction to the four mathematical morphology operations for processing binary image. If F stands for a given binary image in pending, and B is structure element, then:

Erosion can be defined as:

$$F\Theta B=\{x:B+x\subset F\} \tag{1}$$

Dilation can be defined as:

$$F\oplus B=Y\{F+b:b\subset B\} \tag{2}$$

Open and close operation is composed of erosion and dilation algebraic operation and set operation. Open operation is to erode the image first, and then dilate it. Close operation is to dilate the image first, and then erode it. Then:

Open operation can be defined as:

$$F\circ B=(F\Theta B)\oplus B \tag{3}$$

Close operation can be defined as:

$$F\bullet B=(F\oplus B)\Theta B \tag{4}$$

Based on morphological characteristics of computing, the operation of erosion and dilation meets:

$$F\ B\subseteq F\subseteq F\oplus B \tag{5}$$

the operation of open and close meets:

$$F\circ B\subseteq F\subseteq F\bullet B \tag{6}$$

The analysis above indicates that the dilation and close operation can expand the image, while the erosion and open operation can minish the image, but the processed image is similar with the original one. Therefore, in the morphology edge detection field, the following operations are often used to detect image edge [5].

If Ed(F) stands for dilation residue edge detector, it can be defined as the difference between the result of image F which was dilated by structure element B and the original image, namely:

$$Ed(F)= (F \oplus B)-F \tag{7}$$

Similarly, if Ee(F) is the erosion residue edge detector, it can be defined as the difference between the original image and the results of image F which was eroded by structure element B, namely:

$$Ee(F)=F-(F \ominus B) \tag{8}$$

If G(F) is the morphology grads of image F, it can be calculated by dilation and erosion operation, namely:

$$G(F)= (F \oplus B)-(F \ominus B) \tag{9}$$

The processing of erosion and dilation makes the image minishing or enlarging, while the difference between the original image and the changed one is the image edge. However, erosion and dilation can not filter noise very well. The open and the close operation have good filtering effect apparently, which decrease the noise in the concave and convex of the edge [6], but they can not reflect all the edge feature of image. Therefore, we improved the traditional morphology edge detection operator, thus obtaining an edge detection operator with more strong noise proof to get better image edge. That is:

$$F \bullet B \oplus B - F \circ B \ominus B \tag{10}$$

3 The Selection of Multi-structure Elements

Edge is greatly related to structural elements in morphological edge detection. The selection of structural elements is essential to edge detection operators and has direct effect on the results of edge detection. The size and shape of structural elements determine the objects and performance of morphological algorithm. What's more, different structural elements can analyze different images with different results.

In the traditional application of morphological edge detection, single structural element is often used to detect. A certain structural element is simply sensitive to the edge at the same direction, while the edges at different directions are omitted [7]. Therefore, whichever structural element is chosen, some edges information will be lost.

In such case, in view of varied edges at different directions in the image, this article creates structural element at different directions in four degrees 00、450、900、1350, as Figure 1 shows:

0	0	0
1	1	1
0	0	0

0	0	1
0	1	0
1	0	0

0	1	0
0	1	0
0	1	0

1	0	0
0	1	0
0	0	1

0° structural element 45° structural element 90° structural element 135°structural element

Fig. 1. Structural element in different directions

4 Multi-structure Elements Morphology Edge Detection Algorithm

With the anti-noise mathematical morphology edge detection operator (format 10) in this article and the structural elements in Figure 1, we make edge detection on the plate image at four detections, and do image compound depending with different detection results to gain the final image edge.

We compared a binary image, which has salt and pepper noise, with itself after being processed by the edge detection, as shown in figure 2. The test shows that the traditional Sobel operator, Canny, Prewitt and general morphology processing can

(a) With salt and pepper noise plate image

(b) Edge detection by Sobel operator

(c) Edge detection by Canny operator

(d) Edge detection by Prewitt operator

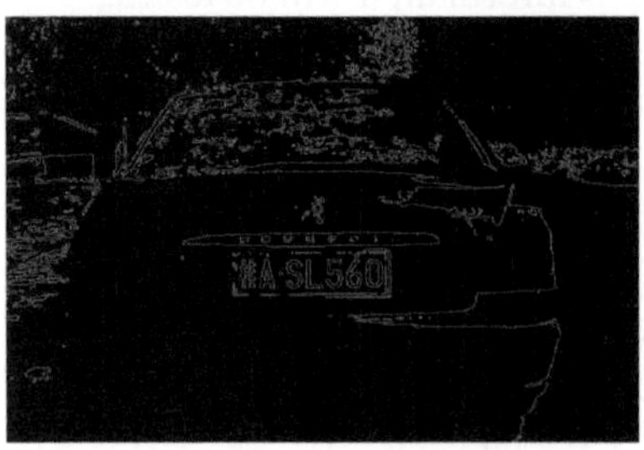

(e) Edge detection by morphological single structure element

(f) Edge detection by this article's operator

Fig. 2. The results of edge detection using different methods detect the plate edge successfully, but can not filter noise very well. The new multi-structure elements mathematical morphology edge detection operator, compared with the traditional edge detection operator based on template and the general morphology edge detection operator, not only detects the plate edge successfully, but also filters noise magnificently.

5 Plate Positioning Algorithm Based on Morphology Multi-structure Element

The procedure of plate positioning is as indicated in Figure 3.

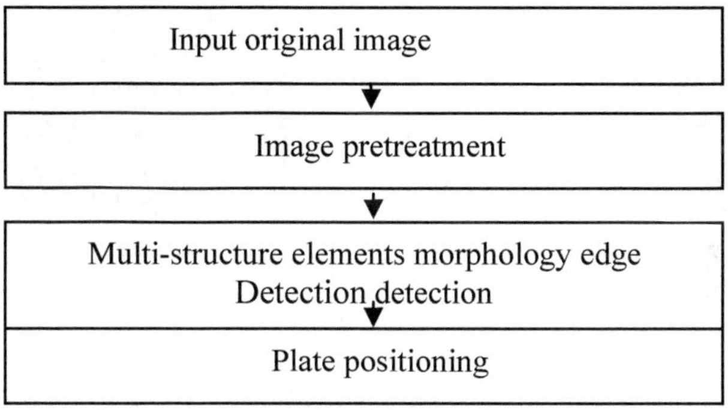

Fig. 3. The procedure of plate positioning

5.1 Image Pretreatment

In practical circumstances, due to various objective causes, e.g. climate when shooting, illumination, machine, blurring of plate, etc, the original plate picture needs taking series of preprocessing before the next operation. The processing includes gray scale transformation, binarization, denoising, filtering, and so on. This article adopts the pretreatment of combining gray scale transformation, gray scale stretching, binarization and median filtering (Figure 4). First, we transformed color image to gray image, because the color information less favorable treatment. and then processed gray scale stretching to make the image more clear, and increase the contrast. Second, we used binary processing to it for only black and white binary image, and maked the plate characteristics prominent. Last, we used median filtering to remove the noise effectively.

(a) Original picture

(b)Gray picture

(c) Gray stretching picture

(d) Binary picture

(e) Median filtering picture

Fig. 4. Image pretreatment

5.2 Multi-structure Element Based on Morphology Edge Detection

In this paper, we chose a new multi-structure element based on morphology operator to detect the edge of Figure 4 (e). The detected edge is not only clear and continuous, but also effective in filtering noise and screening complicated background. Please refer to figure 5 for the result.

5.3 Plate Positioning

On the basis of Figure 5, this article adopts the method of scanning[8,9] to detect plate license area from plate image. As there are seven characters in the license plate area and a certain distance between each other, hopping is defined as from target to background or background to target. It reveals that there are more hopping times beyond license plate area, and more in a certain distance. If there are 7 characters in the plate license, 14 hopping times (about 2 hopping times between every character), the character will be broken, unclear, or the license plate slanting. Thus, we utilize 12 hopping times for safety, choose 30 pixels for character space, and pick the lines satisfied 10 times for use.

Generally speaking, license plate is usually fixed at the lower part of the car, whose similar character interference comes from above. Therefore, we scan it from left to right and down to up. According to this order, we scan every image and write down the location of the hoping point. If there are more than 12 hopping points in some line and the space between one hopping point and the nest point is within 30 pixels, we will write down the location of the initial point and the ending point; if there are 10 conti nual hopping points, we will consider this area as the preselection area of the license plate. With the horizontal scan and vertical scan to identify the boundary between upper and low, and left and right (as in Figure 6), the results show that the identification is more fast and accurate.

Fig. 5. Morphology detection result

Fig. 6. Positioning result

6 Experimental Result and Conclusion

The detailed implementation of above algorithm was simulated in MATLAB, on P4 2G with 2G memory. After the image pretreatment, we took multi-structure element morphology operation to detect the edge of the image, and combined the scanning to position plate region. In order to test the positioning effect of the operation, we applied the above method to 230 sample auto pictures, which contains various backgrounds, illumination conditions and contrasts. The test shows 225 precise positioning, with 97.8% accurate rate. The result states clearly that this method is able to position plate precisely under complicated exterior circumstance and has preferable implementation prospect.

References

1. Serra, J.: Introduction to mathematical morphology. Academic Press, NewYork (1982)
2. Lu, Y.-Q., Wu, L.-C.: Computer Engineering 31, 224 (2005)
3. Wang, H.-F., Zhan, G.-L., Luo, X.-M.: Computer Engineering and Applications 45, 223 (2009)
4. Fang, L.: Edge detection based on mathematics morphology. Kunming University of Sience and Technology, p. 16 (2006)
5. Lian, J., Wang, K.: Computer Engineering and Applications 42, 77 (2006)
6. Li, M., Sun, H., Wu, L.-Y.: Computer Engineering and Applications 46, 160 (2010)
7. Zhang, Q.-L., Kang, B.-S.: Computer Engineering and Applications 45, 166 (2009)
8. Liu, W.-M., Zhao, X.-P.: Computer Engineering and Applications 6, 223 (2004)
9. Lu, C., He, D.-J., He, X.: Computer Engineering and Design 29, 5125 (2008)

Medical Image Segmentation of PACS System Image Post-processing*

Lv Jie, Xiong Chun-rong, and Xie Miao

Department of Professional Technical Institute, Yulin Normal University, Yulin Guangxi 537000, China
lvjie2227@126.com, xiongchunr@163.com, mm.xx@163.com

Abstract. In recent years, the use of computer technology in medical image processing and analysis has been a hot issue of research and application at home and abroad, and the application of the PACS system has created a new era of digital medical treatment. Based on the technology of PACS system for post-processing segmentation, an improved watershed algorithm has been brought up to the medical image segmentation. Compared with the original algorithm, the improved algorithm greatly improves the accuracy of image segmentation; and the problem of over-segmentation of the original algorithm has been solved better. Dynamic merge rule is added in process of implementing watershed algorithm and the formation of over-segmentation regions due to the algorithm itself are restrained directly. Applying the improved algorithm to the PACS system, it has been proved that the algorithm can effectively extract the medical image lesions.

Keywords: PACS system, Medical image segmentation, Watershed algorithm, Dynamic merge.

1 Introduction

Nowadays, the combination of computer technology and medical imaging technology has been the main trend. This combination takes good news to both doctor and patient, because it can raise the accuracy of clinical diagnosis. PACS, Picture Archiving and Communication System, is the typical representative of hospital informatization. PACS is a comprehensive system that can collect, store, manage, diagnose and post process the images generated by kinds of digital medical equipments [1].

As a platform of digital application of medical images, PACS can transfer images that generated by various equipments to one format. Medical Image Segmentation of PACS System Image Post-processing is studied in this paper. Image segmentation is a way by which we can separate the area from the surrounding. Such segmentation can help the doctor to get the specify part from the whole picture so as to eliminate interference from irrelevant parts. This is very essential for doctor to make correct judgment about the disease and decide the correct diagnostic strategy.

* Fund: Supported by Guangxi, Yulin Normal University youth research projects in 2010(No:2010YJQN28).

Y. Zhang (Ed.): Future Communication, Computing, Control and Management, LNEE 141, pp. 495–500.
springerlink.com

Due to characteristic of medical imaging technology, compared with the normal picture, medical image is not so contrasting to background, unclear in edge, with more noise, ect. In recent years, Watershed Algorithm [2] is of great concern due to its effectiveness in operation and precise edge locating. However, serious over-segmentations generally exist. How to overcome such disadvantage has been a major concern in the research. According to the feature of rich detail information in the medical images, many methods are comprehensively used to get the segmentation results before and during the segmentation. First, pre-process the noise in a mathematical morphology method before using Watershed Algorithm. Second, impose restrictions directly on the formation of watershed. The detailed segmentation process is showed in Figure 1.

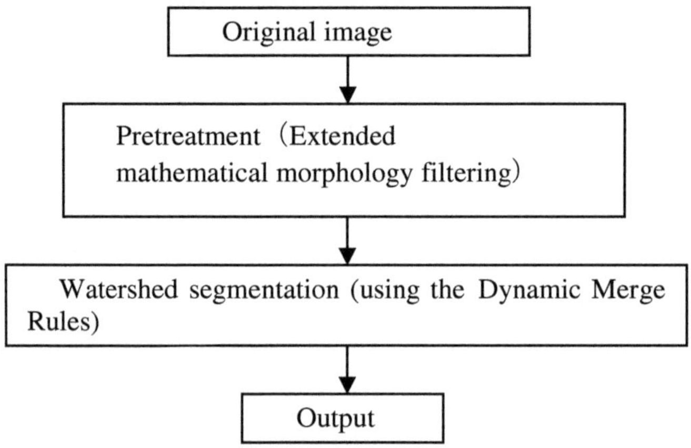

Fig. 1. Segmentation process

2 Mathematical Morphology

Mathematical Morphology is a method applied in image process and pattern recognition. Its basic idea is to measure and extract the shape via some morphological structuring elements, so as to analyze and recognize the image. Mathematical Morphology's basic operation consists of dilation and erosion, and the opening and closing operation based on the extension of the two basic operations above.

Erosion and dilation has most impact in the image edge, so we can get image edge by operation between the treated image and original one. However, erosion and dilation has little filtering effect of noise. Morphological opening operation and Morphological closing operation are obvious more effective at filtering. But since it take use of the complementary between erosion and dilation, the result just reflect image edge concavity and convex [3], but not all the features of the image edge. The image segment method based on the Watershed Algorithm depends largely on the algorithm we use to get the gradient. In view of this, a extended mathematical morphology method to get the image gradient and pre-process the image are suggested in this paper. This algorithm takes use of following operation.

$(M \circ B) \oplus B - M \bullet B$

$(M=(F \circ B) \bullet B$, B is 3×3 square structure element. F is the original image.)

First of all, pre-process the image by morphological opening operation and morphological closing operation so as to filter the noise. Second, use Morphological opening operation to smooth the image .Third, dilate the image. The difference between the original and the one get by dilation is the better image edge.

3 Dynamic Merge Rules

By the step introduced before, we can get the target edge and gradient image reflecting all the edge. Ignore the minute irrelevant object or the details inside. And apply the gradient image processed by noise filter as input image for Watershed Algorithm. However, for the image segmentation based on Watershed Algorithm, minute changes in some factor, such as background, noise, object edge and inside, will result in the division of the same object[4]. In other words, they will be affected by noise and error. Any minimum point will be treated as a valley, and any valley will become a basin while segmented by Watershed Algorithm. Then problems come, the uniform region in image will be divided into many subregions. This makes the result disorganized, and brings many fake edges. This means failure in segmentation. This is what we call over-segmentation.

In this paper, we bring Dynamic Merge rules into Watershed Algorithm, so as to restrain the image divided into subregions unnecessarily, thus eliminating over-segmentation and achieve better effects.

(1) Rearrangement of the gradient graph

As introduced before, we can get the original image gradient graph by extended mathematical morphology operators. The gradient image pixels ordered by spatial relations, but Watershed Algorithm can not process the image in such order. Therefore, we must rearrange the gradient image to meet the algorithm requirement. Watershed Algorithm processes the image from minimum pixel gradient to maximum. Therefore, all the pixels in the gradient map are arranged according to their pixel values in ascending order. The implementation of the algorithm requires repeated access to the neighborhood of each pixel. In order to improve computing speed, we manage to store 8 neighborhood pixel information instead of 4 for each pixel during rearrangement.

(2) Segmentation and merge

After gradient graph rearrangement, algorithm dealing with the image from the pixel with minimum gradient to the one with maximum gradient, this is the entire segmentation and merger process.

First, we must consider which storage solutions we should take for all the pixels to be split. The segmentation is not to process all the pixels in one time, so we don't need to scan the whole image each time. Because of this, we apply First In First Out (FIFO), one kind of circular queue in data structure, in the program to complete the process. This storage structure is not only helpful in storing and finding the pixel quickly, but also helpful in speed popping up the pixel. So the whole segmentation process speeds up.

Based on the algorithm process order, if the pixel with gradient value h disposed, it means that the pixels of gradient value less than or equal to h are all disposed. All the disposed pixels can be treated as a set denoted as M. Then we consider about the pixels with gradient h+1, which are not disposed. We denote it as set N. Then set M and set N may have the following three relationships.

a、 N and M have no value shared by each other. This means all the pixels around set N has have gradient values higher than or equal to h+1. So N is the new valley.

b、 N and M have value shared by each other, and they connected. N belongs to the valley of M.

c、 N and M have value shared by each other, however, they not connected. This means N belong to the other basin [5]. By calculating the distance between the bottom and N, we can find where N should locate.

The following introduce the concrete steps. It is divided into four steps:

① Put the pixels into the queue. First, put the pixels into circular queue under First In First Out (FIFO) rule, then output it at the appropriate time. In other words, if segmentation process request to process pixel with gradient value h, the circular queue response for put the right pixels into the queue. When the process finish, the circular queue should release the disposed one.

② Pixels out of queue. When the input finished, circular queue is responsible for releasing the disposed pixel. We denote the pixel to be released as P. We calculate the basin and the watershed in neighborhood of pixel P, and denote them as NumB and NumW. The affect NumB and NumW have on output of the watershed will be discussed following.

③Judge the output for P based on the NumB and NumW.

If NumB = 1, it means that only one basin adjacent to P. So we should assign the output to P.

If NumB = 0 and NumW> 0, it means that the watershed is P.

If NumB > 1, it means that there are many basins in P's adjacent area. According to the original algorithm, P is the watershed. In this case, there will be over-segmentation. Now we bring the Dynamic Combination Rule to Watershed Algorithm to solve this problem. First, find out the merging mother town for point P, and denote it as W. The principle is to find the point in the adjacent basins. The one in the deepest bottom is W.As for the other basins, calculate the dynamic value for the bottom. Appoint a threshold value first. Then compare the bottom dynamic value with this threshold value. If less than the threshold value, all the pixels of this basin should be incorporated into mother town W of P. Repeat this operation. If all the basins merged into W, output of W will be assigned to P. Otherwise, P is watershed.

④ Judge whether there is a new bottom

If there is no pixel with value lower than h+1, pixel with value h+1 should correspond the bottom of the new basin. At this moment, the algorithm should scan the whole image to judge whether there is a new bottom.

4 Image Segmentation and Analysis

Select tuberculosis CT image from PACS for segmentation, results are shown in Figure 2.

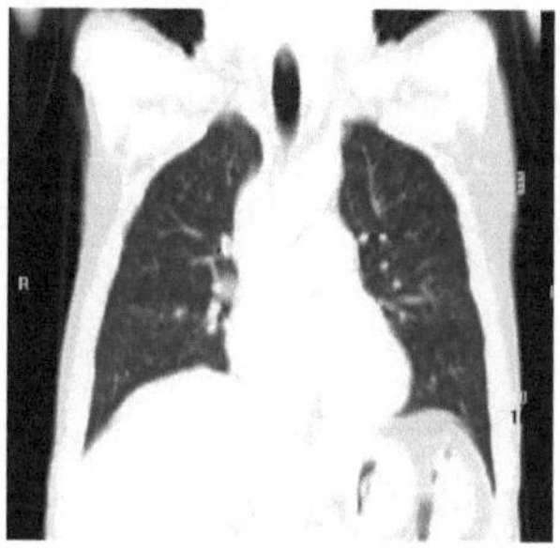

a. Original tuberculosis

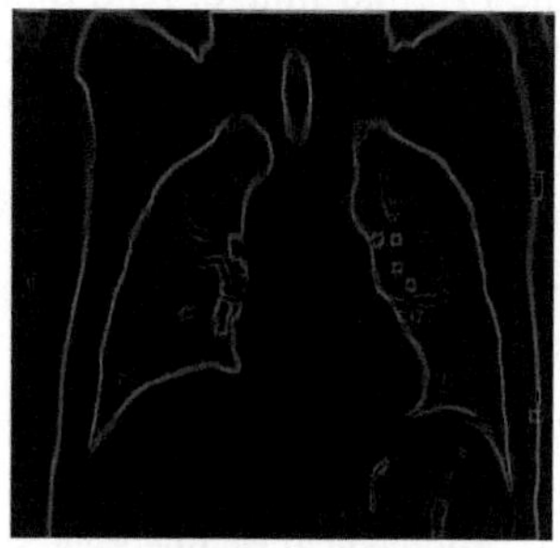

b. Extended operator graph

c. Segmentation results before Improved

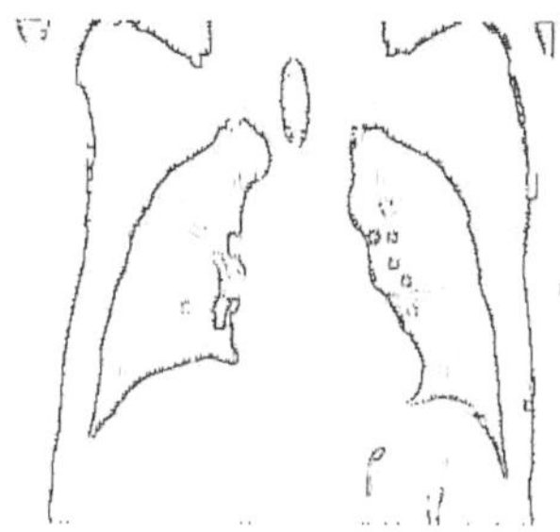

d. Improved segmentation results

Fig. 2. Segmentation results comparison of Tuberculosis CT images before and after using the improved watershed algorithm

Figure 2a is a typical CT image of tuberculosis, the small white dot in the left and right lungs is the tuberculosis. After enlarging and stretching the image, doctor can not make the accurate diagnosis. In order to analyze the image better, we must have the segmentation in the image. After detecting the edge by extended mathematical morphology operators, we get result as Figure 2b. If we do segmentation with traditional Watershed Algorithm, we get result as Figure 2c. You can see that the segmented image makes no sense, because it's over-segmentation. Have segmentation on Figure 2b by improved Watershed Algorithm, we get result as Figure 2d. We can see that segmentation result is very clear, lesion area have been separated from the surrounding successfully. This allows doctor to focus on analyzing lesion area, so doctors can make further diagnosis to the patient.

All operations are running in PACS system in computer with 2.5GHz Intel Pentium IV processor and 1.99GB RAM.

5 Summary

This paper introduces an improved Watershed Algorithm to solve the problem of over-segmentation. First, an extended mathematical morphology operator is proposed to process the original image. And then, Dynamic Merge Rule Is added to Watershed Algorithm to improve it. Applying the improved Watershed Algorithm to the PACS system, we can have segmentation in the image; moreover, extract the lesion area from whole image. The result shows that the improved algorithm has good noise immunity and works well at medical image segmentation. Therefore, it can help to improve the diagnosis and treatment.

References

1. Fang, Q.-L.: Biomedical Engineering Shanghai 1, 58 (2003)
2. Vincent, L., Soille, P.: IEEE Trans. on Pattern Analysis and Machine Intelligence 13, 583 (1991)
3. Zhao, Y.-Q., Gui, W.-H., Chen, Z.-C., Li, L.-Y.: Computer Engineering and Applications 33, 6 (2005)
4. Lui, H.-T., Zhang, Y.-J.: Data Acquisition and Processing 12, 18 (1997)
5. Bieniek, A., Moga, A.: Pattern Recognition 21, 907 (2000)

A New Multi-ant Colony Optimization Algorithm

He Yueshun[1] and Du Ping[2]

[1] No.418, Guanglan Avenue Nanchang Jiangxi, China
hys8418@163.com, pdu@edu.cn

Abstract. This paper introduces the basic ant colony algorithm, the model and its problems in the process of solving the TSP. Because the basic ant colony algorithm to search for to a certain extent, all individuals found the same solutions n in exactly, it can not search the solution space in further, it is not conducive to find better solutions. For the shortcomings of the algorithm, we present a new ant colony algorithm, the new multiple-ant colony algorithm for solving problems; it searches the solution space using a different population, in order to avoid local optimum to obtain the global optimum. Simulation results show that the algorithm in solving TSP problems than other algorithms more efficient, have good practical value.

Keywords: Ant colony algorithm, multiple ant colony optimization algorithm, pheromone, the optimal solution.

1 Introduction

Ant colony algorithm by the Italian scholar M. Dorigo, who in the 20th century and early 90's ants in nature by simulating the behavior of routing group put forward a heuristic based on biomimetic evolutionary systems population. It consists of two basic stages: stage adaptation stage and Collaboration. In the adaptation stage, the candidate solutions based on the information accumulated continuously adjust its structure. In the collaborative stage, through the exchange of information between candidate solutions to generate performance expectations of a better solution, similar to the learning mechanism of learning automata. Ant colony algorithm was first successfully applied to solve the famous traveling salesman problem (TSP) ,The algorithm uses a distributed parallel computer system of positive feedback, easy to combine with other methods, but also has strong robustness.

But ant colony algorithm has slow convergence and stagnation of the lack of prone, that the search carried out to a certain extent, all individuals found in exactly the same solution, not further search the solution space is not conducive to find better solutions. Therefore, for lack of the algorithm, we present a new ant colony algorithm, called Ant Colony Optimization with Multiple Ant Colony (ACOMAC).The concept of multiple ant Colony is affected by parallel genetic algorithm in solving problems using different populations of the search space to avoid local optimal solution to obtain the global optimal set inspired circumstances.

Y. Zhang (Ed.): Future Communication, Computing, Control and Management, LNEE 141, pp. 501–507.
springerlink.com

2 ACO Algorithm

Ant colony algorithm is proposed in recent years, it is mainly used to solve difficult combinatorial optimization problems, and it is a new element Kai French method. It is the decision to imitate the ant foraging ability of the shortest path is designed. Despite the limited capacity of a single ant, but the groups formed by them, you can perform many complex behaviors. Ant colony optimization is a natural way from a moving ant in its path a few left over after a number of pheromones to leave this material in the path of the track. When single random moving ants, ants can detect pheromone on the path of pheromone and decided to move in the direction of high intensity, which further strengthened the intensity of pheromone on the path.

After adding an element, the pheromone on that edge of the volume will be updated to allow the ant pheromone is an edge in the use of scalar exchange with each other. At the side i the information gathered strength elements may be used each time $\tau(i)$ in the beginning. Equation (1) is used to select the next one needs to be added to the solution of the side s, By the Ant K, in step R also need to be added to the solution of a side by the $J_k(r)$ shown. $\eta(s)$ Shown the side s solution will be chosen to be the expectation level. In the TSP problem, $\eta(s)$ is inversely proportional to the distance between two points.

The following will describe the state transition rule, global and local updating rules update rule, they were by the Eq. 1, Eq. 2, and Eq. 3 said. Eq. 1 is a pheromone level and cost that the best height of greedy selection method combined. In order to ensure that other side is incorporated into the solution of each ant, the Eq. 2 the probability of Methods side. This is to avoid premature or converge to a small part of the elite solution is necessary, each equation by the parameter Q to the control.(Which $0 \le q_0 \le 1$).

$$s = \begin{cases} \{\arg\max_{s \in jk(r)} \{\tau(r,u) * \eta(r,u)^{\beta}\} & (q \le q_0) \\ 0 & \end{cases} \tag{1}$$

$$P_k(r,s) = \begin{cases} \dfrac{\tau(r,s)[\eta(r,s)]^{\beta}}{\sum_{w \in J_k(r)} \tau(r,z)[\eta(r,z)]^{\beta}} & (s \in J_k(r)) \\ 0 & \end{cases} \tag{2}$$

For the minimization (maximization) problem, the parameters take negative (positive), which can make each of the minimum (maximum) cost is right. The pheromone of the selected side was partial updated by Eq. 3.

$$\tau(r,s) \leftarrow (1-\alpha) * \tau(r,s) + \alpha * \Delta\tau(r,s) \tag{3}$$

Which :

$$\Delta\tau(r,s)=\begin{cases}(L_{gh})^{-1}, & \text{If } (r,s) \text{ belongs to the global optimum trip}\\ 0 & \end{cases}$$

α express the relative importance of pheromone, $0 \le \alpha \le 1$; $\Delta\tau(r,s)=\tau_0$ express of each edge of the first left on the number of pheromone. L_{gb} express the global optimal journey from the start trajectory. When building a journey TSP, ants visit each side and through, such as Eq. 4 show the local pheromone update rules to change the level.

$$\tau(r,s) \leftarrow (1-\rho)*\tau(r,s)+\rho*\Delta\tau(r,s) \tag{4}$$

Where ρ that the evaporation coefficient of pheromone, $0<\rho<1$, $\Delta\tau(r,s)$ pheromone that each increment of the edge, these edges form the best solution found in Eq. 5.

$$\Delta\tau=\begin{cases}\frac{Q}{L_k} & \text{If } (i,j)\in tabu_k\end{cases} \tag{5}$$

Where Q represents a constant, usually set at 100. L_k represents length of the ants k in this tour in the paths, For each side, 0 times the pheromone intensity $\tau_{ij}(0)$ is usually set to a very small value.

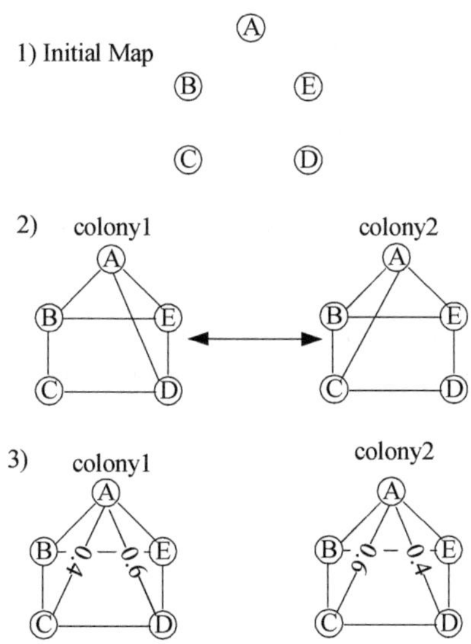

Fig. 1. Multi-ant colony's social rules

3 ACOMAC Algorithm

ACO algorithm is a relatively new use of the communication in ant behavior in the principle of solving the meta-heuristic search optimization methods. Ants can pass through the path of the pheromone left on the method to exchange information on their food source found in the path of information that can guide other ant's prime tracks to find food sources. So far, many studies have concentrated on solving several benchmark problems, such as the TSP problem, JSP issues, and QAP problem. Here, we present ACOMAC algorithm to solve the problem, we use this algorithm not only requires the solution of problems, and requires a solution close to the optimal solution .ACOMAC algorithm through different strategies conducive to the optimal solution obtained by the algorithm of ant colony using different strategies conducive to build mechanism, is to adopt the basic concept of parallel genetic algorithm to search the solution space, this strategy specifically, different groups to avoid solving the problem of local optimum to obtain the global optimum.

Assumes that all the ants are in the same map or area, they are formed according to the different preferences of different groups and communicate with each other, but sometimes a group will obtain information from other groups to get knowledge, so that it can improve the survival competitiveness.Fig.1 shows the social rules of multi-ant colony. Fig. 1(1) describes the initial map, Fig. 1(2) represent two cycles after the colony through the children, their pheromone values in the table each side has different initial pheromone value (each side of the initial value of pheromone equal to 1.0),The equivalent of two different ant pheromone left on the ground, and then left in the path trajectory marking pheromone, that is, the two groups of ants in the group where only know their own knowledge about the shortest path, do not know where the other group of knowledge about the shortest path (local search).Fig.1 (3) describes the circumstances for the information communication group 1, in addition to the length of side equal to 0.4, the other side length is equal to 1.0; For Group 2 is concerned, in addition to the length of side AD is equal to 0.4, the other side length is equal to 1.0.After a few steps, the group and the group will communicate with each other and get all the knowledge about the shortest path or a global search. For more detailed concept of the social rules of colony shown in Fig. 2,Figure 2 describes the concept of local search and global search for the concept, First, all groups will be to search for solutions in each cycle; Second, they will be fixed in a few cycles to build information and communication, At the same time, each group should put it to use the pheromone path mapping the way to their records in, and then using Eq. 6 for information and communication; it is worth noting that the formula represents a group and another about an exchange of information elements of a group situation.

$$\begin{cases} Clan_i = Clan_k * (1-w) + Clan_i * w & (\text{if } i = 1) \\ Clan_i = Clan_{i-1} * (1-w) + Clan_i * w & (\text{otherwise}) \end{cases} \tag{6}$$

Where, i represents the i group, k represents group number, w represents a group of pheromone table weight (local weight), $(1-w)$ on behalf of another group of pheromone table weight (external weight). $0 \le w \le 1$.

In Fig.1(1) is the initial map, Fig.1(2) Suppose there are two ant, After several cycles, which pheromone values in the table is different, each edge of the initial pheromone value is equal to 1.0,Fig. 1(3) describes the situation about information and communication: For group 1 is, AC is equal to the length of side 0.4, AD edge lengths equal to 0.6, the other side of the length 1.0; in it for the group 2, AD is equal to the length of side 0.4, AC edge lengths equal to 0.6, the other side The length is equal to 1.0.

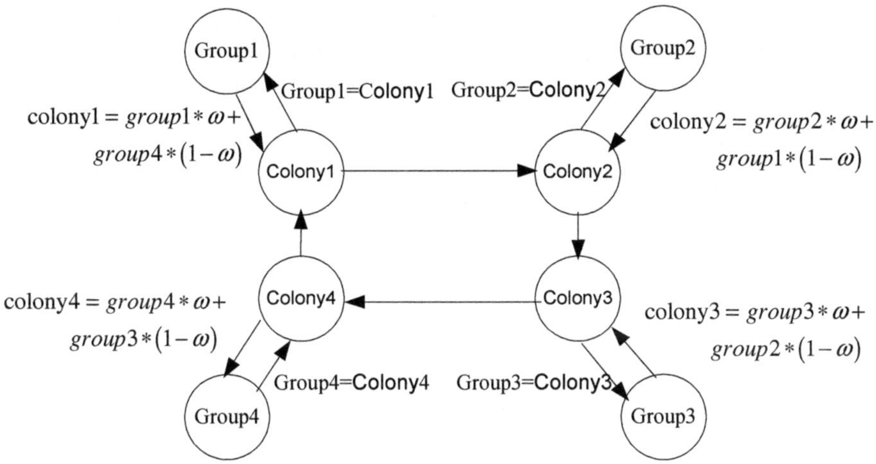

Fig. 2. Multiple ant colonies Society of the rules

ACOMAC algorithm is described as follows:
// For solving TSP problems ACOMAC Algorithm
Set parameters, initialize pheromone track
While (termination condition not met) do
While (termination condition not met) do
Construction Solutions
Application local search
Update local pheromone
After T time adjustment, the local information elements $P_a*(1-w)+(P_i*w)$
Loop
Global pheromone update
Loop
End

4 The Simulation Results

To test the algorithm performance, simulation of the algorithm in the experiment, the algorithm's parameters are: $q_0=0.9,\ \beta=2,\ \rho=\alpha=0.1,\ m=10,\ \tau_0=(n*L_{nn})^{-1}$, L_{nn} represents the nearest neighbor produced by the journey length, n represents the number of sites, the computer simulation run 30 times, local weight w = 0.6, external weight $(1-w)=0.4$, Table 1 shows the result between ACOMAC algorithms and M. Dorigo's ACS algorithm in solving problems of comparing different TSP. Fig. 3 describes the comparison of the evolution curves between ACOMAC algorithm and M. Dorigo's ACS algorithm in solving problems Ei176 (76 nodes, running the number 30.). Fig.4 describes the comparison of the evolution curves between ACOMAC algorithm and M. Dorigo's ACS algorithm in solving problems Kroal00 (100 nodes, running the number 30.);The results show that, ACOMAC algorithm for solving TSP problems in the performance has been significantly improved than the ACS algorithm, we can obtain global optimal or near global optimal solution.

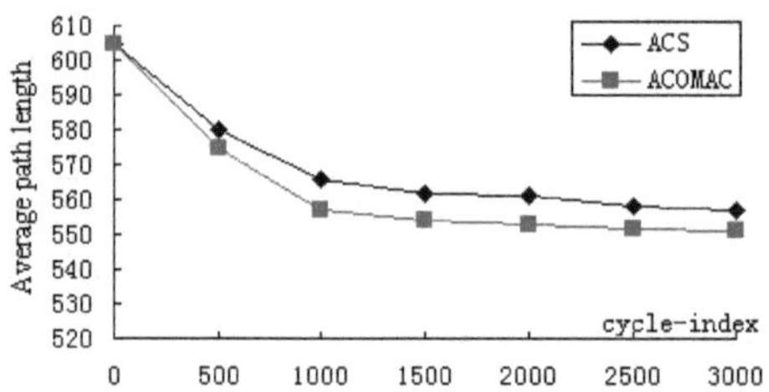

Fig. 3. The Comparison of the evolution curves in solving problems Ei176

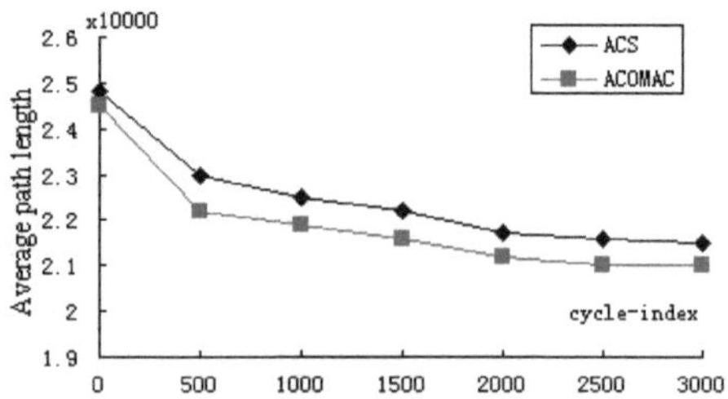

Fig. 4. The Comparison of the evolution curves in solving problems Kroa100

5 Conclusions

Many studies have shown that ant colony algorithm has a strong ability to find better solutions, because the algorithm exploits the positive feedback principle, to a certain extent, can speed up the evolutionary process, but the essence is a parallel algorithm, different ongoing exchange of information between individuals and the transmission, which can work together, is conducive to find better solutions.

This paper presents a new problem to solve TSP algorithm-ACOMAC algorithm, the concept of multiple ant colony is affected by parallel genetic algorithm for solving TSP problems in different populations of the search solution space in order to avoid local optimum conditions to obtain the global optimum inspired and created. Simulation results show that, ACOMAC algorithm for solving TSP problems more effective than the ACS algorithm.

Acknowledgment. This work was supported by the grants from Jiangxi Science and Technology Agency provide financial aid for "an integrated logistics system based on Industrial PDA " and Key Laboratory of Nuclear Resources and Environment Ministry of Education provide financial aid for "The Research on Dynamic Bayesian Network of Intelligent Information Fusion Mechanism for Multi-Source Spatial Based on Ant Colony Optimization".

References

1. Dorigo, M., Di Caro, G., Gambardella, L.M.: Ant algorithms for discrete optimization. Artiticial Life 5(2), 137–172 (1999)
2. Baterina, A.V.: Image edge detection using ant colony optimization. WSEAS Transactions on Signal Processing 6(2), 58–67 (2010)
3. Matthews, D.C.: Improved Lower Limits for Pheromone Trails in Ant Colony Optimization. In: Rudolph, G., Jansen, T., Lucas, S., Poloni, C., Beume, N. (eds.) PPSN 2008. LNCS, vol. 5199, pp. 508–517. Springer, Heidelberg (2008)
4. Jin, Y., Zhao, Y.: Optimization of ship principal parameters with improved ant colony algorithm. Journal of Huazhong University of Science and Technology 36(11), 99–102 (2008)
5. Nonsiri, S.: Modifying ant colony optimization. In: Proceedings of the 2008 IEEE Conference on Soft Computing on Industrial Applications, pp. 95–100 (2008)
6. Yoshikawa, M.: A new pheromone control algorithm of ant colony optimization. In: International Conference on Smart Manufacturing Application, pp. 335–338 (2008)
7. Gao, W.: New continuous ant colony algorithm. In: Proceedings of the World Congress on Intelligent Control and Automation, pp. 1275–1279 (2008)

Design and Realization of Zhejiang Water Environment Information Disclosure System

Yanli Ma[1] and Bin Xie[2]

[1] Institute of Space Information Technique, Zhejiang University, Hangzhou, 310027, China
sandcrazy@163.com
[2] Institute of Remote Sensing and Earth Sciences, Hangzhou Normal University, Hangzhou 310036, China
xiebinbingdiao@gmail.com

Abstract. In this paper, the current management situation of water environment and the demand of the public for the water environment information disclosure are analyzed which indicates the application of WebGIS in the web publishing and disclosing the water environment information is necessary. Take Zhejiang Province as the example, the functional demands, system design, database development, technology development and other of the water environment information disclosure system are discussed. Based on the ArcServer platform, water environment information disclosure WebGIS is realized, which enables managers to easily evaluate, plan and analyze the water environment, while safeguards the public's right to timely know the water environment information, and achieves the effective management and information disclosure for the water environment information.

Keywords: Water environment, information disclosure, WebGIS, ArcServer.

1 Introduction

With the rapid development of the national economy, population growth, accelerated urbanization, water shortages and water pollution has become a major constraining factor in social and economic development, water environment problems have caused more and more attention from the public.

The current water environment information management system mainly realizes the management of the water environmental non-spatial data. Managers cannot intuitively understand the geographical distribution of water environmental factors, which brings out some difficulties to make decisions, therefore, the detection, analysis and evaluation for the water environment temporal and spatial distribution to find out the major factor influencing the water quality and improve water quality of the environment, has become a dispensable part of the modern water environmental protection and management.

In addition, in order to protect the public's right to know information on the water environment, a platform is needed to open the water environment information. Because of the intimate relationship between the water environment information and

Y. Zhang (Ed.): Future Communication, Computing, Control and Management, LNEE 141, pp. 509–516.
springerlink.com

the spatial space[1], GIS has become a powerful instrument to understand, simulate and manage complex water issues[2]. Through WebGIS, the water environment information can be displayed in an intuitive way and also using the Internet to publish can achieve broader regional sharing[3].Relying on the water environment management department, based on the Zhejiang Province's basic geographic data and the water environment data related to the water quality parameters, water polluted companies, water sources, sewage treatment plants, etc., and combining with ASP.NET and ArcServer, this article designs and develops the water environmental information open system using the article designs the three-tier. The purpose is to provide the public with a window to know about the water environment, raise public awareness of water conservation and participation for the water pollution prevention and also to provide the management department an instrument to understand the public's needs, improve management efficiency and scientific decision-making.

2 System Design

2.1 System's Aims

System's goal is to build an information system orienting towards environmental management, visual, dynamic based on the geographic information systems, monitoring data management system and database platform[4]. Main functions of the system requirements are as follows:

(1)Data management requirements: building the basic geographic database (including the data of the province's administrative divisions, river systems, water polluting companies, drinking water sources, sewage treatment plants and so on), realizing the data's hierarchical display based on the administrative Map of Zhejiang Province, and implementing the query of the water quality data and more elements of geographic information.

(2)Information dissemination and sharing requirements: Implementing the online dissemination and sharing of the water environment-based information and its evaluation, meeting the public's needs of scanning, inquiring, and tickling water environment information.

(3)Special Project information's query and mapping: realizing the functions of attribute data query, the time-interval data graphs query, data classified information query, test point query, draw a trend chart or histogram graph, and associate it with the electronic map.

2.2 System Architecture Design

According to the features of the water environment data such as open Geographic Information System data input, remote access and map information dissemination, etc. and as well on the purpose of separateness of the Data logic, business logic and presentation logic to alleviate the pressure of the Client and data server and balance the load better, we adopt the ASP.NET Three-tier architecture(Fig.1).

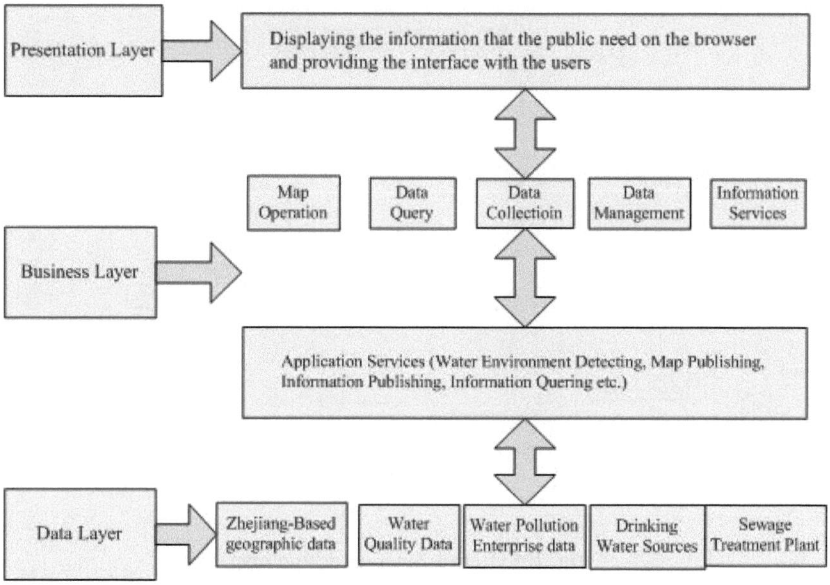

Fig. 1. System's structure design map

1) Presentation layer: displaying the information that the public need on the browser and providing the interface with the users.

2) Business layer: operating the data according to the needs(realizing the operation of adding, deleting, query and changing) and realizing different business functions. To achieve access to visit and transfer the system's information resources and support the service of the analysis on the chart, report forms, distribution and application.

3) Data layer: the base of the whole system is the organization and management layer of the spatial database. Using the database technique realize the management of the spatial data including: Zhejiang basic geographic data, water environment quality data, water polluting plant data, applied water source data and sewage treatment plant data.

2.3 System Function Design

Based on the collection and arrangement of the water environment information, integrate, mine and combine the various information sources to make the date uniformed and integrated. The system provides service about the collecting, reporting, publishing, searching and retrieval, download of the data and exchange and interdynamic. Realize the Processing, conversion and storage of the water environment information dispersed in various departments and integrated the whole information of the province. Its database realizes the communication and sharing of the source from the province, city and county's monitoring and managing departments and display on the portal to increase the transparency of the environmental protection and guarantee the rights for the public to know and participate. The system's overall function is shown in Fig.2.

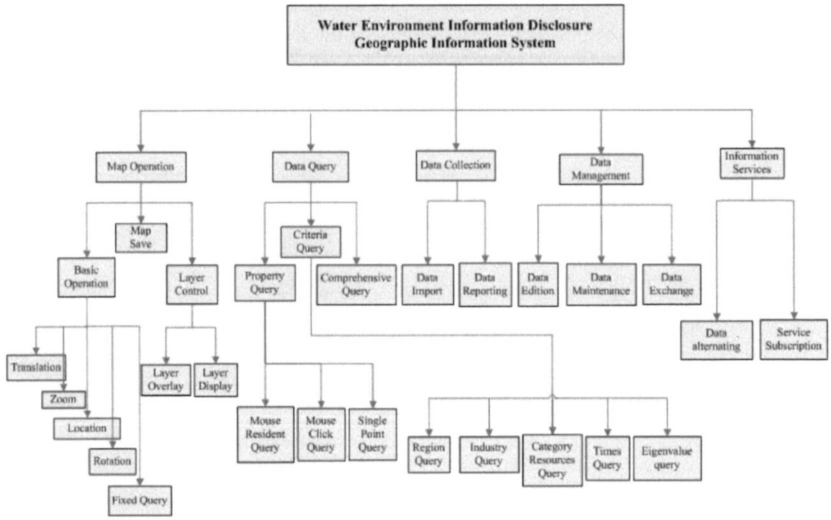

Fig. 2. System's function design map

1) Map operation: Map operation is one of the most basic functions in the geographic information system which can realize the flexible display of the map information[5]. Map information includes the province's river network model, administrative region and other background information. Among them, the map layers include Zhejiang map, water polluting companies, drinking water sources, sewage treatment plants etc.. Each layer is displayed with different legend.

2) Data Query: System's search function is designed to satisfy users' need of the various water environment information surroundings the source and content of the information to provide different kinds of system query platform. From the view of the query types, it mainly includes graphic attribute Query, condition query and integrated query. Graphic attribute query can be divided into mouse staying query, mouse clicking query and single-point query; Condition query can be divided into region query, business query, time query, eigenvalue query and so on; integrated query provides the users with the function of fuzzy query. Users input different key words to do fast search which mainly provide the contrast query about the different index with different time intervals and different regional change based on the Water quantity, water quality and pollution source. The query results can be output in the means of word, chart, conventional diagram, process line, relation line, grid graphic and so on.

3) Data collection: The system's data collection function mainly realizes the collection and management of the water environment data of the water environment quality, water polluting plant, drinking water source , sewage treatment plant and so on which mainly includes functions of automatic input, artificial data report input, reporting query, overdue warn and so on.

4) Data management:

i) Date edition: involving the functions of site's monitoring and maintaining and data editing. Users can artificially adjust the data through this function.

ii) Data maintenance: providing support of the data updating, maintaining in order to automatic manufacture , mine, process, copy, save and so on.

iii) Data exchange: developing the system interface and realizing the interaction with other related systems according to the standards with the purpose of different directions' data sharing.

5) Information service: It mainly provides information query, analysis, statistics and so on in order to provide the water environment management policy and society public service. Also, it provides the service function of various information based on the information sharing database including the information integrated query, changing process contrast, distribution map check, water quality evaluation and analysis, integrating report and so on.

3 System Development and Realization

3.1 Database Construction

Database construction is core of the system development. During the process, it should guarantee the whole structure's flexibility, expandability and maintainability and as well the data's reliability, validity, integrity and safety. According to different stage's need analysis, combining with data's object and its content, the database mainly divided into two kinds: Firstly, , attribute data that is stored in the database of server end of Oracle; Secondly, spatial data that is stored in the server end of Oracle through ArcSDE[6].

Attribute data: system's attribute data charts include water environment quality information chart, water polluting plant information chart, drinking water source information chart, sewage treatment plant and so on.

Spatial data: system's spatial data is special data including administrative Map of Zhejiang Province and the data distribution maps of the water polluting plant, drinking water source, sewage treatment plant and so on. All of them are date in shape format They are stored in the Oracle through ArcSDE and can be visited through ArcServer.

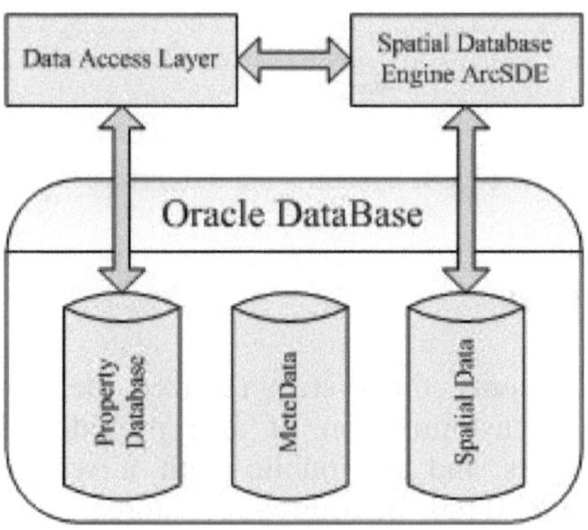

Fig. 3. Database design

In order to meet the need of the system construction, we design the metadatabase. According to the metadatabase standard, its content includes eight major dataset that cannot be used repeatedly and five minor dataset that can be used repeatedly. They are used to comprehensively describe the data sets, data set series, elements and attributes. Major subsets: flag information, data quality information, data log information, data organization information, information referring to system, elements classification information, release information, metadata reference information; minor subset: reference information, time frame, contact information, address.

3.2 System Realization

According to the system design requirements, system realization used the solution based on the ASP.NET + ArcServer + ArcSDE + huge database. Among them, ASP.NET serves as the development tool, ArcServer serves as the WebGIS' server, ArcSDE serves as the spatial data engine and huge relation database serves as the database server of the Oracle.

The system belongs to the classic B/S structure including browser and server end. The user interface is achieved by the browser absolutely, Web server is connected with the server through ArcServer and management tools and spatial service support the server to run in the background. The main interface is like in the following figure.

Fig. 4. System's main interface

4 System's Features

According to the requirements, the system realized the management of the water environment information customization. It also provides the related government departments, research units and the public with a window to know the water environment and its change. It has the following features:

(1) Realizing the digitalization, network for the water environment information and increasing the integrated use ability for the water environment quality and its related data.

(2) Introducing the water source monitoring, pollution source information, sewage treatment information and so on into the water environment system.

(3) Realizing the information disclosure which helps the public know about the surrounding's water environment equality situation in time, monitor pollutant treatment and participate the water environment management positively.

(4) Improving the Environmental Education, increasing the public's awareness to protect the water environment and prevent the water pollution and establishing the environmental and green lifestyle.

(5) Using the currently powerful ArcServer platform, combining the Asp.net Three-tier development model and realizing the management, decision, evaluation, analysis of the water environment information.

5 Conclusion

WebGIS can further develop the GIS functionality in the Internet, which is an effective means to achieve geospatial information web publishing and sharing[7]. The Zhejiang water environment information WebGIS system is based on the ArcServer's powerful spatial date analysis function and the Zhejiang basic geographic data, water environment quality data, water polluting plant data, applied water source data and sewage treatment plant data and combines the water environment management and evaluation system. It realizes the scientific organization and effective management of the water environment spatial information and the sharing of the water environment quality information which help the management workers can effectively develop spatial and visual management. What's more, it guarantees the rights of the public to know the water environment information, to monitor and to participate and also lead and promote the public to participate widely which makes the management and decision more informationized, scientific, digital and efficient. So it plays an instructive role in the water environment protection, management and use.

References

1. Xiong, Y.-J., Tang, B., Liu, H.-O.: Study on GIS of water Resources Based on ASP.NET and ArcServer. Sichuan Environment 8, 131–135 (2009)
2. Fletcher, J.J., Sun, Q., et al.: GIS application for stream water management in West Virginia (4), 113–123 (2001)
3. Wan, L., Chen, Y.-Q., Tan, J.: Information system for on-line publishing and sharing of urban ecological environment quality information. Computer Engineering and Applications 45(6), 25–31 (2009)
4. Lin, H., Gao, J., Xu, J.-Y.: Application of GIS to Environmental Management of Surface Water In Shenyang. Environmental Protection Science 33(1), 26–27 (2007)
5. Jiang, Y., Lei, M., Ma, T.: Study on Flood Information Service System Based on WebGIS. Technology of Soil and Water Conservation (4), 16–19 (2010)

6. Wang, L., Wang, M.-Z., Guo, K.: Design and Realization of Mineral Resources Management System Based on ArcEngine. Computer Technology and Development 20(5), 215–222 (2010)
7. Wang, N.-H., Li, D., Pan, H.: Information service platform of forest pest forecast based on WebGIS. Journal of Forestry Research 20(3), 275–278 (2009)

Matrix Inversion Method for Load Identification in Transfer Paths Analysis

Zhang Fang, Zhou Hong, and Wang Erbing

School of Automobile, Tongji University, Shanghai, China
autozfang@gmail.com, zhouhong@mail.tongji.edu.cn,
wangerbing@gmail.com

Abstract. Transfer Path Analysis (TPA) is fairly well established to study the NVH behavior of a system. It aims to estimate and rank individual noise or vibration contributions to a given target location. Usually, the method involves indirect determination of operational loads. To identify the loads, inverse methods are widely used in which a matrix of measured acceleration data is inverted at each frequency. Due to poor condition of this matrix, however, the results may prone to errors. In this paper, a TPA model is established for an automobile and experiments have been performed to investigate further the matrix inversion problem. Strategies to improve the load identification, over-determination and singular value rejection, are studied. In addition, criteria for omitting singular values and the consequences of using different criterion are discussed to find an appropriate criterion for singular values rejection.

Keywords: Matrix inversion, transfer path analysis, singular value.

1 Introduction

Transfer Path Analysis is fairly well established to study the NVH behavior of a system. It provide the possible ways of energy transfer from sources of excitation to a given target location. For instance, vibration energy of the engine passes through the structures and transmits into panels which radiate into the cabin of an automobile. The operational loads at the interfaces are required to be identified to quantify the contributions from several connection points. Therefore, the accuracy of these loads significantly affects the suitability of the application of transfer path analysis method in evaluating each path's effects on the target, and ultimately affects the insight into the mechanisms responsible for the noise problem at the target location. In general, operational loads cannot be measured directly [2]. As a result, Transfer Path Analysis usually includes the identification of the operational loads from measurement of other quantities. One of the load identification methods which are widely used is the matrix inversion method [1].

Y. Zhang (Ed.): Future Communication, Computing, Control and Management, LNEE 141, pp. 517–524.
springerlink.com

2 The Matrix Inversion Method

In Transfer Paths Analysis, the matrix inversion method involves operational data of indicators $\{X\}$ measured at a series of locations close to the operational loads, in combination with frequency response functions $[H]$ measured between the indicators and the loads.

$$\begin{Bmatrix} X_1 \\ \vdots \\ X_m \end{Bmatrix} = \begin{bmatrix} H_{X_1}/F_1 & H_{X_1}/F_2 & \cdots & H_{X_1}/F_n \\ \vdots & \vdots & \vdots & \vdots \\ H_{X_m}/F_1 & H_{X_m}/F_2 & \cdots & H_{X_m}/F_n \end{bmatrix} \begin{Bmatrix} F_1 \\ \vdots \\ F_n \end{Bmatrix} . \tag{1}$$

A matrix of measured FRFs requires to be inverted due to the following expression:

$$\begin{Bmatrix} F_1 \\ \vdots \\ F_n \end{Bmatrix} = \begin{bmatrix} H_{X_1}/F_1 & H_{X_1}/F_2 & \cdots & H_{X_1}/F_n \\ \vdots & \vdots & \vdots & \vdots \\ H_{X_m}/F_1 & H_{X_m}/F_2 & \cdots & H_{X_m}/F_n \end{bmatrix}^{-1} \begin{Bmatrix} X_1 \\ \vdots \\ X_m \end{Bmatrix} . \tag{2}$$

Where $\{X_1 \quad \cdots \quad X_m\}'$ is the vector of operational data of indicators. H_{X_i/F_i} is the FRF measured between the indicator(e.g. acceleration) and load.(e.g. force) applied at transfer path j

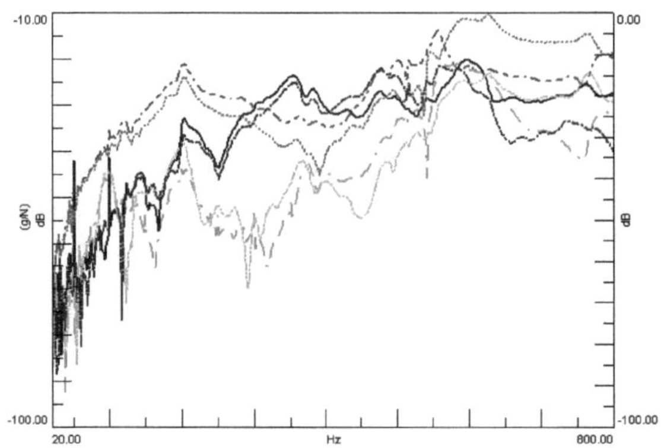

Fig. 1. Similarity of the FRFs belong to the same structure

The matrix, however, are usually ill-conditional because the FRFs in the inversion matrix belong to the same structure. The FRFs belong to the same structure have more or less the same shape since they contain information on the same resonances as shown in Fig.1.

From numerical point of view, the matrix inversion is quite complex because every frequency the rows of the matrix look very similar one to another. This can be explained by a simple example. Inverting a matrix is actually the same as looking for the point of intersection of 'n' n-dimensional surfaces [6]. Take n=2 as an example, estimating two loads using two slave accelerations is the same as looking for the coordinates of the intersection point of 2 lines. In this case, systems have two types: a good conditional system with rather orthogonal rows of the system matrix, and an ill-conditioned system with rather parallel rows of the matrix as shown in Fig.2. The parallel lines express the uncertainty on the operational data (measured slave accelerations) and every point in the area that is marked is a possible solution. The area becomes quite big for a bad conditioned system.

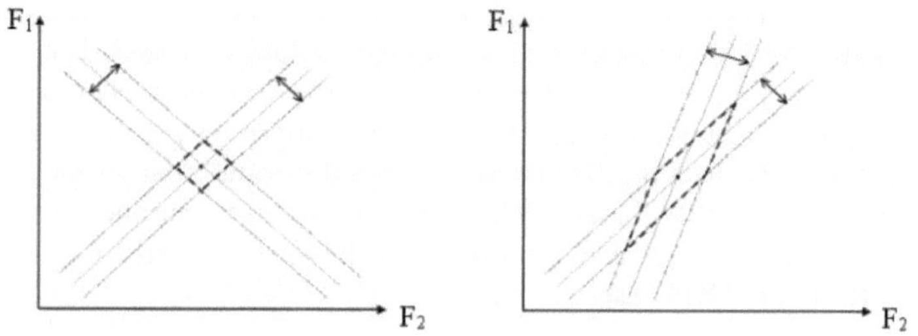

Fig. 2. Point of intersection of surfaces

The identification of indirect loads is, therefore, prone to errors. The loads are overestimated by the matrix inversion, particularly at frequencies where the condition number of the matrix is high. This ultimately results in unreliable estimates of the contributions from relevant sources to the responses via different paths. Hence, solutions for the matrix problem are needed to identify the relevant loads which lie at the heart of reliable results.

3 Over-Determination

One solution for the matrix problem can be achieved by over-determination. All data for Transfer Path Analysis is measured data contaminated with noise. Therefore, over-determination is always necessary which means that the number of indicators are required to be bigger than the number of loads to be identified. As a result of that, the solution of the problem is a 'least squares solution', not an exact solution anymore [3]. In general, condition of matrix improves when more measurements are added to the system. This can be explained by the simple example given before. Estimating two loads using two slave accelerations is the same as looking for the coordinates of the intersection point of 2 lines [6].

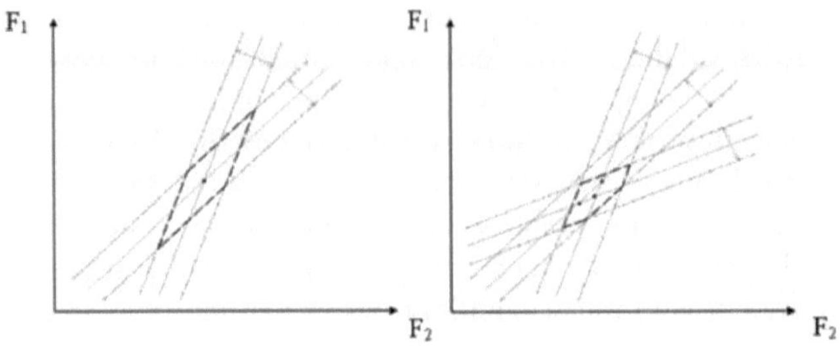

Fig. 3. Least squares solution for the point of intersection of surfaces

Fig.3. shows that when 3 lines, instead of 2 lines, are cutting each other, the intersection point is not exactly defined anymore and hence the point closest to the 3 lines has to be found which means the least squares solution is used. The area in which the solution can vary becomes smaller which can result a more accurate estimate of load identification [5]. In other words, the condition improves greatly. This can also be validated by experiments. Fig. 4(a) shows the condition number of the (7X7) matrix using 7 indicators to identify 7 loads. With the increase of the number of indicators, the peak in the condition number become lower ,as shown in Fig.4(b) for the (9X7) case and Fig. 4(c) for the (14X7) case.

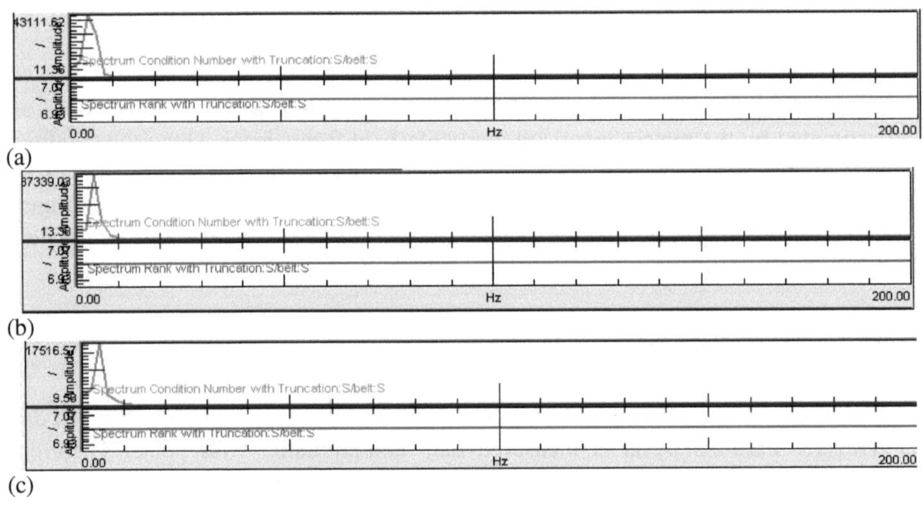

Fig. 4. Condition number of FRF matrix: (a) seven loads and seven indicators; (b) seven loads and nine indicators; and (c) seven loads and fourteen indicators.

In general, the number of indicators, twice the number of loads to be estimated can bring a spectacular improvement of the condition number. The extra effort is bigger than the benefit when adding more indicators.

4 Matrix Inversion Using Singular Value Decomposition

Another solution has considered performing singular value decomposition (SVD) of the FRFs matrix to avoid numerical problems in the matrix inversion [4]. The matrix is rewritten as $[H]=[U][S][V]$ with $[U]$ and $[V]$ unitary matrices and $[S]$ a diagonal matrix which contain the singular values as shown below.

$$[H]_{m\times n}=[U]_{m\times m}[S]_{m\times n}[V]^{*}_{n\times m}. \tag{3}$$

Where

$$[S]=\begin{bmatrix} \sigma_1 & 0 & \cdots & 0 \\ 0 & \sigma_2 & \cdots & \vdots \\ \vdots & \cdots & \cdots & 0 \\ 0 & \cdots & \cdots & \sigma_n \end{bmatrix}$$

The inversion of the matrix:

$$\begin{aligned}[H]^{-1}_{m\times n} &= ([V]^{T}_{n\times m})^{-1}([S]_{m\times n})^{-1}([U]_{m\times m})^{-1} \\ &= [V]_{n\times m}[S^{-1}]_{m\times n}([U]_{m\times m})^{*}. \end{aligned} \tag{4}$$

Where

$$[S^{-1}]_{m\times n}=\begin{bmatrix} \sigma_1^{-1} & 0 & \cdots & 0 \\ 0 & \sigma_2^{-1} & \cdots & \vdots \\ \vdots & \cdots & \cdots & 0 \\ 0 & \cdots & \cdots & \sigma_n^{-1} \end{bmatrix}$$

The system matrix is decomposed in orthogonal parts of the information in it which means the singular values in $[S]$ actually refer to the different independent phenomena acting in the system matrix. The largest singular value corresponds to the leading phenomenon. However, if the diagonal matrix $[S]$ contains small singular value σ_i, the diagonal terms σ_i^{-1} in $[S^{-1}]$ becomes big. Unfortunately, it means that the unimportant part of the system matrix before inverting becomes important after inverting which brings in more noise than information and has a large effect on final result. In this case, rejecting some insignificant singular values is beneficial and hence leads to a more reliable estimate of the loads. However, inappropriate singular value rejection makes the different rows in the matrix to be less linear dependent which means that some significant information is taken away and therefore, induce an under-prediction of the response. As a result, an appropriate criterion is required to abandon singular values.

One criterion is omitting a fixed number of singular values. A fixed number of rows of the FRFS matrix will be taken away after conversion of the system to the singular space. This does not take the matrix condition at each frequency into account, which may result in significant information loss when values are at the same level at some special frequency.

The second criterion is that the number of rows to take away depends on the absolute values of the singular values. That means if the condition number at some specific frequencies is low, but many singular values are small than the threshold, the important information will be taken away.

The last criterion is that the number of rows to take away depends on the ratio of the biggest singular value to the smallest. The omitted singular values vary with frequency and the spread of singular values at each frequency are considered. [6]

A TPA model is established for an automobile and experiments have been carried out. Fig.5 shows the loads that estimated by three different criterion. The estimated loads are quite different from each other. In order to judge the effectiveness of the identification based on three criteria, the final contribution are shown in the Fig.6. The contribution result based on the relative criterion is most similar to the operational data measured from the target location. The results based on the absolute criterion and fixed number criterion are more or less different from the operational data. Therefore, the relative criterion is the most accurate method for singular values rejection to identification of the loads.

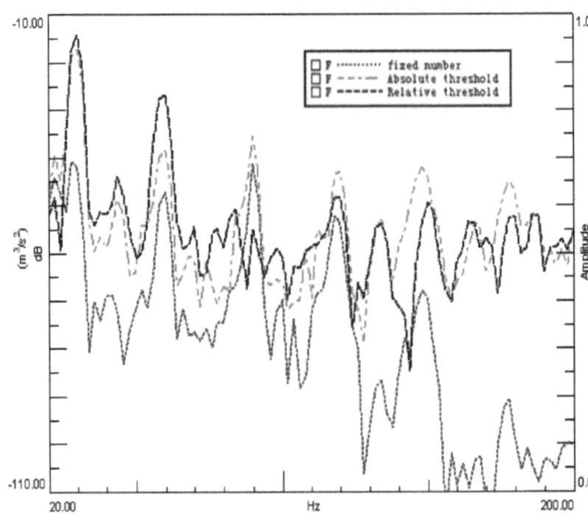

Fig. 5. Loads estimation after omitting singular values

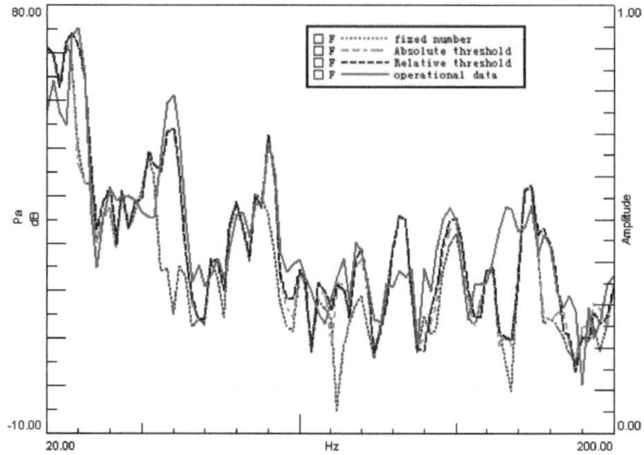

Fig. 6. Contribution result

5 Conclusion

A TPA model is established for an automobile and experiments have been performed to investigate further the problem brought by the matrix inversion method which is for load identification. It has been shown that the estimation of loads by inverse methods can be improved by some techniques. One effective solution is over-determination which yield some improvement in load estimate due to the improvement of the condition number .In addition, singular value rejection is an effective method for reducing error magnification because the matrix system are usually ill-conditioned and discarding some singular values can be beneficial. Three criteria, the fixed number criterion, the absolute criterion and the relative criterion for omitting singular values, are studied. According to the experiments results, the relative criterion is the most accurate method to determine the operational load.

References

1. Thite, A.N., Thompson, D.J.: The quantification of structure-borne transmission paths by inverse methods. part1, improved singular value rejection methods. Journal of Sound and Vibration 264, 411–431 (2003)
2. Thite, A.N., Thompson, D.J.: The quantification of structure-borne transmission paths by inverse methods. part1, improved singular value rejection methods, use of regularization techniques. Journal of Sound and Vibration 264, 433–451 (2003)
3. Mendel, J.M., Harada, G.A.: The Contrained Total Least Squares Technique and its Applications to Harmonic Super resolution. IEEE Transactions on Signal Processing 39(5) (1991)
4. Janssens, M.H.A., Verheij, J.W., Loyau, T.: Experimental example of the pseudo-forces method used in characterisation of a structure-borne sound source. Applied Acoustics 63(1), 9–34 (2002)

5. Tcherniak, D., Schuhmacher, A.P.: Application of Transmissibility Matrix Method to NVH Source Contribution Analysis. In: Proceeding of the IMAC-XXVII, Orlando, Florida USA, February 9-12 (2009)
6. LMS Test_lab: Transfer Paths Analysis manual (2006)

Operational Modal Analysis for Automobile

Wang Jing, Zhou Hong, Xu Gang, Wang Erbing, and Lu Xiang

School of Automobile, Tongji University, Shanghai 201804, China
wangjing_auto@163.com, zhouhong@mail.tongji.edu.cn,
xg@tongji.edu.cn, wangerbing@gmail.com, lux.tongji@gmail.com

Abstract. This paper presents the method of operational modal analysis, which is the identification of modal parameters by only using the response signals under operational conditions or ambient excitations. Giving the actual road spectrum as ambient excitations on vibration table, the operational modal analysis for automobile was conducted by test. The first order natural frequency 14.721Hz of rear axle and the first order torsion frequency 30.928Hz of body were determined in the test. This result implies that the operational modal analysis method is efficient for the modal parameters identification of automobile under operational conditions. It can accurately assess the dynamic characteristics of automobile's structures under actual conditions.

Keywords: Operational Modal Analysis, automobile, PolyMax, operational condition.

1 Introduction

Vehicle vibration and modal identification need to be considered in automotive design and manufacturing process, it relates to NVH performance, strength and overall performance design. Traditional modal analysis method is often controlled in good laboratory conditions; hammer or shakers are used to excite the structure. By both measuring the input and output signals to obtain the Frequency Response Functions (FRFs), the modal parameters can be identified.

However, the real loading conditions to which a structure is subjected often differs considerably from those used in laboratory testing, since pre-stress of suspensions and non-linear factors will be taken into account[1]. Also for some structures it is very difficult to obtain the complete information of input signals. Therefore, the development of operational modal analysis technique is important. Operational modal analysis often refers to the modal analysis under ambient excitations or the identification of modal parameters by only using response signals [2][3].

As the innovation and development of traditional modal analysis method, operational modal analysis has following advantages: it can accurately reflect the dynamic characteristics of automobile's structures under actual operational conditions; it doesn't need to measure input signals, and no expensive equipment is needed to excite the structure; the analysis results can be directly applied to the online health monitoring and damage diagnosis of structures.

Y. Zhang (Ed.): Future Communication, Computing, Control and Management, LNEE 141, pp. 525–532.
springerlink.com

The main methods of operational modal analysis include peak-picking method, poly-reference LSCE (least squares complex exponential), stochastic subspace method [4] and PolyMax method [5]. PolyMax method is a revolutionary modal parameter estimation technique of LMS International Company. It uses discrete-time frequency domain model and fast recursive computing skills, also it has good capability of resisting disturbance, and the stabilization diagram is clean and clear. At present PolyMax method is recognized one of the best experimental modal analysis methods and it has good accuracy for low damping, high damping and dense modes systems.

2 Theoretical Basis of PolyMax Method

2.1 Mathematical Model

PolyMax method uses FRF matrix [H(ω)] as fitting function, while in the operational modal analysis, poly-reference cross-power spectrum matrix is used instead of FRF matrix. In the fitting process, the so-called matrix-fraction model is assumed:

$$\left[H(\omega)\right]_{l\times m}=\left[B(\omega)\right]_{l\times m}\left[A(\omega)\right]^{-1}{}_{m\times m} \tag{1}$$

[H(ω)]l×m is the matrix containing the FRFs between all m inputs and all l outputs or the matrix containing the cross-power spectrums between all m references responses and all l responses. And:

$$\left[B(\omega)\right]=\sum_{r=0}^{p}Z^{r}[\beta_{r}]\ ;\quad \left[A(\omega)\right]=\sum_{r=0}^{p}Z^{r}[\alpha_{r}]\ ;\quad Z=e^{-j\omega\Delta t}\ ;$$

Where [β r]l×m are the numerator matrix polynomial coefficients; [α r]m×m are the denominator matrix polynomial coefficients; p is the model order; Z is derived from a discrete-time model and Δt is the sampling time.

For any frequencies ωk, list the equation (1) according to the measured cross-power spectrum matrix [H(ωk)]; take different frequencies to list a sufficient number of equations (over-determined equations). Then the unknown numerator and denominator matrix polynomial coefficients [β r] [α r] can be found by the Least-Squares solutions.

2.2 Poles and Modal Participation Factors

Once [α r] are determined, the poles and modal participation factors are retrieved as eigenvalues and eigenvectors of their companion matrix:

$$\begin{bmatrix} [O] & [I] & \cdots & [O] & [O] \\ [O] & [O] & \cdots & [O] & [O] \\ \vdots & \vdots & \vdots & \vdots & \vdots \\ [O] & [O] & \cdots & [O] & [I] \\ -[\alpha_0]^T & -[\alpha_1]^T & \cdots & -[\alpha_{p-2}]^T & -[\alpha_{p-1}]^T \end{bmatrix}[V]=[V][\Lambda] \tag{2}$$

The modal participation factors are the last m rows of [V]mp×mp, the matrix [Λ] contains the poles $\lambda_i = e^{-p_i \Delta t}$ on its diagonal. They are related to eigenfrequencies ω_i and damping ratios ξ_i as follows:

$$p_i, p_i^* = -\xi_i \omega_i \pm j\sqrt{1-\xi_i^2}\,\omega_i \tag{3}$$

2.3 Mode Shapes

The mode shapes can be found by considering the so-called pole-residue model:

$$\left[H(\omega)\right] = \sum_{i=1}^{N}\left[\frac{\{\psi\}_i \langle l\rangle_i^T}{j\omega - p_i} + \frac{\{\psi\}_i^* \langle l\rangle_i^H}{j\omega - p_i^*}\right] - \frac{[LR]}{\omega^2} + [UR] \tag{4}$$

Where N is the number of modes; $\{\psi\}_i \in C^l$ are the mode shapes; $\langle l\rangle_i^T \in C^m$ are the modal participation factors; $[LR],[UR] \in R^{l \times m}$ are respectively the lower and upper residuals of the out-of-band modes in the considered frequency band.

Since the mode shapes $\{\psi\}_i$ and the lower and upper residuals are the only unknowns, they are readily obtained by solving (4) in a least-squares sense. This second step is commonly called least-squares frequency-domain (LSFD) method.

3 Operational Modal Analysis for Automobile

The actual working situation was simulated by the vibration table. The modal parameters of automobile could be identified by only measuring output signals of the vehicle system.

3.1 Measurement and Analysis System

The measurement system consisted of three parts, exciting system, acquisition system and analysis system. The exciting system included vibration table; acquisition systems included Accelerometers and LMS SCADAS III SC316W hardware; LMS Test.lab was used for analysis.

The test object was a complete vehicle. Vehicle coordinate system was defined as: the X-axis was aligned with the forward direction of motion. The Y-axis pointed toward the left side of the vehicle, while the Z-axis was directed away from the ground. X, Y, Z coordinates were in accordance with the “right hand rule”.

In order to determine the vehicle dynamic performance under operational conditions, the boundary conditions should match the operational boundary conditions. The four wheels should be put on the ground without any other constrains, in this way the results would be more accurate. To keep the excitation environment be similar with the actual working environment, the collected road spectrum was simulated by vibration table, and then the important operational modals could be excited.

Low frequency modes below 100Hz were mainly considered in the test, so the data was sampled at a rate of 512Hz and the anti-aliasing filter had a cut-off frequency at 128Hz, 20 averages were performed to eliminate the random signal error. Hanning window was used to suppress the leakage of time-domain signals.

3.2 Arrangement of Measurement Points

101 measurement points respectively on body, chassis, engine mounts and rear axles were arranged, table 1 shows the specific distribution of points:

Table 1. Distribution of measurement points

No.	Name	Code	Number of points
1	Left body	L	21
2	Right body	R	21
3	Middle body	M	20
4	Chassis	C	35
5	Engine	E	4

As the given spectrum was along Z-axle, so that the most obvious displacement of each points was also along Z-axle. In order to appropriately reduce the work amounts, only the Z-axle vibration acceleration signals were collected. Two reference locations were selected beneath the A-pillar of left body and above the B-pillar of right body. The geometric model and reference points are illustrated in figure 1 and figure 2.

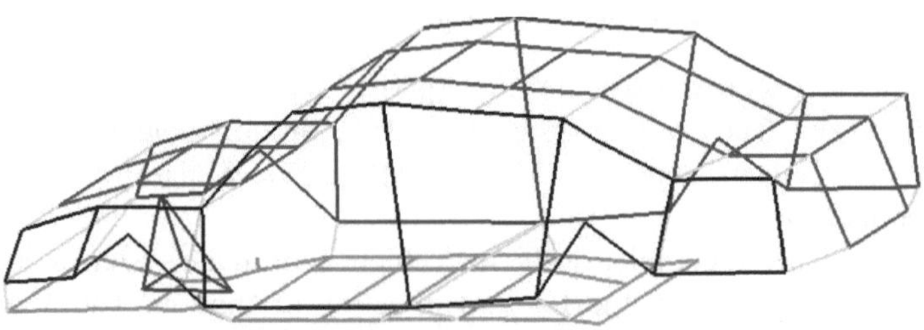

Fig. 1. Geometric model diagram

Fig. 2. Reference points

3.3 Data Processing and Modal Analysis

After the testing of all response signals were completely finished, the modal parameters (such as mode frequencies, damping ratios and mode shapes) can be obtained by means of stabilization diagram. As the drastic movements of engine and rear axle may affect the test results, firstly only the points of body and chassis were selected to identify the modal parameters. The identification of the first 7 modes results are presented in Table 2.

Table 2. Modal parameters of body and chassis (Subjective descriptions)

Mode order	Mode frequency	Damping ratio	Mode shape
1	1.540 Hz	4.90%	first order torsion of body, first order bending of chassis
2	5.340 Hz	4.05%	first order torsion of body
3	8.926 Hz	1.76%	torsion of front body, second order bending of chassis
4	10.668 Hz	0.84%	torsion of body in Y direction
5	14.395 Hz	3.95%	bending of body in X direction
6	20.190 Hz	0.55%	overall bending of body, high order bending of chassis
7	30.928 Hz	0.40%	first order torsion of body

And then the points of engine and rear axles were respectively added to analysis the modal parameters, table 3 and table 4 presents the results:

Table 3. Modal parameters of body, chassis and engine

Mode order	Mode frequency	Damping ratio
1	1.530Hz	4.56%
2	5.348 Hz	3.64%
3	8.885 Hz	1.46%
4	14.046 Hz	3.39%
5	19.543 Hz	0.53%
6	30.878 Hz	0.36%

Table 4. Modal parameters of body, chassis and rear axle

Mode order	Mode frequency	Damping ratio
1	1.523 Hz	1.27%
2	5.478 Hz	2.29%
3	14.721 Hz	2.15%
4	16.343 Hz	1.14%
5	20.545 Hz	0.58%

3.4 Assurance of Mode Determination

In order to estimate the quality of the obtained mode shapes, MAC (Modal Assurance Criterions) values were calculated. MAC values can be used to compare two modal modes obtained from two different modal parameter estimation processes on the same test data or to investigate the validity of the estimated modes within the same modal model. The MAC between two mode shape vectors, $\{\Phi\}_r$ and $\{\Phi\}_s$, is defined as:

$$MAC\left(\{\Phi\}_r,\{\Phi\}_s\right)=\frac{\left|\{\Phi\}_r^T,\{\Phi\}_s\right|^2}{\left(\{\Phi\}_r^T,\{\Phi\}_s\right)\left(\{\Phi\}_s^T,\{\Phi\}_r\right)} \quad (5)$$

When comparing mode shapes, the MAC values for corresponding modes should be closed to 100%, and the MAC values for different modes should be less than 35%. Table 5 shows the MAC values of the first 7 order mode shapes.

Some MAC values are more than 35%, because the measurement points were not enough and maybe some critical points were missed, so that the difference between two modes are not obvious.

Table 5. MAC values of the first 7 order mode shapes

	Mode01	Mode02	Mode02	Mode04	Mode05	Mode06	Mode07
Mode01	100.000	52.671	67.362	24.483	55.026	22.951	7.151
Mode02	52.671	100.000	50.205	41.637	18.751	16.597	1.150
Mode03	67.362	50.205	100.000	48.886	53.296	27.706	3.217
Mode04	24.483	41.637	48.886	100.000	16.130	6.459	0.736
Mode05	55.026	18.751	53.296	16.130	100.000	33.123	1.691
Mode06	22.951	16.597	27.706	6.459	33.123	100.000	6.536
Mode07	7.151	1.150	3.217	0.736	1.691	6.536	100.000

4 Conclusion

It can be observed from the above results:

(1)Limited by the frequency band of road spectrum signal, frequencies below 40Hz were identified in the test. The first order torsion frequency is often near 30Hz for most vehicles' body, and table 2 shows that the 7th mode frequency 30.928Hz is the first order torsion frequency; it can be regarded as the first order torsion of body under operational conditions.

(2)Compared to the modal parameters of body and chassis (in Table 2), the modal parameters of body, chassis and engine (in Table 3) only have a small variation, because the engine has only 4 measurement points so that it can't influence the modal identification results. To further research, more points should be arranged to identify the actual vibration situation of engine.

(3)The mode order is apparently reduced after adding the points of real axle (in Table 4), and the 3rd, 4th and 5th mode shapes are similar—the body doesn't vibrate obviously, while the rear axle vibrates a lot. Reference [6] shows that the first order frequency of rear axle in rear suspension simulation model of this testing automobile is 14.5Hz, it is similar with the 3rd mode frequency 14.721Hz in Table 4. This result means, operational modal analysis effectively identifies the natural frequency of rear axle under operational conditions and this has a lot of engineering significance.

In conclusion, the operational modal analysis method can efficiently extract the automobile's low modes under operational conditions; this provides reliable evidence for the vibration analysis. The automobile's operational modal parameters can accurately assess the dynamic characteristics of automobile's structures under actual conditions.

References

1. LMS Test_lab: Operational modal analysis manual (2006)
2. Ren, W.-X., Zong, Z.-H.: Output-only modal parameter identification of civil engineering structure. Structural Engineering and Mechanics 17(3-4) (2004)

3. Hernans, L., Van der Auweraer, H.: Modal testing and analysis of structures under operational conditions: industrial applications. Mechanical Systems and Signal Processing 13(2), 193–216 (1999)
4. Peeters, B., de Roeck, G., Hermans, L., Wauters, T.: Comparison of system identification methods using operational data of a bridge test. In: Proceedings of ISMA 23, the International Conference on Noise and Vibration Engineering, K.U.Leuven, Belgium, September 1998, pp. 923–930 (1998)
5. LMS Test_lab: LMS PolyMAX—A Revolution in Modal Parameter Estimation
6. Chen, D., Jin, X., Huang, H., Gao, Q.: Dynamics Simulation and Experimental Study on Torsion Bar Rear Suspension of Passenger Car. Automotive Engineering 29(2), 149–152 (2007)

Study on Control Algorithm of the Electro-hydraulic Servo System for Load Simulation Test

Hu Yudong[a], Zhou Hong[b], and Xu Gang[c]

4800 Cao An Road, Shanghai, China, 201804
[a] ajian_1984@126.com, [b] zhouhong@tongji.edu.cn, [c] xg@tongji.edu.cn

Abstract. The process of control algorithm for load reproduction is presented. The procedure of system identification based on frequency response function and target signal iteration based on frequency-domain iteration learning control (ILC) algorithm is illustrated. The frequency-domain iteration learning control algorithm through setting proper overlaps and sections of the signals is proposed. A real electro-hydraulic servo system for load simulation test is chosen, and load reproduction of this system is successfully achieved with the proposed algorithm. The result shows that the algorithm is effective.

Keywords: System identification, iteration learning control, load simulation.

1 Introduction

The electro-hydraulic servo system is widely used in load simulation test because of its compact structure, good performance and easy transfer of energy. The key of load simulation test is to reproduce the operational state of the specimen in laboratory environment. For fatigue and durability performance test, it is essential to get a high precision of load reproduction so as to get reliable results. Therefore, it is really important that the actual load should be as close as possible to the target load. The theory of the method to solve the load reproduction problem was first proposed in 1976[1]. After that, a lot of commercial softwares appear. MTS Systems Corporation proposed RPC（Remote Parameter Control） in 1977. Schenk Corporation proposed ITFC (Iterative Transfer Function Compensation) in 1979. Tiab Corporation proposed MIMIC (Multi-Input, Multi-Output Iterative Control) and Instron Corporation and LMS Corporation proposed TWR (Time Waveform Replication) [2-3]. All these softwares are based on the same mathematic principium which calls frequency-domain iteration learning control (ILC) [4]. It could make the controlled system repeatedly work according to the reference trajectory with high precision.

With the development of computer technology and modern manufacturing industry, it is now easy to get the hardware system for the load simulation test. However, there is sometimes no control software for some load simulation test systems due to the high price of the software. Therefore, it is quite necessary and could also reduce the cost if the control algorithm could be programmed and realized. In this paper, the frequency-domain iteration learning control algorithm through setting proper overlaps and sections of the signals is proposed and programmed in Matlab. Then, a real

Y. Zhang (Ed.): Future Communication, Computing, Control and Management, LNEE 141, pp. 533–541.
springerlink.com

electro-hydraulic servo system for load simulation test is chosen and the control algorithm is successfully used to reproduce the vertical acceleration signal of a specimen.

2 The Procedure of Control Algorithm for Load Reproduction

There are mainly two steps of the control algorithm for load reproduction, system identification and target signal iteration.

2.1 System Identification

In the procedure of load reproduction, the entire system should be indentified as a mathematic model which is a basis of the target iteration. The object of the identification procedure in this paper is the electro-hydraulic servo system for load simulation test which is showed in figure 1. It is composed of controller, electro-hydraulic servo valve, actuator, specimen, displacement sensor and acceleration sensor. It is treated as a whole system in the identification procedure which is showed in figure 2. The input signal of the system is the displacement of the actuator while the output signal of the system is the vertical acceleration signal of the specimen.

Fig. 1. Electro-hydraulic servo system for load simulation test

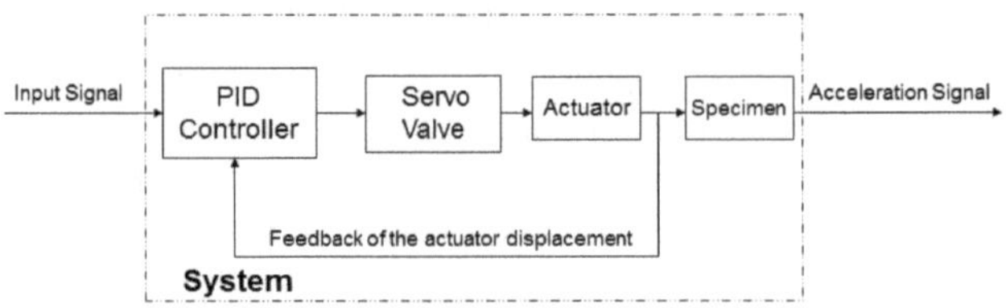

Fig. 2. System diagram of the electro-hydraulic servo system for load simulation test

The identification method is the nonparametric FRF (Frequency Response Function) model identification. The specific diagram of the method is showed in figure 3.

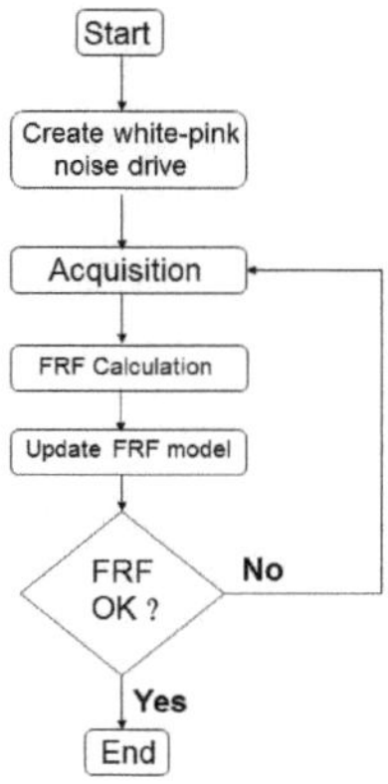

Fig. 3. Flow diagram of the FRF model identification

The drive signal of the identification is usually the white-pink noise which is a combination of a low-frequency white noise and a high-frequency pink noise. The frequency range of the drive signal is selected according to the frequency feature of the target signal and the energy of the drive signal should be the same level of the target signal.

Since the drive signal of the identification is definite, the H1 method is chosen to calculate the FRF model and the coherence function at the same time:

$$H(j\omega) = \frac{G_{yu}(j\omega)}{G_{uu}(j\omega)} \tag{1}$$

$$\gamma = \frac{\left|G_{yu}(j\omega)\right|^2}{G_{uu}(j\omega)\cdot G_{yy}(j\omega)} \tag{2}$$

where $H(j\omega)$ is the FRF of the system, γ is the coherence function of the input signal and output signal, $G_{yu}(j\omega)$ is the cross-power spectrum estimation of the input signal and output signal, $G_{uu}(j\omega)$ and $G_{yy}(j\omega)$ are respectively the auto-power spectrum estimation of the input signal and the output signal.

The value of the coherence function γ is between 0 and 1, which could be used to evaluate the precision or the quality of the calculated FRF model. A low coherence value could be caused by the nonlinearity of the system or the noise of the environment. Basically the FRF model is quite precise if the values of the coherence function are all greater than 0.8 or 0.9.

Usually there are more than one identification processes and several corresponding FRF models. We could finally get a high precise model by averaging the FRF models.

$$FRF_{ave}(j\omega) = \lambda FRF_{new} + (1-\lambda) FRF_{old} \tag{3}$$

where $FRF_{ave}(j\omega)$ is the FRF after averaging, FRF_{new} and FRF_{old} are respectively the FRF of the current identification and the last averaging FRF, λ (0< λ <1) is the weighting coefficient which could be chosen by the experience of the engineer or selected according to the coherence function[5].

2.2 Target Signal Iteration

As the test system always shows more or less some nonlinearity, it is necessary to get the final drive signal through iteration in order to make the drive signal gradually converge which makes the response signals approach its corresponding target signals. The frequency-domain iteration learning control algorithm is used in this paper. The actual iteration flow diagram is showed in figure 4.

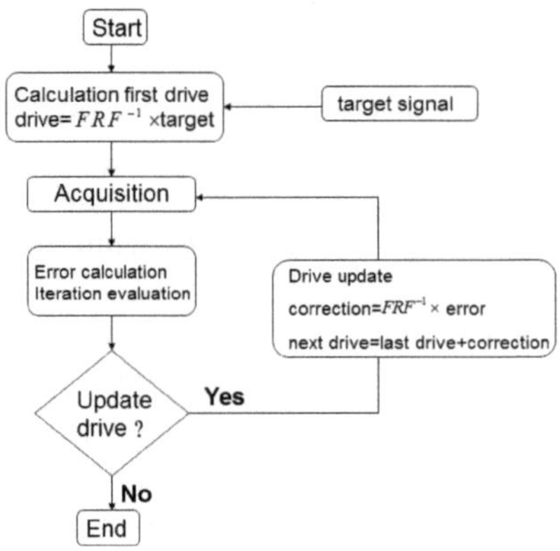

Fig. 4. Flow diagram of iteration

Because of the nonlinearity of the test system, a weighting coefficient α is selected to get the first drive signal so as to prevent the first drive from overloading which will damage the test system. We could get the first drive signal in the frequency-domain according to the formula below:

$$U_1(j\omega) = \frac{1}{H(j\omega)} \times Y_t(j\omega) \times \alpha \tag{4}$$

where $U_1(j\omega)$ is the DFT(Discrete Fourier Transformation) of the first drive signal and we could get $u_1(t)$ through the IDFT(Inverse Discrete Fourier Transform) of it, $H(j\omega)$ is the FRF model got earlier, $Y_T(j\omega)$ is the DFT of target signal, α is the weighting coefficient which range from 0 to 1.

After the drive signal is played, we could get the response signal and the tracking error between the response signal and the target signal. Then we could evaluate the iteration quality by the value of the relative RMS (Root-Mean-Square):

$$\varepsilon = \frac{RMS\{e(t)\}}{RMS\{y_t(t)\}} \times 100\% \tag{5}$$

where $e(t) = y_t(t) - y(t)$ is the tracking error, $y_t(t)$ is the target signal, $RMS\{\}$ is the Root-Mean-Square of the signal.

We could usually end the iteration if ε is less than 10%. Otherwise, we could update the next drive signal according to the tracking error. A weighting coefficient is also chosen in the frequency-domain calculation in order to ensure the convergence of the algorithm.

$$\Delta U(j\omega) = \frac{1}{H(j\omega)} \times E(j\omega) \times \beta \tag{6}$$

$$U_{new}(j\omega) = U_{old}(j\omega) + \Delta U(j\omega) \tag{7}$$

where $\Delta U(j\omega)$ is the update value of the drive signal, $E(j\omega)$ is the DFT of the tracking error $e(t) = y_t(t) - y(t)$, $U_{old}(j\omega)$ and $U_{new}(j\omega)$ are respectively the DFT of the drive signal before updating and after updating and we could get the next drive signal through IDFT of $U_{new}(j\omega)$, β is the weighting coefficient which range from 0 to 1.

3 The Realization of the Frequency-Domain Iteration Learning Control Algorithm

Previously the principle and the process of the frequency-domain iteration learning control algorithm is introduced. Next in this paper, we will have a further study on the realization of this algorithm and propose the iteration algorithm through setting proper overlaps and sections of the signals.

3.1 The Realization of System Identification

It is mentioned previously that the white-pink-noise is used in the system identification procedure. Therefore, in the calculation process, we should separate the drive and response signals of system identification and get the FRF through averaging in order to reduce the effect of random error.

Suppose the sampling frequency of the system is F_s, the point number of each signal part is N (with proper overlap), and if the whole signal could be separated into p sections under these conditions. Then we could get the corresponding FRF of each drive and response signal part by formula (1): $H_i(j\omega) = \frac{G_{yui}(j\omega)}{G_{uui}(j\omega)}$ (i=1,…,p) .

Finally, we can get the FRF model by averaging the p FRF models: $H(j\omega)=\frac{1}{p}\sum_{i=1}^{p}H_i(j\omega)$, whose resolution frequency is $\frac{F_s}{N}$ Hz.

3.2 The Realization of Target Signal Iteration

The target signal should be periodic in the iteration process, but the actual signals usually don't meet this requirement. Therefore we should add a fade-window for the target signal. In this paper we use a custom-defined fade-window to avoid too much energy weaken of the singals, as is showed in figure 5. We just make a sine fade in a short section at the start and end to make it periodic and leave the rest of the signal unchanged.

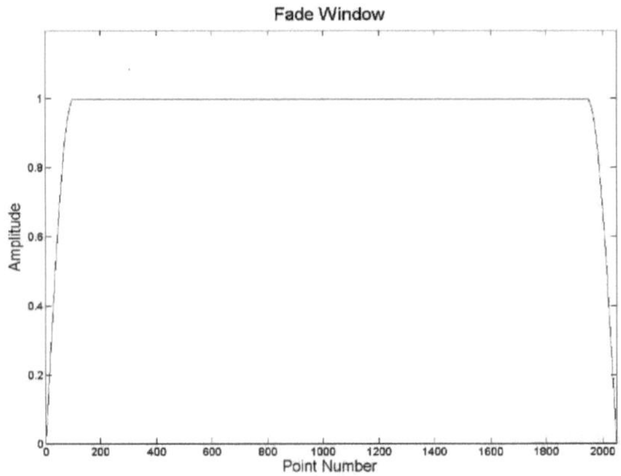

Fig. 5. Custom fade window diagram

Suppose the sampling frequency of the target signal is F_s and the length of the signal is M. Usually M is much greater than the length of the separated section in system identification process. So if we proceed the iteration process with the entire target signal, the resolution of the FRF model should be $\frac{F_s}{M}$ which is actually $\frac{F_s}{N}(>\frac{F_s}{M})$ as showed previously. There are generally two ways to solve this problem:

The first one is that we make the resolution up to $\frac{F_s}{M}$ through interpolation of the FRF model. However, the target signals are sometimes as long as a few hours. The computing complexity of Fourier transform will get a sharp nonlinear rise with the increase of the transform points. So this method is not recommended.

The second one is to separate the target signal into some sections and add fade-window for each section to make them periodic. Nevertheless, the start and end of each section is faded that will lead to problems in iteration. To avoid this problem, a method through setting proper overlaps and sections of the target signal is proposed:
Firstly, separate the target signal, set the length of each section just the same as the length of the section in the system identification process and chose the overlap of 50%. Suppose the whole signal could be separated into q sections under these conditions. (Add zeros if the length is not the multiple of $\frac{N}{2}$). After that add custom-defined fade-window for each section and calculate according to formula (4). At last, splice the calculated drive signals according to the principles below (see figure 6):

1) For the section without overlap (the points from 1 to $\frac{N}{2}$ of the first section and the points from $\frac{N}{2}+1$ to N of the last section), use the calculated result directly.

2) For the section with overlap (take the $\frac{N}{2}$ points of the first section and the second section for example): For the original R points, use the result of the first section. For the last R points, use the result of the second section. For the rest points, use the averaging results of the first section and the second section.

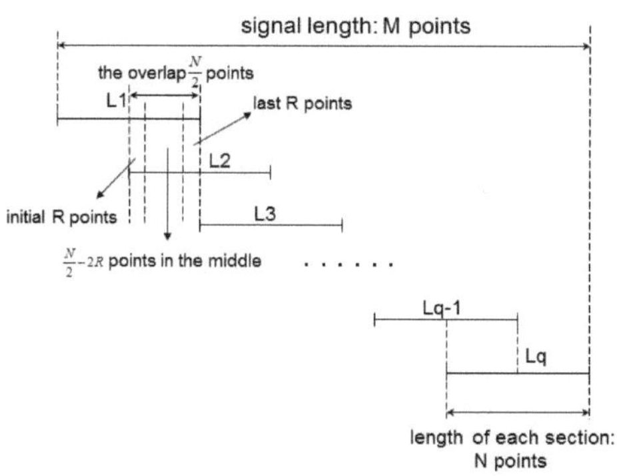

Fig. 6. Improved frequency-domain iterative arithmetic diagram

Then we could get the drive signal with the principles above and continue the iteration process without any problems.

4 Validation of the Algorithm

Next, the electro-hydraulic servo system for load simulation test is selected to validate the proposed algorithm.

The target signal is the vertical acceleration signal of a bracket under operation condition. In the iteration process, we get the drive signal offline through the proposed algorithm, play the drive signal on the test system and record the actual vertical acceleration response of the specimen, then calculate the next drive signal again with the algorithm. The convergence curve of the relative RMS is showed in figure7.

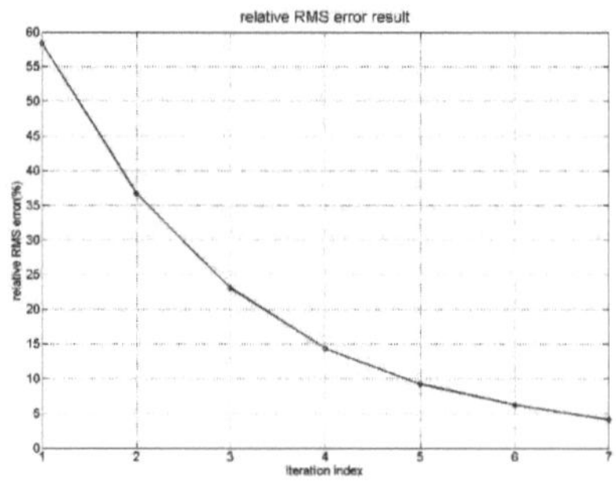

Fig. 7. Relative RMS error convergence curve of the iteration process

As can be seen from the figure, after seven times iteration the relative RMS is less than 5% which is strictly reach the basic requirement of 10%. Figure 8 shows the tracking error after the 7th iteration. Where the red line is the target signal, the blue dash line is the response signal and the black line is the tracking error. It shows that the actual response signal is quite close to the target signal.

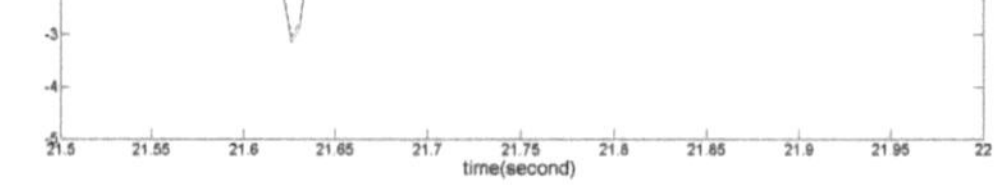

Fig. 8. Tracking error of the 7th iteration (20s-22s)

5 Conclusions

In this paper, the procedure of system identification based on frequency response function and target signal iteration based on frequency-domain iteration learning control algorithm is illustrated. The frequency-domain iteration learning control algorithm through setting proper overlaps and sections of the signals is proposed. At the same time, the algorithm is programmed in Matlab and a real electro-hydraulic servo system for load simulation test is chosen to validate the proposed algorithm. The result shows that the algorithm can get high precision and it could be used as the control algorithm for load simulation test system.

References

1. Cryer, B.W., Nawrocki, P.E., Lund, R.A.: A road simulation system for heavy duty vehicles. SAE 760361 (1976)
2. Dodds, C.J.: A computer system for multi-channel remote parameter control of a test specimen. MTS publication (1977)
3. De Cuyper, J., Dominiek, C.: Service load simulation on multi-axis test rigs. Sound and Vibration 33, 30–35 (1999)
4. Daley, S., Hätönen, J., Owens, D.H.: Hydraulic servo system command shaping using iterative learning control. In: Proceedings of UKACC Control 2004 Mini Symposia, Bath, UK (2004)
5. De Cuyper, J., Coppens, D., Liefooghe, C., Swevers, J., Verhaegen, M.: Advanced drive file development methods for improved service load simulation on multi axial durability test rigs. In: Proceedings of the International Acoustics and Vibration Asia Conference, Singapore (1998)

Research on the Effect of Paths Coupling of Transfer Path Analysis Method

Mu Le[a], Zhou Hong[b], Wang Erbing[c], Xu Haiqing[d], and Su Lang[e]

TongJi University, Cao'an Road 4800, JiaDing District, Shanghai, 201804, China
[a] fushunmule@126.com, [b] zhouhong@mail.tongji.edu.cn,
[c] wangerbing.85061@163.com, [d] coldplay901@126.com,
[e] su171175082@163.com

Abstract. The effect of paths coupling is discussed in transfer path analysis processing. With establishing a TPA model, three different coupling manners were studied: all paths coupling, coupling among three engine mounts and coupling between two exhaust hangers, coupling among three DOFs of each exciting point. The second manner has little effect to the synthesis steering wheel vibration and estimated loads, and the third manner will cause the result overestimated greatly. So the coupling between the engine mounts and exhaust hangers could be ignored. Three DOFs of each exciting point must couple.

Keywords: Paths Coupling, Transfer Path Analysis, Frequency Response Function.

1 Introduction

Experimental transfer path analysis (TPA) is an efficient technique to estimate and rank individual noise or vibration contributions from different paths. The transfer path analysis method designed to solve NVH problems is based on measuring big amount of frequency response functions (FRFs) and known to be extremely laborious and time consuming [1]. The effects of three manners of paths coupling are compared, and FRFs which are unnecessary to measure are analyzed when the coupling do not have much effect.

The paper is organized as following: in Section 2 a brief theoretical introduction to transfer path analysis method is given, Section 3 describes a TPA model, Section 4 analyzes the effects of three paths coupling manners.

2 TPA Theory

Transfer path analysis is based on the superposition principle that is valid for linear, time-invariant systems[2]. The vehicle is described as a source-path-receiver model. If the forces and transfer functions have been determined, partial contributions to as well as the total response at receiver are obtained. This method is effective to identify noise or vibration source, find the main contribution path, and take proper measurement to reduce the noise and vibration.

Y. Zhang (Ed.): Future Communication, Computing, Control and Management, LNEE 141, pp. 543–550.
springerlink.com

To create a TPA model, the global system has to be divided into an active part and a passive part, the former containing the sources, the latter the receiver points where the responses are measured [3]. The two parts are joined at coupling point through hinges, rubber collar and so on. Each degree of freedom (DOF) acting at interface points between active parts and passive parts is a separate transfer path in TPA. Usually only the translational DOFs are included in the measurements [4]. The total response can be expressed as a sum of the path contributions:

$$a_{sum} = \sum_{i=1}^{N} a_i = \sum_{i=1}^{N} H_i(w) \times F_i \tag{1}$$

Where a_{sum} is a vector of total vibration response at receiver, a_i is a vector of partial response, F_i is a vector of load, and $H_i(w)$ is a frequency response function from the load to the receiver.

The determination of operating loads for each DOF can be acquired indirectly, and can be performed in matrix inversion method through measuring FRF matrix between structural response on the receiver side and loads acting at all interfacing DOFs. The forces are obtained as

$$\begin{Bmatrix} F_1 \\ F_2 \\ \vdots \\ F_N \end{Bmatrix} = \begin{bmatrix} H_{11} & H_{12} & \cdots & H_{1N} \\ H_{21} & H_{22} & \cdots & H_{2N} \\ \vdots & \vdots & \vdots & \vdots \\ H_{m1} & H_{m2} & \cdots & H_{mN} \end{bmatrix}^{-1} \times \begin{Bmatrix} a_1 \\ a_2 \\ \vdots \\ a_m \end{Bmatrix} \tag{2}$$

Where a_m is a reference acceleration, F_N is a vector of load, H_{mN} is a frequency response function from load to reference acceleration.

3 TPA Model

The steering wheel vibration problem is studied in idling. Because the transfer paths are multiple and complex, the model should be simplified properly considering the operational situation. In idling, the steering wheel vibration is mainly contributed by engine vibration and exhaust pipe vibration [5]. In TPA, the vehicle system is usually divided into two subsystems. Subsystem A is an active part, including the engine and the exhaust pipe, and subsystem B is a passive part, including car body components. The target is steering wheel vibration, and the joint is the coupling components, including mounts and hangers. The engine connects with car body through three engine mounts. Only the translational DOFs considered in the measurements, engine vibration is transmitted to car body through nine paths. The exhaust pipe connects with car body through two hangers. The exhaust pipe vibration is transmitted to car body through six paths. So there are fifteen paths in all contributing to steering wheel vibration. Fig. 1 shows the transfer paths of steering wheel vibration.

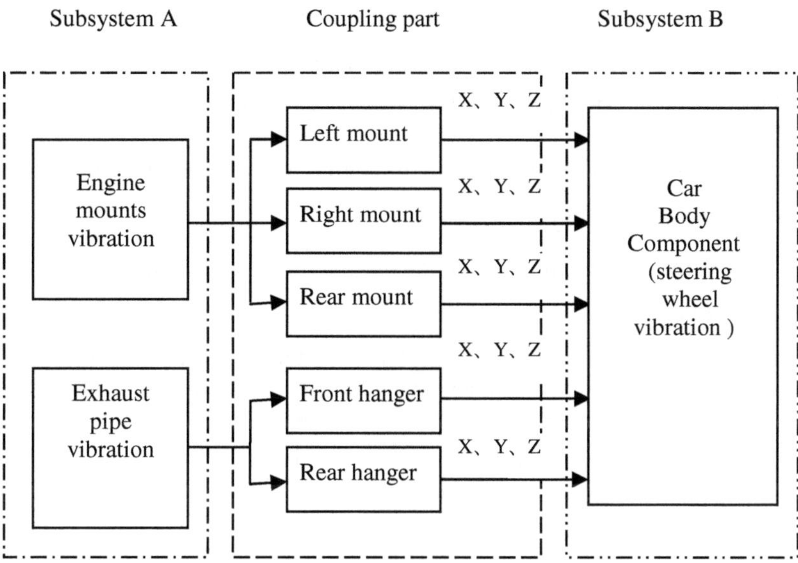

Fig. 1. Transfer paths of steering wheel vibration

4 Paths Coupling Analysis

In TPA model, the active part connects with the passive side through a coupling component, and the loads put on the passive part are transmitted from the active part through coupling components. If there are multiple DOFs of inputs at coupling point, such as DOF 1 and DOF 2, there are corresponding responses at passive side. If the DOF 1 of input at the active side only causes the DOF 1 of response at the passive side, there is no paths coupling. If the DOF 1 of input at the active side causes both the DOF 1 of response and the DOF 2 of response at the passive side, there is paths coupling [4].

Matrix inversion method is used to calculate the loads, and the reference accelerations show the energy of loads at the passive side. If all of the paths couple, the X direction load at the left engine mount not only cause the X direction load at the left engine mount of passive side. Not only X direction reference accelerations at the left engine mount of passive side have responses, but also reference accelerations at other mounts have responses, such as X direction reference accelerations at the right engine mount of passive side and Z direction reference accelerations at the right engine mount of passive side. From equation (2), one can write

$$\begin{bmatrix} F_{lx} & F_{ly} & F_{lz} & F_{rx} & F_{ry} & F_{rz} & F_{rex} & F_{rey} & F_{rez} & F_{fx} & F_{fy} & F_{fz} & F_{bx} & F_{by} & F_{bz} \end{bmatrix}_{1\times 15} =$$

$$\begin{bmatrix} H_{lx,1lx} & H_{lx,1ly} & H_{lx,1lz} & \cdots & \cdots & H_{lx,2bx} & H_{lx,2by} & H_{lx,2bz} \\ H_{ly,1lx} & H_{ly,1ly} & H_{ly,1lz} & \cdots & \cdots & H_{ly,2bx} & H_{ly,2by} & H_{ly,2bz} \\ H_{lz,1lx} & H_{lz,1ly} & H_{lz,1lz} & \cdots & \cdots & H_{lz,2bx} & H_{lz,2by} & H_{lz,2bz} \\ \vdots & \vdots & \vdots & \ddots & & \vdots & \vdots & \vdots \\ \vdots & \vdots & \vdots & & \ddots & \vdots & \vdots & \vdots \\ H_{bx,1lx} & H_{bx,1ly} & H_{bx,1lz} & \cdots & \cdots & H_{bx,2bx} & H_{bx,2by} & H_{bx,2bz} \\ H_{by,1lx} & H_{by,1ly} & H_{by,1lz} & \cdots & \cdots & H_{by,2bx} & H_{by,2by} & H_{by,2bz} \\ H_{bz,1lx} & H_{bz,1ly} & H_{bz,1lz} & \cdots & \cdots & H_{bz,2bx} & H_{bz,2by} & H_{bz,2bz} \end{bmatrix}_{15\times 30}^{-1} \times \begin{bmatrix} a_{1lx} & a_{1ly} & a_{1lz} & \cdots & \cdots & a_{2bx} & a_{2by} & a_{2bz} \end{bmatrix}_{1\times 30} \quad (3)$$

Where $a_{i\,jk}$ is the ith reference acceleration in k direction at the j engine mount. The unit is m/s^2. i is 1 or 2; j is l, r, re, f or b, that respectively mean left engine mount, right engine mount, rear engine mount, front hanger and back hanger. k is X, Y or Z direction.

F_{mn} is the load in n direction at the m engine mount. The meaning of m and n is respectively the same with j and k in $a_{i\,jk}$.

$H_{mn,i\,jk}$ is the FRF from the load F_{mn} to the reference acceleration $a_{i\,jk}$. The unit is $m/(s^2 \cdot N)$.

If the engine mounts are considered far from the exhaust hangers, the couplings between engine mounts and exhaust hangers are ignored. The couplings between three engine mounts and the couplings between two exhaust hangers are considered. In the matrix $H_{mn,i\,jk} = 0$ (j=l, r, re and m=f, b ; j=f, b and m= l, r, re). From equation (2), one can write

$$
\begin{bmatrix} F_{lx} & F_{ly} & F_{lz} & F_{rx} & F_{ry} & F_{rz} & F_{rex} & F_{rey} & F_{rez} & F_{fx} & F_{fy} & F_{fz} & F_{bx} & F_{by} & F_{bz} \end{bmatrix}_{1\times 15} =
$$
$$
\begin{bmatrix}
H_{lx,1lx} & H_{lx,1ly} & H_{lx,1lz} & H_{lx,1rx} & H_{lx,1ry} & H_{lx,1rz} & \cdots & \cdots & 0 & 0 & 0 & 0 & 0 & 0 \\
H_{ly,1lx} & H_{ly,1ly} & H_{ly,1lz} & H_{ly,1rx} & H_{ly,1ry} & H_{ly,1rz} & \cdots & \cdots & 0 & 0 & 0 & 0 & 0 & 0 \\
H_{lz,1lx} & H_{lz,1ly} & H_{lz,1lz} & H_{lz,1rx} & H_{lz,1ry} & H_{lz,1rz} & \cdots & \cdots & 0 & 0 & 0 & 0 & 0 & 0 \\
H_{rx,1lx} & H_{rx,1ly} & H_{rx,1lz} & H_{rx,1rx} & H_{rx,1ry} & H_{rx,1rz} & \cdots & \cdots & 0 & 0 & 0 & 0 & 0 & 0 \\
H_{ry,1lx} & H_{ry,1ly} & H_{ry,1lz} & H_{ry,1rx} & H_{ry,1ry} & H_{ry,1rz} & \cdots & \cdots & 0 & 0 & 0 & 0 & 0 & 0 \\
H_{rz,1lx} & H_{rz,1ly} & H_{rz,1lz} & H_{rz,1rx} & H_{rz,1ry} & H_{rz,1rz} & \cdots & \cdots & 0 & 0 & 0 & 0 & 0 & 0 \\
\vdots & \vdots & \vdots & \vdots & \vdots & \vdots & \ddots & & \vdots & \vdots & \vdots & \vdots & \vdots & \vdots \\
\vdots & \vdots & \vdots & \vdots & \vdots & \vdots & & \ddots & \vdots & \vdots & \vdots & \vdots & \vdots & \vdots \\
0 & 0 & 0 & 0 & 0 & 0 & \cdots & \cdots & H_{fx,2fx} & H_{fx,2fy} & H_{fx,2fz} & H_{fx,2bx} & H_{fx,2by} & H_{fx,2bz} \\
0 & 0 & 0 & 0 & 0 & 0 & \cdots & \cdots & H_{fy,2fx} & H_{fy,2fy} & H_{fy,2fz} & H_{fy,2bx} & H_{fy,2by} & H_{fy,2bz} \\
0 & 0 & 0 & 0 & 0 & 0 & \cdots & \cdots & H_{fz,2fx} & H_{fz,2fy} & H_{fz,2fz} & H_{fz,2bx} & H_{fz,2by} & H_{fz,2bz} \\
0 & 0 & 0 & 0 & 0 & 0 & \cdots & \cdots & H_{bx,2fx} & H_{bx,2fy} & H_{bx,2fz} & H_{bx,2bx} & H_{bx,2by} & H_{bx,2bz} \\
0 & 0 & 0 & 0 & 0 & 0 & \cdots & \cdots & H_{by,2fx} & H_{by,2fy} & H_{by,2fz} & H_{by,2bx} & H_{by,2by} & H_{by,2bz} \\
0 & 0 & 0 & 0 & 0 & 0 & \cdots & \cdots & H_{bz,2fx} & H_{bz,2fy} & H_{bz,2fz} & H_{bz,2bx} & H_{bz,2by} & H_{bz,2bz}
\end{bmatrix}_{15\times 30}^{-1} \tag{4}
$$
$$
\times \begin{bmatrix} a_{1lx} & a_{1ly} & a_{1lz} & \cdots & \cdots & a_{2bx} & a_{2by} & a_{2bz} \end{bmatrix}_{1\times 30}
$$

If the couplings between three engine mounts and the couplings between two exhaust hangers are ignored, only the DOFs at the same mount and the DOFs at the same exhaust hanger are considered. In the matrix $H_{mn,i\,jk} \neq 0$ (m=j). From equation (2), one can write

The paths coupling assumption places some effect on the result. Fig. 2 shows the loads calculated in three coupling manners (e.g. Z direction loads at left engine mount). Fig. 3 shows the synthesis total steering wheel vibration in three coupling manners. We can conclude from the results that the loads of coupling among three engine mounts and coupling between two exhaust hangers is nearly the same with the loads of all paths coupling, and that the synthesis total steering wheel vibrations are almost the same. When only coupling among three DOFs of each exciting point are considered, the load and synthesis total steering wheel vibration are overestimated greatly. So the coupling between the engine mounts and exhaust hangers could be ignored. Three DOFs of each exciting point must couple. Next the FRFs from loads to reference accelerations are analyzed.

$$\begin{bmatrix} F_{lx} & F_{ly} & F_{lz} & F_{rx} & F_{ry} & F_{rz} & F_{rex} & F_{rey} & F_{rez} & F_{fx} & F_{fy} & F_{fz} & F_{bx} & F_{by} & F_{bz} \end{bmatrix}_{1\times 15} =$$

$$\begin{bmatrix} H_{lx,1lx} & H_{lx,1ly} & H_{lx,1lz} & 0 & 0 & 0 & \cdots & \cdots & 0 & 0 & 0 & 0 & 0 & 0 \\ H_{ly,1lx} & H_{ly,1ly} & H_{ly,1lz} & 0 & 0 & 0 & \cdots & \cdots & 0 & 0 & 0 & 0 & 0 & 0 \\ H_{lz,1lx} & H_{lz,1ly} & H_{lz,1lz} & 0 & 0 & 0 & \cdots & \cdots & 0 & 0 & 0 & 0 & 0 & 0 \\ 0 & 0 & 0 & H_{rx,1rx} & H_{rx,1ry} & H_{rx,1rz} & \cdots & \cdots & 0 & 0 & 0 & 0 & 0 & 0 \\ 0 & 0 & 0 & H_{ry,1rx} & H_{ry,1ry} & H_{ry,1rz} & \cdots & \cdots & 0 & 0 & 0 & 0 & 0 & 0 \\ 0 & 0 & 0 & H_{rz,1rx} & H_{rz,1ry} & H_{rz,1rz} & \cdots & \cdots & 0 & 0 & 0 & 0 & 0 & 0 \\ \vdots & \vdots & \vdots & \vdots & \vdots & \vdots & \ddots & & \vdots & \vdots & \vdots & \vdots & \vdots & \vdots \\ \vdots & \vdots & \vdots & \vdots & \vdots & \vdots & & \ddots & \vdots & \vdots & \vdots & \vdots & \vdots & \vdots \\ 0 & 0 & 0 & 0 & 0 & 0 & \cdots & \cdots & H_{fx,2fx} & H_{fx,2fy} & H_{fx,2fz} & 0 & 0 & 0 \\ 0 & 0 & 0 & 0 & 0 & 0 & \cdots & \cdots & H_{fy,2fx} & H_{fy,2fy} & H_{fy,2fz} & 0 & 0 & 0 \\ 0 & 0 & 0 & 0 & 0 & 0 & \cdots & \cdots & H_{fz,2fx} & H_{fz,2fy} & H_{fz,2fz} & 0 & 0 & 0 \\ 0 & 0 & 0 & 0 & 0 & 0 & \cdots & \cdots & 0 & 0 & 0 & H_{bx,2bx} & H_{bx,2by} & H_{bx,2bz} \\ 0 & 0 & 0 & 0 & 0 & 0 & \cdots & \cdots & 0 & 0 & 0 & H_{by,2bx} & H_{by,2by} & H_{by,2bz} \\ 0 & 0 & 0 & 0 & 0 & 0 & \cdots & \cdots & 0 & 0 & 0 & H_{bz,2bx} & H_{bz,2by} & H_{bz,2bz} \end{bmatrix}^{-1}_{15\times 30} \tag{5}$$

$$\times \begin{bmatrix} a_{1lx} & a_{1ly} & a_{1lz} & \cdots & \cdots & a_{2bx} & a_{2by} & a_{2bz} \end{bmatrix}_{1\times 30}$$

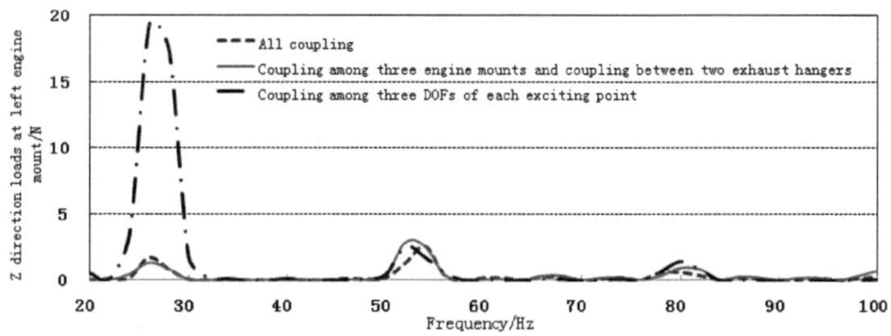

Fig. 2. Z direction loads at left engine mount in three coupling manners

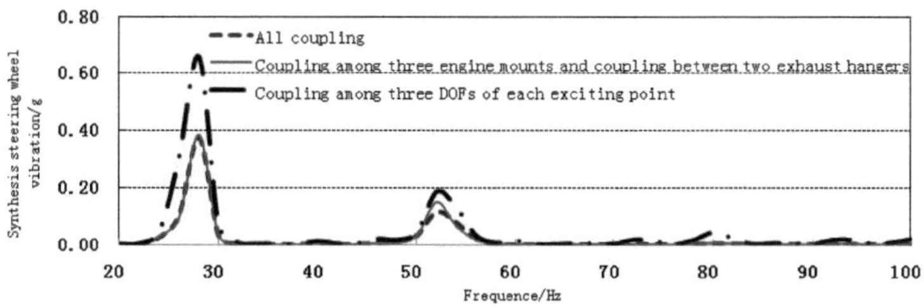

Fig. 3. Synthesis total steering wheel vibration in three coupling manners

Fig. 4 shows the FRFs from the load at right engine mount in Z direction to the first reference accelerations at each engine mount. The maximum of FRF to right engine mount is 3.25 g/N. The maximum of FRF to left engine mount is 0.86 g/N. The maximum of FRF to rear engine mount is 0.07 g/N. Fig. 5 shows the FRFs from the load at rear exhaust hanger in Z direction to the first reference accelerations at each exhaust hanger. The maximum of FRF to front exhaust hanger is 0.16 g/N. The maximum of FRF to back exhaust hanger is 0.78 g/N. Fig. 6 shows the FRFs from the load at right engine mount in Z direction to the first reference accelerations at each exhaust hanger. The maximum of FRF to front exhaust hanger is 0.15 g/N. The maximum of FRF to rear exhaust hanger is 0.12 g/N. The FRFs between engine mounts are far larger than that between engine mounts and exhaust hangers. So the coupling between engine mounts must be considered. The FRFs between exhaust hangers are as much as that between engine mounts and exhaust hangers. Because the estimated loads are influenced by both the FRF matrix and reference accelerations, and the synthesis total steering wheel vibration is influenced by the FRF matrix, reference accelerations and FRF from load to receiver. So the FRF matrix is not the only factor to determine the result. According to the estimated loads and the synthesis total steering wheel vibrations of three manners, the coupling between the engine mounts and exhaust hangers could be ignored. Three DOFs of each exciting point must couple.

In conclusion, paths coupling depends on the real condition. So the coupling between engine mounts and exhaust hangers could be ignored. Three DOFs of each exciting point must couple. The measurement amount of FRFs will be reduced greatly and time will be saved.

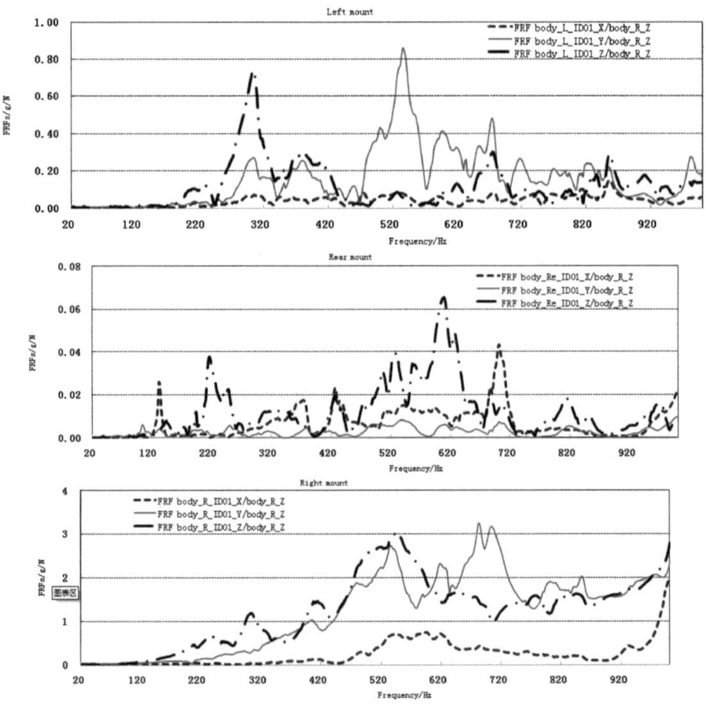

Fig. 4. FRFs from the load at right engine mount in Z to the first reference accelerations at each engine mount

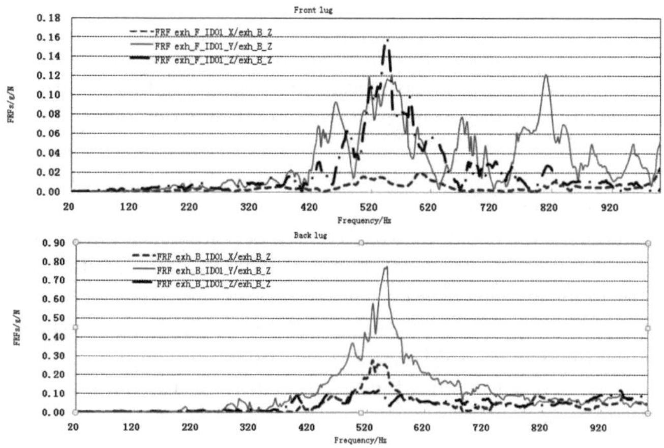

Fig. 5. FRFs from the load at rear exhaust hanger in Z to the first reference accelerations at each exhaust hangers

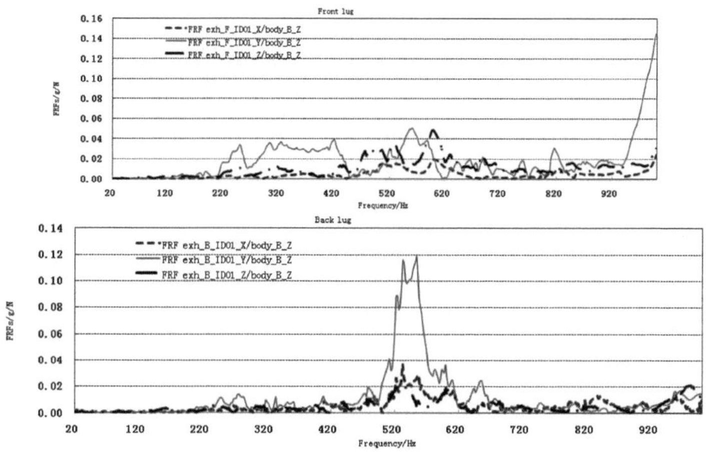

Fig. 6. FRFs from the load at right engine mount in Z to the first reference accelerations at each exhaust hangers

5 Conclusions

A TPA model is established based on the fundamental principal of transfer path analysis. Three coupling manners, all paths coupling, engine mounts and exhaust hangers decoupling , engine mounts decoupling and exhaust hangers decoupling, were discussed. The second manner has little effect to the result, and the third manner will cause the result overestimated greatly. So the engine mounts and exhaust hangers could decouple. Engine mounts should couple and exhaust hangers should couple.

References

1. Tcherniak, D.: Application of Transmissibility Matrix method to structure borne path contribution analysis. In: Proceedings of NAG/DAGA Conference, Rotterdam, Holland (2009)
2. Jian, P., Gang, C., Hua, H.: Automobile noise and vibration-theory and application. Beijing Institute of Press, Beijing (2006)
3. Gajdatsy, P., Janssens, K., Giele, L., Mas, P., Van der Auweraer, H.: Critical assessment of Operational Path Analysis: effect of coupling between path inputs. In: Proceedings of Acoustics 2008, Paris, France (2008)
4. Plunt, J.: Strategy for transfer path analysis (TPA) applied to vibro-acoustic systems at medium and high frequencies. ISMA 23, Leuven, Belgium (1999)
5. Liang, Y.: Interior Noise Research Based on Transfer Path Analysis. Tongji University, Shanghai (2010)

Network Security Analysis Based on Graph Theory Model with Neutral Network

Mo Ming-zhong

Department of Mathematics and Computer Science, Liuzhou Teachers College, Liuzhou, Guangxi 545004, P.R. China
momingzh@163.com

Abstract. In order to improve the safety of network and grasp the safety situation of network correctly, the graph theory and neutral network was combined to have safety analysis for the network security. Firstly, the relating studies was summarized; and then basic theory of the graph theory model of neutral network system was analyzed; and then procedure of network attack situation analysis based on graph theory was discussed, and the simple algorithm was designed; and finally a case study was carried out to verify the effective of this method.

Keywords: Network security, Graph theory, Neutral network.

1 Introduction

The study on the network security analysis based on model was still in the original step, and the systematical theory method had not formed. The main studies related to it concluded the following researches: the attack tree model was put forward firstly by a scientist and the tree structure in the form of AND-OR shape was used in the model construction of attack activities, and the safety of system was evaluated, but the limitation of tree structure, the attack tree could not construct the model of multi trying to attack, time dependent and visit controlling scene. And other scientists applied the conception of right improving graph, and used different traces leading to attack object to express the different attack procession of attacker, and the procession of improving the right of attacker was expressed, and the mean attack price of attack activities was computed according to the experience, but this method lacked the theory basis during computing mean attacking value. Attack figure model was established by some scientists, the network topology information was considered in the procession of security analysis, but the disadvantages of this method was that the attack figure was made by hand, and could not suit for the a large scale network environment. Some scientists used amended model tester to construct the attack figure [1].

The above methods only considered the network security problem from the single angle of attacker, and the factors that would were harmful to the safety of network system, such as the topology structure of network, the weakness of system, the trust relations of main engine, topology structure of network, detail information assignment of system, the usage of software and hardware and other problems could not be input into the model of evaluation, and these methods could not suit for the a scale network

Y. Zhang (Ed.): Future Communication, Computing, Control and Management, LNEE 141, pp. 551–557.
springerlink.com

system, and with the increasing of network scale, the algorithm could form the "status burst" phenomena.

2 Graph Theory Model of Network System

2.1 Define 1 the Graph Theory Model of Network System

The whole network could be abstracted as a directional graph $G = (V, \{VR\})$, the top points of graph denoted the nodes of network, such as server, PC main engine, where V denoted the limited non blank collection of the top point, the top points were linked by directional arc, VR denoted the relation collection between the two nodes [2].

If C and D denoted the nodes of network and the six kinds of visit relation were divided as follows:

$K_1 : C$ could be executed in the capacity of manager on D, and all resources were controlled.

$K_2 : C$ could be executed in the capacity of general client on D, the part resources were controlled.

K_3: C could get or distribute the personal and public information in the capacity of register client of serve software of D, and the system command could not be executed.

K_4: C could get or distribute the personal and public information in the capacity of anonymous client of serve software of D, the connectivity of input layer could be denoted only, such as visiting inner service through firewall.

$K_5 : C$ only visit D on the IP network layer, this relation embodied the connectivity of IP layer.

K_6: C only visit D on link layer, this relation embodied the connectivity of link layer.

The visit relation between node C and node D $(C, D) \in VR$, and $VR = \{K_1, K_2, K_3, K_4, K_5, K_6\}$, which was shown in Figure 1.

2.2 Graph Theory Algorithm Based on Neutral Network

Much importance had always been related to the graph isomorphism problem by science staffs in mathematics and engineering circles, particularly about assigning the models of system, and all kinds of constructing algorithms of graphs. Some scientists put forward the algorithms of graph isomorphism by applying Hopfield neural networks, and based on this method some scientists amended their algorithm, through the construction of relating theory model, the energy equations were amended and

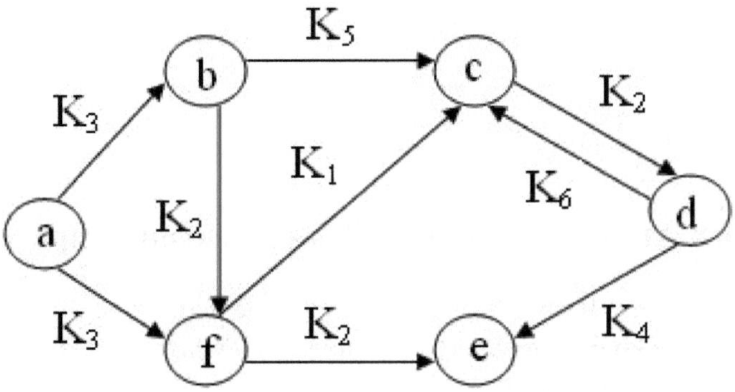

Fig. 1. Model figure of graph theory

reduced thus reducing the complicacy and improving the convergent efficient of networks. An expression was listed as follows [2].

G and G' was defined as two isomorphic graphs with vertex-settings of $V(G)=\{g_1, g_2, \cdots, g_n\}$ and $V(G')=\{g_1', g_2', \cdots, g_n'\}$. (g_i, g_j') was assumed as a unit, which expressed the isomorphic-graph β that mapped the vertex g_i in G to the vertex g_j' in G', Obviously, therefore neural network produced that was made up of $n \times n$ units. When it was steady, the output result of the unit was expressed as w_{ij}, which was expressed as follows [3]:

$$w_{ij} = \begin{cases} 1 & if, \beta(g_i) = g_i' \\ 0 & otherwise \end{cases} \tag{1}$$

The energy equation could be expressed in the following forms:

$$W = \frac{A}{2}\sum_{i=1}^{n}(\sum_{i=1}^{n} w_{ik} - 1)^2 + \frac{B}{2}\sum_{i=1}^{n}(\sum_{i=1}^{n} w_{ik} - 1)^2 + C\sum_{i=1}^{n}\sum_{k=1}^{n}\sum_{j=1}^{n}\sum_{m=1}^{n} | a_{ij} - b_{ij} | w_{ik} w_{jm} \tag{2}$$

where $A = \{a_{ij}\}_{n \times n}$, $B = \{b_{ij}\}_{n \times n}$ was the linkage variable respectively, the corresponding dynamical functions could be expressed as follows:

$$\frac{dw_{ik}}{dt} = -A(\sum_{i=1}^{n} w_{ik} - 1)^2 + B(\sum_{i=1}^{n} w_{ik} - 1) + C(\sum_{i=1}^{n}\sum_{k=1}^{n} | a_{ij} - b_{km} | w_{km}) \tag{3}$$

where $w_{ik}(t)$ was the status of unit (g_i, g_j'), t was time, A, B, C denoted the testing indexes respectively. And the weight of network $\omega_{ik,jm}$ and the threshold value I_{ik} could be expressed as follows:

$$\omega_{ik,jm} = -A\delta_{ij} - B\delta_{km} - C\,|\,a_{ij} - b_{km}\,| \tag{4}$$

$$I_{ik} = -A - B \tag{5}$$

where $i, j, k, m = 1, 2, \cdots, n$

3 Network Attack Situation Analysis Based on Graph Theory

Define 2 attack price: If one attacker attack into one node of network successful, and the price of attacking from node i to node j, which included time, cost and technology and other factors [3].

$A(i \to j)$ was defined as attack price of node i and j.

There were four factors in $A(i \to j)$ in network information system:

(1) Serious degree of leak in node j was defined by $S(j)$;

(2) Confidence level between node i and node j was defined by $C(i \to j)$;

(3) Protect measures strengthen of node j was defined as $P(i)$;

(4) Safe degree of link between node i and node j was defined by $D(i \to j)$.

The efficient value of attack price was listed in Table 1.
Choosing rule of weight value of network was listed as follows
(1) The sum of all properties equal to 1;
(2) The property of high attack price must over 0.25;
(3) The property of low attack price must less 0.25;

The graph theory model was integrated into network system based on multi data origin, the object dynamic procession was achieved under the condition of the original network system security status by using graph theory, and the safety risk procession of attack main body illegal obtained the visit right of the critical information capital was described formally. The basic structure of the network analysis method based on graph theory was given in Figure 2.

Table 1. Evaluating value of four factors

$S(j)$		$C(i \to j)$		$P(i)$		$D(i \to j)$	
label	level	label	level	label	level	label	level
Very low	1	Very little	1	Very high	5	Very high	5
Low	2	Little	2	High	4	High	4
Middle	3	Middle	3	Middle	3	Middle	3
High	4	Big	4	Weak	2	Weak	2
Very high	5	Very big	5	Very weak	1	Very weak	1

The main function of every model was description as follows [4]:

(1) Network parameters abstracted. The parameters such as network main body, attack main body, protect main body and weakness was abstracted, and offering the preparations for constructing model.

(2) Data preprocessing. The formula processing could be carried out for the network parameters obtained, and the weakness information could be changed into atomic collection, at the same time, the simply procession could be applied into the graph theory.

(3) The graph theory model was constructed. The formatting network parameters and weakness knowledge was used as input, using the corresponding algorithm.

(4) Making the graph theory model figure. The graph theory model was embodied as figure, and the present safety situation could be embodied completely, and the safety analysis could be carried out based the relating theory of graph theory.

The main idea of constructing the graph theory model of network security was expressed as follows [4]:

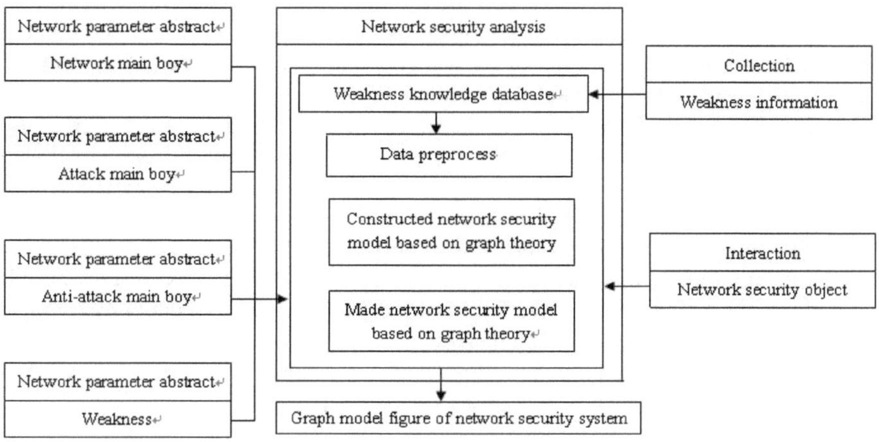

Fig. 2. The basic structure of the network analysis method based on graph theory

(1) Atomic penetrating collection of objective network system was defined according to the results of the network abstract parameter and data preprocessing.

(2) The mapping was carried out for different atomic originated from the status of system, and the network penetrating activity was found out.

(3) When there were no useful activities that made the status of system continue to change, and whether the final status and attack main body had same objection was checked, if not consistence, the relating system status and penetrating activity could be deleted.

(4) If the final status and attack main body had same objection, the lasting status and activity would consist of the model base on graph theory.

The simple algorithm of graph theory was listed as follows:

Begin proc

When the original operation was carried out, assumed $D'=D$

For every $d_i \in D$ do

For every $d_j \in D(j \neq i)$ do

If pre(d_i)=pre(d_j) and $R_f \subseteq (R_p \cup R_d) - post(d_i)$ and $R_f \subseteq (R_p \cup R_d) - post(d_j)$ then

$D' = D' - d_j$

End if

End for

End for

End proc

4 Application Case Study

In order to verify the effective of graph theory model, the network was constructed for test. The network was made up of six business parts and one information center, and every business part was made up of 24 working stations, and information center was made up of 5 severs. Firewall set the serious visit controlling strategies, and the Web Server, File Server and Mail Server of the inner network could only be linked by outer visitors. VLAN was set between the every test and management thorough exchanging machine, and the visit could be carried out. DB Server could not be visit directly by Web Server, File Server and Mail Server.

Security object of network system was to protect DB Server, and ensured the attack main body not obtain the root right. The parameters of network system could be obtained and the 145 weakness was found out; during the stage of data processing, the forma of network parameters was changed, and the algorithm mentioned above was carried out to find 20 atomic penetrating. And the security situation of network was decrypted by graph theory.

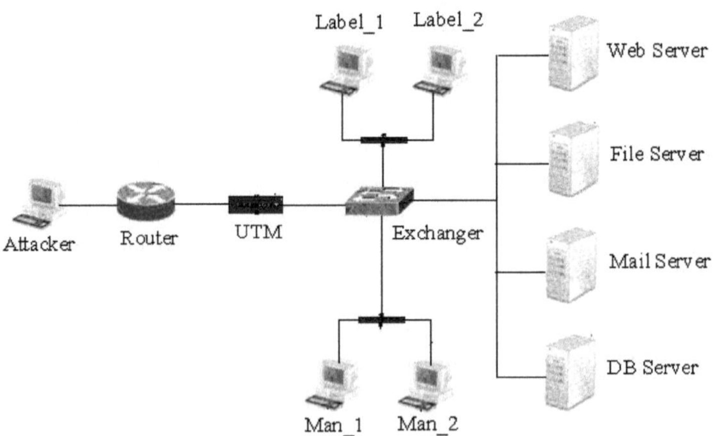

Fig. 3. Topology figure of network security

5 Conclusions

The safety status of network was carried out based on graph theory, and the corresponding effective and optimal measurements were carried out. The new graph theory method was put forward, and the network safety situation was described correctly, and the safety of network was improved.

References

1. Li, N., Xia, J.B., Feng, K.S.: Journal of Naval University of Engineering 22(2), 54–59 (2010)
2. Gan, Z.B., Wu, P., Lu, S.F.: Application Research of Computers 20(11), 20–21 (2007)
3. Zhang, X., Hu, C.Z., Liu, S.H., et al.: Computer Engineering 33, 10–12 (2007)
4. Tang, Y.L., Xu, G.A., Niu, X.X., et al.: Journal of Beijing University of Posts and Telecommunications 31(2), 50–53 (2008)

Fault Diagnosis of Gearboxes Based on the Local Mean Decomposition Method

Jinshan Lin

School of Mechanical and Electronic Engineering, Weifang University,
Weifang 261061, P.R. China
jslinmec@yahoo.cn

Abstract. Fault diagnosis of gearboxes is always a central topic for the maintenance of mechanical systems. However, the common methods are unfit for processing the faulty signals from gearboxes which are usually non-stationary and nonlinear. Empirical mode decomposition (EMD) is suitable for non-stationary and nonlinear signals. Nonetheless, the EMD method still suffers from some shortcomings and these problems are in process. Recently, local mean decomposition (LMD) is put forward for exploring non-stationary and nonlinear signals and reveals some better performance compared with the EMD method in some extent. The paper applies the LMD method to fault diagnosis of a gearbox and successfully identifies the faults of the gearbox. Consequently, the paper points out that the LMD method is seemingly a useful method for fault diagnosis of gearboxes.

Keywords: Fault diagnosis, local mean decomposition (LMD), gearbox.

1 Introduction

Gearboxes are key parts of mechanical systems and their condition monitoring is always a topical issue for the maintenance of mechanical systems [1]. The vibration-based signal processing technique is one of major tools for diagnosing faults of gearboxes and has been proved to be effective in fault diagnosis of gearboxes [2], [3].

Traditional signal processing techniques cover time-domain method and Fourier transform [4]-[6]. Nevertheless, these techniques usually assume that the process generating data is stationary and linear. Considering that the vibration signals captured from the faulty gearboxes are strongly non-stationary and nonlinear, the conventional signal processing techniques are inevitable to cause serious errors [7]. To explore the non-stationary signal, many methods, such as short time Fourier transform (STFT), Wigner-Ville distribution (WVD), and wavelet transform (WT), have been proposed and receives a lot attention recently. However, all the mentioned above methods still depend on Fourier transform, is short of the self-adaption for investigated signals and, consequently, their practicalities are constrained greatly. Empirical mode decomposition (EMD), a novel time-frequency analysis technique, has been introduced to explore the non-stationary signals of gearboxes and achieves a rapid development by

Y. Zhang (Ed.): Future Communication, Computing, Control and Management, LNEE 141, pp. 559–563.
springerlink.com

reason of its excellent performance [8], [9]. Nevertheless, the EMD method still suffers from some weaknesses and these problems are under way.

Recently, local mean decomposition (LMD) is proposed to process non-stationary and nonlinear signals [10]. The LMD method can relieve some imperfections of the EMD method and reveals better performance compared with the EMD method in some aspects. The paper utilizes the LMD method to examine the faulty signal from a gearbox and succeeds in identifying the faults of the gearbox. Also, the paper manifests that the LMD method is apparently a feasible method for fault diagnosis of gearboxes.

2 Local Mean Decomposition (LMD)

The LMD method can decompose a signal into the sum of a set of AM-FM product functions (PF) and a residual. More specifically, the LMD method is stated as follows:

(1) Find all the maximum and minimum points of the signal.

(2) The ith mean value mi and the ith local magnitude ai of each two successive extrema ni and ni+1 are expressed as, respectively,

$$m_i = \frac{n_i + n_{i+1}}{2} \tag{1}$$

$$a_i = \frac{|n_i - n_{i+1}|}{2} \tag{2}$$

(3) Plot straight lines between two successive extrema with the corresponding local mean values and form the piecewise mean lines; similarly, the piecewise envelope lines can be obtained.

(4) Smooth the piecewise mean lines and the piecewise envelope lines using moving averaging to construct a smoothly varying continuous local mean curve m(t) and a smoothly varying continuous local envelope curve a(t).

(5) Subtract the local mean curve m(t) from the signal, then the remaining signal h(t) is divided by the envelope curve a(t) and derive the nominal FM signal s(t); if s(t) is not a truly pure FM signal, repeat the previous steps until s(t) is a truly pure FM signal. Suppose m iterations are required for obtaining ith AM-FM component, and the iterative process can be written as

$$\begin{cases} h_{i1}(t) = x_i(t) - m_{i1}(t) \\ h_{i2}(t) = h_{i1}(t) - m_{i2}(t) \\ \quad\vdots \\ h_{im}(t) = h_{i(m-1)}(t) - m_{im}(t) \end{cases} \tag{3}$$

where

$$\begin{cases} s_{i1}(t) = \dfrac{h_{i1}(t)}{a_{11}(t)} \\ s_{i2}(t) = \dfrac{h_{i2}(t)}{a_{i2}(t)} \\ \vdots \\ s_{im}(t) = \dfrac{h_{im}(t)}{a_{im}(t)} \end{cases} \tag{4}$$

Thus, the ith envelope is given as

$$a_i(t) = a_{i1}(t)a_{i2}(t)\ldots a_{im}(t) = \prod_{q=1}^{m} a_{iq}(t) \tag{5}$$

with the objective of

$$\lim_{m\to\infty} a_{im}(t) = 1 \tag{6}$$

Finally, the ith AM-FM component can be expressed as

$$\mathrm{PF}_i(t) = a_i(t)s_{im}(t) = a_i(t)\cos(\varphi_i(t)) \tag{7}$$

Consequently,

$$\varphi_i(t) = \arccos(s_{im}(t)) \tag{8}$$

$$\omega_i(t) = \frac{d(\varphi_i(t))}{dt} \tag{9}$$

where $\omega_i(t)$ is the instantaneous frequency of the PFi(t).

3 Application of the LMD Method to Fault Diagnosis of a Gearbox

In this part, the LMD method was utilized to diagnose the faults of a gearbox. The gearbox is double-stage driving and contains three shafts with the rotation frequency 24.32Hz, 15.38Hz, and 6.15Hz, respectively. The mesh frequencies of the first shaft and the third shaft are 615.36Hz and 338.56Hz, respectively. The signal in Fig. 1(a), with the sampling frequency 3690Hz and the length 5904 points, was captured from the acceleration sensor around the output shaft.

Furthermore, the signal was examined using the LMD method. The principal PFs, holding the most typical information, are displayed in Fig. 1. The PF1 in Fig. 1(b) is a modulation signal with the carrier frequency of 1225Hz and the modulation frequency of 15 Hz. The carrier frequency 1225Hz is approximate to double the mesh frequency of the input shaft and the modulation frequency 15 Hz is nearly close to the rotation frequency of the input shaft. The PF2 in Fig. 1(c) embodies two modulation components: one with the carrier frequency of 337Hz and the modulation frequency of 6 Hz and another with the carrier frequency of 670Hz and the modulation frequency of 7 Hz. The frequency 337Hz approximates to the rotation frequency of the output shaft and the frequency 670Hz is approximate to double that. The modulation frequency 6Hz or 7Hz approaches to the rotation frequency of the output shaft. The PF3 in Fig. 1(d) is largely made up of the modulation signal with the carrier frequency 337Hz and the modulation frequency 12Hz. The carrier frequency 337Hz is approaching to the mesh frequency of the output shaft and the modulation frequency 12 Hz is roughly equivalent to double the rotation frequency of the output shaft. The PF4 in Fig. 1(e) is virtually governed by the component with the frequency 105Hz, which is approximate to 1/3 time of the rotation frequency of the output shaft. The PF5, PF6, and PF7 seem to stand for the rotation frequency of the input shaft, middle shaft, and output shaft, respectively. Since that the mesh frequency and their harmonics are the carrier frequency and the rotation frequency and their harmonics are the modulation frequency is the characteristic symptom of shaft misalignment, it could sound plausible that there is a misalignment fault arising in the gearbox. Additionally, 1/3 time the rotation frequency of the output shaft hints a rub-impact fault. Actually, the misalignment fault is induced by the abnormal flywheel mounted on the output shaft and the rub-impact results from the bottom bolt looseness. Hence, the LMD method effectively extracts the multiple-fault information that is rather difficult for the conventional methods including the EMD method.

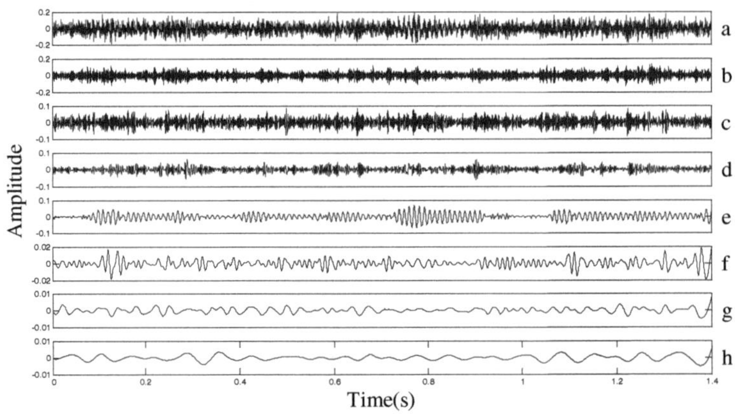

Fig. 1. The decomposition results of the gearbox vibration signal using the LMD method: (a) the vibration signal; (b)-(h) the PF1-PF7.

4 Conclusions

The LMD method is introduced to diagnose the faults of gearboxes, which effectively overcoms the shortcomings of the EMD method in some extend. Next, the LMD method is employed to process the signal collected from a gearbox and successful to extract the multiple-fault information from the captured signal. The results show that the LMD-based method is apparent to be a promising method for fault diagnosis of gearboxes.

References

1. Bartelmus, W., Zimroz, R.: A new feature for monitoring the condition of gearboxes in non-stationary operating conditions. Mechanical System and Signal Processing 23, 1528–1534 (2009)
2. Ocak, H., Loparo, K.A., Discenzo, F.M.: Online tracking of bearing wear using wavelet packet decomposition and probabilistic modeling: A method for bearing prognostics. Journal of Sound and Vibration 302, 951–961 (2007)
3. Lei, Y.G., He, Z.J., Zi, Y.Y., Chen, X.F.: New clustering algorithm-based fault diagnosis using compensation distance evaluation technique. Mechanical Systems and Signal Processing 22, 419–435 (2008)
4. Jardine, A.K.S., Lin, D., Banjevic, D.: A review machenical diagnostics and prognostics implementing condition-based maintenance. Mechanical Systems and Signal Processing 20, 1483–1510 (2006)
5. Tandon, N., Choudhury, A.: A review of vibration and acoustic measurement methods for the diction of defects in rolling element bearing. Tribology International 32, 469–480 (1999)
6. Peng, Z.K., Chu, F.L.: Application of the wavelet transform in mechanical condition monitoring and fault diagnostics: a review bibliography. Mechanical Systems and Signal Processing 18, 199–221 (2004)
7. Loutridis, S.J.: Damage detection in gear systems using empirical mode decomposition. Engineering Structures 26, 1833–1841 (2004)
8. Huang, N.E., Shen, Z., Long, S.R.: The Empirical mode decomposition and the Hilbert spectrum for nonlinear and non-stationary time series analysis. Proceedings of the Royal Society of London A454, 903–995 (1998)
9. Liu, B., Riemenschneider, S., Xu, Y.: Gearbox fault diagnosis using empirical mode decomposition and Hilbert spectrum. Mechanical Systems and Signal Processing 20, 718–734 (2006)
10. Smith, J.S.: The local mean decomposition and its application to EEG perception data. J. R. Soc. Interface 2, 443–454 (2005)

Study on Stability of Continuous-Time Hopfield Network

Li Tu[1] and Juanjuan Song[2]

[1] Department of Computer Science, Hunan City University, Yiyang, Hunan, 413000, China
tulip1907@163.com
[2] Department of Chemical and Environmental Engineering, Hunan City University, Yiyang, Hunan, 413000, China
s-jj-18@163.com

Abstract. The dynamics of the asymmetric continuous-time Hopfield networks are discussed, and the sufficient conditions for the global and local stability of the network are proposed. Furthermore, two system designing methods for endowing the network with retrieval properties are proposed based on the matrix decomposition method, respectively. And the applications of the network in pattern recognitions and information retrieval are also studied by numerical simulations.

Keywords: Hopfield neural network, matrix decomposition, dynamics, eigenvalue.

1 Introduction

In the traditional studying,continuous-time Hopfield network connection weight matrix is symmetric,but biological studies have shown that synaptic connections in the human brain has a natural asymmetry, and it is difficult to achieve an absolute symmetry the Hopfield network with hardware, so the study of dynamic behavior of asymmetric Hopfield network is important.

Memory storage is one of the important applications of Hopfield neural network, analysis of the dynamic behavior of the network, especially the memory recovery characteristics analysis, is the premise of network applications. In this paper we studied dynamic behavior of asymmetric continuous-time Hopfield network,and developed Hopfield network storage.

2 Network Modle

Asymmetric continuous-time Hopfield networks can be described with the followingdifferential equation:

$$\mathrm{X} = -\mathrm{A}X + WF(X) + \Theta \tag{2.1}$$

$X = [x_1, x_2 ... x_n]^T \in R^n$ is the state vector of neurons , $\Theta = [\theta_1, \theta_2 ... \theta_n] \in R^n$ is a Constant vector, $\mathrm{A} = \mathrm{diag}[\mathrm{a}_1, \mathrm{a}_2 ... \mathrm{a}_n] \in \mathrm{R}^{n \times n}$ is a Positive definite diagonal matrix, $W \in R^{n \times n}$ is a asymmetric matrix,it is connection weights among synapses

Y. Zhang (Ed.): Future Communication, Computing, Control and Management, LNEE 141, pp. 565–570.
springerlink.com

of neurnols. $F:R^n \rightarrow R^n$ is a continuous function, and $\mathrm{F}(0)=0, \mathrm{F}(\mathrm{X})=[f_1(x_1), f_2(x_2)...f_n(x_n)]^T$ is a transfer function of neurnols, $f_j(x)$ is differentiabl and meets the following conditions:

$$0 < \frac{f_j(x)-f_j(y)}{x-y} \le k, \forall x, y \in R, j=1,2,...n \qquad (2.2),$$

so function $f_j(\cdot)(j=1,2,...,n)$ is monotonically increasing, and only if x = 0, $f_j(x)=0$ .furthermorr, if function $F(\cdot)$ is a linear transfer function, network (2.1) is a linear system. A linear system can not have the ability of recovering its memory,, so assume that function $F(\cdot)$ is a nonlinear transfer function. And function $\mathrm{X}^{(\mathrm{i})}=[x_1^{(i)}, x_2^{(i)},..., x_n^{(i)}], i=1,2,...,m, m \ge 1$ is the pattern of asymmetric network (2.1) need to be recognized.

In order to make the network have the ability of restoring its memory, a set of given states need to be designed to balance of the network, when the initial state is close to a balance enough, the network can converge to it.

3 System Design

Recognition of binary pattern is one important application of Hopfield network, in order to make the network have the ability of restoring its memory, We need to design parameter A, W and Θrationally, Make function $\mathrm{X}^{(\mathrm{i})}(i=1,2,...,m)$ to be a balance of network (2.1)

For simplicity, We make an assumption in the system design based on matrix decomposition :

$$a_1 = ... = a_n \ge 0 \qquad (2.3)$$

$$f_1(x) = f_2(x) = ... = f_n(x), \forall x \in R \qquad (2.4)$$

$$\left|x_{j_1}^{(i_1)}\right| = \left|x_{j_2}^{(i_2)}\right|, i_1, i_2 = 1,2,...,m, j_1, j_2 = 1,2,...n \qquad (2.5)$$

and $x_j^{(i)}$ is a Component of $x^{(i)}$.

Set $\varsigma^{(i)} = F(x^{(i)})$, $\overline{Y} = [\varsigma^{(1)} - \varsigma^{(m)}, \varsigma^{(2)} - \varsigma^{(m)},..., \varsigma^{(m-1)} - \varsigma^{(m)}]$, make singular value decomposition on $\overline{Y}$,we get a function $\overline{Y} = \hat{V}\Lambda Z^T$, In the function $\hat{V}$ and Z are V orthogonal arraies, $\Lambda \in R^{n\times(m-1)}$ is a diagonal matrix, The following conclusions can be drawed:.

About i = 1,2,...,m, $\mathrm{W}\varsigma^{(i)} + \Theta = \overline{a}\varsigma^{(i)}$, $\overline{a}$ is a real constant, and $W = \overline{a}\hat{V}L\hat{V}^T$ (2.6)

$$\Theta = \overline{a}\varsigma^{(m)} - W\varsigma^{(m)} \quad (2.7)$$

In these equations $L = [l_{ij}] \in R^{n \times n}$ is a under diagonal matrix,its elements l_{ij} are defined as follows:

$$l_{ij} = \begin{cases} 1, \forall i = j, i \le m-1 \\ \tau_i, \forall i = j, i > m-1 \\ rand(\gamma_{ij}, \hat{\gamma}_{ij}), \forall i = j, i \ge m, j \ge m \\ 0, other \end{cases} \quad (2.8)$$

$\gamma_{ij}, \hat{\gamma}_{ij}, \tau_i \in R, rand(\gamma_{ij}, \hat{\gamma}_{ij})$ are random numbers in area $(\gamma_{ij}, \hat{\gamma}_{ij})$.

The following is designing method of continuous-time Hopfield network based on matrix decomposition:

Step 1) calculate $\overline{Y} = [\varsigma^{(1)} - \varsigma^{(m)}, \varsigma^{(2)} - \varsigma^{(m)}, \ldots, \varsigma^{(m-1)} - \varsigma^{(m)}]$

Step 2) make singular value decomposition on $\overline{Y}$, we got $\overline{Y} = \hat{V}\Lambda Z^T$

Step 3) set under diagonal matrix L, in the matrix $\tau_i \le 0$

Step 4) alculate $\overline{a} = a_1 x_1^{(1)} / f_1 x_1^{(1)}$

Step 5) alculate $W = \overline{a}\hat{V}L\hat{V}^T, \Theta = \overline{a}\varsigma^{(m)} - W\varsigma^{(m)}$

Because $W = \overline{a}\hat{V}L\hat{V}^T = \hat{V}\overline{a}L\hat{V}^T$ •can be reguard as Schur decomposition of matrix W, $\overline{a}l_{jj} (j = 1,2,\ldots,n)$ is eigenvalues of the matrix W, obviously, by adjusting the parameter $\overline{a}$ and parameter l_{jj} ,the eigenvalue of matrix W can be adjusted easily.

The algorithm shows, $\mathrm{WF}(\mathrm{x}^{(i)} + \Theta = \overline{a}F(x^{(i)}), i = 1,2,\ldots,m)$

Moreover, Equation (2.3), (2.4) and (2.5) shows $\overline{a} = \dfrac{a_i x_j^{(i)}}{f_j(x_j^{(i)})}, 1 \le i \le m, 1 \le j \le n$,

so $\mathrm{WF}(\mathrm{x}^{(i)} + \Theta = \dfrac{a_i x_j^{(i)}}{f_j(x_j^{(i)})} F(x^{(i)}) = AX^{(i)}$

And we can set vector $X^{(i)} (i = 1,2,\ldots,m)$ as a balance of network (2.1). In system designing,set $\tau_i \le 0$, so $\lambda_1(W) = \overline{a}$, and $\lambda_1(\cdot)$ is the maximum eigenvalue of a matrix.

From step 2 perform the designing, set $\overline{Y}=[\varsigma^{(1)},\varsigma^{(2)},...,\varsigma^{(m)}]$, an asymmetric network $\Theta=[0,0,...,0]^T$ can be designed, and a symmetrical network can be designed by setting $l_{ij}=0,(i>j,i\geq m,j\geq m)$.

4 Simulation of the System

Assume that there are four models need to be identified:

$$X_{(1)}=[-0.8,0.8,0.8,0.8,0.8,-0.8,\ -0.8,0.8]^{\mathrm{T}}$$

$$X_{(2)}=[0.8,-0.8,0.8,-0.8,0.8,\ 0.8,\ 0.8,-0.8]^{\mathrm{T}}$$

$$X_{(3)}=[0.8,0.8,-0.8,-0.8,0.8,\ 0.8,\ -0.8,-0.8]^{\mathrm{T}}$$

$$X_{(4)}=[-0.8,-0.8,0.8,-0.8,-0.8,\ 0.8,\ 0.8,-0.8]^{\mathrm{T}}$$

So $\hat{S}=\{X^{(1)},X^{(2)},X^{(3)},X^{(4)}\}$, Set the parameters of this function a1=...=a8=1, $f_1(\cdot)=...=f_8(\cdot)=f_8(\cdot)$

According to system designing based on matrix decomposition , results can be calculated:

$$\overline{a}=\frac{a_1x_1^{(1)}}{f_1(x_1^{(1)})}=1.205,$$

$$\overline{Y}=\begin{vmatrix} 0 & 1.33 & 1.33 \\ 1.33 & 0 & 1.33 \\ 0 & 0 & -1.33 \\ 1.33 & 0 & 0 \\ 1.33 & 1.33 & 1.33 \\ -1.33 & 0 & 0 \\ -1.33 & 0 & -1.33 \\ 1.33 & 0 & 0 \end{vmatrix}$$

Set $\tau_{\mathrm{i}}=0(3<i\leq 8),\gamma_{ij}=-0.5,\overline{\gamma}_{ij}=0.5$, variable $l_{ij}(i>j,i\geq 4,j\geq 4)$ generated randomly in area (-0.5,0.5), W and Θ can be calculated from function (2.6) and (2.7) . Specific results are shown below:

Θ= [0.32, 0.07, 0.17, 0.36, 0.32, 0.16, 0.24, 0.03]T.

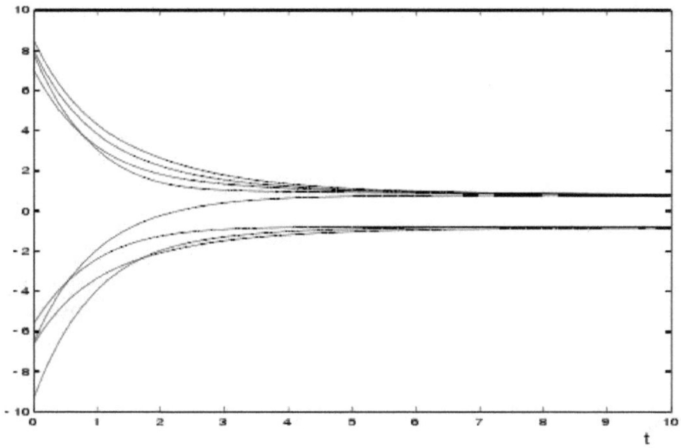

Fig. 1. Network state trajectory of initial state X1 (0)

By calculating these results is easy to get $\lambda_1(H - A\Phi(X^{(i)}))= -1.58 < 0, (i = 1,2,3,4)$, This result shows , $X^{(i)}, (i = 1,2,3,4)$ has been designed as local balance.

Given an initial state X1(0) = [7.0, 6.6, 8.5, 5.6, 8.0, 6.5, 7.8, 9.3]T, Network will converge eventually to X(2). Figure 1. shows the operating status on time of the network.

There are balance points that don't belongs to the set $\hat{S}$ in the network, , This balance is called pseudo-state. The pseudo-state will affect performance of the network .

Set $X_s^1 = [0,\ 0.8, 0.8,\ 0.8, 0, 0.8, 0.8,\ 0.8]^{\mathrm{T}}$, it is easy to prove that X_s^1 is a balance point. Given an initial state $X_2(0) = [0.3,\ 0.6, 1,\ 0.7,\ 0.3, 0.7, 0.7,\ 0.9]^{\mathrm{T}}$,network will converge eventually to X_s^1, so X_s^1 is a pseudo-state.

Network (2.1) is a non-linear coupling network, it is difficult to calculate all of the balance. But if the network's initial state is close to a pseudo-state enough, network will converge to the pseudo-state. The balance point of network (2.1) exists only in the area of a certain limited size of the origin. so all the pseudo-states can be found through the initial point. In collection $\mathrm{B}_s = \{X : -4 \le x_i \le 4, i = 1,2,...,8\}$ 7140 uniformly distributed vectors are provided to the network as the initial state, All balance are in collection Bs（Memory model, the unstable equilibrium point and the pseudo-state） .

Only three pseudo-states were found by numerical simulation, there were

$$X_s^1 = [0,\ -0.8, 0.8, -0.8, 0, 0.8, 0.8,\ -0.8]^{\mathrm{T}},$$

$$X_s^2 = [0,\ 0.8, -0.8, -0.8, 0, 0.8, -0.8,\ -0.8]^{\mathrm{T}}$$

$$X_s^3 = [-0.8,\ 0.8, -0.8,\ -0.8, -0.8, 0.8, -0.8, -0.8]^T$$

It is easy to prove that :

$$X_s^1 = (X^{(2)} + X^{(4)})/2,$$

$$X_s^2 = -X^{(2)} + X^{(3)})/2 + X^{(4)}/2,$$

$$X_s^2 = -X^{(2)} + X^{(3)} + X^{(4)}$$

So all the pseudo-state can be written inlinear combination of vectorX(1), X(2), X(3) and X(4).There are a large number of initial state, therefore, all the pseudo-states were found.

5 Conclusion

In this study, there are a minimum number of pseudo-states in the asymmetric Hopfield networks. In method based on the matrix decomposition, binary mode can easily be stored as locally balance point of network (2.1) ,however, the algorithm can not process non-binary problems.

References

1. Cheng, X.Z., Dasgupta, C., Singh, M.P.: Retrieval properties of a Hopfieldmodel with random asymmetric interactions. Neural Comput. 12, 865–880 (2000)
2. Jin, T., Zhao, H.: Pattern recognition using asymmetric attractor neural networks. Phys. Rev. E 72, 066111 (2005)
3. Zhang, H.G., Wang, Z.S., Liu, D.R.: Global asymptotic stability and robust stability of a class of Cohen-Grossberg neural networks with mixed delays. IEEE Trans. Circuits Syst. 56(3), 616–629 (2009)
4. Wang, Z.S., Zhang, H.G., Yu, W.: Robust stability of Cohen-Grossberg neural networks via state transmission matrix. IEEE Trans. Neural Networks 20(1), 169–174 (2009)
5. Zuo, Z., Yang, C., Wang, Y.: A new method for stability analysis of recurrent neural networks with interval time-varying delay. IEEE Trans. Neural Networks 21(2), 339–344 (2010)
6. Zhao, H.: Designing asymmetric neural networks with associative memory. Physical Rev. E 70, 066137 (2004)
7. Krauth, W., Mezard, M.: Learning algorithms with optimal stability in neural networks. Journal of Physics A: Mathematical and General 20(11), L745–L752 (1987)

An Economic Model for the Evaluation of the Economic Value of Cloud Computing Federation

Zehua Zhang and Xuejie Zhang

School of Information Science and Engineering, Yunnan University
North Cuihu Road2, 650091 Kunming, P.R. China
zehuazh@ynu.edu.cn email, xjzhang@ynu.edu.cn

Abstract. With the widely adoption and further study of cloud computing technologies, there is an imperative demand to incorporate the resources from different cloud computing platforms to provide service for the users. Cloud Computing Federation is a computing mechanism which can federate the resources from different Cloud Computing Service Providers to provide a uniform resource interface for the user. We think that CCF is the only way to get the cloud computing be widely used and realize the greatest value of it, in order to manifest this fact, we propose an economic model for the valuation of the Cloud Computing Federation, and a Cloud Computing Federation architecture based on P2P and market is also given in this paper. Our work provides a valuable reference for the realization of Cloud Computing Federation and the design of the cloud computing standards.

Keywords: Cloud Computing federation, Economic Model, Economic Value, Evaluation, Architecture.

1 Introduction

Cloud computing is an emerging new distributed computing paradigm with a great hope that it will provide computing as the 5th utility (after water, electricity, gas, and telephony) to meet the everyday needs of the general community [1]. By the use of the automate systems management, load balancing, and virtualization technologies, cloud computing intend to integrate distributed resources into an infinite computing resource pool. Provide the computational power and data storage facilities to users, and hidden the technical detail at the same time. Cloud computing is considered has the advantages including infinite scalability, lower the TCO (Total Cost of Ownership), reduce the upfront investment and risk for the user, system automation and promote the realization of green IT.

With the widely adoption and further study of cloud computing technologies, there is an imperative demand to incorporate the resources from different cloud computing platforms to provide service for the users. The conception of Cloud Computing Federation (CCF) is proposed by many researchers from industry and academe [2,3,4]; Cloud Computing Federation is a computing mechanism which can federate the resources from different CCSPs (Cloud Computing Service Provider) to provide a

Y. Zhang (Ed.): Future Communication, Computing, Control and Management, LNEE 141, pp. 571–577.
springerlink.com

uniform resource interface for the user. Cloud Computing Federation is also called as Inter-cloud in some cases [5,6].

Cloud computing announced that it had transferred the user's upfront investment risk to the CCSPs, and the users needn't worry about that how much IT facilities should they invest at the beginning of their projects. Actually, the CCSPs now face the problem that how much cloud computing facilities should they invest to meet the peak requirements of the users. We think that CCF is the only way to get the cloud computing be widely used and realize the greatest value of it, since the main value of cloud computing is brought by the Scale Economy Effect which is caused by the share and reuse of resources by lots of users. In order to manifest this fact, we propose an economic model for the valuation of the Cloud Computing Federation, and a Cloud Computing Federation architecture based on P2P and market is also given in this paper. Our work provides a valuable reference for the realization of Cloud Computing Federation and the design of the cloud computing standards.

The rest of this paper is organized as follows: Section 2 analyses the advantages of Cloud Computing Federation. Section 3 introduces an economic model for the evaluation of the economic value of Cloud Computing Federation. Section 4 presents a P2P and market based Cloud Computing Federation architecture. Section 5 describes a prototype of the Cloud Computing Federation with performance comparison between insulated cloud computing platforms and Cloud Computing Federation. Finally, a conclusion of the work is discussed in Section 6.

2 Advantages of Cloud Computing Federation

Many problems of cloud computing today are attributed to the lack of portability (user's application can spans over multiple CCSPs) and interoperability (user's application can deploy on multiple CCSPs), with reference to some viewpoints on cloud computing in the white paper of University of California, Berkeley [7], we analyze the problems in cloud computing and the solutions to these problems in Cloud Computing Federation.

Limited scalability. Most of the CCSPs claim that they provide infinite scalability for the user, in fact, with the rapid increase of the users and the widely adoption of cloud computing technologies, none of the CCSPs can meet all the resource requirements of all the users, only a Cloud Computing Federation can provide a real infinite computing resources pool for all the users.

Unreliable availability. Actually, shutdown events have happened recently with most of the main CCSPs, including Amazon, Google and Microsoft. If a user uses only one CCSP's specific service, once the CCSP breakdown, the user's service becomes unusable. In Cloud Computing Federation, if one CCSP's service is shutdown, the user can use other CCSP's service, which offers a good availability of service.

Service provider lock-in. Currently, if you use one CCSP's specific service, it's impossible for you to move your data and application from one CCSP to another, you are locked to one CCSP. A Cloud Computing Federation will democratize the whole cloud computing market.

Unable to deployment service over multiple CCSPs. At the present time, Applications in cloud computing can't scale over multiple CCSPs since there is no interoperability between CCSPs. Cloud Computing Federation make it possible for an provider to integrate different CCSP's service to provide more valuable service for the user.

High TCO (Total Cost of Ownership). The main value of cloud computing is brought by the scale economy which caused by the share and reuse of resources by lots of users, but now, the CCSPs have to prepare extra expensive resource for the peak request from the user, driven by cost, Cloud Computing Federation is a better choice for the CCSPs to provide a high cost performance service for the user.

Cloud Computing Federation is one of the main trench of cloud computing, the above-mentioned analyses indicates that only the Cloud Computing Federation can realize the goal to provide computing as the 5th utility.

3 An Economic Model for the Evaluation of the Economic Value of Cloud Computing Federation

Aims to compare the economic difference between Cloud Computing and Cloud Computing Federation, we propose the following economic model for the evaluation of the economic value of these two computing mechanisms.

Here are some parameters will be used in the model:

NP: the CCSP's net profit

C: the resource capacity that a cloud computing platform can provide

D: the actual average resource demand of the user

α: the price for the CCSP to sell one computing unit

β: the price for the CCSP to buy one computing unit

γ: the cost for the CCSP to provide one computing unit (such as the power and the human resources, etc.)

Then, the net profit is computed as:

$$NP=\alpha D-\beta C-\gamma D. \tag{1}$$

i.e. ,

$$NP=(\alpha-\gamma)D-\beta C. \tag{2}$$

In an actual cloud computing market, when the CCSP continuously increase the resource capacity C, the user's average resource demand D will not linearly increase with C. We estimate D as:

$$D=C1/k. \tag{3}$$

Where k>1. So we can get:

$$NP=(\alpha-\gamma)C1/k-\beta C. \tag{4}$$

$$\frac{\partial NP}{\partial C}=(\alpha-\gamma)k\text{-}1C1/k\text{-}1-\beta. \tag{5}$$

When k→∞,

$$\frac{\partial NP}{\partial C}=-\beta. \tag{6}$$

From equation 5 and 6, we can see that with the continuously increase of the resource capacity C, the increase of the average resource demand D will be less compare with C; the net profit NP will decrease, and even become a negative value at last.

To solve this problem, the only way is to federate different cloud computing platform's resources, thus form a huge resource pool, and provide high quality service for the user with low cost.

In Cloud Computing Federation, the CCSP needn't to prepare excessive cloud computing facilities to meet the peak requirements of the users. In case of the peak requirements, a CCSP can lease resources from other CCSPs, the resource capacity C will grow linearly with the average resource demand D, we can consider: D≈C, then equation 2 is simplified as:

$$NP \approx (\alpha-\gamma-\beta)C = hC. \quad (7)$$

Where $h=(\alpha-\gamma-\beta)$, h is the earning coefficient. There is a linear relation between net profit NP and resource capacity C, this means that the CCSP needn't to prepare excessive resource for the peak requirements, and the net profit NP can grow linearly with the average resource demand D.

From the economic point of view, Cloud Computing Federation can bring greater benefits to CCSP than cloud computing, and provides more cost-effective cloud computing services for users. Economic studies have shown that if there are no other barriers, then more economical technology will inevitably replace the noneconomic technology. Apparently now Cloud Computing Federation is more economical than cloud computing.

4 A P2P and Market Based Architecture for Cloud Computing Federation

The goal of Cloud Computing Federation is to ingrate multiple cloud computing platforms' resource into a huge resource pool, and provides computing as a kind of utility. To reach this goal, two problems need to be solved, the first one is the granularity in the partition of the whole resource pool, and the second one is that how the components of the Cloud Computing Federation interact with each other to realize collaboration under certain granularity, these two problems are issues of the architecture of the Cloud Computing Federation. We propose a P2P and market based architecture for Cloud Computing Federation here.

4.1 The Resource Management Granularity of Cloud Computing Federation

We think cloud computing management region (abbreviated as CCMR) is a reasonable resource management granularity for Cloud Computing Federation's resource management. CCSP's data centers are often located in different areas (such as Amazon's EC2 elastic computing cloud instances locate in different domains, and a domain is divided into multiple regions), and these data centers are interconnected to form a cloud computing platform. A CCMR is partitioned geographically or by the management strategies, it can be a data center of a CCSP, or there may be many CCMRs in a data center.

Cloud Computing Federation is a very huge system which has numerous of computing resources and data resources, it's impossible to management the resources centrally. If adopt the distributed resource sharing and collaboration mechanisms, it's unreasonable to define the resource management granularity as small as a virtual machine or as big as a cloud computing platform (i.e. a CCSP). In the former case, the particle size is too small, which will inevitably lead to significant network traffic load, and taking up a lot of additional computing resources to achieve resource sharing and collaboration. In the latter case, because most of the CCSP's data and data flow converge to the facilities which take charge of the resource sharing and collaboration, thus undoubtedly become a bottleneck in Cloud Computing Federation, and decreases the efficiency of resource allocation to a very low level.

At the same time, considering the CCMR (such as a data center) can provide cloud computing services in a self-autonomy way, and the probability will be high for a CCMR to has a more close distance with other CCMRs (may belong to other CCSPs), thus has more advantages to carry out resource sharing and collaboration with low cost. Therefore, choose CCMR as the resource management granularity for Cloud Computing Federation is a reasonable.

4.2 A P2P and Market Based Architecture for Cloud Computing Federation

The resources in a Cloud Computing Federation is huge and the states of the resources change very quickly, so if the CCMRs can only collaborate with each other through the specific facilities in the CCSPs, the efficiency of resource allocation will be greatly reduced, and these facilities will become the bottleneck of the whole Cloud Computing Federation, and greatly weaken the whole system's flexibility and extensibility. So, aims to get a high efficiency in resource allocation, let the CCMRs interact with each other directly in a P2P way is reasonable. Moreover, take into account that the market mechanism is the best way for large scale resource allocation, we propose a p2p and market based architecture for Cloud Computing Federation, this architecture is depictured as in fig .1.

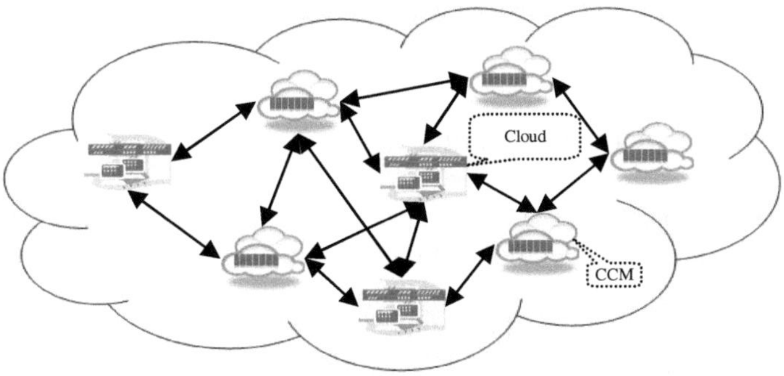

Fig. 1. A P2P and Market Based Architecture for Cloud Computing Federation

In this architecture, the CCMR and the Cloud Computing Market (like a E-commerce market, the CCMR can register its resources to this market for sell, or buy resources from this market) are all interconnected to form a overlay network, and interact with each other in a P2P mechanism to trade the resources, once a CCMR meets the peak resource requirements from the user, and its resources can't satisfy these requirements, then, it can buy resources from other CCMRs or from the Cloud Computing Markets. Contrarily, when a CCMR has idle resources, it can sell these resources to the market or to other CCMRs. Therefore, this architecture integrates resources from different CCSP to realize a P2P and market based Cloud Computing Federation.

In order to enable resource sharing and collaboration among CCSPs, there must be large amounts of data exchange between the CCSPs, let the CCMRs interact with each other in a P2P and market based mechanism, a large proportion of data exchange will be done locally or closely, thus brings the advantages including low latency, high bandwidth, and low cost. This is similar with the bottleneck problem between different ISPs (Internet Service Provider), which is greatly improved after the addition of connections among the local network of different ISPs.

5 Performance Analysis

We construct a prototype call PMBACCF1.0 in a LAN to compare the average user satisfaction and the average utilization ratio under the CCF and non-CCF situation. There are 5 PCs (CPU: Intel Pentium4 530 3.0G, RAM: 2G, hard disk: 250G, OS: Widows XP Professional) in the LAN, we setup two virtual machines on each computers by VMware Workstation 9.

Initially, 1,000 tasks with random size are generated and distributed evenly in number on 10 nodes, each node represents a CCMR which has a random Execute Capability EC=0~220 tasks/iteration, after each iteration step, another 1,000 tasks are added randomly to these ten CCMRs. In CCF, all CCMRs share its resources with other CCMRs to deal with the peak requirements, contrarily, in non-CCF, the CCMR deal with all the requirements deliver to it by itself. After 5,000 iteration steps, the experiment results manifests that compare with non-CCF, the average user satisfaction and the average utilization ratio of CCF is considerably enhanced 61.5% and 53.1%, respectively.

6 Conclusion and Future Work

In order to manifest the value of CCF, we propose an economic model for the valuation of the Cloud Computing Federation, and a Cloud Computing Federation architecture based on P2P and market is also given in this paper.

In the Future work, based on this architecture, we will concentrate on the resource management mechanism in this large P2P overlay network, and the standard of cloud computing. These two problems will have fundamental influence on the construction of Cloud Computing Federation in the future.

Acknowledgments. This work is supported by the National Natural Science Foundation of China (NSFC) (No.60573104) and the Natural Science Foundation of Education Department of Yunnan Province (No.09Y0050).

References

1. Buyya, R., Yeo, C.S., Venugopal, S., Broberg, J., Brandic, I.: in: Cloud computing and emerging IT platforms: Vision, hype, and reality for delivering computing as the 5th utility. Future Generation Computer Systems 25, 599–616 (2009)
2. Rochwerger, B., Breitgand, D., Levy, E., Galis, A., Nagin, K., Llorente, I., et al.: in: The Reservoir model and architecture for open federated cloud computing. IBM Journal of Research and Development 53 (2009)
3. Ranjan, R., Buyya, R.: in: Decentralized Overlay for Federation of Enterprise Clouds (2010), Information on, `http://arxiv.org/ftp/arxiv/papers/0811/0811.2563.pdf`
4. Rochwerger, B., Breitgand, D., Levy, E., Galis, A., Nagin, K., Llorente, I., et al.: The Reservoir model and architecture for open federated cloud computing. IBM Journal of Research and Development 53, 1–11 (2009)
5. Buyya, R., Chee, Y.S., Venugopal, S.: in: Market-Oriented Cloud Computing: Vision, Hype, and Reality for Delivering IT Services as Computing Utilities. In: Proc. of the HPCC, pp. 5–13 (2008)
6. Use Cases and Functional Requirements for Inter-Cloud Computing (2010), Information on, `http://www.gictf.jp/doc/GICTF_Whitepaper_20100809.pdf`
7. A Berkeley View of Cloud Computing (2010), Information on, `http://www.eecs.berkeley.edu/Pubs/TechRpts/2009/EECS-2009-28.pdf`

Color Calibration and Color Management in Digital Camera

Songhua He[1,2,a], Zhen Liu[1,b], and Lijie Wang[2,c]

[1] Institute of light Science and technology, NanJing Forestry University, NanJing, 10037, China
[2] Institute of Media and Communication, Shenzhen Polytechnic, Shenzhen, 518055, China
[a] hdh1818@szpt.edu.cn, [b] cehuiliu@163.net, [c] wlj6205@szpt.edu.cn

Abstract. Digital camera is being widely used in people's day life. In this paper, the corresponding relationship between device independent color and device depend color is discussed, the making and use of digital camera's are analyzed, and color calibration and color management under manual white balance are studied.

Keywords: color correction, color management, white balance.

1 Introduction

Along with rapid development of the high-tech industry in the domestic and foreign countries, the digital camera has come to people's day life, while packing industry and printing industry are using manuscripts from the digital camera. Because of the more and more highly increasing requirement of image color, whether the digital camera can meet the standard of the packing and printing industry has become a very important task. In fact the digital camera has produced lots of color tolerances when you take photos, the same object shows different colors under different shooting conditions. But the industry has a very high requirement for the color, the same product should present the real color when shoot under different conditions; The same product shoot by digital carve of different camera should provide save result, etc. Otherwise, it will cause consumer misunderstanding and unnecessary loss, so it highly demanded that the digital camera can perform the color correction and color management.

2 The Color Profile of Imaging Equipment

In the work flow of color management, the color profile is the key content, its generation, storage, selection and transfer are necessary in the color management.

When choosing L*a*b as the color space which is independent to the device, characteristic of device color can be presented as: the corresponding relationship between device independent color and device depend color [1,2].For different devices, the relationship shows different characteries.

Y. Zhang (Ed.): Future Communication, Computing, Control and Management, LNEE 141, pp. 579–585.
springerlink.com

Digital camera: the color of the captured will be transfered to red, green and blue signals (RGB), which in turn are related to photographed object. From this relation and the three signal values, we can computed actual the color of the shot object.

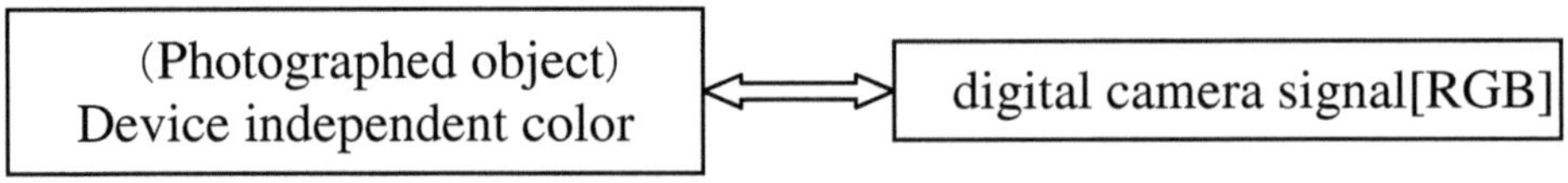

Monitor: From the three RGB signal output, the color be displayed on the screen, the color on the screen and the three RGB signal also has a certain relationship, according to which, we can revert the requested color to red, green and blue signals.

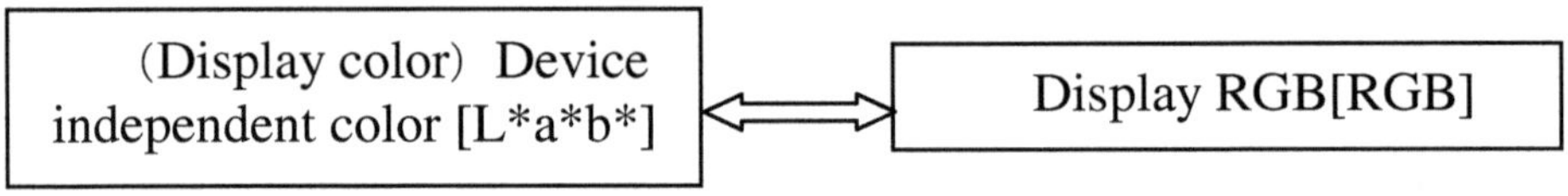

Printing: After drying, proofing and printing, the dot area rate[C%,M%,Y%,K%] in the films will get the color printing proof. The color on the prints has a certain data relation with the dot area. We can revert process color to the corresponding dot area[1,3].

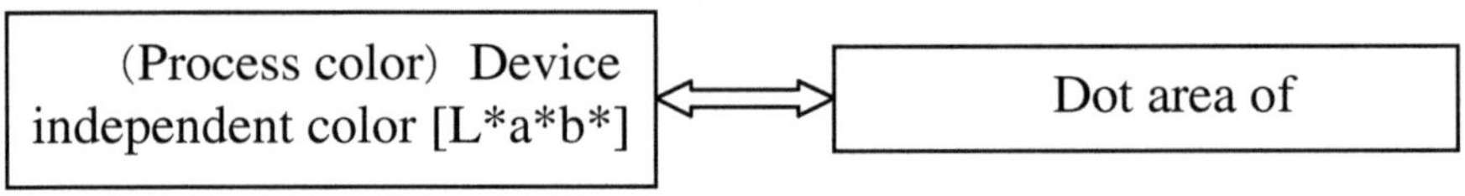

Actually, corresponding relation mentioned above is the relationship between device color and CIE1976 L*a*b. This is necessary when the color match is concerned. It needs to be stored in the color management system as a color profile description document, which is called "profile".

3 The Making and Use of Digital Camera's Profile

\The making of digital camera's profile is similar to other imaging input device such as scanner, and no matter what special software used, the methods the profiles are generated are similar. The specific process is first to take a photo of a standard color chart. The most common chart is GretagMacbeth ColorChecker DC, which used in this experiment. The size of the chart is: 21.59cm x 31.56cm with a length and width ratio of 2:3. The chart is composed of 237 small color patches which arranged as a 12 x 20 matrix. In the middle of the chart is a 2.8cm x 2.8cm white patch, which is usually used to do manual white balance. The surrounding of the white patch are 177 color patches of size 1.3cm x 1.3cm, which reflect the color gamut of the digital camera. All of the patches listed are representatives of color of natural world such as skin color, green color of leaves, blue color of sky and water, etc. On the right hand side of the chart is a column of patches with high brightness, which are used to enlarge the color gamut of the digital camera. In order to reflect the uniform of the light source, the edge of the chart is a set of white, gray and black patches. The four corners of the chart are marked with cross lines in order to accurately define the location in the ProfileMaker. All the colors in the chart are from the Munsell color system.

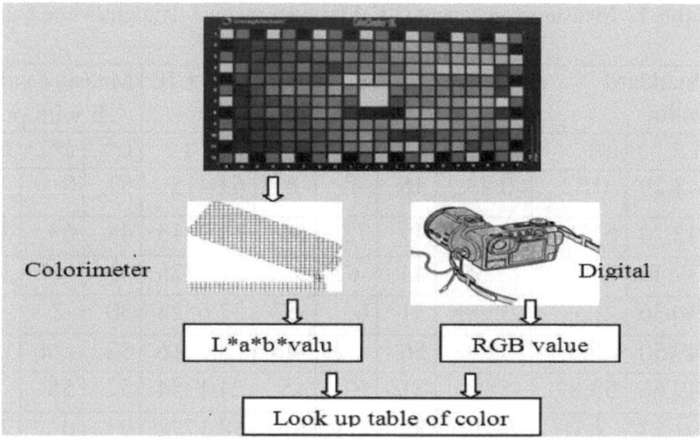

Fig. 1. The generation and use of digital camera's color profile

The process of making profile is showed in Figure 1. The L*a*b* value of each color patch is first measured using a calibrated spectrophotometer, and then the chart's L*a*b* table is generated referring to manufacture table provided. This parameter table is usually provided by manufacture, but as time going, there is a difference between the provided and actual data, thus it is necessary to measure the L*a*b* values using a spectrophotometer. When making digital camera's profile under a certain condition, we should take am image of the whole chart and then measure the RGB values of each color patch. Therefore, a lookup table can be generated for the transformation between RGB and L*a*b*, which is used to image a single point from camera's RGB space to L*a*b* space. This is the basic theory and structure of the digital camera's profile.

4 The Analysis of Calibration Using Camera Profile

It is necessary to analyze a new generated profile to see if the color of an image taken by the digital camera can match to the actual color transformed by the lookup table.

Firstly, within the GretagMacbeth ColorCheckerDC, pick up 9 usual color patches Gray(I7), R(D2), G(D7), B(D10),C(H9), M(G11), Y(M5), White(J6), Black(L8), and measure their CIE L*a*b* values as a standard reference. Secondly, under illuminant condition of daylight and fluorescent lamp respectively, switch the white balance mode to manual setting, then take a picture of GretagMacbeth ColorCheckerDC chart, and open the two images with Photoshop (color space should be set Adobe RGB). A set of L*a*b values will be obtained by using sucker tool, and after using Image\Mode\Assign profile command, the profile can be generated under the corresponding condition. The same process can be applied to other data sets.Then using color difference formula(Eq. 1) to calculate color difference values, table 1 and table 2 can be made for comparing.

$$\Delta E_{ab} = [(\Delta L^*)^2 + (\Delta a^*)^2 + (\Delta b^*)^2]^{\frac{1}{2}} \tag{1}$$

Table 1. Measure values and CIE ΔE under manual daylight/white balance

Color patches	Standard reference value			Measure values and CIE △E without profile				Measure values and CIE △E with profile			
	L*	a*	b*	L*	a*	b*	△E 1	L*	a*	b*	△E 2
Gray	48.29	-0.52	-0.43	46	-1	-6	6.0415	52	0	-1	3.7894
R	42.70	61.39	29.18	49	73	64	37.2413	48	63	30	5.5995
G	39.44	-41.60	12.35	44	-62	9	21.1725	41	-44	18	6.3248
B	30.36	-2.37	-35.96	31	6	-57	22.6528	30	-2	-41	5.0664
C	49.50	-25.67	-32.88	56	-32	-46	15.9516	52	-24	-36	4.3328
M	49.66	53.89	-5.22	51	67	15	24.1354	52	55	-2	4.1323
Y	91.17	-2.51	74.20	82	-11	61	18.1772	93	0	71	3.8064
White	95.13	-1.42	1.38	85	-2	-3	11.0516	96	2	1	3.5493
Black	20.83	-0.12	-0.67	10	2	-7	12.7221	23	1	-1	2.4697

Table 2. Measure values and CIE △E under manual daylight lamp/white balance

Color patches	Standard reference value			Measure values and CIE △E without profile				Measure values and CIE △E with profile			
	L*	a*	b*	L*	a*	B*	ΔE 1	L*	a*	b*	△E 2
Gray	48.29	-0.52	-0.43	55	0	-2	6.9108	52	0	-1	3.7894
R	42.70	61.39	29.18	52	77	58	34.0698	43	68	31	6.8625
G	39.44	-41.60	12.35	50	-74	29	37.9275	40	-42	9	3.42
B	30.36	-2.37	-35.96	34	-11	-44	12.3437	30	-2	-41	5.0664
C	49.50	-25.67	-32.88	63	-56	-19	35.9666	54	-23	-32	5.3059
M	49.66	53.89	-5.22	58	66	-3	14.8706	54	51	-6	5.2777
Y	91.17	-2.51	74.20	91	-13	68	12.1864	93	-1	74	2.4554
White	95.13	-1.42	1.38	92	0	1	3.5488	98	1	0	3.9997
Black	20.83	-0.12	-0.67	17	2	-7	7.6336	23	0	-1	2.2742

After comparing the color difference △E*ab using and without using profile, we can get the following figures(fig.2 and fig.3).

From the tables and figures shown above, it can be found that the image captured with manual setting white balance and then calibrated using the profile shows a better result with small color differences, most of which are less than 6. By comparing the image displayed on the screen and the real object, we can see that the pair is visually matched. Thus color calibration is achieved.

Why using digital camera's profile can achieve color calibration? Because the profile connects camera RGB values and device independent values[L*a*b*]. The colors of captured objects give responses as RGB signals, which have a certain relationship with the color of object. Using this relationship and the three signals, the captured object's color can be calculated and the color can be calibrated automatically.

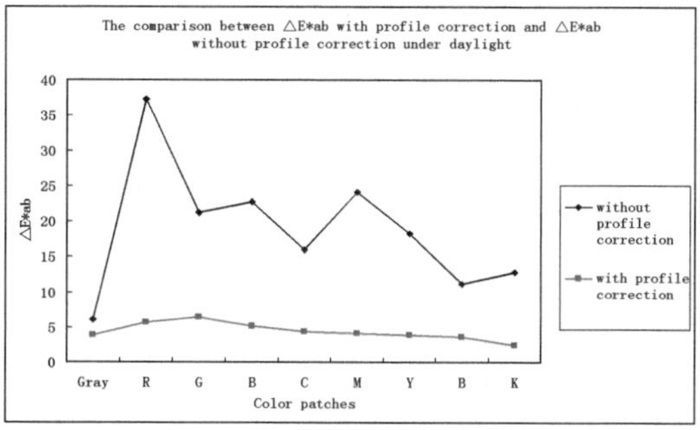

Fig. 2. The comparison between △E*ab with profile correction and △E*ab without profile correction under daylight

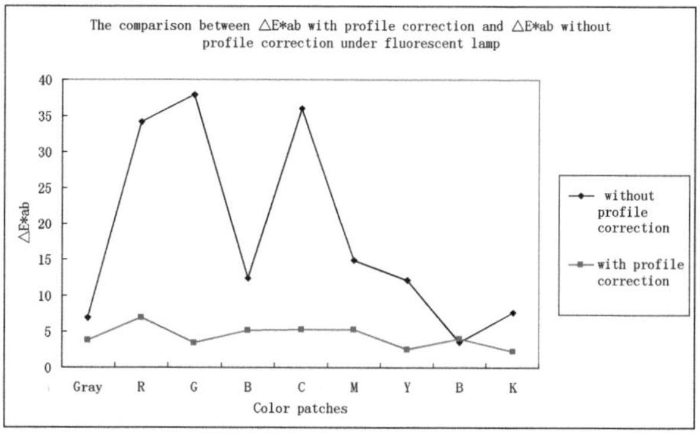

Fig. 3. The comparison between △E*ab with profile correction and △E*ab without profile correction under fluorescent lamp.

5 Color Management System Based on Profiles

From the analysis above, we can see that the device's profile established a relationship between a certain device's color appearance and CIE 1976 L*a*b* values[5]. Since L*a*b* is device independent, as soon as L*a*b* is defined, the corresponding color will be decided. Different devices provide the same L*a*b* will show the same color. So by using device profile, we can consider CIE L*a*b* to be a bridge connecting various devices to perform color management[6].

There are 4 common device profiles used in color management system: Scanner Profile, Camera Profile, Monitor Profile and Printer Profile. In this experiment, we made the profiles for digital camera, display, and Epson 7000. via these files, we can analyze each device's characteristic, Figure 4 shows the gamut of the above three devices:

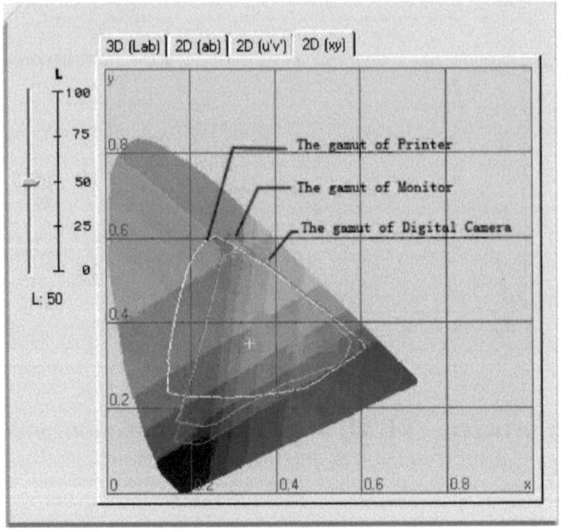

Fig. 4. The color gamut of instruments

It is obvious from the above figure that the gamut of display is fairly the same as the gamut of digital camera. While the reproduction of printer is worse than the display and the digital camera: most color of images captured by digital camera can be reproduced by display, but the highly saturated blue can not be printed accurately by the printer. Actually the blue outside the gamut of printer has to be transformed by a certain algorithm before being reproduced by the printer. Thus when the different color gamut is not matched with each other, color management system can only map the most color within the gamut, and the color outside the gamut needs to be replaced by other colors. The most common color gamut mapping algorithms are divided in four types: perceptual, saturation, relative colorimetric and absolute colorimetric [7].

From the previous analysis, we can conclude that considering the involvement of color management system, the necessary requirement before converting the color captured by digital camera is: the certain profiles of camera, display and printer are obtained and selected.

Supposing the colors of all captured objects are normal, images are captured by digital camera, whose Profiler provides the transformation from camera RGB to Lab, then the whole system can calculate Lab of object color accordingly.

The image captured then needs to be displayed on color screens. Because the system already has the converting information between Lab and display RGB, instead of using RGB values of digital camera directly, the Lab calculated by previous step should be converting to display RGB, which is supposed to accurately represent the color of object captured, thus ensure the color of display to match that of the captured object.

The operator will then process color separation on the display. Color separation will result in the accurate dot area of CMYK according to the color transformation information of the printer Profile. The reproduction of printer is finally obtained by the procedure of imposition, RIP processing, output recording, plate burning and proofing. The printing reproductions are matched with the captured objects, which achieves color consistency in the whole process. This experiment fulfilled the color management system based on characterization profiles. The results show that the captured image gives the same appearance on diffident displays, printers and displays.

6 Conclusions

Although using manual white balance can greatly improve the automation calibration function of digital camera, the color differences between the captured images and real objects are very large. Because the requirement of color reproduction in the industry is quite high, which needs color difference be less than 6 or even smaller, further color calibration needs to been done in addition to white balance process. And the good results will be obtained by make characterization profiles for digital cameras. Because the light sources of digital camera are various, different camera profiles should be made under different conditions. So for the industry productions which have relatively stable light source and require highly accurate color reproduction, color management system and color calibration based Profile will contribute a lot.

References

1. Yang, J., Song, Y.-H.: The theory and methods of color management, vol. Printing Technology, Beijing (2), 15–17 (2000)
2. Hu, C.-W., Wan, X.-X.: Color management in the image system. Packaging Engineering, Chongqin, China (5), 27–29 (2004)
3. Hu, T.: Color management system based on ICC criterions. Printing Technology, Beijing, China (2), 13–15 (2002)
4. Xu, D., Meng, S.-Y., Shi, J.-Y.: The Research of Color management based on ICC criterions. Soft Transaction, Beijing (10), 740–746 (1998)
5. Tang, S.-Q.: The chromatology. Publishing company of Beijing Institute of Technology, Beijing Institute of Technology (1988)
6. Zhang, X.-M.: A Spatial Extension to CIELAB for Digital Color Image Reproduction. Society of Information Display Sumposium Technical Degest, Beijing, China (27), 731–734 (1996)
7. Zhang, X.-M., Wandell, B.A.: Color Image Fidelity Metrics Evalubted Using Image Distortion Maps. Signal Processing, Beijing, China (3), 201–214 (1998)

A New Method for CT/MR Image Fusion Based on Bandelets Transform

Huimin Lu[1, a], Yujie Li[1,2,b], Lifeng Zhang[1,c], Shiyuan Yang[1,d], and Seiichi Serikawa[1,e]

[1] Department of Electrical Engineering and Electronics, Kyushu Institute of Technology, Japan
[2] College of Information Engineering, Yangzhou University, China
[a,b] {luhuimin,yzyjli}@boss.ecs.kyutech.ac.jp,
[c,d,e] {zhang,yang,serikawa}@elcs.kyutech.ac.jp
http://www.boss.ecs.kyutech.ac.jp

Abstract. In recent years, as the bandelets transform has some benefits, many scholars interest in this field. They proposed many methods to solve different problems in different fields. In this paper, we propose a new method maximum local energy method to calculate the low coefficients of images. And then adopt the absolute maximum method to select the high coefficients of images. Later, we compare the results with wedgelets transform and wavelet transform. Beside the human vision, we also compare the results by quantitative analysis. In our experiments, we take wedgelet transform and bandelets transform for comparing the results. The numerical experiments explain that the maximum local energy is an effect way for image fusion, which can get well performance.

Keywords: computed tomography, magnetic resonance image, wedgelets transform, bandelets transform, maximum local energy, medical image processing.

1 Introduction

Imaging sensors are an important resource in today's world. Because of that a single sensor cannot provide a complete view of the scene in many applications. The result of fused images, if suitably obtained from a set of source sensor images, can get a better view of the sight than the view provided by any of the individual source images. In resent decades, there has been growing interest in the use of multiple sensors to increase the capabilities of intelligent machines and systems. As a result, multisensor fusion has become an area of intense research and development activity in the past few years.

A variety of image fusion techniques have been developed. Generally speaking, we can roughly divide them into two groups, multiscale decomposition based fusion methods such as pyramid algorithm [1], wavelet transform method [2], wedgelet transform [3],bandelet transform [4] etc., and nonmultiscale decomposition based fusion methods, for example, weighted average method, nonliner method, estimation theory based methods and so on.

Y. Zhang (Ed.): Future Communication, Computing, Control and Management, LNEE 141, pp. 587–592.
springerlink.com

The weighted average method is one of the most simple image fusion methods. The sourcc images do not be transformed and decomposed and fused image directly averages the gray level of defocused images' pixels. This method is suitable for real-time processing, but will decrease the signal to noise ratio of the image. These are certified by some researchers. The pyramid method firstly constructs the input image pyramid, and then takes some feature selection approach to form the fusion value pyramid. By the inverter of the pyramid, the pyramid of image can be reconstructed, to produce fusion images. This method is relatively simple, but it also has some drawbacks. The themes of classical wavelets include terms such as compression and efficient representation. These features which play an important role in analysis of functions in two variables are dilation, translation, spatial and frequency localization and singularity orientation. However, classical wavelets have drawbacks in representing images, such as the problem of efficient representation in two dimensions. Recently, several theoretical papers have called attention to the benefits of Bandelets transform [5,6]. So this paper will introduce this method and propose a new method to apply in these transforms in CT/MRI images fusion.

We briefly introduce the structure of this paper. In Section 2, we primitively introduce the principle of Bandelets transform for image fusion. As a solution, we propose in Section 3 a new method --- Maximum Local Energy method in multi-focus images fusion. Numerical experiments are presented in Section 4 to confirm that our proposed method is useful for image fusion. At last, we conclude the paper in Section 5.

2 Bandelets Transform for Image Fusion

The Bandelets transform [5,7] are defined as anisotropic wavelets that are warped along the geometric flow, which is a vector field indicating the local direction of the regularity along edges. The dictionary of bandelet frames is constructed by using a dyadic square segmentation and parameterized geometric flows. The ability to exploit image geometry, makes its approximation error decay optimal asymptotically for piece-wise regular images.

For image surfaces, the geometry is not a collection of discontinuities, but rather areas of high curvature. The Bandelet transform recasts these areas of high curvature into an optimal estimation of regularity direction. Figure 1 shows an example of bandelets along the geometric flow in the direction of edges. In real applications, the geometry is estimated by searching for the regularity flow and then for a polynomial to describe that flow.

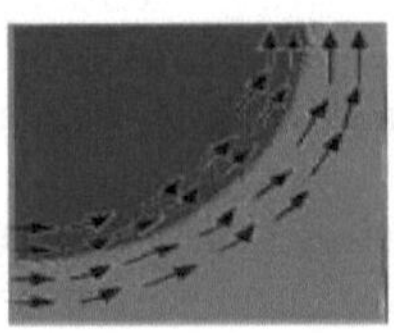

Fig. 1. An illustration of bandelets with geometric flows in the direction of the edge

Let us suppose the image support is $S = \cup_i \Omega_i$. Ω_i is the region in depth i. In each Ω_i, the flow is either parallel horizontally or vertically. The image is devised as dynamic square regions, each region Ω_i includes only one contour. If the region does not contain any contour, the image intensity is uniformly regular and the flow is not defined.

The Bandelet transform is first implement by reordering the two-dimensional Wavelet coefficients in multiscale analysis and then performing a one-dimensional wavelet transform in geometric directional analysis. In using Bandelets transform to process images, we suppose the image f(x1,x2) is uniformly regular, C^{α} is α times continuously differentiable, the regions of an image are approximated in the separable wavelet basis of $L^2(\Omega)$ in

$$\begin{Bmatrix} \phi_{j,m_1}(x_1)\psi_{j,m_2}(x_2) \\ \psi_{j,m_1}(x_1)\phi_{j,m_2}(x_2) \\ \psi_{j,m_1}(x_1)\psi_{j,m_2}(x_2) \end{Bmatrix}_{(j,m_1,m_2)\in I_\Omega} \quad (1)$$

where I_Ω is an index set that depends upon the geometry of the boundary of Ω. x1, x2 are the location of pixel in the image. And $\phi_{j,m_1}(x_1)\psi_{j,m_2}(x_2)$, $\psi_{j,m_1}(x_1)\phi_{j,m_2}(x_2)$, $\psi_{j,m_1}(x_1)\psi_{j,m_2}(x_2)$ are the modified wavelets at the boundary. If a geometric flow is calculated in Ω, this suggests replacing the orthogonal family $\{\phi_{j,m_1}(x_1)\psi_{j,m_2}(x_2)\}_{j,m_1,m_2}$ by the family $\{\psi_{l,m_1}(x_1)\psi_{j,m_2^{'}}(x_2)\}_{j,l>j,m_1,m_2}$ which generates the same space. The functions $\{\psi_{l,m_1}(x_1)\psi_{j,m_2^{'}}(x_2)\}$ are called Bandelets because their support is parallel to the flow lines and is more elongated (2l>2j) in the direction of the geometric flow. The above expression (1) can be replaced by

$$\begin{Bmatrix} \psi_{l,m_1}(x_1)\psi_{j,m_2}(x_2 - c(x_1)) \\ \psi_{j,m_1}(x_1)\phi_{j,m_2}(x_2 - c(x_1)) \\ \psi_{j,m_1}(x_1)\psi_{j,m_2}(x_2 - c(x_1)) \end{Bmatrix}_{j,l>j,m_1,m_2} \quad (2)$$

Since the flow is parallel vertically, a flow line associated to a fixed translation parameter x2 is a set of point $(x_1, x_2 + c(x_1)) \in \Omega$ for x1 varying, with $c(x) = \int_{x_{\min}}^{x} c^{'}(u)du$. l is the direction of geometric flow which is more elongated (2l>2j). The above function (2) is got by inserting Bandelets in the Warped Wavelet basis in

$$\begin{Bmatrix} \phi_{j,m_1}(x_1)\psi_{j,m_2}(x_2 - c(x_1)) \\ \psi_{j,m_1}(x_1)\phi_{j,m_2}(x_2 - c(x_1)) \\ \psi_{j,m_1}(x_1)\psi_{j,m_2}(x_2 - c(x_1)) \end{Bmatrix}_{(j,m_1,m_2)\in I_{W\Omega}} \tag{3}$$

where $W\Omega$ is the warped region, $W\Omega = \{(x_1, x_2) : (x_1, x_2 + c(x_1)) \in \Omega\}$.

Bandelets transform can adaptively track the geometric direction of the image. It also can process the different changes in different regions. It abandoned the concept of ``edge", which is not easy to define in mathematic. Instead of this, it adopted the concept of ``geometric flow" to reflect the continuous transformation in the image.

3 Image Fusion Algorithm

Bandelets transform based image fusion is completed primarily by the way that different type images are respectively bandelets transformed, and then according to certain criteria for selecting the appropriate low-frequency and high-frequency coefficients. And through Bandelet inverse transform, the two type images are fused as a clear, more information image. In this paper, we take the following steps; firstly, use Bandelets transform in two images to get the coefficients. After that, we process the coefficients of the low frequency and high frequency of these images, and fuse these images according to some fusion method. Finally, through inverse Bandelets transform, we get a clear image.

4 Principle of Low Frequency Fusion

This paper uses the maximum local energy (LE) [8,9] as a measurement in low frequency domain. Select the maximum energy of two source images as output. Due to the partial human visual perception characteristics and the relationship of decomposition about local correlation coefficients, the statistical characteristics of neighbor should be considered. Therefore, the statistic algorithm is based on the 3X3 window. The algorithm is described as follows:

$$\mathrm{LE}_\xi(i,j) = \sum_{i' \in M, j' \in N} p(i+i', j+j') \bullet f_\xi^{(0)2}(i+i', j+j') \tag{4}$$

where p is the local filtering operator. M, N is the scope of local window. $\xi \in$ A or B (A, B is the window for scanning two images) . $f_\xi^{(0)}(i,j)$ is low frequency coefficients.

Local bandelets energy (LBE) is

$$LBE_\xi^{l,k}(i,j) = E_1 * f_\xi^{(0)2}(i,j) + E_2 * f_\xi^{(0)2}(i,j) + \cdots + E_K * f_\xi^{(0)2}(i,j). \tag{5}$$

where E1,E2,…, EK-1 and EK are the filter operators in K different directions.

5 Principle of High Frequency Fusion

In order to compare the results easily, we take the Absolute values (ABS) as a treatment. Under the assumption that image details are contained in the high-frequency subbands. The high-frequency components of different images have larger absolute values in the same level and position than those of original images. Input two images, image A and image B. Then after Bandelets transform, we get the high frequency coefficients separately, C_H^A and C_H^B. In order to choose the high frequency of the fused image, we adopt the max value of $|C_H^A|$ and $|C_H^B|$ for the fused high frequency.

6 Experimental Results and Discussion

In this section, we compare the LE-Bandelets transform with other methods. From the visual analysis, we can find that, the proposed method is better than other methods. Beside visual analysis, we compare the results in numerical analysis. The quantitative analysis result is shown in Table 1. We use the evaluation functions in [10,11] to measure our results.

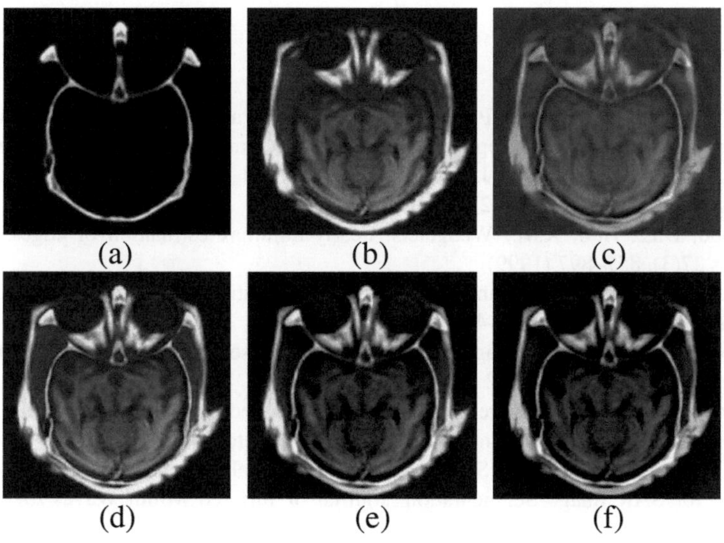

Fig. 2. (a) CT image. (b) MR image. (c) Wedgelet transform fused image. (d) LE-Wedgelet transform fused image. (e) Bandelet transform fused image. (f) LE-Bandelet transform fused image.

Table 1. Quantitative analysis

Methods	Wedgelets	Bandelets	LE-Wedgelets	LE-Bandelets
PSNR	17.703	18.229	23.531	**23.549**
Q	0.8132	0.8876	0.9149	**0.9226**
Q_W	0.7903	0.8890	0.8789	**0.9081**
Q_E	0.5201	0.5303	0.6788	**0.6903**
SSIM	0.7015	0.6563	**0.8364**	0.5820
MS-SSIM	0.8237	0.9105	0.8398	**0.9306**

7 Conclusion

In this paper, we present Local Bandelets Energy fusion method, a new algorithm that gives the number of benefits for image fusion. From the above table, we can see that, in these experiments, maximum Local Energy (LE) improved the results of traditional methods. Meanwhile, LBE fusion method gives a best performance. We can conclude that LBE perform a better result. It is useful for medical image fusion.

There are several areas in which our method can be improved or extended. We currently consider improving the LBE method. In addition, we also plan to apply it in other beyond wavelet transforms.

References

1. Yu, Z.M., Gao, F.: Laplacian Pyramid and Contrast Pyramid Based Image Fusion and Their Performance Comparison. Application Research of Computers 21(10), 96–98 (2004)
2. Pajares, G., de la Cruz, J.M.: A Wavelet-based Image Fusion Tutorial. Pattern Recognition 37(9), 1855–1872 (2004)
3. Donoho, D.L., Huo, X.M.: Wedgelets: nearly-minimax estimation of edges. Annals of Statics 27(3), 857–897 (1999)
4. Le Pennec, E., Mallat, S.: Bandelet Image Approximation and Compression. IEEE Trans. Image Processing 14(4), 423–438 (2005)
5. Qu, X., Yan, J.: A Novel Image Fusion Algorithm based on Bandelet Transform. Chinese Optics Letters 5(10), 569–572 (2007)
6. Maalouf, A., Carre, P., Augereau, B.: Bandelet-based Anisotropic Diffusion. In: IEEE International Conference on Image Processing, San Antonio, TX, pp. (I)289–(I)292 (2007)
7. Biao, H., Xia, Z., Licheng, J.: Second Generation Bandelet-domain Hidden Markov Tree-3S Model for SAR Image Segmentation. Journal of Infrared Millin Waves 29(2), 145–149 (2010)
8. Lu, H., Zhang, L., Serikawa, S., et al.: Local Energy based Image Fusion in Sharp Frequency Localized Contourlet Transform. Journal of Computational Information Systems 6(12), 3997–4005 (2010)
9. Lu, H., Li, Y., Serikawa, S., et al.: Local Energy based Multi-focus Image Fusion Method on Curvelet Transforms. In: IEEE 10th International Symposium on Communications and Information Technologies, pp. 1154–1157 (2010)
10. Gemma, P., Heijmans, H.: A New Quality Metric for Image Fusion. In: International Conference on Image Processing, vol. 2, pp. (III)173–(III)176 (2003)
11. Wang, Z., Li, Q.: Information Content Weighting for Perceptual Image Quality Assessmen., IEEE Trans. on Image Processing (accepted)

Color Image Segmentation Using Fast Density-Based Clustering Method

Yujie Li[1,2,a], Huimin Lu[1,b], Lifeng Zhang[c], Shiyuan Yang[c], and Seiichi Serikawa[1,c]

[1] Department of Electrical Engineering and Electronics, Kyushu Institute of Technology, Japan
[2] College of Information Engineering, Yangzhou University, China
[a,b] {yzyjli,luhuimin}@boss.ecs.kyutech.ac.jp,
[c] {zhang,yang,serikawa}@elcs.kyutech.ac.jp

Abstract. Color image segmentation is an important research topic in the field of computer vision. In this paper, we propose a method for image segmentation by computing similarity coefficient in RGB color space. Then, we apply the density-based clustering algorithm TI-DBSCAN on regions growing rules that in turn speeds up the process. This new method has three advantages. First, this method can reduce the disturbance of noise and get the segmentation numbers more accurately. Second, it needn't to change the RGB color space to other space. Third, it uses a triangle inequality property to quickly reduce the neighborhood search space. The experimental results illustrate that the new approach segmentation method can efficiently segment image.

Keywords: Density-based clustering, Color image segmentation, Clustering analysis, RGB color space, TI-DBSCAN.

1 Introduction

Image segmentation is one of the important image analysis technologies. One of the most basic definitions is: divide image into each with distinct features of regional and extract interested goals. Many different segmentation approaches [1-7] have been proposed such as: thresholding, region extraction and so on. But all of them cannot be generalized under signal scheme. In this paper, we bring forward an approach based on clustering.

Region growing and clustering are two representative methods of region based segmentation. Main drawbacks of this method is that it is difficult to make the growing or stop growing criteria for different images and the method is sensitive to noise. Recently, most researchers focus on the segmentation problem as an unsupervised classification problem.

In 1996, Martin Ester etc. proposed a density-based clustering method, called DBSCAN algorithm [8], for data clustering. The DBSCAN algorithm is recognized as a high quality scalable algorithm for clustering low dimensional data. However, the DBSCAN algorithm is time consuming. During some research, we can found that the time consuming operation in DBSCAN is the calculation of a neighborhood for each data point. In 2010, Marzena Kryszkiewicz etc. proposed a new method, called

Y. Zhang (Ed.): Future Communication, Computing, Control and Management, LNEE 141, pp. 593–598.
springerlink.com

TI-DBSCAN [9], for speed up the performance by use the triangle inequality property to quickly reduce the neighborhood search space.

In this paper, we combine the region growing and TI-DBSCAN algorithm for image segmentation in RBG color space. The paper is organized as follows. We discuss DBSCAN and TI-DBSCAN algorithm in section 2. In Section3, we describe our segmentation in detail. Section4 presents the experimental results. The conclusions of this paper are summarized in section 5.

2 DBSCAN and TI-DBSCAN

2.1 DBSCAN

In this section, we briefly introduce the basic ideas of density-based clustering involve a number of new definitions. DBSCAN [8] is central around the key idea of "to form a new cluster or grow an existing cluster the eps-neighborhood of a point should contain at least a minimum number of points (MinPts)". A detailed theoretical formulation is given in [8]. As the parameters Eps and Minpts are set, DBSCAN starts to cluster data points form an arbitrarily chosen point q. Firstly, it searches the neighbor points of the point q. Decide the neighbor points require a region query or not. If the neighborhood is sparsely populated, point q is marked as noise. Otherwise, a cluster is initiated and all points in neighborhood of point q are labeled by the new cluster ID. Next, the neighborhoods of all q's neighbors are examined iteratively to check if new candidates can be added into the cluster. If a cluster cannot be expanded anymore, DBSCAN chooses another arbitrary unlabeled point, if any such point exists, and repeats the same procedure to form new cluster. DBSCAN progresses until all data points in the dataset have been labeled either as noise or with a cluster ID. In image processing, an image can be considered as a special spatial dataset in which each pixel has a spatial location and a color value.

2.2 TI-DBSCAN

As we all know, the DBSCAN algorithm is recognized as a high quality scalable algorithm for clustering low dimensional data. The most time consuming operation of it is the calculation of a neighborhood for each data point. In order to speed up this operation in DBSCAN, Marzena Kryszkiewicz etc proposed TI-DBSCAN algorithm [9]. In this section, we briefly introduce this algorithm.

Property 1. (Triangle Inequality Property) For any three points p, q, r:

$$\text{distance}(p,q) \geq \text{distance}(p,r) - \text{distance}(q,r). \quad (1)$$

Property 2. Let D be a set of points. For any two points p, q in D and any point r:

$$\text{distance}(p,r) - \text{distance}(q,r) > Eps \Rightarrow q \notin NEps(p) \wedge p \notin NEps(q). \quad (2)$$

Property 3. Let r be any point and D be a set of points ordered in a non-decreasing way with respect to their distances to r. Let p be any point in D, qf be a point following point p in D such that distance(qf,r) – distance(p,r) > Eps, and qb be a point preceding point p in D such that distance(p,r) – distance(qb,r) > Eps . Then:

qf and all points following qf in D do not belong to NEps(p).

qb and all points preceding qb in D do not belong to NEps(p).

The pseodучode of TI-DBSCAN is shown as follows.

Algorithm TI-DBSCAN(set of points D, *Eps*, *MinPts*);

/* *Eps* – the radius of the point neighborhood*/

/* *MinPts* – the required minimal number of point *MinPts* within *Eps*-neighborhood*/

/* *r* – a reference point assumed to be fixed*/

/* *p.ClusterId* – label of a cluster to which *p* belongs; Initially assigned the UNCLASSIFIED label*/

/* *p.NeighborsNo* – the number of neighbors of *p* already found; Initially assigned 1 to indicate that a point itself belongs to its own *Eps*-neighborhood */

/* *Border* – the information about neighbors of point *p* that turned out noncore points for which it is not clear temporary if they are noise ones or border ones; Initially assigned an empty set.*/

D' = empty set of points; /* D'– the result of clustering of D */

for each point *p* in set D **do** /* D – the set of points that is subject to clustering */

p.ClusterId = UNCLASSIFIED;

p.dist = *Distance*(*p*,*r*); /* *p.dist* – the distance of point *p* to reference point *r* */

p.NeighborsNo = 1;

p.Border = ∅;

endfor

sort all points in D non-decreasingly w.r.t field dist;

ClusterId = label of first cluster;

for each point p in the ordered set D starting from the first point until last point in D **do**

if *TI-ExpandCluster*(D, D', *p*, *ClusterId*, *Eps*, *MinPts*) **then**

ClusterId = *NextId*(*ClusterId*); /**TI-ExpandCluster* calculates *Eps*-neighborhood for each point */

endif

endfor

return D';

hborhood for each point onlyntation region.ghest similarity in color sir them.l neighborhood According to some experiments in [9], the proposed TI-DBSCAN algorithm speeded up DBSCAN's performance by 50%. The improved algorithm also suit for high dimensional data processing.

3 Proposed Segmentation Method

By reference to the definitions for DBSCAN in settion 2 [8]. We present these definitions and specifications for image segmentation and follow up with the process of our segmentation method.

SpatialEps The spatial neighborhood of a given pixel is called the spatial Eps of the pixel.we name it SpatialEps. SpatialEps is set as a circle in this paper. Figure 1. illustrates the SpatialEps of a pixel p. The size of SpatialEps is the number of all pixels in the circle. Black points in the circle represent pixels which are color similar with p.

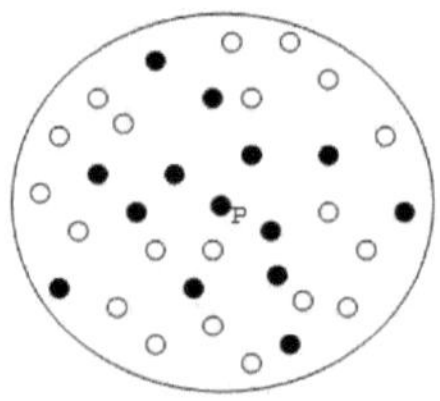

Fig. 1. SpatialEps of a pixel p

Color Similarity In this paper, we use the following steps. Suppose f and h are respectively color vectors in RGB space. Where, f = (f1,f2,f3)', h = (h1,h2,h3)'. The correlation coefficient between the color vectors is defined as

$$KC(f,h) = \lambda R KR(f,h) + \lambda B KB(f,h). \tag{3}$$

where, KR(f,h) is the hue and saturation similarity coefficients of two color vectors. KB(f,h) is the intensity similarity coefficients of two color vectors.

$$\boldsymbol{KR}(\boldsymbol{f},\boldsymbol{h}) = \frac{\langle \boldsymbol{f},\boldsymbol{h} \rangle}{\| \boldsymbol{f} \| \cdot \| \boldsymbol{h} \|} \tag{4}$$

$$\mathrm{KB}(f\,,\,h) = 1 - \frac{| f_1 + f_2 + f_3 - h_1 - h_2 - h_3 |}{765} \tag{5}$$

where λR and λB are the weight coefficients of hue and saturation similarity coefficient and intensity similarity coefficient. In pactive, we choose λR = 0.85, λB = 0.15.

To determine whether the two colors are similar, according to the value of KC(f,h) and pre-set T. If KC(f,h) > T, then f, h is similar in colors, otherwise they are not similar.

Core pixel In the spatial neighborhood of point p, if the number of pixels with the similarity color with point p is not less than the threshold MinPts, then p is the core pixel.

Based on the above definitions, we prescribe our segmentation method as follows:

Search the unlabeled pixels in an image in order for current core pixel and current core pixel neighborhood. From the top left corner to bottom down corner to search the image.

If a core pixel p is found, a new cluster is created. Then iteratively collect unlabeled pixels that are density-connected with p, and label these pixels with the same cluster label. For other pixels, use triangle inequality density-connected method to cluster them.

If there are still existing core pixels in the image, return to step 2).

For pixels that are not labeled with any clusters, merge them with the cluster that is adjacent to them and has the highest similarity in color similarity.

Label all of clusters as a segmentation region.

Determining the parameters

There are two parameters for SpatialEps in each image, the size of the SpatialEps circle and the MinPts. We simply select the size of SpatialEpt circle size according to the size of the image. MinPts is choosed by SpatialEps. That is, MinPts is the half number of the point p neighborhood. According to [10], we choose T = 0.985.

4 Experimental Results and Discussion

To verify the proposed segmentation method, experiments are performed on images with different complexity on Core 2 2.0GHz PC. It takes less than 0.5 second to segment a 349×522 image. From the results, we can found that, our method is fast than original methods. Meanwhile, the segment result is performing than former methods.

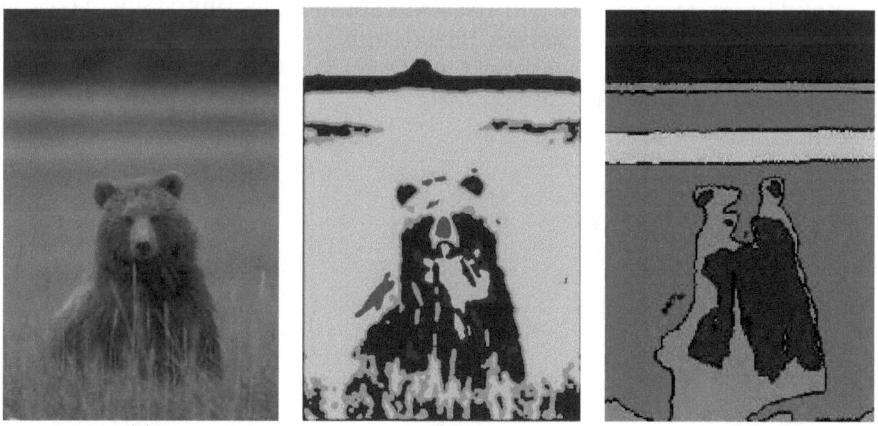

(a) (b) (c)

Fig. 2. Segmentation results of our algorithm. (a) Input image. (b) Traditional segmentation result. (c) Our method segmentation result.

5 Summary

In this paper, we present a new color image segmentation method. The use of density-based clustering and the RGB color space. The proposed method speeds up the computation complexity. As follows from the experiments, our algorithm is much more efficient than original DBSCAN algorithm. The new algorithm also has highly ability of denoising.

References

1. Chen, C.W., Luo, J., Parker, K.J.: IEEE Trans. on Image Processing 7(12), 1673–1683 (1998)
2. Parvati, K., Prakasa, R.S., Mariya, D.M.: Discrete Dynamics in Nature and Society, pp. 1–8 (2008)
3. Otman, B., Hongwei, Z., Fakhri, K.: Fuzzy Based Image Segmentation. Springer, Berlin (2003)
4. Lu, H., Zhang, L., Serikawa, S., et al.: A Method for Infrared Image Segment Based on Sharp Frequency Localized Contourlet Transform and Morphology. In: ICICIP 2010, pp. 79–82 (2010)
5. Gonzalez, R.C., Woods, R.E.: Digital Image Processing. Publishing House of Electronics Industry, Beijing (2004)
6. Han, J., Kamber, M.: Data mining: Concepts and Techniques. Morgan Kaufmann Publishers, San Francisco (2001)
7. Hung, M., Yang, D.: An Efficient Fuzzy C-Means Clustering Algorithm. In: IEEE Intel. Conf. on Data Mining, pp. 225–232 (2001)
8. Ester, M., Kriegel, H.P., Sander, J., Xu, X.W.: A Density-based Algorithm for Discovering Spatial Databases With Noise. In: Proc. of 2th ICKDDM (1996)
9. Kryszkiewicz, M., Lasek, P.: TI-DBSCAN: Clustering with DBSCAN by Means of the Triangle Inequality. LNCS. Springer, Berlin (2010)
10. Han, X., Li, J., et al.: An Approach of Color Object Searching for Vision System of Soccer Robot. In: Proc. ICRB, pp. 535–539 (2004)

Modeling High-Speed Network Traffic with Truncated α-Stable Processes

Lin Bai[1], Wenyang Guan[2], Chen Chen[3,*], Jianhua He[2], and Rui Wang[3]

[1] School of Electronic and Information Engineering,
Beihang University, Beijing, China
[2] School of Engineering, Swansea University,
Swansea, United Kingdom
[3] School of Electronics Engineering and Computer Science
Peking University, Beijing, China
chen.chen@pku.edu.cn

Abstract. It has been reported that high-speed communication network traffic exhibits both long-range dependence (LRD) and burstiness, which posed new challenges in network engineering. While many models have been studied in capturing the traffic LRD, they are not capable of capturing efficiently the traffic impulsiveness. It is desirable to develop a model that can capture both LRD and burstiness. In this letter, we propose a truncated α-stable LRD process model for this purpose, which can characterize both LRD and burstiness accurately. A procedure is developed further to estimate the model parameters from real traffic. Simulations demonstrate that our proposed model has a higher accuracy compared to existing models and is flexible in capturing the characteristics of high-speed network traffic.

Keywords: System modeling, network traffic modeling, truncated α-stable processes.

1 Introduction

From the seminal work in [1] it has been widely recognized that aggregated network traffic in a variety of wired and wireless networks exhibits burstiness and long-range dependence (LRD) [2-4]. The burstiness and LRD rule out the use of conventional Markov models. Many self-similar models have been proposed in the past decade to capture the LRD and burstiness properties.

According to the two major features, burstiness and LRD, network performance may be optimized through efficient traffic engineering. In the literature LRD-based optimization has been well developed previously in some self-similar models, while the burstiness, however, has not been characterized well. Recently proposed α-stable distribution models [2][3] and gamma distribution [4] have been found performing better in capturing traffic burstiness. Nevertheless, these two models showed some limitations as well: 1) the α-stable distribution has infinite variance and its tail is

* Corresponding author.

Y. Zhang (Ed.): Future Communication, Computing, Control and Management, LNEE 141, pp. 599–603.
springerlink.com

factually too heavy for modeling real traffic; 2) the gamma distribution is not flexible enough in capturing real network traffic.

In this paper, we propose a truncated α-stable LRD process model, which capture both traffic LRD and burstiness accurately. Treating the α-stable distribution and gamma distribution as special cases, we show that our proposed model can capture a larger variety of network traffic with higher flexibility.

2 Mathematical Preliminaries

In this section, after a brief introduction of the LRD, we present the statistical distribution of our proposed truncated α-stable LRD process model. In [1] and [5], LRD is characterized by autocorrelation function that decays in a power-law fashion. Let $\mathbf{X} = \{X_{(n)}, n = 0, 1, 2, \dots\}$ be a covariance stationary stochastic process with mean m, variance σ^2, where its autocorrelation function is given by $r(k)$, $k \geq 0$. Note that $\mathbf{X}$ is defined to exhibit the LRD with its autocorrelation function satisfies the following asymptotic approach [1][5]:

$$r(k) \sim k^{-2(1-H)} L(k). \tag{1}$$

Here, the coefficient $H \in (1/2, 1)$ denotes the Hurst parameter which represents the degree of LRD and $L(k)$ is a slowly varying function (e.g., constants) satisfying that

$$\lim_{t \to \infty} L(tx)/L(t) = 1, \quad \forall x > 0. \tag{2}$$

Truncated α-stable distribution introduces a cutoff in the power-law tail to the α-stable distribution [4]. Such a cutoff leads to a finite variance. Under the assumption that network traffic is nonnegative, our proposed truncated α-stable distribution is carried out to be positively skewed, which is defined by the following cumulant function (log of the characteristic function):

$$\Phi(\omega) = a\Gamma(-\alpha)[(\lambda - j\omega)^{\alpha} - \lambda^{\alpha}]. \tag{3}$$

Here, $\alpha \in (0, 1)$, $a > 0$, and $\lambda > 0$ denote the characteristic exponent, scale parameter, and truncation parameter, respectively. Note that the proposed distribution is defined on positive x-axis as the one shown in [4].

3 Proposed Truncated α-Stable LRD Process Model

In this paper we use the total packets per unit time to measure aggregated network traffic [1][2]. Denote by $\mathbf{S} = \{S_{(n)}, n = 0, 1, \dots\}$ the network traffic process, where $S_{(n)}$ represents the number of packets arrived within the duration of the nth

unit time. Based on the distribution represented in (3), a stochastic process is called a truncated α-stable LRD process if it satisfies the two following conditions: 1) $S_{(n)}$ follows truncated α-stable distribution defined by (3); 2) Process $\{S_{(n)}\}$ exhibits LRD.

It has been widely accepted that burstiness is inherently related to non-Gaussian and positively skewed distributions [2]. As a joint consideration of the α-stable distribution and the gamma distribution, factually, the α-stable distribution and gamma distribution can be regarded as special cases of the truncated α-stable distribution: 1) when $\alpha \to 0$ in (3), it results in a gamma distribution; 2) when $\lambda \to 0$ in (3), it leads to a positively skewed α-stable distribution. In the sequel, we can show that the proposed model outperforms the existing models in [2] and [4]. The proposed model demonstrates a higher flexibility than the existing models and can capture a larger variety of real network traffic.

With a proper truncation parameter λ[1], the truncated α-stable distribution behaves like (but not necessarily equals) an α-stable distribution over several orders of magnitude and is able to capture the traffic burstiness. However, its tail decays more rapidly than that of α-stable distribution, which is approximated by

$$f_{T\alpha}(x) \sim Cx^{-\alpha-1}e^{-\lambda x}, \quad x \to \infty, \tag{4}$$

where C is a constant. Accordingly, a truncated α-stable variable has a finite variance [4], where the aggregation of its variables converges to Gaussian distribution in a certain limit as provided by the central limit theorem.

4 Parameters Estimation

Our proposed model can be characterized mainly by two parameters, H and θ, where H represents the degree of LRD and $\theta = \{\alpha, \lambda, a\}$ characterizes the distributional property. In this section, we present a method to estimate these parameters from real traffic.

Since parameter H can be estimated by using the variance-time plot method in [2], we focus on the estimation of parameter $\theta = \{\alpha, \lambda, a\}$. A cumulant-based approach is proposed to estimate θ, where the least squares (LS) criterion is employed to minimize the summation of squared errors between the theoretical and empirical cumulants.

For $m \geq 1$, from (3), the cumulants can be represented in terms of the parameter $\theta = \{\alpha, \lambda, a\}$ as

$$\mathrm{cum}_m(\mathbf{X}) = a\lambda^{\alpha-m}\Gamma(m-\alpha). \tag{5}$$

Factually, there is a one-to-one correspondence between the moments and cumulants. Denote the central moment of kth order by $m_k = E[(\mathbf{X} - E[\mathbf{X}])^k]$. We have

[1] The value of λ will be discussed in the following section.

$$\begin{aligned}&\mathrm{cum}_1(\mathbf{X}) = E[\mathbf{X}]; \quad \mathrm{cum}_2(\mathbf{X}) = m_2; \quad \mathrm{cum}_3(\mathbf{X}) = m_3;\\&\mathrm{cum}_4(\mathbf{X}) = m_4 - 3m_2^2; \quad \mathrm{cum}_5(\mathbf{X}) = m_5 - 10m_3m_2;\\&\mathrm{cum}_6(\mathbf{X}) = m_6 - 15m_4m_2 - 10m_3^2 + 30m_2^3...\end{aligned} \tag{6}$$

The estimated k^{th} order central moment can be obtained as

$$\widehat{m}_k = \overline{(\mathbf{X} - \overline{\mathbf{X}})^k}. \tag{7}$$

Combining (6) with (7), the empirical cumulants $\widehat{\mathrm{cum}}_m(\mathbf{X})$ can be obtained from the real data. Thus, the optimal estimation of the parameter θ based on the LS criterion is represented by

$$\widehat{\theta} = \arg\min_{\theta}\Big\{\sum_m \big|a\lambda^{\alpha-m}\Gamma(m-\alpha) - \widehat{\mathrm{cum}}_m(\mathbf{X})\big|^2\Big\}. \tag{8}$$

5 Numerical Results and Discussions

In this section, we evaluate the accuracy of the proposed model in capturing the real network traffic behavior and compare it with existing models. In order to provide a realistic evaluation of the traffic models, we used real traffic trace named "Drexel trace 0226" studied by Drexel University[2]. Due to the limited space, other traffic traces based models are not included in this letter.

The model inaccuracy, which characterizes the mismatch between analytical results and real traffic, is used as the performance metric for the evaluation of our proposed traffic model.[3]. Denote by $\bar{F}(x) = P(X > x)$ the probability that X exceeds the value x, which gives the definition of complementary cumulative distribution function (CCDF) for process X. In Fig. 1, we present results of absolute differences between the CCDF for the studied traffic models and the empirical CCDF (ECCDF) for real traffic (i.e. $\mathrm{abs}(\mathrm{CCDF} - \mathrm{ECCDF})$) versus the received packets per unit time, x. Since the α-stable distribution exhibits a heavy tail, where its CCDF deviates obviously from ECCDF even for a large x, it can not provide a good modeling performance. More importantly, it shows that the tail (i.e., $\mathrm{abs}(\mathrm{CCDF} - \mathrm{ECCDF})$) of our proposed truncated α_stable distribution is thinner than that of the other two distributions at a broad range of x. Therefore, we can conclude that our proposed model shows a higher accuracy than existing ones.

---

[2] Available: http://www.ece.drexel.edu/CSPL/DUTrace-2003/

[3] We use a search algorithm to find the optimal parameter vector $\theta = (\alpha, \gamma, a)$ of our proposed model numerically based on [8].

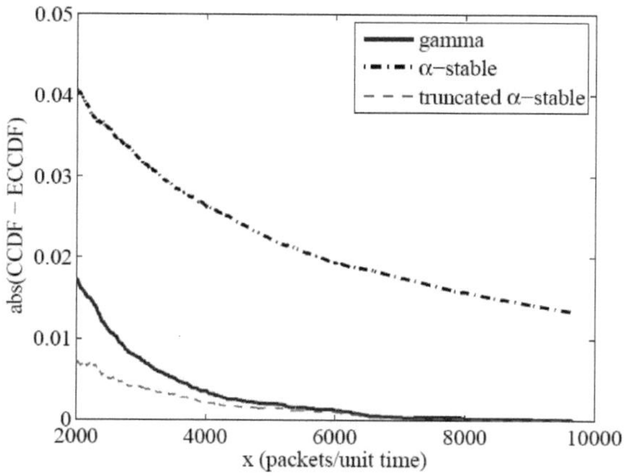

Fig. 1. Absolute differences between CCDFs of models and ECCDF

6 Conclusion

In this paper, we proposed a truncated α-stable distribution based traffic model for high-speed network traffic, which exhibits both strong LRD and burstiness. It has been demonstrated by simulations that our proposed traffic model can efficiently capture both LRD and burstiness properties. It overcomes the problem of overestimating traffic burstiness in α-stable distribution based traffic models and can achieve better accuracy in characterizing an observation of non-Gaussian burstiness. Furthermore, the proposed model is flexible in characterizing network traffic.

References

1. Leland, W.E., Taqqu, M.S., Willinger, W., Wilson, D.V.: On the self-similar nature of Ethernet traffic. IEEE/ACM Trans. Networking 2, 1–15 (1994)
2. Karasaridis, A., Hatzinakos, D.: Network heavy traffic modeling using α-stable self-similar processes. IEEE Trans. Commun. 49, 1203–1214 (2001)
3. Qin, W., Wang, Q., Sivasubramiam, A.: An [graphics object to be inserted manually]-stable model-based linear-parameter-varying control for managing server performance under self-similar workloads. IEEE Trans. Control Systems Technology 17, 123–134 (2009)
4. Scherrer, A., Larrieu, N., Owezarski, P., Borgnat, P., Abry, P.: Non-Gaussian and long memory statistical characterisations for internet traffic with anomalies. IEEE Trans. Depend. Secur. Comput. 4, 56–70 (2007)
5. Beran, J., Sherman, R., Taqqu, M.S., Willinger, W.: Long-range dendence in variable-bit-rate video traffic. IEEE Trans. Comm. 43, 1566–1579 (1995)
6. Ge, X., Yang, Y., Wang, C., Liu, Y., Liu, C., Xiang, L.: Characteristics analysis and modeling of frame traffic in 802.11 wireless networks. Wirel. Commun. Mob. Comput. 10, 584–592 (2010)

Author Index

Ameur, Sonia Ait 227
Ascar, Bouabdellah 227

Bai, Lin 599
Bekkal, Amel 227
Bhargava, R.R. 285
Bouden, T. 241

Chen, Chen 599
Chen, Hua 173
Chen, Huanlin 125
Chen, Jie 329, 337
Chen, Shin-Bo 257
Chen, Tie-jun 487
Chen, Xiaoping 17
Cheng, Mingzhi 117
Cheng, Shihui 349
Chu, Ying 479

De, Tian 43, 473
Deng, Yi-Min 265
Dib, S. 241
Djerfi, F. 241
Du, Ping 501
Du, Xianbin 329, 337

Erbing, Wang 517, 543

Fang, Zhang 517
Feng, Wang Pan 149

Gan, Shuoqiu 451
Gao, Hailin 99
Gao, Junxiang 117
Gu, Changzhou 17
Gu, Min 131
Gu, Wenyang 125
Guan, Jiaqing 383
Guan, Wenyang 599

Hao, Shiru 165
He, Jianhua 599
He, Peng 441
He, Qian 383
He, Songhua 579
He, Yueshun 501
Hong, Liang Zhi 207
Hong, Zhou 517, 543
Hsu, Chien-Yeh 257
Hu, Ruihua 457
Hu, Yongcai 89, 405
Hu, Yudong 533
Huang, Chi-Hung 257
Huang, Ju-hua 373

Ján, Duplák 421
Ji, Xiaojuan 35
Jiang, Wei 35
Jirib, Kala 365
Jun, Lu 25

Kan, Yuan 35
Ke, Huang 271
Koga, Toshio 131
Kong, Xianwang 271

Łach, Łukasz 291
Lang, Jia 157
Lang, Su 543
Le, Mu 543
Li, Bing 441
Li, Cunbin 235

Li, Ding-wei 441
Li, Ge 427
Li, Gechen 357
Li, Heping 415
Li, Junhuai 181
Li, Lemin 83
Li, Qi 389, 397
Li, Qiong 343
Li, Ran 357
Li, Xianghai 457
Li, Xin 109
Li, Yujie 587, 593
Liang, Tian 479
Liao, Qun 373
Liao, Xiaoju 307
Lin, Jinshan 559
Liu, Xinhua 389, 397
Liu, Youhui 389, 397
Liu, Zhen 579
Long, Jie 17
Lu, Hai 307
Lu, Huimin 587, 593
Lu, Jun 29
Lu, Xiang 525
Lv, Jie 487, 495

Ma, Bingbing 329, 337
Ma, Haifeng 467
Ma, Wenjing 213
Ma, Yanli 509
Ma, Yongsheng 265
Man, Zhao Xue 149
Michalik, Peter 421
Min, Zhang Li 149
Ming, Li 207
Ming-zhong, Mo 551

Nibouche, M. 241
Niu, Changxi 83
Nowak, Jarosław 291

Pan, Wei-feng 441
Pei, Baohao 329, 337
Pivovarnik, Alexander 421

Qin, Qin 427
Qiu, Lei 157

Serikawa, Seiichi 587, 593
Shao, Pengfei 321
Sharma, Kuldeep 285
Shi, Changxiu 451
Shi, Liang 265
Singamneni, Sarat 279
Song, Haihui 43, 473
Song, Jingfeng 467
Song, Juanjuan 565
Su, Jianyuan 221
Sun, Chaoli 249
Sun, Qiming 173
Svyetlichny, Dmytro S. 291

Taghezout, Noria 227
Tan, Ying 249
Tang, Wen-xian 343
Tang, Zhiwei 141
Tu, Li 565

Viktord, Kanicky 365
Vlastislavc, Salajka 365

Wang, Erbing 525
Wang, Jing 525
Wang, Junwei 271
Wang, Lijie 579
Wang, Meiyan 265
Wang, Ping 1, 9
Wang, Rui 599
Wang, Weiqing 65, 73, 433
Wang, Xiao 329
Wang, Xiaoming 221
Wei, Rencan 321
Wen, Dongge 467
Wen, Youmei 49
Wenping, Wu 197
Wu, Junfeng 357
Wu, Wenping 189
Wu, Xiang 125
Wu, Yang 207
Wu, Yaohui 321
Wu, Yixuan 141

Xia, Feng 189
Xie, Bin 509
Xie, Miao 495
Xiong, Chun-rong 487, 495
Xiong, Jian-qiang 373
Xu, Du 83
Xu, Gang 525, 533
Xu, Haiqing 543

Xu, ShiWu 1, 9
Xuelan, Liang 197

Yan, Qingyou 235
Yang, Shiyuan 587, 593
Yang, Yongkui 271
Yao, Jiajun 451
Yin, Jilin 389
Yin, Mingyong 307
Ying, Guo 17
Yu, Fan 451
Yu, Mingjiu 157
Yue, Xin 467

Zajac, Jozef 421
Zdeneka, Cada 365
Zeng, Jianchao 249
Zhang, Jiajian 125
Zhang, Jian 343
Zhang, Jing 157, 181
Zhang, Lifeng 587, 593
Zhang, Liping 271
Zhang, Ping 213
Zhang, Shilin 415
Zhang, Shuwu 415
Zhang, Tiejun 213
Zhang, Weichun 329, 337
Zhang, Xuejie 571
Zhang, Yunqiang 249
Zhang, Zehua 571
Zhao, Yanwei 297, 313
Zhou, Hong 525, 533
Zhou, Xiaoxiong 57
Zhu, Lili 235
Zhu, Shi-Song 131
Zhu, Yangpeng 181
Zhuang, Hong 343

MIX
Papier aus verantwortungsvollen Quellen
Paper from responsible sources
FSC® C105338

If you have any concerns about our products,
you can contact us on
ProductSafety@springernature.com

In case Publisher is established outside the EU,
the EU authorized representative is:
Springer Nature Customer Service Center GmbH
Europaplatz 3, 69115 Heidelberg, Germany

Printed by Libri Plureos GmbH
in Hamburg, Germany